Lecture Notes in Networks and Systems 2014

The series "Lecture Notes in Networks and Systems" publishes the latest developments in Networks and Systems—quickly, informally and with high quality. Original research reported in proceedings and post-proceedings represents the core of LNNS.

Volumes published in LNNS embrace all aspects and subfields of, as well as new challenges in, Networks and Systems.

The series contains proceedings and edited volumes in systems and networks, spanning the areas of Cyber-Physical Systems, Autonomous Systems, Sensor Networks, Control Systems, Energy Systems, Automotive Systems, Biological Systems, Vehicular Networking and Connected Vehicles, Aerospace Systems, Automation, Manufacturing, Smart Grids, Nonlinear Systems, Power Systems, Robotics, Social Systems, Economic Systems and other. Of particular value to both the contributors and the readership are the short publication timeframe and the world-wide distribution and exposure which enable both a wide and rapid dissemination of research output.

The series covers the theory, applications, and perspectives on the state of the art and future developments relevant to systems and networks, decision making, control, complex processes and related areas, as embedded in the fields of interdisciplinary and applied sciences, engineering, computer science, physics, economics, social, and life sciences, as well as the paradigms and methodologies behind them.

Indexed by SCOPUS, EI Compendex, INSPEC, WTI Frankfurt eG, zbMATH, SCImago.

All books published in the series are submitted for consideration in Web of Science.

For proposals from Asia please contact Aninda Bose (aninda.bose@springer.com).

Wojciech Zamojski · Jacek Mazurkiewicz ·
Jarosław Sugier · Tomasz Walkowiak ·
Janusz Kacprzyk
Editors

Dependable Computer Systems: Methods, Models, and Applications

Proceedings of the Twenty First International Conference on Dependability of Computer Systems DepCoS-RELCOMEX, June 29 – July 2, 2026, Brunów, Poland

Springer

Editors
Wojciech Zamojski
Department of Computer Engineering
Wrocław University of Science and Technology
Wrocław, Poland

Jacek Mazurkiewicz
Department of Computer Engineering
Wrocław University of Science and Technology
Wrocław, Poland

Jarosław Sugier
Department of Computer Engineering
Wrocław University of Science and Technology
Wrocław, Poland

Tomasz Walkowiak
Department of Computer Engineering
Wrocław University of Science and Technology
Wrocław, Poland

Janusz Kacprzyk
Systems Research Institute
Polish Academy of Sciences
Warsaw, Poland

ISSN 2367-3370 ISSN 2367-3389 (electronic)
Lecture Notes in Networks and Systems
ISBN 978-3-032-27926-2 ISBN 978-3-032-27927-9 (eBook)
https://doi.org/10.1007/978-3-032-27927-9

This Springer imprint is published by the registered company Springer Nature Switzerland AG
The registered company address is: Gewerbestrasse 11, 6330 Cham, Switzerland

Preface

In this volume we would like to present proceedings of the 21st International Conference on Dependability of Computer Systems DepCoS-RELCOMEX which is scheduled to be held on-site in Brunów Palace, Poland, from June 29 to July 2 2026. The beautiful Brunów Palace is our traditional venue—close to many palaces and castles located in the charming Lower Silesia's Valley of Palaces and Gardens.

The scope of the DepCoS-RELCOMEX conference has always encompassed the wide range of challenges that arise in the performability and dependability analysis of contemporary computer systems and networks. This year, we are pleased to welcome a strong group of contributions related to **quantum computing**—a rapidly developing paradigm of data processing that has the potential to transform the way computers are used today. Dependable computer processing requires the delivery of reliable, accurate, and timely results under conditions involving both quantitative and qualitative data, and it relies on a combination of precise models and algorithms as well as approaches based on approximation or imitation. It is increasingly evident that **artificial-intelligence** methods and tools play a central role in modern information technology and computer engineering. **Cybersecurity**—both its tools and its standards—is becoming increasingly important in modern information systems, especially in light of the rapidly growing threats posed by attacking systems based on artificial intelligence. In this year's proceedings, we present also a normative approach to the cybersecurity of information systems dedicated, among others, to public-sector and local-government administration. These three thematic directions—artificial intelligence, quantum technologies, and cybersecurity—together with research on dependability, including both theoretical foundations and practical tools, constitute the core areas of the DepCoS2026 conference.

For similar reasons, we are also broadening our perspective on system dependability to include systems that make extensive use of cognitive techniques and deep-learning methods. In our view, this understanding of dependability—as the credibility of systems, artificial-intelligence tools, and computer applications—addresses the key challenges currently faced by computer science, both in theoretical research and in engineering practice. The growing number of research methods developed for such analyses draws heavily on the latest advances in artificial intelligence and computational intelligence. **The topical diversity of the papers included in these proceedings reflects the broad spectrum of multidisciplinary issues that must be considered in contemporary dependability studies.**

The conference is organized annually since 2006 by Department of Computer Engineering at the Faculty of Information and Communication Technology, Wrocław University of Science and Technology. This year DSW Ideis University from Wrocław has joined us as a co-organizer, and we hope that this cooperation will be continued in the future.

The idea and roots of DepCoS series go back to the heritage of two much older events: RELCOMEX (1977–1989) and Microcomputer School (1985–1995) which were organized by the Institute of Engineering Cybernetics (predecessor of the Department) under the leadership of prof. Wojciech Zamojski, now also the DepCoS chairman. Since 2006 the proceedings were printed, chronologically, first by the IEEE Computer Society (till 2009), then by Wrocław University of Science and Technology Publishing House (2010–2012) and in 2011–2021 by Springer Nature in the "Advances in Intelligent Systems and Computing" series (volumes no. 97, 170, 224, 286, 365, 479, 582, 761, 987, 1173, and 1389). Since 2022 DepCoS proceedings are a part of the "Lecture Notes in Networks and Systems" series and so far appeared in volumes 484 (2022), 737 (2023), 1026 (2024), and 1427 (2025). Springer Nature is one of the largest and most prestigious scientific publishers, with the LNNS titles being submitted for indexing in CORE Computing Research & Education database, Web of Science, SCOPUS, INSPEC, DBLP and other indexing services.

It is both a pleasure and an honor to express our gratitude to everyone who contributed to the organization of the Conference and to the preparation of this volume: the authors, the members of the Program and Organizing Committees, and all individuals whose efforts supported the creation of this book. Special recognition is due to the reviewers, whose assessments and comments were invaluable in selecting and improving the submitted papers. In alphabetical order, our sincere thanks this year go to: Koorosh Aslansefat, Wojciech Bożejko, Frank Coolen, Wiktor Daszczuk, Zbigniew Gomółka, Paweł Gora, Radosław Idzikowski, Michał Jaroszczuk, Łukasz Jeleń, Karol Kaluga, Karol Kozak, Andrzej Kucharski, Urszula Kużelewska, Alexey Lastovetsky, Jan Magott, Jacek Mazurkiewicz, Marek Młyńczak, Teodor Niżyński, Marcin Rudziński, Ścibór Sobieski, Robert Sobolewski, Jarosław Sugier, Kamil Szyc, Przemysław Śliwiński, Mariusz Uchroński, Tomasz Walkowiak, Marek Woda, Wojciech Zamojski, Wiktor Ziętara, and Ewa Żesławska. Their work, although not mentioned elsewhere in this volume, deserves to be acknowledged prominently in this introduction.

At the conclusion of this preface, we would like to express our gratitude to all authors who chose DepCoS-RELCOMEX as *the* forum for presenting and discussing their research results. We hope that the papers included in this volume will contribute to further advances in the design, analysis, and engineering of dependable computer systems and networks and that they will serve as a valuable resource for scientists, researchers, engineers, and students working in these fields.

Wojciech Zamojski
Jacek Mazurkiewicz
Jarosław Sugier
Tomasz Walkowiak
Janusz Kacprzyk

Twenty First International Conference on Dependability of Computer Systems DepCoS-RELCOMEX

Brunów, Poland, June 29–July 2, 2026

Programme Committee

Wojciech Zamojski (Chairman)	Wrocław University of Science and Technology, DSW Ideis University, Wrocław, Poland
Michael Affenzeller	Upper Austria University of Applied Sciences, Austria
Johnson Agbinya	Upper Austria University of Applied Sciences, Austria
Ali Al-Dahoud	Al-Zaytoonah University, Amman, Jordan
Patricio García Baez	Universidad de La Laguna, San Cristóbal de La Laguna, Spain
Andrzej Białas	Research Network ŁUKASIEWICZ—Institute of Innovative Technologies EMAG, Katowice, Poland
Ilona Bluemke	Warsaw University of Technology, Poland
Magdalena Bogalecka	Gdynia Maritime University, Poland
Wojciech Bożejko	Wrocław University of Science and Technology, Poland
Dariusz Caban	Wrocław University of Science and Technology, Poland
De-Jiu Chen	KTH Royal Institute of Technology, Stockholm, Sweden
Jacek Cichoń	Wrocław University of Science and Technology, Poland
Frank Coolen	Durham University, UK
Wiktor B. Daszczuk	Warsaw University of Science and Technology, Poland
Mieczysław Drabowski	Cracow University of Technology, Poland
Francesco Flammini	University of Linnaeus, Sweden
Peter Galambos	Óbuda University, Hungary
Manuel Gill Perez	University of Murcia, Spain
Franciszek Grabski	Polish Naval Academy, Gdynia, Poland
Aleksander Grakowskis	Transport and Telecommunication Institute, Riga, Latvia

Organizing Committee

Chair

Members

Jacek Mazurkiewicz	.
Jarosław Sugier	.
Tomasz Walkowiak	.
Tomasz Zamojski	.
Mirosława Nurek	Wrocław University of Science and Technology, Poland
Marcin Szydłowski	.
Wiktor Ziętara	DSW Ideis University, Wrocław, Poland

Contents

About the Editors

Wojciech Zamojski has been a professor of computer engineering at the Wroclaw University of Science and Technology (WUST) and at the University of Zielona Góra, the Collegium Witelona in Legnica, and the DSW Ideis University of Lower Silesia in Wrocław. His research focuses on the theory of system dependability, computer systems and networks, artificial intelligence, and modeling of transport systems. He is the founder of an academic school of dependability of computer systems. Apart from DepCoS Conferences, he was the main organizer of cyclic conferences RELCOMEX and MICROCOMPUTER Schools. He is the (co)author and editor of scientific publications (articles, books, notebooks, reports, proceedings) related to computer engineering and dependability.

Jacek Mazurkiewicz is an associate professor at Faculty of Information and Communication Technology at WUST, Poland. He is a member of Polish Neural Network Society and Polish Safety and Reliability Association and the (co)author of over 180 papers in international journals, book chapters, and proceedings. His fields of interest include soft computing, computer architecture, wireless systems, dependability, functional analysis, and modeling of systems.

Jarosław Sugier is an assistant professor at the Faculty of Information and Communication Technology at WUST. His research interests include programmable digital circuits, hardware implementation of cryptographic algorithms, and modeling of human behavior in computer simulators. He is an author or co-author of about 90 scientific publications and has been an editor of 20 monographs and conference proceedings.

Tomasz Walkowiak is a scientist working in the field of machine learning. He is one of the leading designers of the concept and architecture of the CLARIN-PL Language Technology Centre, which is built upon parallel, distributed, and efficient text processing. He consistently serves as a member of program committees and as a reviewer for machine learning and NLP-related international journals and scientific conferences. Additionally, he is an editor of 10 monographs and has (co)authored over 240 papers.

Janusz Kacprzyk is a professor of computer science at the Systems Research Institute of Polish Academy of Sciences and also at The Warsaw School of Information Technology, PIAP—Industrial Institute of Automation and Measurements, and Cracow University of Technology. He has been a visiting professor at universities in USA, England, Italy, UK, and Mexico. He is a full member of the Polish Academy of Sciences, a fellow of IEEE and IFSA, and a foreign member of the Spanish Royal Academy of Economic and Financial Sciences. His research interests include intelligent systems, soft computing, fuzzy logic, decision-making, decision analysis and decision support, database querying, information retrieval, data analysis, data mining, etc. He has (co)authored 5 books, 50 volumes, and ca. 400 papers.

About the Author

Exploring the New Edge AI Computer Based on Raspberry Pi CM5

Ali Al Dahoud[1](✉), Mohamed Fezari[2], Ahmad Al Dahoud[3], Ala' M. Al-Zoubi[1], and Hani Al-Mimi[1]

[1] Faculty of Science and IT, Al-Zaytoonah University of Jordan, Amman, Jordan
{aldahoud,ah.aldahoud,a.alzoubi,Hani.Mimi}@zuj.edu.jo
[2] Faculty of Technology, Badji Mokhtar Annaba University, Annaba, Algeria
mouradfezari@yahoo.fr
[3] Faculty of Architecture and Design, Al-Zaytoonah University of Jordan, Amman, Jordan

Abstract. AI is being utilized by companies differently than previously. Companies now operate at greater speed, through AI's capability to automate numerous processes, and at a much greater reliance upon data to guide decision-making than at any other point in history. To understand what impacts AI will bring to business and to know how it will be utilized, we have developed a future generation computing system, based upon the Raspberry Pi CM5 computer module called "Alpon X5." Alpon X5 represents an advanced, AI-based platform for the large commercial environment, which combines several forms of AI to function as a single cognition in complex industrial environments. These forms of AI include: computer vision; natural language processing; machine learning; and predictive analytics. These forms of AI are utilized for the following primary applications: Predictive Maintenance-Using sensor data to identify when equipment may fail and replace or fix the equipment before it fails; Computer Vision for Real-Time Quality Control and Anomaly Detection to enable the automation of manufacturing processes and optimize manufacturing production; Smart Supply Chain Management-Utilizing accurate demand forecasting, optimizing logistics and managing inventory to create high levels of precision; and Seamless Integration into Existing Industrial Internet of Things (IIoT) Systems, enabling the development of Autonomous Factories and Data-Centric Business Models. As demonstrated by the results of implementing Alpon X5, there is a significant reduction in cost of operation, improvement of product quality and increase in worker safety. Our paper provides the results of and discussion regarding our research and we feel that Alpon X5 represents a major technological advancement for the Fourth Industrial Revolution and IIoT, and that Alpon X5 is a scalable and powerful means of utilizing the capabilities of AI to stay competitive in today's global marketplace.

Keywords: Edge AI · Raspberry Pi compute module 5 (CM5) · ALPON X5 · Industrial edge computing · Neural processing unit (NPU) · AI acceleration · Industrial internet of things (IIoT)

W. Zamojski et al. (Eds.): DepCoS-RELCOMEX 2026, LNNS 2014, pp. 1–10, 2026.
https://doi.org/10.1007/978-3-032-27927-9_1

1 Introduction

There are many needs for processing data and making intelligent decisions at the point of creation—at the "edge"—due to the rising number of Internet of Things (IoT) and Artificial Intelligence (AI) devices [1, 2]. The ALPON X5 addresses these needs through its powerful design. The ALPON X5 is far more than just another single-board computer (SBC); it is an industrial-grade platform that can serve as the brain for future generations of smart machines, autonomous systems, and intelligent visual inspection systems [3, 4].

The move from the current Raspberry Pi Compute Module 4 (CM4) to the upcoming Compute Module 5 (CM5) should be a massive jump, much like the architectural leap from the Pi 4 to the Pi 5. In addition to building upon the solid foundation provided by the CM4's custom BCM2711 SoC and flexible PCIe support, the CM5 will significantly enhance the overall architecture by replacing the BCM2711 with the more powerful and modern BCM2712 SoC. These enhancements include improved CPU performance utilizing ARM Cortex-A76 cores, significantly greater GPU power, and other features such as the ability to utilize LPDDR5X memory, two HDMI 2.1 outputs supporting up to dual 8K displays, and a native PCIe 2.0 interface for increased storage and peripheral speeds. The advancements provided through the transition will give industrial and embedded applications a great deal more performance for high-resolution vision systems, complex AI inference, and digital signage, cementing the Compute Module's position as a top-tier system-on-module for professional developers.

The progression from the Alpon X4 to the Alpon X5 AI is also a large step forward in terms of generational improvements, transitioning from a capable perception-based system to a powerful, comprehensive cognitive-based platform. With the Alpon X4, most likely, some fundamental computer vision-based models were used to perform perception-based tasks such as object detection and basic recognition. The Alpon X5, however, was designed to manage a complex multi-model pipeline and integrate the real-time perceptual information obtained from models like SCRFD and YOLO with the advanced reasoning and natural language capabilities of Large Language and Vision-Language Models (LLMs/VLMs). With a new hardware base that includes improved processing cores and neural engine capabilities, the system transitions from a passive sensor to a fully functional intelligent analytics engine capable of real-time scene understanding, natural language interaction, and generating actionable, descriptive intelligence directly at the edge.

The ALPON X5 is a small industrial-grade SBC designed specifically for demanding Edge AI and industrial IoT applications. It is not a consumer-grade SBC like a typical Raspberry Pi board, but rather a heavy-duty SBC built around the powerful Raspberry Pi Compute Module 5 (CM5).

The ALPON X5 builds upon the impressive Raspberry Pi Compute Module 5 (CM5) to take the well-known Raspberry Pi ecosystem and elevate it to an industrial-grade level. It uses a dedicated NPU to greatly improve efficiency when using the AI workload and serves as the best example of a synergistic relationship between a general-purpose high-performance computer and a specialized AI accelerator that is ready for use in real-world deployments. [1], in this paper we describe the main components of ALPON X5 in Chap. 2.

2 Core Technology ALPON X5 "Main Components"

The ALPON X5 serves as a high-performance edge ai computing system with an integrated architecture combining four technology building blocks that provide the necessary elements for robust, efficient, and intelligent edge computing. These four technology building blocks provide the foundation for the Alpon x5's capabilities that come from the design of the intelligent device through its integration of several key technologies.

2.1 High-Performance Compute Core: Raspberry Pi Compute Module 5 (CM5)

The CM5 serves as the general-purpose processing brain.

Key Components:

- SoC: Broadcom BCM2712 with a CPU Quad-core Arm Cortex-A76 @ up to 2.4GHz. Provides the raw power for the operating system, data management, and complex application logic.
- RAM: LPDDR4X (4GB to 16GB). Ensures smooth handling of multiple processes and large datasets. Figure 1 is a photo of the module.

This foundation offers a modern, powerful, and software-compatible base from the renowned Raspberry Pi ecosystem [10].

Fig. 1. Pi Compute module 5

2.2 Dedicated AI Acceleration: Neural Processing Unit (NPU)

Fig. 2. DEEPX DX-M1

DEEPX is a specialized co-processor for ultra-efficient AI Inference.

Key Capability: Delivers Several TOPS (Trillion Operations Per Second) of performance for running pre-trained neural network models (e.g., for object detection, classification, segmentation). AI Performance: 25 TOPS. support S-O-T-A Algorithms: ResNet, MobileNet V1/v2/v3 SSD, EfficientNet, EfficientDet, YOLOv5, YOLOv7, YOLO8, DeepLabv3, PIDNet and the latest YOLO Models, VLM (CLIP etc.), the presentation of the module is in Fig. 2.

This is the "AI engine." It offloads intensive ai tasks from the cpu, enabling real-time, low-power ai inference that is orders of magnitude faster and more efficient than cpu-based processing. this is the definitive feature for edge AI workloads [5, 6].

2.3 Industrial-Grade I/O and Connectivity Fabric

Grade I/O provides robust and versatile interfacing for sensors, networks, and peripherals.

Key Components: Dual Gigabit Ethernet: For network redundancy, high-bandwidth camera streams, or network segmentation. PCIe Interface: Enables high-speed expansion with NVMe SSDs (for fast local storage) or other specialized cards. Wide-Voltage DC Input (12-24V): Allows direct integration into industrial, automotive, and outdoor power systems. And USB 3.0, GPIO, CSI/DSI: Maintains compatibility with a vast ecosystem of peripherals and HATs.

2.4 Robust Thermal and Mechanical Design

The mechanical design supports reliable, long-term operation.

Key Features: Metal housing: Provides a solid barrier against physical damage or environmental extremes, and is an efficient method for removing excess heat from the system, as illustrated in Fig. 3. This housing provides the ability to operate in extended temperature ranges, as shown in Fig. 3, thereby allowing the CPU and NPU to perform at their maximum level during extended periods of time when operating continuously under full load. This feature is very important for continuous operation in 24/7 industrial environments (Table 1).

Fig. 3. Alpon X5 AI metal enclosure

Table 1. ALPON X5 vs. standard Raspberry Pi 5[3]

Feature	ALPON X5 (with CM5)	Raspberry Pi 5
Form factor	Industrial carrier board + CM5	All-in-one consumer SBC
CPU	CM5's Quad-core Cortex-A76	Raspberry Pi 5's Quad-core Cortex-A76
AI Acceleration	Yes, dedicated NPU (3–6 TOPS)	No (relies on CPU/GPU, much slower)
I/O Robustness	Dual GbE, Wide-range DC input, PCIe	Single GbE, USB-C Power, No PCIe
Applications	Industrial, Edge AI, Robotics	Hobbyists, Education, Prototyping
Ruggedness	High (wide temp range, metal enclosure)	Standard

3 ALPON X5 as "Intelligent Edge"

The ALPON X5 integrates processing capabilities with sensors and input/output interfaces, allowing it to perform tasks such as signal processing, pattern recognition, and lightweight machine-learning inference directly on the device. This local intelligence enables faster decision-making in real time, which is essential for applications such as smart monitoring systems, industrial automation, and IoT devices. By combining embedded processing power with on-device data analysis, the ALPON X5 exemplifies the concept of intelligent edge computing, where computation happens close to the data source instead of in centralized cloud infrastructure.

The true power of the ALPON X5 lies on the integration of the components:

- The CM5 handles the complex operating system and application flow.
- The NPU dedicates itself to parallel AI computations upon request.
- The Industrial I/O seamlessly brings data in from high-speed sensors and cameras and connects to enterprise networks.

This synergy creates a unified platform that is greater than the sum of its parts, purpose-built for deploying sophisticated intelligence directly where data is generated—at the edge.

In addition, the following add-ons provide the Alpon X5 with communication flexibility, an extension using the PCI slot, and acceptance of Pi HATs:

- **Rugged I/O**: Dual Gigabit Ethernet ports (to provide network redundancy or to support multiple video feeds), wide-voltage 12–24 V DC input, and other features that are rugged enough for a variety of industrial settings, in addition to providing a level of reliability and consistency for a variety of different uses.
- **Expansion Capability**: A PCIe slot is provided as part of the design, which allows users to expand the device's capabilities by adding high-speed NVMe storage for data logging or a variety of expansion cards based on specific user needs.

- **Compatibility**: The standard 40-pin GPIO connector is retained, so users can continue to draw upon the large number of Raspberry Pi HATs and sensors available.

4 Applications: Upgrading Industries with On-Device Intelligence

The ALPON X5 is built for scenarios where low latency, data privacy, and high reliability are paramount. Its applications span numerous industries:

4.1 Smart Manufacturing and Automated Optical Inspection (AOI)

Real-time inspection using AI on production lines. The ALPON X5, with its cameras, can perform real-time inspection of a product for defects and/or improper measurements and verify that components are assembled correctly; it also has the ability to sort products as well. Its NPU can run complex vision algorithms instantly and reject defective products so the production process does not slow down [7]. For example, it can be used for pipe control and leak detection; see Fig. 2 in [8] and [9].

In reference [8] and [9], The Raspberry Pi AI Kit, a recent collaboration between Raspberry Pi and Hailo, has been launched to make artificial intelligence more accessible to enthusiasts and developers alike. Priced at $70, this kit integrates a powerful Hailo-8L AI accelerator module into the Raspberry Pi ecosystem, specifically designed for the Raspberry Pi 5. It can be integrated in Oil Leak detection on Pipeline and based on the wireless communication of the ALPON X5 using LTE, the operator can determine problem location.

4.2 Smart Retail and Security

Video analysis for security and customer insights. For example, in a retail environment, it could be used to count people, create heat maps to show foot traffic through stores, and check inventory levels on shelves, and in a security environment, it could be used to identify intrusion, read license plates, etc. It would do this by analyzing video feeds and sending alerts based on what was detected without having to send the actual video feed to the cloud.

4.3 Intelligent Robotics and Autonomous Systems

Serve as the central processing unit for Autonomous Mobile Robots (AMRs), robotic arms, etc. The ALPON X5 will process sensor data from cameras, LiDAR, etc. to allow the AMR/robotic arm to navigate around objects in real time and avoid obstacles. The NPU is especially important for tasks such as "vision-based picking," where an object must be identified and then located.

4.4 Agricultural Technology (AgriTech)

Precision agriculture and automated harvesting. Mounted on drones or farm machinery, the ALPON X5 can analyze crops to detect diseases, monitor health, and identify ripe produce for automated harvesting. This enables data-driven decisions to optimize yield and reduce waste.

4.5 Machine Vision and Automated Optical Inspection (AOI)

The product can be used in Identifying defects on a production line. In sorting objects based on visual characteristics. And reading barcodes and OCR. However, it executes less Tops than the NVIDIA Jetson AGX 32GB which reaches 200Tops while ALPON X5 it reaches 26 Tops.

4.6 Edge Computing and Networking

Acting as a powerful, customizable edge server or network appliance. With its dual Ethernet ports, PCIe slot (for adding a second network card), and powerful CPU, it can be deployed as a firewall, a VPN gateway, or a local server for processing IoT sensor data from a factory floor.

5 Tests

Some tests were done using Alpon X5 and other SBC with less computation power, the tests were selected from some works cited in references:

In [10] we noticed that testing YOLOv7 on the ALPON X5 demonstrates the capability of this edge-AI platform to perform real-time computer vision tasks directly on-device. The ALPON X5 integrates a Raspberry Pi Compute Module 5 with a DEEPX DX-M1 neural accelerator delivering about 25 TOPS of AI performance, enabling efficient inference for deep-learning models such as YOLOv7 without relying on cloud processing. During testing, YOLOv7 can be used to perform object detection on live camera streams, identifying and tracking multiple objects in real time while maintaining low latency thanks to on-device inference. This setup is particularly suitable for edge applications such as smart surveillance, industrial monitoring, robotics, and traffic analysis, where immediate processing of visual data is required. The combination of the AI accelerator, local processing capability, and compatibility with frameworks like PyTorch, ONNX, and TensorFlow Lite makes the ALPON X5 an effective platform for deploying and evaluating YOLO-based detection models at the intelligent edge. Moreover, the YOLOPose Pose Detection Model, developed within a Kickstarter-supported project, is a deep-learning model designed for real-time human pose detection and estimation using the YOLO (You Only Look Once) framework. The model extends traditional object detection by predicting key points representing human body joints—such as shoulders, elbows, knees, and ankles—allowing the system to reconstruct the posture or movement of a person in an image or video stream. Unlike traditional multi-stage pose estimation methods, YOLOPose performs object detection and pose estimation simultaneously in a single forward pass, enabling faster processing and making it suitable for real-time edge-AI applications.

Always in [10] YOLOv5 20-Channel Object Detection refers to a configuration of the YOLOv5 model designed to detect objects belonging to an image or 20-channelvideo stream at the same time. Each channel in the model's output corresponds to a specific object category, allowing the neural network to classify and localize multiple types of objects simultaneously. During inference, YOLOv5 processes the input image in a single pass and outputs bounding boxes, class probabilities, and confidence scores for each detected object. This multi-class detection capability makes the model suitable for applications such as smart surveillance, autonomous systems, and industrial monitoring, where recognizing several object categories at the same time is required while maintaining real-time performance. These performances show the power of the ALPON X5.

In [12] authors included an AI module on Raspberry Pi and tested face recognition model, Modern face recognition systems rely heavily on deep learning models to perform these tasks. The pipeline begins with face detection, where specialized models identify faces and their key landmarks. These landmarks are used to align faces, ensuring that variations in pose, orientation, or size do not affect subsequent steps. The aligned faces are then passed through embedding models, which generate high-dimensional vectors representing unique facial features. Finally, these embeddings are compared using similarity metrics such as cosine similarity to identify or verify individuals. Face detection is the first step in a face recognition pipeline, and using models that detect key facial landmarks ensures accurate alignment for subsequent steps. Below is an example of using the SCRFD model with the DeGirum PySDK to perform face detection and extract key points. The code demonstrates how to load a face detection model, run inference on an image, and obtain both bounding boxes and key points, the results were good and they will be better with ALPON X5 AI module.

Finally, in [13], the authors include information about the existing datasets of road defects, which are commonly used for detection and segmentation. The presented work is based on deep learning approach with the use of synthetic generated training data for segmentation of cracks in driver-view image. The novelty of their approach lies in creating synthetic dataset for training state-of-the-art deep learning frameworks. The relevance of the research is emphasized by processing of wide-view images in which heterogeneous pixel intensity, complex crack topology, different illumination condition and complexity of background make the task challenging. They used an embedded U-Net model to segment image on an SBC with less characteristics than ALPON X5, which proves that we can implement segmentation models also on this device for detecting faults industrial machines.

6 Discussion

The "good results" are not from any single model, but from their orchestration.

Running multiple modern DNN models simultaneously is extremely demanding. The Alpon X5 provides the necessary CPU/GPU (likely NPU) power to do this without lag. If the Alpon X5 is an edge device, it allows for real-time processing without constant cloud dependency, ensuring low latency, data privacy, and reliability. Moreover, the platform is likely optimized to run these specific models efficiently, perhaps using TensorRT or similar SDKs to maximize throughput.

The success of the testing of the Alpon X5 is an example of a developing trend in AI: moving from standalone AI models to a multimodal, integrated approach for achieving AI [14–16]. The combined strength of the low-level perception of the models (SCRFD, VGG16, and YOLO) and the higher-level cognition (LLMs and VLMs) has produced a solution that does not simply "see" but also "understand and report." The Alpon X5 provides the necessary "muscle" to allow this complex brain to operate in a useful, real-world environment. The "good results" demonstrate both the effectiveness of each model individually and, most importantly, the benefit derived through their deliberate combination.

7 Conclusion

The ALPON X5 is a complete Edge AI solution that allows companies to build complex intelligence in products and in processes by using a powerful AI accelerator and the robustness of the community supported Raspberry Pi CM5 platform. It is a solution representing the next generation of computing–Intelligent Computing Distributed at the Edge.

It is therefore a big step from prototype to production-ready, as it is a revolutionary new approach in the industrial world. By combining the robust general-purpose computing capabilities of the Raspberry Pi CM5 and the high-efficiency special purpose processing capabilities of a dedicated NPU, the ALPON X5 has delivered the industrial capabilities required to realize real-time on-device intelligence. With the ALPON X5 addressing the primary need of the industry for higher productivity, quality, and autonomous performance, the advantages of the ALPON X5 are evident across multiple applications. Examples include: the automation of optical inspections on production lines, which will revolutionize quality control; enabling intelligent robotics for precision material handling; and enabling predictive maintenance and enhanced security through intelligent video analytics. In total, the ALPON X5 is an essential tool for realizing Industry 4.0 by providing companies the ability to embed advanced AI capabilities within their operational processes, thereby reducing latency, improving data privacy, and enhancing reliability at the edge.

References

1. Shi, W., Cao, J., Zhang, Q., Li, Y., Xu, L.: Edge computing: vision and challenges. IEEE Internet Things J. **3**(5), 637–646 (2016)
2. Satyanarayanan, M.: The emergence of edge computing. Computer **50**(1), 30–39 (2017)
3. Xu, L.D., He, W., Li, S.: Internet of Things in industries: a survey. IEEE Trans. Industr. Inf. **10**(4), 2233–2243 (2014)
4. Schwab, K.: The Fourth Industrial Revolution: What it means and how to respond. Foreign Affairs (2016)
5. Chen, Y., Krishna, T., Emer, J., Sze, V.: Eyeriss: an energy-efficient reconfigurable accelerator for deep convolutional neural networks. IEEE J. Solid-State Circuits **52**(1), 127–138 (2017)
6. Jouppi, N.P.: et al.: In-datacenter performance analysis of a tensor processing unit. In: Proceedings of the 44th Annual International Symposium on Computer Architecture (ISCA) (2017)
7. Saxena, A., Goebel, K.: Turbofan engine degradation simulation data set. NASA Ames Prognostics Data Repository (2008)
8. CNX Software.: Alpon X5 Raspberry Pi CM5 edge AI computer features 25 TOPS accelerator, 4G LTE connectivity. CNX Software (2025). https://www.cnx-software.com/2025/07/30/alpon-x5-raspberry-pi-cm5-edge-ai-computer-25-tops-accelerator-4g-lte-connectivity/

9. LinuxGizmos.: Raspberry Pi AI Kit: 13 TOPS AI performance with Hailo-8L and M.2 HAT. *LinuxGizmos* (2024). https://linuxgizmos.com/raspberry-pi-ai-kit-13-tops-ai-performance-with-hailo-8l-and-m-2-hat/
10. ALPON.: ALPON X5: Powerful edge AI computer powered by Raspberry Pi. Kickstarter (2025). https://www.kickstarter.com/projects/alpon/alpon-x5-powerful-edge-ai-computer-powered-by-raspberry-pi
11. Running local LLMs and VLMs on the Raspberry Pi.: Medium (n.d.). https://medium.com/data-science/running-local-llms-and-vlms-on-the-raspberry-pi-57bd0059c41a
12. Hailo Community.: A comprehensive guide to building a face recognition system. Hailo Developer Community Forum (2024). https://community.hailo.ai/t/a-comprehensive-guide-to-building-a-face-recognition-system/8803
13. Kanaeva, I.A., Ivanova, J.A.: Road pavement crack detection using deep learning with synthetic data. In: IOP Conference Series: Materials Science and Engineering, vol. 1019, no. 1, Article 012036. IOP Publishing (2021)
14. Aldahoud, A., Fezari, M., Al-Dahoud, A., Aqel, D.: A systematic review on applying machine learning and deep learning on SBCs. WSEAS Trans. Inf. Sci. Appl. (2024). https://doi.org/10.37394/23209.2024.21.26
15. Al-Dahoud, A., Fezari, M., Al-Dahoud, A., Aqel, D., Mimi, H., Daoud, M.S.: Revolutionizing space: The potential of artificial intelligence. WSEAS Trans. Comput. Res. (2024). https://doi.org/10.37394/232018.2024.12.40
16. Al-Dahoud, A., Fezari, M., Al-Dahoud, A., Dahoud, A.: Embedding more autonomous safety in MCU-based embedded systems using project-based learning (Case study). Int. J. Educ. Inf. Technol. **18** (2024). https://doi.org/10.46300/9109.2024.18.6

A Dependable Explainable NLP Framework for Clinical Personality Trait and Subtype Analysis

Aleksandra Ata[1(✉)], Stanisław Lota[2], and Karol Kozak[1]

[1] DSW Ideis University, Wrocław, Poland
{aleksandra.ata,karol.kozak}@ideis.pl
[2] WSB Merito University in Wrocław, Wrocław, Poland
stanislaw.lota@wroclaw.merito.pl

Abstract. The application of natural language processing models in clinical and mental health contexts requires not only high predictive performance but also interpretability, robustness, and well-calibrated uncertainty. This paper presents a dependable and explainable NLP framework for fine-grained personality trait analysis, grounded in clinical theory. The proposed approach combines a parameter-efficient DistilBERT architecture with lightweight adapter modules, resulting in only 1.91% trainable parameters, thereby reducing overfitting risks in domain-specific datasets. To address severe class imbalance, the framework integrates data augmentation with a Focal Loss objective. Explainability is achieved through a multi-layer approach combining local post-hoc explanations with class-level statistical validation. The framework is evaluated on a hold-out test set of 1,600 social media posts, achieving 82.06% accuracy, 0.8985 ROC-AUC, and a well-calibrated ECE of 0.042. The model demonstrates statistically significant differences between grandiose and vulnerable narcissistic expressions, consistent with established psychological theory.

Keywords: Clinical NLP · Trustworthy AI · Dependable XAI · Mental health NLP · Personality trait analysis · Narcissism detection

1 Introduction

The deployment of artificial intelligence in clinically-relevant domains, such as mental health, presents a distinct set of challenges that transcend standard performance metrics. For a computational tool to be considered dependable, it must satisfy a triad of requirements: predictive accuracy, ensuring reliable classifications; robust interpretability, allowing clinicians to validate the model's reasoning against established diagnostic frameworks; and calibrated uncertainty, providing an honest measure of the model's confidence. The task of identifying narcissistic traits from text, a phenomenon of growing interest in online contexts [19], serves as a critical case study for this challenge.

Narcissistic Personality Disorder (NPD) [1], as defined in the DSM-5, is not a monolithic construct. It manifests across a spectrum, most notably through the clinically

W. Zamojski et al. (Eds.): DepCoS-RELCOMEX 2026, LNNS 2014, pp. 11–19, 2026.
https://doi.org/10.1007/978-3-032-27927-9_2

distinct Grandiose (overt, exhibitionist) and Vulnerable (defensive, hypersensitive) presentations [5, 15, 16]. However, many existing computational models oversimplify this complexity, treating narcissism as a binary classification problem. Such a reductionist approach limits clinical utility, as the distinction between subtypes is crucial for accurate diagnosis and treatment planning.

Furthermore, existing systems face two significant technical hurdles. First, large pre-trained language models (e.g., BERT [10]) introduce a substantial risk of overfitting when fine-tuned on the smaller, specialised datasets typical of clinical research. Second, these models often operate as 'black boxes,' providing predictions without clear explanations, which prevents their integration into evidence-based clinical workflows.

The present research directly addresses the dependability gap in clinical NLP [30]. The framework makes the following specific contributions:

1. **A Parameter-Efficient and Dependable Architecture**: A novel architecture is introduced combining a frozen DistilBERT encoder with lightweight Adapter Modules [2, 21]. This approach reduces trainable parameters to just 1.91% of the total model, significantly mitigating overfitting on small clinical datasets while maintaining high performance.
2. **A Clinically Grounded, Multi-Layer Explainability Framework**: The framework moves beyond isolated post-hoc explanations by implementing a two-layer validation process. Local explanations are generated using LIME [6], while class-level linguistic patterns are validated using Chi-Square statistical tests, providing a robust mechanism for clinical verification.
3. **A Fine-Grained Subtype Analysis with Empirical Validation**: The framework is, to the best of the authors' knowledge, the first to explicitly model and empirically distinguish between Grandiose and Vulnerable narcissism subtypes from text, revealing statistically significant linguistic markers consistent with clinical theory [15–17].

2 Related Work

Computational analysis of personality traits from text is an established research area. Early work relied on dictionary-based methods (e.g., LIWC [29]) and psychometric instruments such as the Narcissistic Personality Inventory [18], which are transparent but lack the contextual understanding of modern neural networks. More recent approaches have employed deep learning [8, 9], achieving accuracies of up to 79% in binary narcissism classification [8, 9, 11]. However, these approaches treat narcissism as a monolithic construct and lack fine-grained subtype modelling.

The concept of parameter-efficient fine-tuning (PEFT) has gained traction as a solution to the problem of overfitting large models. Adapter Modules, introduced by Houlsby et al. [2], provide a method for fine-tuning large models without retraining all parameters. Methods like AdapterFusion [21] further extend this idea. The present work adopts this philosophy by inserting lightweight adapter modules into the frozen DistilBERT encoder, constituting a highly effective PEFT strategy.

In the realm of explainable AI (XAI), post-hoc methods like LIME [6], SHAP [29], and gradient-based visualisation techniques such as Grad-CAM [28] are widely used to explain individual predictions. However, these local explanations can be unstable and may not reflect the model's global behaviour [26]. The proposed framework addresses this limitation by supplementing LIME with a statistical analysis of class-level linguistic features, providing a more robust and clinically verifiable form of explanation.

3 Methodology

The methodological framework is designed for dependability and reproducibility. The following sections detail the dataset curation, model architecture, training strategy, and the multi-layer explainability framework. The entire pipeline is designed to be transferable to other mental health and personality assessment [18, 20] tasks.

3.1 Dataset Curation and Annotation

The study utilises an expanded version of the cleaned Mastodon dataset from prior work [11], supplemented to a total of 8,000 posts. A structured, semi-automated annotation protocol was employed using a gpt-4.1-mini model with a rigorous, clinically-informed, hierarchical strategy: (1) Binary Classification using DSM-5 criteria for NPD [1]; (2) Fine-Grained Subtyping using definitions from seminal works [15–17]. The final dataset comprised 86.1% Non-Narcissistic, 9.9% Grandiose, and 3.9% Vulnerable posts. A validation subset yielded a Cohen's Kappa of 0.78 [3, 14] for binary classification and 0.71 for subtyping.

Limitations of LLM Annotation. To scale the annotation process, an LLM-based system (GPT-4) was used to label a portion of the dataset. To validate this approach, human annotators manually reviewed a subset of 500 LLM-labelled posts. Inter-annotator agreement between human raters and the LLM was substantial (Cohen's Kappa $\kappa = 0.78$ for binary, $\kappa = 0.71$ for subtypes), suggesting the LLM provides a reliable, though not infallible, proxy for human judgment. It is acknowledged that LLM-based annotation is not a substitute for formal clinical diagnosis, and the process is susceptible to model biases. Labels should be considered a high-quality proxy, not definitive ground truth. Future work should involve direct validation by qualified clinicians.

3.2 Model Architecture and Training

Model Architecture. The model is based on DistilBERT [7], a distilled version of BERT [10], which builds upon the Transformer architecture [12]. All pre-trained layers of the encoder were frozen to preserve general language understanding. Lightweight Adapter Modules [2] were inserted after each transformer block, projecting the 768-dimensional input to a bottleneck of 64, applying a non-linear activation, and projecting back to 768. This parameter-efficient approach results in only 1.27 million trainable parameters (1.91% of the total model parameters), dramatically reducing overfitting potential on the specialised clinical dataset.

Training Strategy. To address class imbalance, a two-pronged strategy was employed: (1) Data Augmentation, replicating minority classes three times; and (2) Focal Loss [4] with hyperparameters $\alpha = 0.25$ and $\gamma = 2.0$. The model was trained for 8 epochs using the AdamW optimiser [13] with a learning rate of 5×10^{-4} and a batch size of 32. A 5-fold stratified cross-validation strategy was employed to ensure robust evaluation, with the best-performing model selected based on validation accuracy.

3.3 Explainability Framework

A multi-layer explainability framework was designed to ensure clinical trustworthiness [26]. It operates on two levels: (1) Local, Instance-Level Explanations via LIME [6], which generates approximately 1,000 perturbed text variations and fits a simple logistic regression to approximate the local decision boundary, revealing which words contributed most to a classification; and (2) Global, Statistical Validation via Chi-Square tests to confirm that the association of LIME-identified features with a given class is statistically significant ($p < 0.05$). This combination provides a more robust framework than using isolated post-hoc methods like LIME, SHAP [29], or Grad-CAM [28] alone.

4 Experimental Results

The empirical evaluation covers primary classification metrics, an ablation study, and a linguistic analysis of narcissism subtypes. In a clinical screening context, recall (correctly identifying potential cases) and precision (minimising false alarms) are particularly meaningful metrics.

4.1 Binary Classification Performance

On a held-out test set of 1,600 posts, the full model achieved an overall accuracy of 82.06%. The precision and recall were well-balanced, and the ROC-AUC was 0.8985, indicating a strong ability to distinguish between narcissistic and non-narcissistic posts. Performance was stable across 5-fold cross-validation, suggesting reasonable generalisation to unseen data.

A confidence calibration analysis [24, 25] was conducted to assess the reliability of the model's confidence scores. For predictions where the model was 90–100% confident, its accuracy was 96.2%. The overall Expected Calibration Error (ECE) [24] was a low 0.042, suggesting the model is well-calibrated and its confidence estimates can be trusted in a clinical setting, building on uncertainty estimation principles [22, 23](Fig. 1).

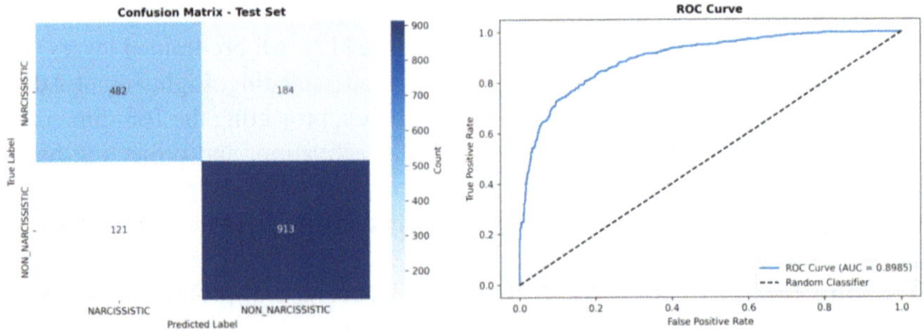

Fig. 1. Binary Classification Performance. The confusion matrix (left) provides a breakdown of the model's classifications (913 True Negatives, 184 False Positives, 121 False Negatives, 482 True Positives). The ROC curve (right), with an AUC of 0.8985, demonstrates a strong capability to distinguish between narcissistic and non-narcissistic posts

4.2 Ablation Study

An ablation study was conducted to systematically evaluate the contribution of data augmentation and Focal Loss. The results, detailed in Table 1, highlight the trade-offs involved in the model design.

A baseline model (DistilBERT without augmentation or Focal Loss) achieved the highest accuracy (87.25%) but at the cost of a very low recall (18.47%), indicating it simply predicted most samples as non-narcissistic. In a clinical context, false negatives (missed cases) are often considered more costly than false positives. The full model sacrifices a small amount of accuracy to achieve a much higher recall of 62.16%. Each component makes a meaningful contribution: data augmentation primarily improves recall, while Focal Loss primarily improves precision (Fig. 2).

Table 1. Ablation study results. All metrics are reported for the positive (narcissistic) class.

Configuration	Accuracy (%)	Precision (%)	Recall (%)	F1-Score (%)
Baseline (Cross-Entropy)	87.25	64.06	18.47	28.67
+ Data Augmentation	86.94	52.97	52.25	52.61
+ Focal Loss	87.44	83.87	11.71	20.55
Full Model	**84.00**	**44.52**	**62.16**	**51.88**

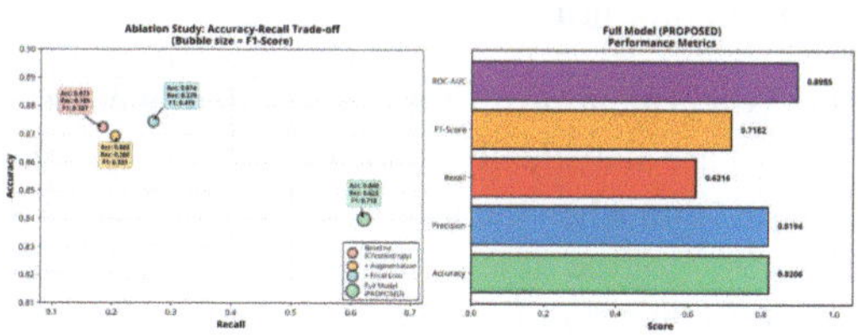

Fig. 2. Ablation Study Results. The bubble chart visualises the accuracy-recall trade-off. Bubble position shows Accuracy (y-axis) vs. Recall (x-axis), and bubble size represents F1-Score. The Full Model achieves the highest recall, prioritising clinical detection over raw accuracy.

4.3 Subtype Analysis and Explainability

A key contribution of this work is the empirical analysis of linguistic differences between Grandiose and Vulnerable narcissism. A Chi-Square test revealed that the word frequency distributions of the two subtypes are statistically distinct ($\chi 2(49) = 89.7$, $p < .001$). As shown in Table 2, Grandiose narcissism is associated with self-promoting and exhibitionist language, while Vulnerable narcissism is characterised by themes related to art and aesthetic cultivation.

Table 2. Most discriminative hashtags per Narcissism subtype (Chi-Square, $p < 0.05$).

Grandiose subtype	Freq. Ratio	Vulnerable subtype	Freq. Ratio
#naughty	23.1 ×	#photography	5.6 ×
#asshole	7.6 ×	#visualart	6.8 ×
#shaved	13.8 ×	#doubleexposure	14.8 ×
#sexy	2.5 ×	#aesthetic	4.9 ×
#erotic	3.8 ×	#composition	5.8 ×

The Grandiose subtype is characterised by exhibitionist and self-promoting language, consistent with its overt clinical presentation [15]. The Vulnerable subtype exhibits language associated with aesthetic cultivation and artistic expression, aligning with its defensive, hypersensitive clinical profile [16, 17]. These findings provide empirical linguistic evidence for the theoretical distinction between the two subtypes.

5 Discussion

The empirical results highlight the framework's dependability, but their interpretation requires a nuanced discussion of the model's calibration, its position relative to prior work, and the practical implications of its explainability features.

5.1 Calibration and Dependability

The low ECE of 0.042 is a critical result for clinical deployment. It indicates that the model's confidence scores are statistically 'honest' and can be used as a reliable triage signal. A clinician can trust that when the model reports 90% confidence, the prediction is correct approximately 96% of the time. This property, grounded in calibration theory [22–25], is a prerequisite for any AI tool intended for use in a clinical support role.

5.2 Comparison with Prior Work

Compared to previous computational work in narcissism detection [8, 9], which often reported accuracy around 79%, the proposed model achieves a higher accuracy of 82.06% while also providing subtype classification grounded in established psychological theory [11] and robust explainability. Unlike models that require full fine-tuning, the parameter-efficient approach presented here is more scalable and less prone to overfitting. The combination of LIME-based local explanations with Chi-Square statistical validation [6, 26] represents a qualitative advance over prior work that relied on isolated post-hoc methods.

5.3 Explainability and Clinical Utility

The two-layer explainability framework provides a mechanism for clinical validation that goes beyond simple feature attribution. By combining LIME [6] with Chi-Square statistical testing, the framework can distinguish between features that are merely correlated with a class and those that are statistically associated at the population level. This distinction is crucial for clinical trustworthiness [26, 27] and reduces the risk of clinicians acting on spurious correlations identified by purely local explanation methods.

5.4 Ethical Considerations

The automated assessment of personality traits raises significant ethical concerns that must be addressed before any clinical deployment. First, the risk of misuse is substantial: a system capable of identifying narcissistic traits from text could be weaponised for profiling, discrimination, or surveillance. Second, the use of social media data introduces privacy concerns and the risk of identifying individuals without their consent. Third, the model's labels are derived from an LLM annotation protocol, not from clinical diagnosis, and must not be used as a substitute for professional assessment.

To mitigate these risks, the framework is designed to operate exclusively in a human-in-the-loop model, where all predictions are reviewed by a qualified clinician before any action is taken. The code and model weights are made publicly available at: https://gitlab.com/aleksandra.ata/dependable-xai to ensure transparency and to allow independent auditing of the system's behaviour.

5.5 Limitations

The study has several limitations. The dataset, while carefully annotated, is derived from a single social media platform and may not generalise to other online communities or demographic groups. Social media language is often performative, and the model may capture elements of online self-presentation rather than underlying clinical pathology. The LLM-based annotation, while validated against human raters, is not a substitute for formal clinical diagnosis. Finally, the explanations provided by LIME are post-hoc and may not perfectly reflect the model's internal logic.

A significant limitation arises from the class imbalance inherent in the dataset. Although data augmentation was used to create a more balanced training set, the validation and test sets retained the natural distribution where 86.1% of posts were non-narcissistic. When deployed in real-world social media [8, 19] environments, the prevalence of narcissistic content may be substantially lower, potentially degrading the model's positive predictive value. Future work should test the model on datasets with more extreme class imbalance to assess robustness.

6 Conclusion

A dependable and explainable NLP framework for fine-grained clinical personality trait analysis has been presented. By combining parameter-efficient adapter-based fine-tuning with a multi-layer explainability framework and rigorous calibration analysis, the proposed approach addresses the key limitations of existing systems. The framework achieves 82.06% accuracy and 0.8985 ROC-AUC on a challenging, imbalanced dataset, while maintaining a well-calibrated ECE of 0.042. The linguistic analysis provides empirical evidence for the theoretical distinction between Grandiose and Vulnerable narcissism subtypes, offering a novel contribution to both computational linguistics and clinical psychology.

Future work will focus on: (1) validation with clinical experts and standardised psychometric instruments [3, 14, 18]; (2) extension to other personality disorders and mental health conditions; and (3) longitudinal analysis of personality trait expression in online discourse [19, 20].

Acknowledgments. Assistance with the preparation of the manuscript was provided by Manus, an AI assistant. Manus was involved in data processing and annotation, generating and refining visualisations [27], implementing the statistical analysis, and assisting with the preparation and formatting of the manuscript.

Disclosure of Interests The authors have no competing interests to declare that are relevant to the content of this article.

References

1. American Psychiatric Association: Diagnostic and Statistical Manual of Mental Disorders, 5th edn. American Psychiatric Publishing, Arlington (2013)
2. Houlsby, N., et al.: Parameter-efficient transfer learning for NLP. In: Proceeding of the 36th International Conference on Machine Learning, pp. 2790–2799. PMLR (2019)
3. Ames, D.R., Rose, P., Anderson, C.P.: The NPI-16 as a short measure of narcissism. J. Res. Pers. **40**(4), 440–450 (2006)
4. Lin, T.Y., et al.: Focal loss for dense object detection. In: Proceeding of the IEEE International Conference on Computer Vision, pp. 2980–2988 (2017)
5. Campbell, W.K., et al.: Does self-love lead to love for others? A story of narcissistic game playing. JPSP **83**(2), 340–354 (2002)
6. Ribeiro, M.T., Singh, S., Guestrin, C.: Why should I trust you?: Explaining the predictions of any classifier. In: Proceeding KDD, pp. 1135–1144 (2016)
7. Sanh, V., et al.: DistilBERT, a distilled version of BERT (2019). arXiv:1910.01108
8. Sumner, C., et al.: Predicting dark triad personality traits from Twitter usage and a linguistic analysis of tweets. In: Proceeding ASONAM, pp. 560–565 (2012)
9. Volkova, S., Bachrach, Y., Armstrong, M., Sharma, V.: Inferring latent user properties from texts published in social media. In: Proceeding EMNLP, pp. 1996–2002 (2015)
10. Devlin, J., et al.: BERT: Pre-training of deep bidirectional transformers for language understanding (2018). arXiv:1810.04805
11. Ata, A., Kozak, K.: Lightweight NSM: A clinical-linguistic approach to narcissismdetection on social media (this is preprint research article): https://open-research-europe.ec.europa.eu/articles/6-74/v1
12. Vaswani, A., et al.: Attention is all you need. arXiv:1706.03762 (2017)
13. Kingma, D.P., Ba, J.: Adam: A method for stochastic optimisation. arXiv:1412.6980 (2014)
14. Konrath, S., et al.: Development and validation of the Single Item Narcissism Scale (SINS). PLoS ONE **9**(8), e103469 (2014)

15. Pincus, A.L., Lukowitsky, M.R.: Pathological narcissism and narcissistic personality disorder. Annu. Rev. Clin. Psychol. **6**, 421–446 (2010)
16. Wink, P.: Two faces of narcissism. JPSP **61**(6), 867–877 (1991)
17. Hendin, H.M., Cheek, J.M.: Assessing hypersensitive narcissism: A reexamination of Murray's Narcism Scale. J. Res. Pers. **31**(4), 588–599 (1997)
18. Raskin, R., Hall, C.S.: A narcissistic personality inventory. Psychol. Rep. **45**(2), 590 (1979)
19. Twenge, J.M., Campbell, W.K.: The Narcissism Epidemic: Living in the Age of Entitlement. Free Press (2009)
20. Petty, R.E., et al.: Attitude Strength: Antecedents and Consequences. Soc. J. Psychol. **26**, 1–13 (2023)
21. Pfeiffer, J., et al.: AdapterFusion: Non-destructive task composition for transfer learning. arXiv:2005.00247 (2020)
22. Gal, Y., Ghahramani, Z.: Dropout as a Bayesian approximation: Representing model uncertainty in deep learning. In: Proc. ICML, pp. 1050–1059 (2016)
23. Lakshminarayanan, B., Pritzel, A., Blundell, C.: Simple and scalable predictive uncertainty estimation using deep ensembles. In: Proc. NeurIPS, pp. 6402–6413 (2017)
24. Guo, C., et al.: On calibration of modern neural networks. In: Proc. ICML, pp. 1321–1330 (2017)
25. Niculescu-Mizil, A., Caruana, R.: Predicting good probabilities with supervised learning. In: Proc. ICML, pp. 625–632 (2005)
26. Montavon, G., Samek, W., Müller, K.R.: Methods for interpreting and understanding deep neural networks. Digit. Signal Process. **73**, 1–15 (2017)
27. Simonyan, K., Vedaldi, A., Zisserman, A.: Deep inside convolutional networks: Visualising image classification models and saliency maps. arXiv:1312.6034 (2013)
28. Selvaraju, R.R., et al.: Grad-CAM: Visual explanations from deep networks via gradient-based localisation. In: Proc. ICCV, pp. 618–626 (2016)
29. Lundberg, S.M., Lee, S.I.: A unified approach to interpreting model predictions. In: Proc. NeurIPS, pp. 4765–4774 (2017)
30. Molnar, C.: Interpretable Machine Learning: A Guide for Making Black Box Models Explainable, 3rd edn. (2025)

Road-Lighting Variant Search via Similarity and Conformity Graphs

Artur Basiura(✉)

Department of Applied Computer Science, AGH University of Science and Technology, Al.Mickiewicza 30, 30-059 Kraków, Poland
abasiura@agh.edu.pl

Abstract. Urban road-lighting modernization at city scale demands fast, repeatable evaluation of large configuration spaces across thousands of heterogeneous segments, which makes standards-compliant photometric optimization impractical to run exhaustively. In professional practice, redesign is therefore often executed in a manual, segment-by-segment workflow: each road segment is transferred into a dedicated tool, geometrically modeled, parametrized, and iteratively verified against the applicable standard, which becomes prohibitive at network scale. This paper demonstrates the practical use of Similarity and Conformity Graphs as an acceleration layer that replaces segment-by-segment analysis with an aggregation-first workflow. Comparable segments are grouped into conforming aggregates, and optimization and high-fidelity photometric evaluation are performed primarily for representative cases, reducing the number of required evaluations by several- to over an order-of-magnitude depending on the adopted strategy. Experiments on a real municipal investment dataset and three planning variants show that the conformity limit provides tunable accuracy and drives the estimated installed power toward the reference optimum, while substantially reducing the number of segments that require manual processing.

Keywords: graph · graph methods · LED lighting · roadway lighting · similarity graph · conformity graph

1 Introduction

Public road lighting is a large-scale municipal infrastructure, with an estimated 317 million streetlights worldwide and growth projected to about 363 million by 2027 [17]. It can account for roughly 40% of municipal electricity use and 1–3% of global electricity demand, while LED retrofits and control strategies can reduce energy consumption by 50–70% relative to legacy systems [4,13,16].

In practice, redesign is still often performed manually, segment by segment, in tools such as RELUX and DIALux by transferring each road layout, modeling geometry and mounting, and iteratively verifying compliance [5,6,10]. Repeating this workflow and evaluation across thousands of segments and multiple retrofit

W. Zamojski et al. (Eds.): DepCoS-RELCOMEX 2026, LNNS 2014, pp. 20–30, 2026.
https://doi.org/10.1007/978-3-032-27927-9_3

alternatives becomes a major bottleneck in early-stage planning and investment screening.

This paper addresses this bottleneck by using Similarity and Conformity Graph techniques [2] to enable *aggregation-first* variant search. Instead of designing and evaluating every segment independently, we group comparable segments into a substantially smaller set of conforming aggregates and perform optimization and high-fidelity evaluation primarily at the aggregate (representative) level. In practical city-scale settings, this can reduce the analysis workload from several thousand segments to only a few dozen representative cases, while retaining a controlled path to standards-compliant verification for the final candidates.

We validate the approach on a real municipal investment dataset and consider three practically relevant scenarios: new infrastructure with flexible pole spacing, constrained design with fixed pole spacing, and retrofit with unchanged pole locations and mounting height.

The remainder of the paper is organized as follows. The first part reviews related optimization and assessment approaches and motivates graph-based reuse, then summarizes the computational workflow and the similarity/conformity-graph construction. The subsequent part reports comparative results for the three scenarios and analyzes the accuracy–optimality behavior. The final section concludes and outlines directions for further research.

2 Related Work and Motivation

Road lighting design is typically validated through standards driven high fidelity photometric computations, which are essential for final compliance but become prohibitive for city scale variant search across thousands of lighting segments and many retrofit options. Previous work accelerates this process mainly by improving the search procedure (e.g. metaheuristics), imposing discrete decision structures (e.g. mixed-integer formulations) or reducing evaluation cost through surrogates and control oriented dimming strategies; however, these approaches still tend to require many expensive segment-level evaluations when applied at scale [3,7,9,11]. In parallel, graph-based and agent-based computational models have been explored to support the design, planning, and modernization of energy-efficient street-lighting, highlighting the benefits of explicit relational representations and distributed reasoning in practical deployments [12,15].

In this paper, we adopt the graph-centric reuse mechanism introduced in [2] and adapt it to our representation of the infrastructure. We model the road network as a Base Graph (BG), where the vertices represent homogeneous lighting situations (segments) described by geometric and contextual attributes, and where the admissible configuration space (liturgy catalog, mounting parameters, and operating regimes) is kept explicit and separable from the infrastructure state. This modeling choice is consistent with previous work on graph-based street lighting optimization and with broader foundations in distributed graph transformations, which motivate scalable transformations and reuse across large, structured problem instances [8,14]. On top of the BG, we construct a Similarity Graph (SG) that connects comparable segments using a selected similarity measure over segment descriptors, and we derive a Conformity Graph (CG) by applying a conformity filter that removes similarity relations deemed inadmissible for safe transfer. Beyond lighting-specific studies, our reuse pipeline is aligned with the general idea of obtaining a common graph representation to allow transfer and interoperability between heterogeneous structured descriptions [1]. The connected structure of CG enables optimizer-agnostic acceleration: instead of optimizing every segment independently, we optimize representatives and propagate (retrieve-and-refine) candidate lighting variants within conforming components, thereby reducing the number of high-cost evaluations. This positions SG/CG-driven variant search as a practical tool for rapid screening and early-stage prediction of power/energy impact, while reserving full standards-compliant verification for the final candidate set.

3 Case Study Results

To ensure complete comparability with previous results, we use the same data set as in [2], originating from the Kraków road-lighting investment implemented within the ISE project. The investment comprised the replacement of lighting equipment for 3,741 luminaires and covered 662 lighting segments that were modeled for computational analysis. Among them, 602 segments exhibit unique infrastructure descriptors (e.g., roadway width, segment length, and geometric layout), which implies that a large fraction of the network cannot be reduced to a small set of identical instances.

For the set of considered segments, we look for installation and operating parameters that minimize the total installed electrical power of the selected luminaires, under the photometric constraints of the CEN road-lighting framework (EN 13201).

In practical redesign and retrofit workflows, the decision space is driven by a small set of infrastructure and luminaire parameters. In the simplest setting (a single roadway), the optimization reduces to selecting the optimal combination of the following variables: the pole setback from the edge of the roadway, pole spacing, mounting height H, the length of the arm a, the type of luminaire F, the lumen flux ratio (LFR), the horizontal angle of mounting, the vertical angle of tilt and the rotation of the luminaire on the arm axis. More complex roadway cross-sections (e.g., parallel carriageways, bicycle lanes, or footpaths) extend this parameterization with additional geometric and photometric coupling constraints.

Even under conservative discretizations, the combinatorial burden is substantial. If each of the nine parameters is restricted to only ten admissible values, the resulting design space already exceeds the candidate solutions 10^9 for a single segment. Restricting the decision set to six parameters still yields 10^6 variants. In realistic scenarios, the effective space is larger due to richer admissible sets and expanding luminaire catalogs. For example, luminaire families can grow rapidly over time, which directly increases the branching factor of the search and makes exhaustive evaluation infeasible without dedicated acceleration strategies.

3.1 Evaluation Variants

The data set was analyzed under three investment driven variants, reflecting typical constraints encountered in municipal practice:

1. **Variant I (new infrastructure, flexible spacing):** design of a new lighting infrastructure where pole spacing is allowed to vary between 30 and 50 m.
2. **Variant II (existing electric infrastructure, new poles):** design of a lighting system with a fixed pole spacing of 20 m.
3. **Variant III (fixture retrofit):** retrofit scenario in which pole locations and height are not changed and the redesign focuses on fixtures (and potentially minor mounting adjustments).

As a result, the three variants differ in both feasible solution structure and computational complexity, making them suitable for assessing the practical benefits of similarity graph driven variant search under heterogeneous investment conditions.

3.2 Variant I: Full Infrastructure Optimization

Variant I considers a greenfield design in which the road-lighting infrastructure is planned from scratch, as in major reconstructions or new road investments. This setting offers the greatest optimization freedom and the largest potential for power reduction, but typically entails higher capital expenditure because pole locations and mounting geometry can be changed. The decision variables and admissible ranges are summarized in Table 1; under this discretization, the search space exceeds 5.4×10^9 variants per segment, motivating accelerated variant-search strategies.

Table 1. Variant I—optimization parameters.

Parameter	Min	Max	Step	# combinations
LFR [%]	1	100	1	100
Arm length a [m]	0	2	0.5	4
Inclination [deg]	0	30	1	30
Setback [m]	0	2	0.5	4
Mounting height H [m]	6	12	1	6
Pole spacing S [m]	30	60	1	30
Luminaire type F		–		625

Analyzing Variant I, we obtain the characteristics shown in Fig. 1. The figure illustrates how the estimated installed power varies with the comfortability limit; in the limit case, the method converges to the reference optimum of 99.8 kW for the analyzed infrastructure. Detailed numerical results are reported in Table 2.

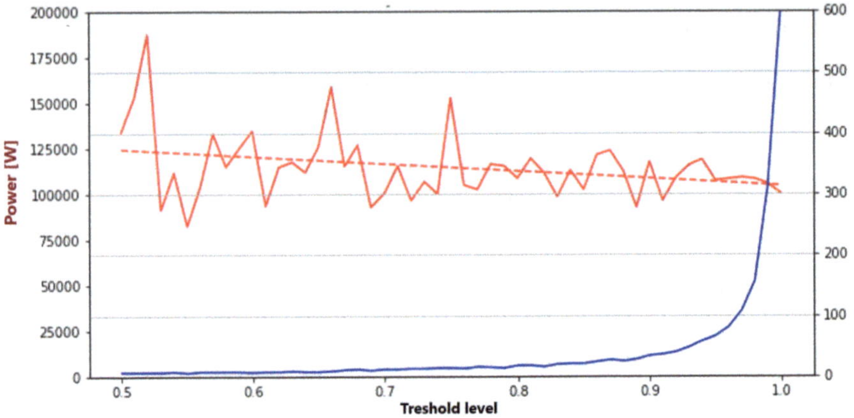

Fig. 1. Estimated power and optimization workload (segments optimized) in Variant I.

Table 2. Variant I—results

Conformity limit	# conformity subgraphs	Estimated power [kW]	Accuracy [%]
0.50	6	133.6	33.9
0.60	6	134.3	34.3
0.70	11	100.1	0.2
0.80	17	108.2	8.2
0.90	33	117.3	17.2
0.95	65	107.0	7.2
0.97	107	108.7	8.9
0.98	155	107.9	8.1
0.99	310	105.4	5.6
1.00	602	99.8	0.0

The results show clear convergence: as the comfortability limit increases, the Conformity Graph becomes more fine-grained, reuse decreases, and the estimated power approaches the baseline optimum. Low limits (0.5–0.6) yield coarse estimates with large errors (about 34%), whereas near the upper end accuracy improves (7.2% at 0.95) and reaches exact agreement at 1.0. Overall, lower limits reduce workload via stronger reuse but sacrifice accuracy, while higher limits improve accuracy by restricting transfer at the cost of more high-fidelity photometric evaluation.

3.3 Variant II: Optimization with Fixed Pole Spacing

Variant II differs from Variant I in that it reflects the design of a lighting installation for existing roads under constrained pole placement conditions. In such settings, pre-existing infrastructure is typically present in the right-of-way, including

Table 3. Variant II—optimization parameters.

Parameter	Min	Max	Step	# combinations
LFR [%]	1	100	1	100
Arm length a [m]	0	2	0.5	4
Inclination [deg]	0	30	1	30
Setback [m]	0	2	0.5	4
Mounting height H [m]	6	12	1	6
Pole spacing S [m]	20	20	1	1
Luminaire type F		–		625

Table 4. Variant II—results

Conformity limit	# conformity subgraphs	Estimated power [kW]	Accuracy [%]
0.50	6	100.47	26.2
0.60	6	88.60	11.3
0.70	11	75.90	-4.6
0.80	17	78.10	-1.9
0.90	33	84.70	6.4
0.95	65	79.90	0.4
0.97	107	82.20	3.3
0.98	155	82.20	3.3
0.99	310	81.50	2.4
1.00	602	79.60	0.0

electrical networks and underground utilities (e.g., gas and water pipelines). For safety and construct ability reasons, lighting poles cannot be placed arbitrarily close to these installations, restricting admissible pole locations to a limited set of feasible points. In this variant, we model this constraint by fixing the pole spacing to $S = 20$ m, representing a practically relevant case with limited freedom in pole placement.

The decision variables and admissible ranges used in Variant II are summarized in Table 3. Fixing S reduces the search space relative to Variant I to approximately 1.8×10^8 candidate variants. Analyzing Variant II, we obtain the characteristics shown in Fig. 2.

For the design space defined by fixed pole spacing ($S = 20$ m) and the remaining admissible parameters, the figure shows how the estimated installed power varies with the conformity limit. The corresponding detailed numerical results are reported in Table 4.

The results again show convergence toward the reference optimum. As the conformity limit increases, the Conformity Graph splits into more subgraphs, reuse is progressively reduced, and the estimated power approaches the baseline

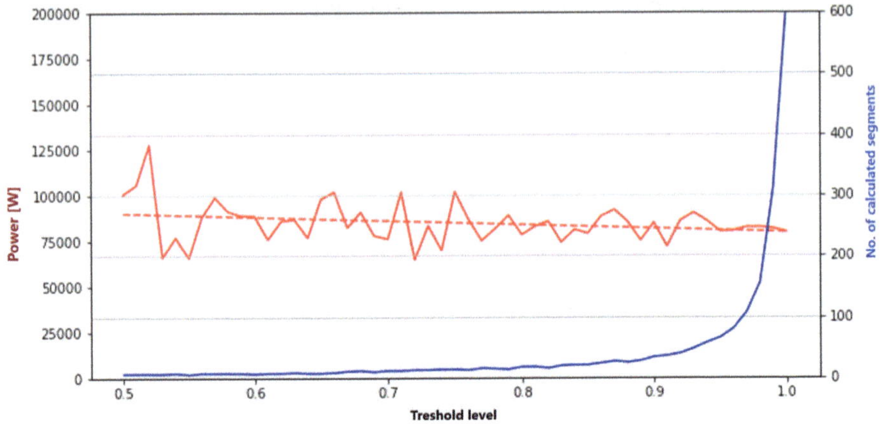

Fig. 2. Estimated power and optimization workload (segments optimized) in Variant II.

Table 5. Variant III—optimization parameters.

Parameter	Min	Max	Step	# combinations
LFR [%]	1	100	1	100
Arm length a [m]	0	2	0.25	8
Inclination [deg]	0	30	1	30
Setback [m]	fixed (as observed)			1
Mounting height H [m]	6	12	1	6
Pole spacing S [m]	fixed (as observed)			1
Luminaire type F		–		625

value of 79.6 kW. For low limits (0.50–0.60), the approximation is coarse and overestimates power (26.2% and 11.3% error). For tighter limits (0.95–0.99), the estimates stabilize within a few percent of the optimum (0.4%–3.3%). Overall, Variant II confirms that the conformity limit provides an effective accuracy control and that, in a constrained design space with fixed spacing, high-accuracy power estimates can be obtained without fully disabling reuse.

3.4 Variant III: Fixture Retrofit Optimization

Variant III represents a retrofit scenario in which an existing lighting installation is preserved and only fixture and mounting related parameters are modified. In this case, pole locations are not relocated; instead, the goal is to improve energy performance by selecting more suitable luminaires and adjusting the mounting configuration. Consequently, parameters that are determined by the existing infrastructure, in particular, the pole spacing and setback, are treated as fixed and are not included as decision variables. This variant corresponds to the retrofit setting analyzed in [2] and is included here as a reference case to ensure con-

Table 6. Variant III—results

Conformity limit	# conformity subgraphs	Estimated power [kW]	Accuracy [%]
0.50	6	133.6	30.6
0.60	6	130.3	27.4
0.70	11	89.5	-12.5
0.80	17	100.6	-1.7
0.90	33	131.3	28.3
0.95	65	103.5	1.2
0.97	107	116.7	14.1
0.98	155	114.2	11.6
0.99	310	106.4	4.0
1.00	602	102.3	0.0

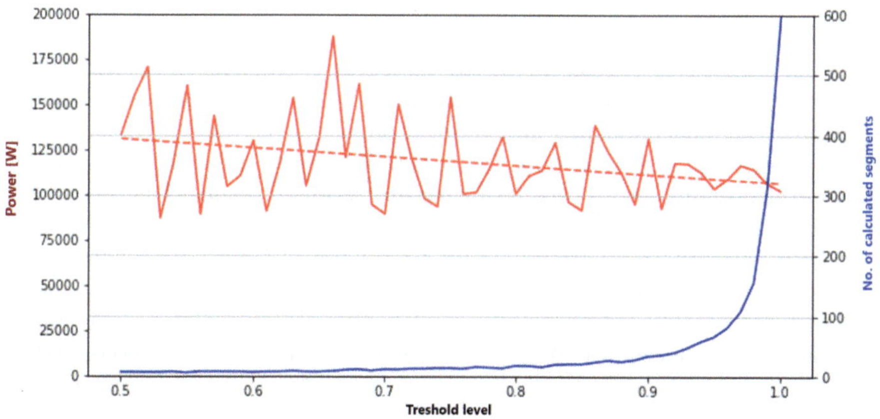

Fig. 3. Estimated power and optimization workload (segments optimized) in Variant III.

tinuity and comparability with prior results. The admissible parameter set for Variant III is summarized in Table 5. Compared to Variant II, this variant yields a smaller search space, with approximately 9.1×10^7 candidate variants.

Analyzing Variant III, we obtain the characteristics shown in Fig. 3. In this retrofit oriented setting, pole locations (and thus pole spacing and setback) are fixed to their observed values, and the search is limited to fixture selection and mounting related parameters. The figure shows the dependence of the estimated installed power on the conformity limit, while detailed numerical results are reported in Table 6. The reference optimum for this variant equals 102.3 kW.

The results indicate that the conformity limit again controls the accuracy of the estimate, but the convergence is less monotonic than in the previous variants. For low limits (0.50–0.60), strong reuse yields substantial overestimation of installed power (27.4%–30.6% error).

For tighter limits, the estimates generally move toward the optimum; in particular, at 0.95 the error drops to 1.2% and at 0.99 it remains within 4.0%.

Overall, Variant III shows that increasing the conformity limit drives the estimated power toward the reference optimum, even under strict retrofit constraints. At the same time, Similarity/Conformity Graphs enable fast, tunable power estimates that are suitable for rapid *what-if* screening and can support higher-level decision-making workflows before full compliance-grade photometric evaluation.

4 Conclusion

This paper studied a similarity-graph-driven variant search for road-lighting optimization, where conformity filtering reduces the number of required computations by transferring solutions within connected components of the Conformity Graph, i.e., across similar branches of the network. The results confirm that the Similarity/Conformity Graph concept provides an effective acceleration mechanism for large-scale studies in which comparable segment-level optimization problems recur across a network, substantially reducing the amount of manual segment-by-segment modeling and verification.

From a practical perspective, the results support a clear recommendation: a conformity limit of 0.95 provides a highly favorable trade-off between effort reduction and solution quality. Compared to the limit of 1.0, where transfer is effectively disabled and segments must be treated independently, the number of required representative cases drops by approximately one order of magnitude (from 602 to 65), while the corresponding power overhead remains limited to about 7% for Variant I, 0.4% for Variant II, and 1.2% for Variant III. In practical terms, this means that a threshold of 0.95 can be recommended as a default screening setting in professional workflows, especially in large network studies where photometric calculations, model preparation, and manual verification dominate the total effort. At the same time, the results also show that the final choice of threshold should remain variant-aware: Variant II tolerates transfer particularly well, Variant III remains robust with a small penalty, whereas Variant I is more sensitive and may require a more conservative setting when strict energy accuracy is prioritized.

Future work will focus on improving stability at intermediate thresholds and on systematic calibration of similarity descriptors, distance functions, and conformity parameters. Adaptive, context-dependent thresholding and uncertainty estimates for transferred solutions are also promising directions toward risk-aware screening and more robust decision support.

References

1. Basiura, A., Kotulski, L., Ziembinski, D.: Common graph representation of different XBRL taxonomies. In: Nguyen, N.T., Tran, T.K., Tukeyev, U., Hong, T., Trawinski, B., Szczerbicki, E. (eds.) Intelligent Information and Database Systems - 14th Asian Conference, ACIIDS 2022, Ho Chi Minh City, Vietnam, November 28–30, 2022, Proceedings, Part I. Lecture Notes in Computer Science, vol. 13757, pp. 507–515. Springer, Cham (2022). https://doi.org/10.1007/978-3-031-21743-2_40
2. Basiura, A., Sędziwy, A., Komnata, K.: Similarity and conformity graphs in lighting optimization and assessment. In: Paszynski, M., Kranzlmüller, D., Krzhizhanovskaya, V.V., Dongarra, J.J., Sloot, P.M.A. (eds.) Computational Science – ICCS 2021: 21st International Conference, Krakow, Poland, June 16–18, 2021, Proceedings, Part IV. Lecture Notes in Computer Science, vol. 12745, pp. 145–157. Springer, Cham (2021)https://doi.org/10.1007/978-3-030-77970-2_12
3. Castillo-Martinez, A., et al.: Particle swarm optimization for outdoor lighting design. Energies **10**(1), 141 (2017). https://doi.org/10.3390/en10010141
4. Clean Energy Ministerial (SEAD): Street lighting (2026). https://www.cleanenergyministerial.org/efficient_products/street-lighting/
5. DIAL GmbH: DIALux evo (lighting design software)
6. European Committee for Standardization (CEN): EN 13201 road lighting (parts 1–5) (2015)
7. Gómez-Lorente, D., Rabaza, O., Espín Estrella, A., Peña-García, A.: A new methodology for calculating roadway lighting design based on a multi-objective evolutionary algorithm. Expert Syst. Appl. **40**(6), 2156–2164 (2013). https://doi.org/10.1016/j.eswa.2012.10.026
8. Kotulski, L.: Rozproszone transformacje grafowe: teoria i zastosowania. Wydawnictwa AGH, Kraków. monografia; tytuł ang. spotykany w literaturze: Distributed Graph Transformations: Theory and Applications (2013)
9. Nourkojouri, H., Zomorodian, Z.S., Tahsildoost, M., Shaghaghian, Z.: A machine-learning framework for daylight and visual comfort assessment in early design stages. In: Proceedings of the 17th IBPSA Conference, pp. 1262–1268. Bruges, Belgium (2021). https://doi.org/10.26868/25222708.2021.30235
10. RELUX Informatik AG: ReluxDesktop (lighting and sensor planning software)
11. Rodríguez, C.C., Romero Quete, A.A., Suvire, G.O., Rivera, S.R.: Optimization of multi-period investment planning in street lighting systems by mixed-integer linear programming. Int. J. Simul. Multidisc. Design Optim. **14**, 14 (2023). https://doi.org/10.1051/smdo/2023017
12. Sędziwy, A.: Sustainable street lighting design supported by hypergraph-based computational model. Sustainability **8**(1), 13 (2016). https://doi.org/10.3390/su8010013
13. Sędziwy, A., Basiura, A.: Energy reduction in roadway lighting achieved with novel design approach and LEDs. LEUKOS **14**(1), 45–51 (2018). https://doi.org/10.1080/15502724.2017.1330155
14. Sędziwy, A., Kotulski, L.: Graph-based optimization of energy efficiency of street lighting. In: Rutkowski, L., Korytkowski, M., Scherer, R., Tadeusiewicz, R., Zadeh, L.A., Zurada, J.M. (eds.) Artificial Intelligence and Soft Computing: 14th International Conference, ICAISC 2015, Zakopane, Poland, June 14–18, 2015, Proceedings, Part II. Lecture Notes in Computer Science, vol. 9120, pp. 515–526. Springer, Cham (2015). https://doi.org/10.1007/978-3-319-19369-4_46

15. Sedziwy, A., Kotulski, L., Basiura, A.: Multi-agent support for street lighting modernization planning. In: Nguyen, N.T., Gaol, F.L., Hong, T., Trawinski, B. (eds.) Intelligent Information and Database Systems - 11th Asian Conference, ACIIDS 2019, Yogyakarta, Indonesia, April 8–11, 2019, Proceedings, Part I. Lecture Notes in Computer Science, vol. 11431, pp. 442–452. Springer, Cham (2019). https://doi.org/10.1007/978-3-030-14799-0_38
16. Valetti, L., Piccablotto, G., Taraglio, R., Pellegrino: long-term monitoring campaign of LED street lighting systems: focus on photometric performances, maintenance and energy savings. Sustainability **15**(24), 16910 (2023). https://doi.org/10.3390/su152416910
17. Zissis, G., Bertoldi, P., Ribeiro Serrenho, T.: Update on the status of LED-lighting world market since 2018. Tech. rep., European Commission, Joint Research Centre (JRC) (2020), https://publications.jrc.ec.europa.eu/repository/handle/JRC122760

Reinforcement Learning Based Control of Quantum Key Distribution Protocol

Aleksandra Birut(✉) and Paweł Ksieniewicz

Department of Systems and Computer Networks, Wrocław University of Science and Technology, Wybrzeze Wyspianskiego 27, 50-370 Wrocław, Poland
aleksandrabirut@gmail.com, pawel.ksieniewicz@pwr.edu.pl

Abstract. This article explores the intersection of quantum cryptography and machine learning, addressing the challenges of real-world applications of the E91 Quantum Key Distribution protocol. Standard approaches utilize fixed measurement bases for both sides of the communication, which are optimal only under ideal conditions, causing inefficiency in high-noise scenarios. This article proposes the implementation of the reinforcement learning agent capable of adjusting measurement bases to improve parameters - the CHSH inequality parameter S and the Quantum Bit Error Rate (QBER) without knowledge of noise model. The problem was formulated as a control task in a custom Gymnasium environment with a simulation of the E91 protocol using Qiskit, targeting phase flip noise with error probabilities up to 0.25. Obtained results showed that the application of reinforcement learning can improve communication compared to the standard implementation. The final agent matched or exceeded the theoretical baseline across the entire range of noise levels tested. Furthermore, it extended the operational range of communication: while the standard protocol failed at an error probability of $p_{error} = 0.175$, the agent maintained the parameter $S > 2$ up to $p_{error} \approx 0.23$. Additionally, in high-noise conditions, the agent successfully reduced the QBER from 0.09 to 0.03.

Keywords: E91 Protocol · Deep Q-Network · Quantum Communication

1 Introduction

The emerging field of quantum cryptography offers novel cryptographic solutions, especially in a field of Quantum Key Distribution (QKD) [20]. Unlike classical systems, security of exchanged information is guaranteed by fundamental laws of physics and not by computational complexity. Additionally, obtained key consists of purely random bits due to the probabilistic nature of quantum measurements [12].

The E91 protocol is based on the quantum entanglement phenomenon to generate a shared secret key [11]. The procedure of communication between

W. Zamojski et al. (Eds.): DepCoS-RELCOMEX 2026, LNNS 2014, pp. 31–42, 2026.
https://doi.org/10.1007/978-3-032-27927-9_4

Alice and Bob can be divided into four distinct stages. Firstly, a sequence of N entangled qubit pairs is generated either by independent source or by side of communication and distributed to both parties. For every incoming qubit, Alice and Bob independently and randomly choose a measurement basis to measure the spin projection. Alice chooses her basis $A \in \{X, W, Z\}$ and Bob chooses basis $B \in \{W, Z, V\}$. Then, using a public authenticated classical channel, they announce which basis they used for each measurement, looking for instances where their choices coincided. If both chose the same basis, they know their results are perfectly anti-correlated due to entanglement. These matching instances form the key.

The last stage is security check – the measurement results from the non-matching bases are utilized to calculate the Clauser-Horne-Shimony-Holt inequality (CHSH) correlation parameter S defined as follows:

$$S = \langle X, W \rangle - \langle X, V \rangle + \langle Z, W \rangle + \langle Z, V \rangle, \tag{1}$$

where $\langle A, B \rangle$ represents the expectation value of the correlation between results obtained in bases A and B.

In ideal conditions, the theoretical maximum for each correlation is $|\langle A, B \rangle| = \frac{1}{\sqrt{2}}$. Consequently, observing value of S that exceeds the classical limit ($S > 2$) confirms that the qubit pairs were entangled and the channel remains secure and, on the other hand, if $S \leq 2$, the communication must be aborted, as the generated key is not secure.

However, the quantum channel in real-world implementations is never ideal [19]. Noise processes reduce the degree of entanglement between qubits, degrading the CHSH correlation parameter, introducing errors in final keys [26]. Since security check relies exclusively on CHSH parameter, its noise-induced degradation causes premature termination of communication under noise conditions even without eavesdropping.

Noise in a quantum channel generally manifests in two ways: as unwanted unitary operations on qubit or as energy relaxation – formally known as amplitude damping. Unlike the non-unitary measurement performed by an eavesdropper – which collapses the wavefunction and destroys entanglement – a unitary error preserves the quantum correlations but changes the basis in which they are maximally observable. This leads to the hypothesis – since noise acts as a rotation of the quantum state, its effects can be mitigated by applying an adequate operation, which is not possible for unwanted measurement introduced by eavesdropper.

Whereas standard implementation of E91 protocol utilizes fixed measurement bases, which are optimal only under ideal conditions, this article proposes a modification to the protocol, where measurement bases are adjusted by reinforcement learning (RL) agent without direct observations of the channel parameters. The research aims to optimize the performance of E91 protocol in presence of noise in terms of CHSH parameter and Quantum Bit Error Rate compared to the standard implementation of the protocol.

The research operates within limitations due to the use of a simulation environment. This choice was dictated by the need for a reproducible environment to establish a baseline for the agent's performance without the fluctuations of current Noisy Intermediate-Scale Quantum (NISQ) devices. However, it is important to acknowledge that results obtained via simulation may differ from those on real quantum circuits due to hardware-specific noise. Optimization focuses strictly on raw communication metrics at the key exchange stage, excluding post-processing steps such as privacy amplification and information reconciliation. Additionally, the goal is to demonstrate that an RL approach can improve performance, rather than to identify the optimal RL algorithm. Therefore, the study focuses on a single RL architecture and does not compare other approaches.

2 Literature Review

The E91 protocol was proposed by Artur Ekert in 1991 [10]. It was the second most prominent Quantum Key Distribution protocol following the BB84 scheme proposed earlier by Charles Bennett and Gilles Brassard in 1984 [4] (republished in 2014 [5]).

In 1964, John Stewart Bell proposed an experiment demonstrating that quantum mechanics is incompatible with local realism [3], which had the possibility to prove the phenomenon of entanglement âĂŞ security guarantee of E91 protocol. Subsequently, in 1969, Clauser, Horne, Shimony, and Holt derived a specific inequality (CHSH) suitable for experimental verification [8]. This verification was carried out successfully by independent teams led by John Clauser, Alain Aspect, and Anton Zeilinger, for which they were awarded the Nobel Prize in Physics in 2022 [1,7,27].

Since these foundational discoveries, numerous attempts have been made to physically implement QKD protocols. A comparative study of various QKD protocols is presented in [9]. Additionally, noise models and their impact on transmission parameters, as well as comparisons between eavesdropping and non-eavesdropping scenarios, are discussed in [15,28]. In [22], the author analyzes the degradation of the signal with increasing communication distance and compares these effects to error correction using Shor's algorithm. Furthermore, [14] outlines various attack scenarios in QKD protocols, including those that use deep machine learning techniques. Artificial intelligence algorithms find applications beyond attack simulations, offering accelerated training, optimization of transmission parameters and key generation, eavesdropping detection, and error correction. A comprehensive review of current solutions in this domain is provided in [21].

Reinforcement learning has emerged as a promising technology for agent optimizing in complex control tasks. In 2015, Mnih et al. of DeepMind introduced the Double Deep Q-Network (DDQN) agent, which achieved human-level performance across a diverse set of Atari 2600 games [17,18]. Furthermore, AlphaGo and AlphaGo Zero agents successfully mastered the game of Go, defeating the world champion and demonstrating that RL agents can discover new strategies in environments with high state-space complexity [23].

The success of deep reinforcement learning in classical domains has inspired research at the intersection of artificial intelligence and quantum information science. It involves both applying classical RL to optimize quantum control problems [16], as well as exploiting quantum properties to improve learning algorithms themselves. For example, Quantum Neural Networks (QNNs) implemented via Parametrized Quantum Circuits (PQC) (also called Variational Quantum Circuits (VQC)) have been proposed as quantum analogs to classical neural networks for the approximation of the value function [6,24].

Direct application of reinforcement learning to Quantum Key Distribution protocols was presented in a recent study by Behera et al. [2]. The authors took a hybrid approach that combined reinforcement learning with quantum neural networks to optimize the BB84 and B92 protocols. Furthermore, they proposed a novel protocol relying on measurements in arbitrary bases selected by an RL agent. While this article also explores the use of a reinforcement learning agent to adjust measurement bases, it diverges by employing classical deep reinforcement learning on the standard E91 protocol as a noise mitigation strategy. To the best of our knowledge, this specific application of classical deep reinforcement learning to the E91 protocol remains unexplored in the existing literature.

3 Implementation

The purpose of this section is to present the preliminary experiment, including a detailed description of the simulation of the communication protocol 3.1 and a formal definition of the reinforcement learning task 3.2.

3.1 Simulation of Communication Protocol

he simulation was developed in Python (v3.10.12) using the Qiskit framework (v2.2.1) and the `AerSimulator` backend (`qiskit-aer` v0.17.2) [13]. It was implemented as a custom environment within the Gymnasium framework (v1.2.2) [25].

In the `AerSimulator` environment, custom noise models were constructed using Qiskit `NoiseModel` class. The core implementation relies on the `pauli_error` function, which defines a quantum error as stochastic Pauli operators. In this model, Pauli Z error is applied with probability p_{gate} after each gate operation, while the identity operator I is applied with probability $(1 - p_{gate})$. This type of error rotates the state around the Z-axis, altering the phase relationship within a superposition (changing $|0\rangle + |1\rangle$ to $|0\rangle - |1\rangle$) and frequently occurs in optical fibers due to temperature fluctuations or physical bending.

With developed simulation of communication under noise conditions, the impact of the error probability p_{gate} on the parameter S and the Quantum Bit Error Rate was simulated for later comparison with developed method. The security parameter S dropped below the classical limit at error probability of 0.175, corresponding to QBER of 0.07.

Additionally, with help of the simulation, relationship between the CHSH parameter S and measurement angles θ used during key exchange was analyzed for justification of hypothesis that phase flip error changes the optimal bases for measurement. The results are presented in Fig. 1 and demonstrate that the maximum value of the S parameter is achieved for different angles under changing noise conditions and differs for theoretically optimal angles used in standard implementation of protocol $(-0.25\pi, 0.25\pi)$. For error probabilities $p_{error} \geq 0.15$ the value of S at the standard angle drops below 2, while a global maximum exceeding this value exists for different angles.

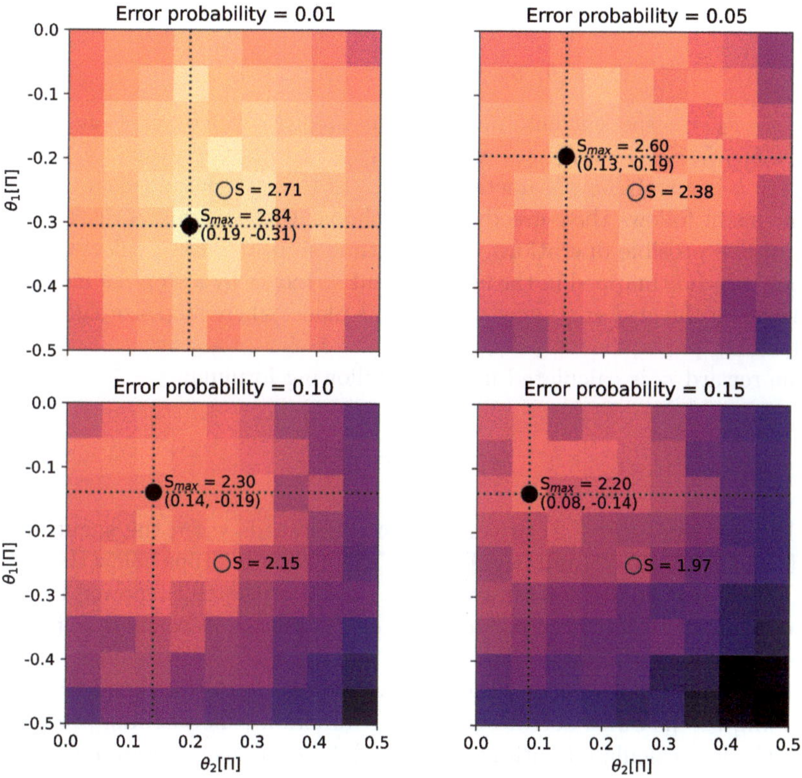

Fig. 1. Relationship between the CHSH parameter S and measurement angles θ under different noise conditions.

This confirms that changing angles can achieve higher values of S than the theoretically optimal ones. Therefore, the objective of the reinforcement learning agent could be formulated. Specifically, the goal was to improve the S and QBER parameters under phase flip noise conditions (with error probabilities up to 0.25, targeting QBER < 0.11) by adjusting the measurement angles in a quantum circuit simulation using the Qiskit `AerSimulator`.

3.2 Agent

The reinforcement learning agent architecture follows the one proposed in [18] implementing Double Deep Q-Network approach.

The observation space was defined as a Box space consisting of two controlled angles that replace gates W and V. Recent history of parameter S was added to the observation vector to address the problem of partial observability. In addition, to help the agent distinguish statistical noise from the consequences of its own actions, the last action taken is included in the observation. The observation vector O_t consists of $2 + k + 1$ elements:

$$O_t = [\theta_1, \theta_2, S_t, S_{t-1}, \ldots, S_{t-k+1}, a_{t-1}],$$

where θ_1, θ_2 are the current measurement angles divided by π in the range $[-0.5, 0.5]$, $S_t, S_{t-1}, \ldots, S_{t-k+1}$ is a history of the last k values of parameter S, where k is a training parameter and a_{t-1} is the last action taken.

The agent follows the ϵ-greedy policy, choosing action from a discrete action space of five possible operations. At each time step, the agent selects an action number, which is mapped in the environment as moves by a step $\Delta\theta$ in 4 possible directions on the grid or staying at the same place, where step size $\Delta\theta$ is defined as a training parameter.

The reward r_t is calculated using the following formula:

$$r_t = \left(\frac{|S_t| - 2}{2\sqrt{2} - 2} \right)^3. \tag{2}$$

The base term $(|S| - 2)$ produces positive rewards when the agent achieves quantum effects ($S > 2$) and penalizes the agent when S drops below the classical limit. The term is normalized to 1 and raised to the third power to further promote maximal S and introduce greater differences in settings approaching maximum to ensure stability.

The Online and Target networks consist of two fully connected layers, each containing 128 neurons, activated by ReLU function. The learning process minimizes the loss calculated with `HuberLoss()` function using `Adam` optimizer.

Additionally, two generalization strategies were implemented to help with adaptation to different noise conditions:

- *Domain Randomization* – the error probability is sampled from a uniform distribution at the beginning of each episode to force the agent to actively look for maximum, rather than memorizing specific angle values for each noise level.
- *Curriculum Learning* – due to the wide range of potential noise and the stochastic nature of quantum simulation results, the learning process required facilitation. Exposing the agent to high noise levels at the beginning of the training caused convergence issues, as the reward became too chaotic. Therefore, the training was divided into two phases. In the first one, the range of error probability was gradually increased. This allows the agent to first

Table 1. Phases of Curriculum Learning.

Phase	Episodes range [% of total]	Error probability range
I	[0,10)	[0, 0.05]
	[10, 20)	[0, 0.1]
	[20, 30)	[0.05, 0.15]
	[30, 40)	[0.1,0.2]
	[40, 50)	[0.15,0.3]
II	[50, 100)	20% probability: [0,0.1] 60% probability: [0,0.3] 20% probability: [0.2,0.3]

adapt to the environment dynamics and stochastic rewards. In the second phase, the training was aimed at stabilizing the policy with a greater focus on extreme scenarios to prevent forgetting. The specific error ranges for each training phase are presented in Table 1.

The complete set of hyperparameters used in the final training is presented in Table 2. The training was executed on Google Colab Pro using Nvidia L4 GPU and a 12-core CPU.

Table 2. Hyperparamteres used in the final training.

Hyperparameter	Value	Hyperparameter	Value
Number of episodes	9000	Number of steps in each episode	200
Batch size	128	Buffer size	$1e^5$
Update rate to the target network	0.05	Learning rate	$1e^{-5}$
Discount factor	0.99	Start epsilon	1
Final epsilon	0.05	Epsilon decay end episode	5000
Burn-in phase	1000	Number of historical S in observation	7
Possible angle adjustment $\Delta\theta$	$0.05\ \pi$		

4 Results and Discussion

Developed algorithm requires assessment of both dynamic properties for the velocity of calibration and optimization relative to the theoretical baseline. Hence, two tests of the trained model were performed.

In the auto-calibration test, the agent was expected to achieve a level of S parameter above the classical limit (if possible) within the first 10% steps of episode. The results are presented in Fig. 2. The black solid line represents the parameter S with the classical limit $S = 2$ marked to show the acceptance criterion as a black dashed line. The agent consistently exceeds the classical limit within the first 20 steps of the episode. After this initial convergence phase, the S parameter stabilizes and remains above the threshold until the end of the episode with fluctuations caused by the stochastic nature of the simulation measurements. The Quantum Bit Error Rate (magenta line) stays at an acceptable level during the test.

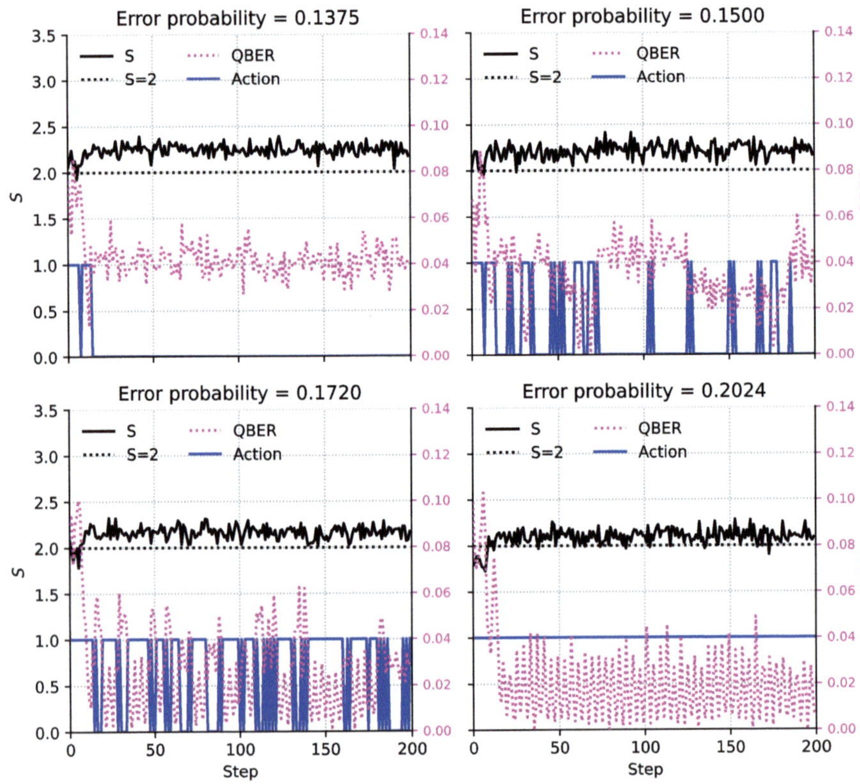

Fig. 2. Agent's performance in auto-calibration test for increasing error probabilities.

In terms of stability, the agent's behavior depends on the noise level. For error probabilities $p_{error} \leq 0.13$ agent is perfectly stable. As shown by the blue line representing taking action (1 for step in any direction, 0 for staying at the same place), once the agent finds the optimal angles, it stops moving completely. In higher noise levels, the agent gradually increases the number of adjustments, and for $p_{error} \geq 0.2$ it moves continuously. However, this is a desired behavior since S oscillates near the classical limit and the agent tries to adjust the angles to maintain connection, but higher values cannot be achieved because of noise.

In the second scenario, the agent was evaluated for its ability to optimize the standard E91 protocol implementation. To ensure a fair comparison, each episode was initialized at the theoretical maximum to avoid wasting time for calibration, which was assessed in the previous test. The comparison between the agent and the theoretical baseline revealed interesting insights. Initially, when tested with the same step size $\Delta\theta = 0.05\pi$ (Fig. 3) used during training, the agent performed slightly worse than the baseline in the range of $p_{error} \in [0.05, 0.125]$. To check if that was a precision issue and the agent overshot maximum, the step size was reduced to $\Delta\theta = 0.01\pi$ for the testing phase (Fig. 3). With this adjustment, the agent performed same or better than the theoretical baseline under all conditions. This suggests that further experimentation with step size can be performed, or a dynamic step adjustment would be beneficial.

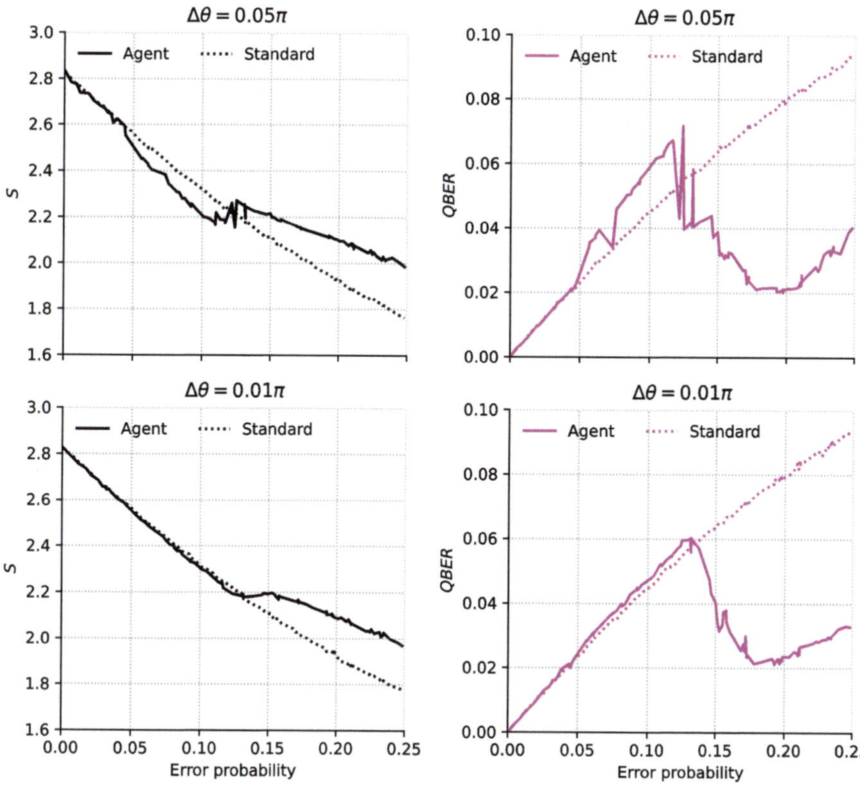

Fig. 3. Comparison between communication parameters achieved by the agent and the theoretical baseline with step size $\Delta\theta = 0.05\pi$ and $\Delta\theta = 0.01\pi$.

The highest increase in performance was observed at noise levels where $p_{error} > 0.15$. The agent extended the operational range for secure communication from $p_{error} = 0.175$ for standard implementation up to $p_{error} \approx 0.23$.

Additionally, the agent drastically reduced the QBER from 0.09 to 0.03 for high noise levels.

However, an analysis of the performance trend suggests space for further optimization. Although the agent matches the baseline in the range $p_{error} \in [0.1, 0.15]$, the trend indicates the potential for better performance. Additionally, the training curve dropped exactly when these error rates were introduced, suggesting that the Curriculum Learning scheme could be improved. Extending the training duration for this noise range or adding more samples during the stabilization phase could help the agent in these scenarios.

5 Conclusion

The experiments confirmed that the proposed agent effectively restores the S parameter above the classical threshold. In scenarios where the standard static protocol failed due to noise, the proposed agent successfully identified measurement angles that maintained $S > 2$ for error probabilities up to $p_{error} = 0.23$ compared to $p_{error} = 0.175$ for the standard protocol.

The trained agent operates effectively in unknown environmental conditions, eliminating the need for explicit channel modeling or waste of key bits for parameter estimation. The proposed method of angle control can benefit the broader range of control problems. This solution can be easily adapted to different problems in the quantum information science domain. However, agent operates in discrete action space due to limitations of Deep Q-Network architecture, which limits the precision of angles settings.

Acknowledgments. This work was supported by the statutory funds of the Department of Systems and Computer Networks, Faculty of Information and Communication Technology, Wrocław University of Science and Technology.

Disclosure of Interests. The authors have no competing interests to declare that are relevant to the content of this article.

References

1. Aspect, A., Grangier, P., Roger, G.: Experimental realization of Einstein-Podolsky-Rosen-Bohm Gedankenexperiment: a new violation of Bell's inequalities. Tech. Rep. 2 (1982)
2. Behera, B.K., Al-Kuwari, S., Farouk, A.: QNN-QRL: quantum neural network integrated with quantum reinforcement learning for quantum key distribution (2025). http://arxiv.org/abs/2501.18188
3. Bell, J.: On The Eistein Podolsky Rosen Paradox. Physics **1**(3), 195–200 (1964)
4. Bennett, C., Brassard, G.: Quantum cryptography: public key distribution and coin tossing. In: International Conference on Computers, Systems & Digital Processing (1984)
5. Bennett, C.H., Brassard, G.: Quantum cryptography: public key distribution and coin tossing. Theor. Compt. Sci. **560**, 7–11 (2020). https://doi.org/10.1016/j.tcs.2014.05.025

6. Chen, S.Y.C.: An Introduction to Quantum Reinforcement Learning (QRL) (2024). http://arxiv.org/abs/2409.05846
7. Clauser, J., Freedman, S.: Experimental test of local hidden-variable theories (1972)
8. Clauser, J.F., Horne, M.A., Shimony, A., Holt, R.A.: Proposed experiment to test local hidden-variable theories. Phys. Rev. Lett. **23**(15), 880–884 (1969) https://doi.org/10.1103/PhysRevLett.23.880
9. Dhakal, P., Dawadi, B.R., Adhikari, N.B.: Performance analysis of different quantum key distribution protocols for optimised security and efficiency. IET Quantum Commun. **6**(1) (2025). https://doi.org/10.1049/qtc2.70015
10. Ekert, A.K.: Quantum cryptography based on Bell's theorem. Phys. Rev. Lett. **67**(6), 661–663 (1991). https://doi.org/10.1103/PhysRevLett.67.661
11. Gisin, N., Ribordy, G., Tittel, W., Zbinden, H.: Quantum cryptography. Rev. Mod. Phys. **74**(1), 145–195 (2002). https://doi.org/10.1103/revmodphys.74.145
12. Herrero-Collantes, M., Garcia-Escartin, J.C.: Quantum random number generators. Rev. Modern Phys. **89** (2016). https://doi.org/10.1103/RevModPhys.89.015004
13. Javadi-Abhari, A., et al.: Quantum computing with Qiskit (2024). https://doi.org/10.48550/arXiv.2405.08810
14. Lejeune, T., Damanet, F.: Deep-learning-based continuous attacks on quantum key distribution protocols (2025). http://arxiv.org/abs/2408.12571
15. Li, L., et al.: The security analysis of E91 protocol in collective-rotation noise channel. Int. J. Distrib. Sens. Networks **14**(5) (2018). https://doi.org/10.1177/1550147718778192
16. Mackeprang, J., Dasari, D.B., Wrachtrup, J.: A reinforcement learning approach for quantum state engineering. Quantum Mach. Intell. **2**(1) (2020). https://doi.org/10.1007/s42484-020-00016-8
17. Mnih, V., et al.: Playing Atari with Deep Reinforcement Learning (2013)
18. Mnih, V., et al.: Human-level control through deep reinforcement learning. Nature **518**(7540), 529–533 (2015). https://doi.org/10.1038/nature14236
19. Pan, J.W., Chen, Z.B., Lu, C.Y., Weinfurter, H., Zeilinger, A., Żukowski, M.: Multiphoton entanglement and interferometry. Rev. Mod. Phys. **84**(2), 777–838 (2012). https://doi.org/10.1103/revmodphys.84.777
20. Pirandola, S., et al.: Advances in quantum cryptography. Adv. Optics Photonics **12**(4), 1012 (2020). https://doi.org/10.1364/aop.361502
21. Purohit, K., Vyas, A.K.: Quantum key distribution through quantum machine learning: a research review. Front. Quantum Sci. Technol. **4**(5) (2025). https://doi.org/10.3389/frqst.2025.1575498
22. Reji Kurien, T.: Quantum communication through entangled particles in an alternate reality-breaking the speed-of-light barrier and beyond. Int. J. Modern Eng. Res. **14**(5), 88–98 (10 2024). www.ijmer.com
23. Silver, D., et al.: Mastering the game of Go with deep neural networks and tree search. Nature **529**(7587) (2016). https://doi.org/10.1038/nature16961
24. Skolik, A., Jerbi, S., Dunjko, V.: Quantum agents in the Gym: a variational quantum algorithm for deep Q-learning. Quantum **6** (2022). https://doi.org/10.22331/Q-2022-05-24-720
25. Towers, M., et al.: Gymnasium: a standard interface for reinforcement learning environments. arXiv preprint arXiv:2407.17032 (2024)
26. Ursin, R., et al.: Entanglement-based quantum communication over 144km. Nature Phys. **3**(7), 481–486 (2007). https://doi.org/10.1038/nphys629

27. Weihs, G., Jennewein, T., Simon, C., Weinfurter, H., Zeilinger, A.: Violation of Bell's inequality under strict Einstein locality conditions. Tech. Rep. (1998)
28. Zulfa, M.F., Anwar, K.: Development of quantum key distribution (QKD) with E91 protocol for future secure quantum networks. In: Journal of Physics: Conference Series, vol. 2980. Institute of Physics (2025). https://doi.org/10.1088/1742-6596/2980/1/012038

On Quantum Annealing Chain Strength Related Parameters Control in D-Wave for the Single Machine Total Weighted Tardiness Problem

Wojciech Bożejko[1], Filip Górajewski[2(✉)], Yana Baloshenko[2], Oskar Siemieniuk[2], and Jarosław Wielgus[2]

[1] Department of Control and Quantum Computing, Wroclaw University of Science and Technology, Wyb. Wyspiańskiego 27, 50-370 Wrocław, Poland
wojciech.bozejko@pwr.edu.pl

[2] Faculty of Information and Communication Technology, Wroclaw University of Science and Technology, Wyb. Wyspiańskiego 27, 50-370 Wrocław, Poland

Abstract. With the emergence of Noisy Intermediate-Scale Quantum (NISQ) devices, quantum annealing (QA) has become a viable approach for solving NP-hard combinatorial optimization problems, such as the Single Machine Total Weighted Tardiness Problem (SMTWTP), by mapping them to Quadratic Unconstrained Binary Optimization (QUBO) formulations. This paper investigates the impact of hardware-specific parameters on the performance of D-Wave Advantage systems when solving SMTWTP instances. The influence of chain strength tuning, chain break fractions, embedding size, and physical qubit connectivity across generations of the Advantage architecture is analysed. The study provides practical insights for parameter tuning in scheduling problems, highlighting the trade-offs between physical constraints and optimization accuracy.

Keywords: combinatorial optimization · scheduling algorithms · NISQ computing · QUBO · minor-embedding · chain strength

1 Introduction

Combinatorial optimization serves as the backbone of industrial logistics and manufacturing. The SMTWTP remains a benchmark for testing scheduling algorithms due to its non-linear penalty structure. As traditional solvers approach their scaling limits, quantum technologies offer a paradigm shift by utilizing quantum tunnelling to navigate complex energy landscapes. The current state of quantum hardware is characterized by the NISQ era, where devices are limited by qubit decoherence and sparse connectivity. Quantum computing primarily operates through two distinct paradigms: the gate-based model and Quantum Annealing (QA). The gate-based model utilizes a sequence of quantum logic gates

W. Zamojski et al. (Eds.): DepCoS-RELCOMEX 2026, LNNS 2014, pp. 43–54, 2026.
https://doi.org/10.1007/978-3-032-27927-9_5

to manipulate qubits and perform general-purpose computations. In contrast, QA is a specialized approach that exploits quantum fluctuations to navigate an energy landscape and identify its global minimum. D-Wave Systems has been at the forefront of this latter approach since its founding in 1999, releasing the first commercially available quantum computer in 2011. The company's hardware has since evolved through multiple architectural generations, progressing to the current Advantage and Advantage2 generations.

The primary objective of this paper is to evaluate how specific D-Wave QPU parameters affect the feasibility and quality of SMTWTP solutions. This paper analyses chain behaviour and the role of chain strength tuning, chain break fractions, and embedding size across different QPU topologies. The work contributes to the field by providing empirical evidence on parameter sensitivity and QPU-dependent hardware constraints for scheduling tasks on D-Wave architectures.

The remainder of this paper is structured as follows: Sect. 2 provides a literature review of quantum annealing applications in scheduling. Section 3 defines the SMTWTP and details its mathematical formulation. Section 4 describes the methodology, including the complete transformation of the SMTWTP into a QUBO model. Section 5 presents experimental results from the D-Wave Advantage systems, analysing parameter sensitivity. Finally, Sect. 6 summarizes our conclusions and proposes avenues for future research.

2 Literature Review

The application of quantum annealing to scheduling has evolved from theoretical proofs-of-concept to empirical performance studies. Early work focused on the Chimera architecture, but recent literature has shifted towards the Pegasus topology, which provides higher connectivity (K_{24}), as presented by Pelofske [5] and included in D-Wave documentation [4]. Comparative studies across three generations of D-Wave hardware–Advantage, Advantage2, and earlier models–reveal a consistent improvement in sampling capability, with newer Zephyr-based architectures demonstrating superior approximation ratios and lower chain break frequencies for complex combinatorial problems [5,8]. Bożejko et al. [1,2] demonstrated the potential of QA-driven Branch and Bound methods, showing that quantum heuristics can effectively prune search trees for tardiness-related problems.

Recent studies by Pelofske [5] and Park et al. [11] have highlighted that minor embedding–the process of mapping logical variables to physical qubits–introduces 'chains' of qubits that must behave as a single logical unit through ferromagnetic coupling. If the coupling strength within these chains (chain strength) is insufficient, the chain breaks, leading to invalid solutions. Conversely, excessive chain strength can 'freeze' the system prematurely, obscuring the original problem's energy landscape and reducing accuracy, as observed by Pelofske and Weinberg et al. [6,11]. While the D-Wave Ocean SDK provides a default heuristic known as uniform torque compensation described in the documentation [4], empirical evidence suggests it often fails to reach the 'optimal' coupling necessary for maximizing the probability of observing the global minimum, as also mentioned by Park et al. [11]. Research by Giergiel et al. [3] and Park [11] indicates that an optimal chain strength typically resides near 1.2 times the critical value required to prevent breaks, suggesting that problem-aware calibration is essential for hybrid workflows to surpass classical meta-heuristics.

The physical performance of these devices is also influenced by diabatic effects and noise mechanisms. Scaling studies of Weinberg et al. [6] on transverse-field Using models have identified an optimal annealing rate where the competing effects of reduced quantum fluctuations and increasing environmental noise are balanced.

Despite these advancements, a research gap exists regarding the systematic tuning of annealing parameters specifically for the SMTWTP on the Advantage generations v4.1 and v6.4. Most literature focuses on the Maximum Clique, Max-Cut, or Travelling Salesperson Problem (TSP), such as Pelofske and Sharma et al. [5,9], leaving scheduling-specific constraints like precedence and non-overlapping penalties under-explored in the context of physical QPU noise and diabatic effects.

3 Problem Formulation

The SMTWTP involves a set of jobs $\mathcal{N} = \{1, 2, ..., n\}$ to be processed on a single machine. Each job $i \in \mathcal{N}$ has a processing time p_i, weight w_i, and due date d_i. Let C_i be the completion time of job i. The tardiness T_i is defined as:

$$T_i = \max\{C_i - d_i, 0\}. \tag{1}$$

The goal is to minimize:

$$\min_S \sum_{i=1}^{n} w_i T_i, \tag{2}$$

subject to non-overlapping constraints:

$$C_i \leq S_j \vee C_j \leq S_i, \quad \forall i, j \in \mathcal{N}, i \neq j$$

where S is a set of starting times $S = \{S_1, S_2, ..., S_n\}$, $S_i, C_i \geq 0$, $i = 1, 2, \ldots, n$. A job i cannot be interrupted: $C_i = S_i + p_i$. [12]

4 Methodology

The SMTWTP is a combinatorial scheduling problem that must be expressed as a QUBO model to be compatible with quantum annealing. As described by Hammer et al. [12] QUBO is a pattern-matching mathematical formulation where the objective is to find a binary vector x that minimizes the following energy function:

$$f(x) = \sum_{i=1}^{n} Q_{ii} x_i + \sum_{i<j}^{n} Q_{ij} x_i x_j, \tag{3}$$

where:

- $x = (x_1, x_2, \ldots, x_n)$ is a vector of binary decision variables,
- Q is an $n \times n$ upper-triangular matrix of real weights,
- Q_{ii} (the diagonal elements) represent the linear coefficients (biases) for each variable,
- Q_{ij} (the off-diagonal elements) represent the quadratic coefficients (couplings) between variables x_i and x_j.

In the context of QA [9], constraints from the original problem are not handled through boundaries but are instead transformed into penalty terms. These penalties are added to the objective function such that any violation of a constraint increases the 'energy' (cost) of the solution, making that state unfavourable for the solver.

While established benchmark sets for the SMTWTP exist in the classical scheduling literature [3], they typically begin at sizes of 40 jobs or more; thus, a custom dataset was generated for research purposes. The data-generation script utilizes job counts $N \in \{3, 4, 5, 6\}$ to systematically scale the complexity of the task. This limitation is a direct result of the physical qubit connectivity and minor-embedding bottlenecks inherent to the D-Wave system, an aspect that is analysed in detail in Sect. 5. Despite the reduced problem size (N $\leq$ 6), the generation methodology strictly mirrors the established statistical strategies used to create classical scheduling benchmarks. Both processing durations and job weights are constrained to small integer values within the interval [1,5]. Restricting the initial problem data to small integers limits the required dynamic range of the QUBO coefficients, mitigates analogue precision errors, prevents constraint-penalty violations, and provides a cleaner assessment of the QPU's fundamental optimization capabilities. For each generated instance, given the small problem size ($N \leq 6$), a brute-force algorithm was employed to search all $n!$ permutations. This provides an exact benchmark for the subsequent performance analysis of the D-Wave system. While classical methods exist for larger instances, a direct algorithmic comparison is not the objective of this hardware-focused study. Detailed benchmarking against classical lower bounds and heuristics for single-machine scheduling on D-Wave can be found in complementary research (e.g. [13]). While the scale of these experiments is strictly constrained by current NISQ hardware limitations, analysing these small instances remains crucial for understanding the fundamental behaviour of quantum annealing in scheduling contexts. Specifically, this scale allows for the exact identification of optimal parameter configurations and the precise measurement of how hardware constraints influence the energy landscape.

The difficulty of each instance is governed by two dimensionless parameters that define the distribution of due dates relative to the total processing time: the Tightness Factor (TF) and the Relative Due Date Range (RDD). The TF determines the average proximity of due dates to the expected completion time; a high TF increases the likelihood of tardiness and forces the solver to prioritize weight-heavy jobs more aggressively. The RDD Range specifies the breadth of the window in which these due dates are distributed.

The experimental setup adopted different configurations for the TF and RDD parameters depending on the experimental context. These parameters were either systematically scaled to evaluate their direct influence on the solver's performance or maintained as static control variables during the analysis of quantum hardware configurations, such as chain strength. The integer due dates (d_i) for each job were generated using a uniform distribution between a minimum and maximum limit, calculated based on the total processing time $\sum p_i$ as follows:

$$D_{\text{min}} = \max\left\{0, \left\lfloor \sum_{i=1}^{n} p_i \left(1 - \text{TF} - \frac{\text{RDD}}{2}\right)\right\rfloor\right\}, \tag{4}$$

$$D_{\text{max}} = \max\left\{D_{\text{min}}, \left\lfloor \sum_{i=1}^{n} p_i \left(1 - \text{TF} + \frac{\text{RDD}}{2}\right)\right\rfloor\right\}. \tag{5}$$

Hardware embeddings were dynamically recomputed between runs utilizing D-Wave's default minor-embedding heuristic. For each instance submitted to the QPU, 1000 reads were performed. The success probability of the solver was empirically measured as the relative frequency of obtaining a valid solution that exactly matched the pre-calculated optimal objective cost out of the total number of reads performed for a given instance.

5 Experiments and Results

For the implementation and execution of the research, the D-Wave Ocean SDK was utilized [4]. The software stack relies on several specialized libraries:

- `dimod` (v0.12.21): Serves as the primary API for quadratic models. It was used to define the Constrained Quadratic Model (CQM) and manage its transformation into Binary Quadratic Models (BQM).
- `dwave-system` (v1.34.0): Provided the interface to access the physical Quantum Processing Units (QPU), specifically managing the minor-embedding process [10,11] and communication with D-Wave Leap.
- `dwave-cloud-client` (v0.14.3): Python implementation of the REST interface used to communicate with D-Wave Solver API (SAPI) servers.
- `dwave-inspector` (v0.5.5): Provides a graphic interface in your web browser for visualizing and analysing how problems are mapped (minor-embedded) onto D-Wave quantum processors.

Python version 3.14.0 was used with standard libraries such as `datetime`, `itertools`, `json`, `os`, `pickle`, `re`, `and` `sys`. Other additional libraries included `dotenv` v0.9.9, `matplotlib` v3.10.7, `numpy` v2.2.6, `networkx` v3.5, and `seaborn` v0.13.2.

D-Wave Advantage2 System v1.11, Advantage System v4.1, and v6.4 were used. In Table 1, the characteristics of the utilized systems are presented. [4]

The job count was capped at $N \leq 6$ due to the physical limitations of the quantum hardware's connectivity. For problem size $N = 7$, the minor-embedding

Table 1. Advantage System comparison.

Characteristics	Advantage v4.1	Advantage v6.4	Advantage2 v1.11
Topology	Pegasus (P16)	Pegasus (P16)	Zephyr (Z12)
Number of qubits	5627	5612	4582
Number of couplers	40 279	40 088	41 596
Qubit temperature	16.0 ± 0.1 mK	15.4 ± 0.1 mK	20 ± 1.0 mK
Readout time range	17.0 to 235.0 µs	18.0 to 173.0 µs	17.0 to 101.0 µs
Programming time	$\sim$ 14100 µs	$\sim$ 14200 µs	$\sim$ 33600 µs
QPU-delay-time per sample	20.5 µs	20.5 µs	60.6 µs
Readout error rate	≤ 0.001	≤ 0.001	≤ 0.001

process – which maps the logical variables of the SMTWTP to the physical topologies – failed to find a valid configuration. The required interaction density of the scheduling constraints exceeds the available routing paths on the chip. That prevents the generation of a functional BQM for larger problem instances.

The aforementioned interaction density implies a quadratic relationship between the problem size (N) and the number of physical qubits required to map the problem instance onto the QPU. This assertion is corroborated by the experimental data presented in Fig. 1. As the problem size scales, the spatial dispersion of both logical and physical qubits increases significantly. The nominal qubit counts listed in Table 1 represent the theoretical graph size; in practice, not all qubits and couplers on a given chip are functional at any given time. Manufacturing defects and calibration drift result in a *working graph* that is a sub-graph of the ideal topology [4]. As the required physical qubit count approaches the total capacity of a given QPU, the minor-embedding process is forced to utilize peripheral routing paths along the boundaries of the QPU fabric, where the probability of encountering inactive qubits or missing couplers increases. These effects contribute to the growing variance in solution quality observed for larger instances ($N = 5, 6$) and are particularly pronounced in the Advantage2 system due to its lower qubit count compared to the Advantage 6.4 and 4.1 architectures. Notably, despite the enhanced internal connectivity of the Advantage2 architecture, certain instances for $N = 6$ remain unmappable.

To establish a performance benchmark, a baseline experiment using the default configuration provided by the D-Wave Ocean SDK was conducted. A critical component of this setup is the automatic determination of the chain strength. In the D-Wave architecture, a chain is a set of physical qubits programmed to represent a single logical variable that is physically enforced through the application of ferromagnetic coupling, where the quadratic interaction term J_{ij} between adjacent qubits in a chain is set to a negative value. The magnitude of this negative coupling is defined as the chain strength. The automatic determination of the chain strength utilizes the *uniform_torque_compensation* function found within the *dwave.embedding.chain_strength* module, which is designed to

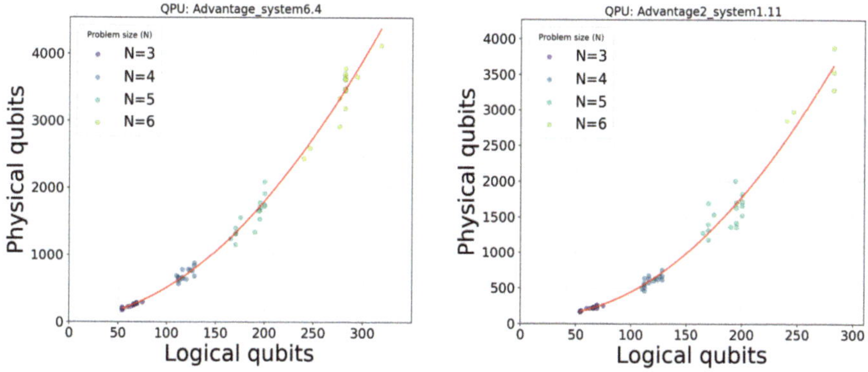

(a) Scaling of physical qubit requirements relative to problem size N Quadratic trend-line: $y = 0.0407x^2 + 0.66x + 43.8$ ($R^2 = 0.9902$).

(b) Scaling of physical qubit requirements relative to problem size N. Quadratic trend-line: $y = 0.0505x^2 - 1.82x + 126.1$ ($R^2 = 0.9770$).

Fig. 1. Comparison of physical qubit requirements relative to problem size N on different architectures.

balance the energy scales of the problem's quadratic terms with the ferromagnetic coupling required to maintain the integrity of the qubit chains [4].

For the smallest instances, the solver successfully identifies several feasible solutions. However, for all instances where $N > 3$, the system fails to return any feasible results. Chain strengths provided by the *uniform_torque_compensation* function are approximately an order of magnitude lower than the values experimentally determined to be optimal for the investigated problem sizes. Subsequent experiments demonstrate that a significant disparity exists between the chain strength values derived from the heuristic function and the values required for peak performance, as illustrated in Figs. 2a and 2b.

Driven by the high frequency of broken chains observed in the baseline results, a systematic investigation into the chain strength parameter was conducted to stabilize the logical variables on the physical hardware. Given the significant variance in recommendations found in the literature and community documentation, an expansive search space ranging from 10^2 to 10^8 was explored.

The initial broad sweep, visible in Fig. 3, is evaluated using the relative tardiness gap – defined as the ratio of the best feasible solution's tardiness to the known optimum (C_{best}/C_{opt}).

The physical mechanism underlying these observations is revealed by Fig. 4a, which plots the mean chain break fraction against chain strength. At low chain strength values, no feasible solutions are returned, and the chain break fraction remains critically high, confirming that the ferromagnetic coupling is too weak to maintain logical consistency under the competing biases of the SMTWTP formulation. As the chain strength increases beyond 10^5, feasible solutions begin to appear, and the chain break fraction drops to values approaching zero across

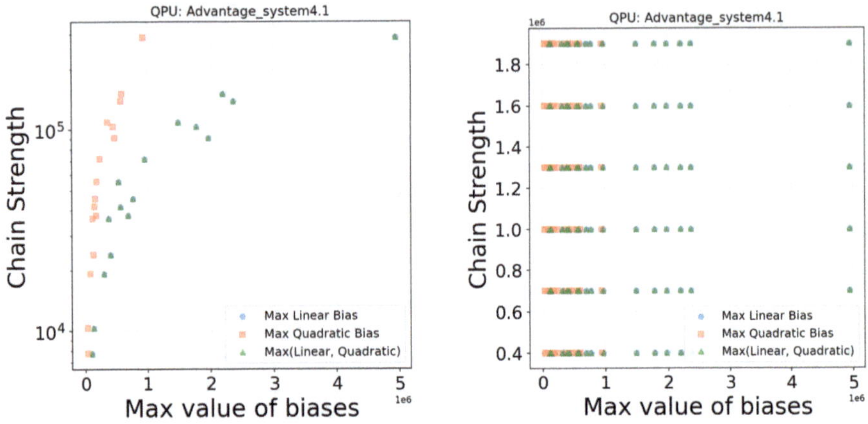

(a) Chain strengths derived via the uniform_torque_compensation heuristic.

(b) Manually tuned chain strengths optimized for maximum success probability.

Fig. 2. Comparative analysis of chain strength scaling for SMTWTP instances, compared to linear and quadratic biases ($|h|_{max}$, $|J|_{max}$) in QUBO.

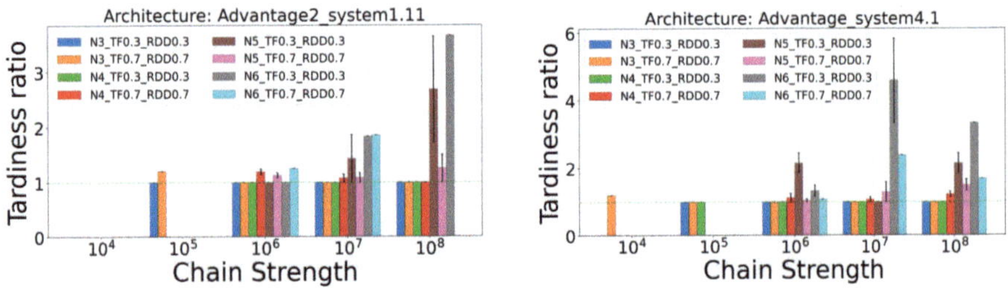

(a) Relative tardiness gap as a function of chain strength for dataset instances.

(b) Relative tardiness gap as a function of chain strength for dataset instances.

Fig. 3. Relative tardiness gap as a function of chain strength for dataset instances on different architectures. A value of 1.0 indicates an optimal solution.

all instance sizes and architectures. This behaviour is consistent across all three systems. Excessively high values (beyond 10^7) degrade solution quality: the ferromagnetic energy dominates the Hamiltonian, preventing the annealer from exploring the solution space adequately. Based on these results, a chain strength of 10^6 is identified as the threshold at which the chain break fraction stabilizes near zero across all tested configurations.

The causal link between chain integrity and solution quality is further corroborated by Fig. 4b, which directly relates the mean chain break fraction to the number of feasible and optimal solutions returned per run. A strong inverse relationship is observed: the vast majority of feasible and optimal solutions are concentrated in the narrow region where the chain break fraction is near zero

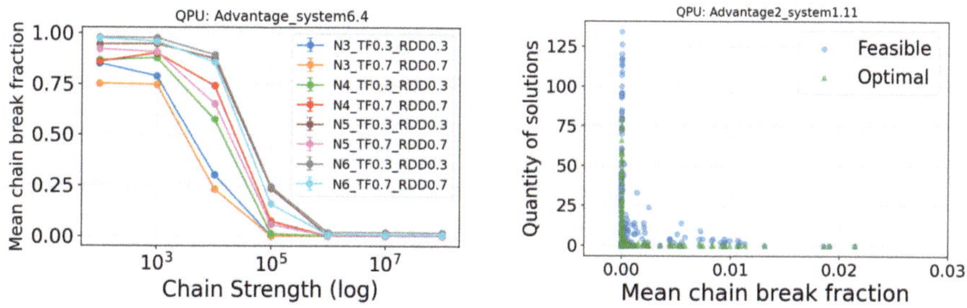

(a) Mean chain break fraction as a function of chain strength for instances.

(b) Number of feasible and optimal solutions as a function of mean fraction.

Fig. 4. Chain integrity analysis: relationship between chain strength, chain break fraction, and solution quality.

(below 0.01). The distribution of optimal solutions is even more tightly concentrated near the zero chain break fraction, suggesting that even marginal chain degradation disproportionately affects the solver's ability to locate the global optimum rather than just any valid schedule. This confirms that maintaining chain integrity is a prerequisite for the solver to produce any meaningful output.

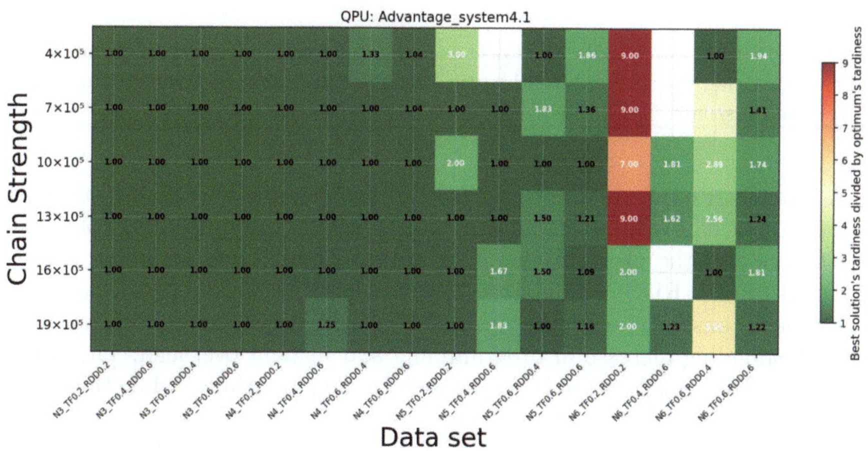

Fig. 5. Heat-map of the relative tardiness gap as a function of chain strength and dataset instance. Blank spaces mean no feasible solution was found.

To refine the chain strength selection around the identified 10^6 region, a fine-grained sweep from 4×10^5 to 1.9×10^6 was conducted. The results, presented as a heat-map in Fig. 5, reveal the detailed interplay between chain strength and instance-specific difficulty. For smaller instances, the solver consistently returns solutions that match the known optimum (gap = 1.0, dark green) across the entire tested range. As the instance size increases to $N = 5$ and $N = 6$, the relative tardiness gap rises significantly, with some configurations yielding feasible solutions whose cost is up to 10 times the known optimum. This is most pronounced for instances with high TF values, which impose tighter deadlines and amplify the penalty terms in the QUBO formulation. The results confirm that no single chain strength value universally produces optimal solutions, underscoring the need for adaptive or instance-aware parameter tuning.

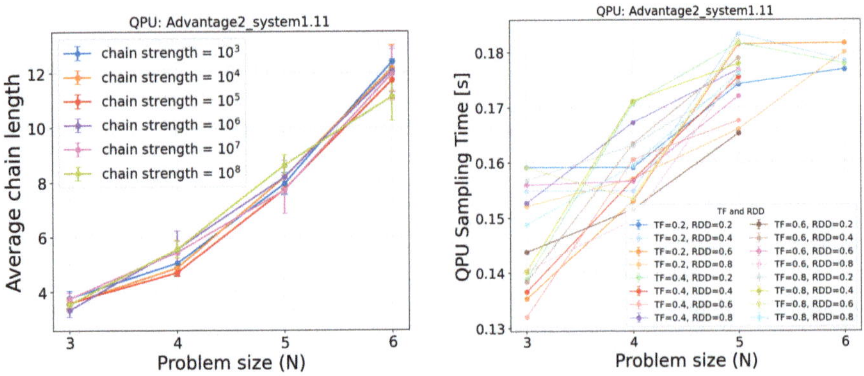

(a) Mean chain length as a function of problem size N.

(b) QPU sampling time as a function of problem size N for tested TF and RDD.

Fig. 6. Embedding complexity and QPU execution characteristics as a function of problem size N. Other QPUs exhibited identical trends, thus these plots are representative of all tested systems.

As context for interpreting the effects of problem parameters on solver performance, Fig. 6 presents two complementary views of the embedding complexity as a function of the problem size N. Figure 6b shows QPU sampling time for all tested TF and RDD combinations. A clear increase in sampling time with N is observed, consistent with the quadratic growth in physical qubit count and the corresponding increase in programming and readout overhead. The Advantage v4.1 system exhibits the highest absolute sampling times, while the Advantage2 and Advantage v6.4 systems demonstrate lower and comparable ranges. The spread of sampling times across different TF/RDD configurations remains relatively narrow for a given N, indicating that the problem parameters have a limited direct effect on QPU execution time, which is primarily governed by the embedding size and the fixed annealing schedule. A modest increase in time variance is visible at $N = 5$ and $N = 6$, likely reflecting the greater variability in embedding quality for more complex instances.

Figure 6a presents the mean chain lengths of the minor embeddings as a function of N for various chain strength values. The scaling behaviour of the maximum chain length exhibits an identical trend, albeit operating over a broader numerical range (from approximately 10 to 60) compared to the mean values

(4 to 12). Both metrics grow monotonically with N. This directly reflects the increasing difficulty of embedding the dense SMTWTP constraint graph into the sparse QPU topology, as each additional job introduces $\mathcal{O}(N)$ new pairwise interactions. The error bars, which widen substantially at $N = 5$ and $N = 6$, indicate that the embedding quality becomes increasingly variable for larger instances. Importantly, chain strength has no significant effect on chain length, confirming that it is a post-embedding parameter that influences the annealing dynamics rather than the embedding structure itself. These longer chains at higher N are inherently more susceptible to breakage, providing a direct physical explanation for the elevated chain break fractions and degraded solution quality observed for larger problem instances in the preceding analysis.

6 Conclusions

The analysis of quantum annealing for the SMTWTP demonstrates that hardware-specific parameters are the primary determinants of solution quality in NISQ devices. A precise balance in chain strength is critical, as insufficient coupling leads to invalid solutions through chain breaks, while excessive strength prematurely freezes the system and obscures the optimization landscape. Furthermore, the experimental analysis underscores how embedding size and QPU-dependent connectivity heavily influence the performance of these devices. Ultimately, this research highlights key considerations for parameter tuning, showing that the transition from abstract mathematical formulations to physical execution hinges on understanding the intricate trade-offs between hardware constraints and optimization accuracy. Future research could focus on developing adaptive, instance-aware parameter tuning algorithms that automatically adjust parameters, such as chain strength, based on specific problem characteristics like the Tightness Factor or job count. Additionally, investigating the integration of advanced minor-embedding heuristics or hybrid quantum-classical decomposition methods may alleviate the connectivity constraints found in current QPU architectures.

References

1. Bożejko, W., Pempera, J., Uchroński, M., Wodecki, M.: Quantum annealing-driven branch and bound for the single machine total weighted number of tardy jobs scheduling problem. Future Gener. Comput. Syst. **155**, 245–255 (2024)
2. Bożejko, W.: On quantum computing formulations for the total weighted tardiness and total weighted earliness optimization. In: Burduk, A., et al. (eds.) International Symposium on Precision Engineering and Manufacturing (ISPEM), Springer, Cham (2025). https://doi.org/10.1007/978-3-032-14163-7_26
3. Giergiel, K., Sam Yang, Y., Murphy, A.B.: Hybrid quantum-classical scheduling with problem-aware calibration (2025). arXiv:2509.04808 arXiv preprint
4. D-Wave Documentation. https://docs.dwavequantum.com/en/latest/index.html. Accessed 20 Feb 2026

5. Pelofske, E.: Comparing three generations of D-wave quantum annealers for minor embedded combinatorial optimization problems. Quantum Sci. Technol. **9**(1), 015010 (2023)
6. Weinberg, P., et al.: Scaling and diabatic effects in quantum annealing with a D-Wave device. Phys. Rev. Lett. **124**(9), 090502 (2020)
7. Pelofske, E., Hahn, G., Djidjev, H.N.: Parallel quantum annealing. Sci. Rep. **12**(1), 4392 (2022)
8. Mehta, V., De Raedt, H., Michielsen, K., Jin, F.: Unraveling reverse annealing: a study of D-Wave quantum annealers. Phys. Rev. A **112**(1), 012414 (2025)
9. Sharma, R., Katukam, R., Nagulapally, A.: Quantum annealing for combinatorial optimization: foundations, architectures, benchmarks, and emerging directions. arXiv preprint arXiv:2602.03101 (2026)
10. Grant, E., Humble, T.S.: Benchmarking embedded chain breaking in quantum annealing. Quantum Sci. Technol. **7**(2), 025029 (2022)
11. Park, H., Lee, H.: Determination of optimal chain coupling made by embedding in D-wave quantum annealer. AVS Quantum Sci. **6**(3), 033804 (2024)
12. Hammer, P.L., Johnson, E.L., Korte, B.H., Nemhauser, G.L.: Annals of Discrete Mathematics, 1st edn. North-Holland Publishing Company, Amsterdam (1977)
13. Bożejko, W., Pempera, J., Uchroński, M., Wodecki, M.: Determination of the lower bounds of the goal function for a single-machine scheduling problem on D-wave quantum annealer. In: Mikyška, J., de Mulatier, C., Paszynski, M., Krzhizhanovskaya, V.V., Dongarra, J.J., Sloot, P.M. (eds) International Conference on Computational Science, pp. 317–324. Springer, Cham (2023). https://doi.org/10.1007/978-3-031-36030-5_16

Trustworthy EEG-Based Epilepsy Classification: Evaluating Explanation Consistency with the E_cons Index

Flavia Costi(✉), Emanuel Covaci, and Darian Onchis

West University of Timisoara, Timisoara, Romania
{flavia.costi,emanuel.covaci,darian.onchis}@e-uvt.ro

Abstract. Automatic detection of epileptic activity in scalp EEG can reach high accuracy, yet many neural models remain opaque and provide limited guidance on when predictions should be trusted. We introduce a simple, model-agnostic index of explanation consistency, E_{cons}, for EEG-based epilepsy classification. Using the BEED dataset (8,000 twenty-second segments; four conditions; 16 features), we train three architectures: an MLP, a 1D CNN, and an LSTM. For each test sample and predicted class, we compute two attribution profiles: intrinsic saliency via gradient×input and post-hoc Integrated Gradients (IG) over multiple baselines. E_{cons} combines these signals by measuring rank alignment and penalising baseline-induced variance of IG attributions, so higher scores indicate explanations that are both consistent and stable. We evaluate E_{cons} through calibration diagnostics and selective prediction, comparing consistency-based selection with confidence-based selection. On BEED, the MLP achieves around 91% test accuracy and serves as the main case study. Higher E_{cons} is associated with more coherent explanations, slightly improved local accuracy and better confidence–reliability behaviour, while thresholding on E_{cons} yields selective regimes with increased accuracy. Similar trends are observed for the CNN and LSTM models. Overall, E_{cons} offers a lightweight way to quantify explanation quality and to flag predictions supported by stable, convergent explanatory evidence. In addition, an ablation analysis shows that both the alignment and stability components contribute to the observed reliability stratification. Risk–coverage evaluation further indicates that E_{cons} acts as a complementary signal to softmax confidence, supporting post-hoc auditing and selection under coverage constraints.

Keywords: Epilepsy · Electroencephalography · Explainable artificial intelligence · Consistency index

1 Introduction

Epilepsy is a chronic neurological disorder characterized by recurrent unprovoked seizures and can strongly affect quality of life and daily functioning. EEG is one

W. Zamojski et al. (Eds.): DepCoS-RELCOMEX 2026, LNNS 2014, pp. 55–67, 2026.
https://doi.org/10.1007/978-3-032-27927-9_6

of the main clinical tools for diagnosis, seizure assessment, and patient monitoring, but manual review is slow, subjective, and dependent on expert availability. This has motivated the development of automatic EEG-based classification systems to support faster and more reproducible clinical decisions [1]. Recent machine learning and deep learning models, including MLP, CNN, and LSTM, have improved EEG classification for seizure detection and epilepsy diagnosis. However, most studies still focus on predictive performance, such as accuracy, sensitivity, and specificity, while paying less attention to how predictions are formed and whether their explanations are reliable [2].

This limitation is particularly important in healthcare, where high accuracy alone is insufficient. Clinicians also need to understand why a model makes a prediction and when that prediction can be trusted. In EEG classification, this means identifying which features or patterns drive the decision. For this reason, XAI has become increasingly relevant. Post-hoc methods such as gradient-based saliency, IG, Layer-wise Relevance Propagation, and SHAP can explain individual predictions [3]. However, their reliability is rarely assessed systematically, and different explainers may highlight different features for the same sample [2]. In many EEG studies, explanations are still judged mainly by visual inspection. Two issues remain central: post-hoc explanations are rarely compared with the model's own input sensitivity [4], and their stability is often ignored. To address this gap, we propose E_{cons}, a per-sample consistency index for EEG-based epilepsy classification. The idea is that a prediction is easier to trust when two explanation sources identify similar important features and when the post-hoc explanation remains stable under small changes. In our framework, E_{cons} combines agreement between an intrinsic attribution signal from the trained model, computed with gradient$\times$input for the predicted class, and a post-hoc explanation produced by IG, together with a penalty for IG instability across multiple realistic baselines. Thus, E_{cons} captures both alignment and stability. We evaluate the framework on the BEED dataset using three neural architectures trained under the same protocol: an MLP, a one-dimensional CNN, and an LSTM. The MLP achieves the best test performance and is therefore used as the main case study, while the CNN and LSTM serve as complementary baselines [5]. Beyond standard classification metrics, we examine how E_{cons} relates to prediction confidence and local accuracy. For the MLP, higher E_{cons} values are associated with higher confidence and a greater likelihood of correct prediction. At the same time, the results show that higher predictive accuracy does not necessarily imply higher E_{cons}, indicating that predictive performance and explanation reliability are related but distinct.

The main contributions of this work are threefold. We introduce E_{cons}, a lightweight per-sample index that measures both agreement between intrinsic and post-hoc explanations and the stability of IG under baseline variation. We evaluate this index on the BEED dataset across three neural architectures trained under the same protocol. We analyse the relation between E_{cons}, confidence, local accuracy, and selective prediction, showing that explanation consistency provides useful information beyond standard performance metrics [6].

2 Materials and Methods

The study uses the BEED dataset, a publicly available EEG collection for epilepsy research released through the UCI Machine Learning Repository [7]. The recordings were acquired at a neurological centre in Bangalore, India, from adult participants. EEG was recorded with 16 scalp channels following the international 10–20 system at 256 Hz and segmented into 20-second epochs. The full dataset contains 16,000 segments distributed across four clinical conditions. In the tabular version used here, the data are represented as 8,000 instances, each described by 16 integer-valued features and one class label. The dataset is balanced, with an equal number of instances per class. The features $X_1, \ldots, X_{16}$ encode channel-level EEG characteristics extracted from the 16 scalp channels. Although the exact electrode labels are not provided, each feature corresponds consistently to one anonymised channel location across subjects. In this work, we treat $X_1, \ldots, X_{16}$ as channel-level descriptors of EEG activity, without adding anatomical assumptions. No missing values are present. The target variable y encodes the clinical condition of each segment as follows:

$$y = \begin{cases} 0 & \text{healthy controls without a clinical diagnosis of epilepsy,} \\ 1 & \text{generalized epileptic seizures with diffuse bilateral onset,} \\ 2 & \text{focal epileptic seizures with localized cortical onset,} \\ 3 & \text{seizure events with observable behaviours.} \end{cases}$$

In all subsequent analyses, we use the full BEED dataset of 8,000 instances represented in the 16-dimensional feature space $(X_1, \ldots, X_{16})$ and annotated with the four-class label y. Panel (1a) illustrates the pipeline from multichannel EEG acquisition, through segmentation into 20 s epochs and feature extraction, to the final tabular representation with $y \in \{0, 1, 2, 3\}$. Panel (1b) shows the class distribution and confirms that the dataset is balanced across healthy controls, generalized seizures, focal seizures, and seizure events with concurrent behaviours.

We formulate the task as a four-class supervised classification problem. The 8,000 labelled instances are split into training and test sets using a stratified 80%/20% partition with a fixed random seed. Before training, all features are standardised using the training-set mean and standard deviation, and the same transformation is applied to the test set. No additional feature engineering, dimensionality reduction, or data augmentation is used, so the analysis remains focused on model behaviour and the proposed explainability index [8].

To study predictive performance and the behaviour of the explainability index across model families, we consider three neural architectures: MLP, 1D CNN, and LSTM. All use the same 16-dimensional input but impose different inductive biases. The MLP is the central model, as it processes the feature vector directly through fully connected layers with nonlinear activations and dropout, making it well suited to tabular data without explicit spatial or temporal structure; it also achieves the highest test accuracy on BEED and yields an intrinsic explanation signal that is easier to interpret with gradient-based saliency [9]. The

1D CNN and LSTM serve as complementary baselines: the CNN captures local feature interactions through convolution and pooling, while the LSTM models ordered dependencies through recurrent processing of the same sequence [10–12].

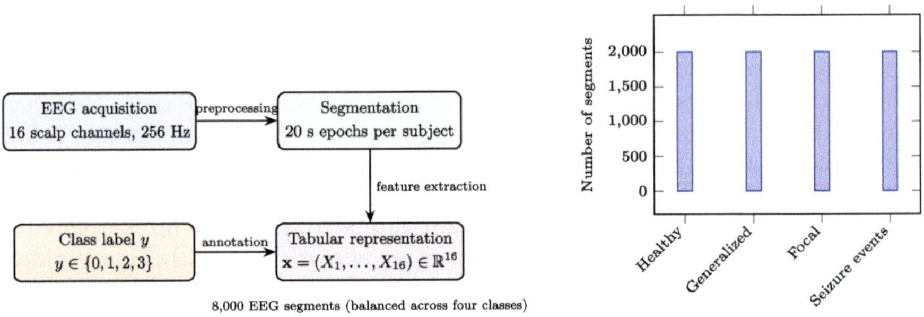

(a) Schematic representation of the EEG material. (b) Class-wise distribution of labelled EEG segments.

Fig. 1. Overview of the BEED material and the modelling setup used in this work. Panel a shows the pipeline from multichannel EEG acquisition to the 16-dimensional feature vector $\mathbf{x}$ and the four-class label y. Panel b illustrates the balanced distribution of the 8,000 segments across the four clinical categories.

To ensure comparability, all three architectures share the same training protocol. We use the same stratified train-test split, the same standardised inputs, cross-entropy loss, the Adam optimiser, a batch size of 64, and 30 training epochs [13]. This common setup makes performance differences easier to interpret and helps link variations in the explainability index primarily to architectural differences rather than preprocessing or optimisation choices.

For explainability analysis, we use two complementary attribution methods: an intrinsic saliency map derived directly from the model and a post-hoc explainer based on IG. Both operate on the same 16-dimensional input and produce feature-level importance scores from different perspectives. This allows us to compare internal model sensitivity with post-hoc attribution and then quantify their agreement through the proposed consistency index [14] (see Fig. 2).

For a test sample $\mathbf{x} \in \mathbb{R}^{16}$, the intrinsic saliency is computed with gradient $\times$ input. We first obtain the predicted class score from the trained model, then compute its gradient with respect to the input, and multiply it element-wise by $\mathbf{x}$. This yields a saliency vector $S_{\text{model}} \in \mathbb{R}^{16}$, where each component reflects the contribution of the corresponding feature to the prediction. The method is simple, efficient, and can be applied to all three architectures considered in this study [15]. As a post-hoc explainer, we use IG, which integrates the model gradient along a path from a baseline input to the actual sample. Instead of a single fixed baseline, we sample several baseline points from the training set, so that each explanation starts from a realistic EEG-like input. For each sample, this produces K attribution vectors $IG_1, \ldots, IG_K \in \mathbb{R}^{16}$.

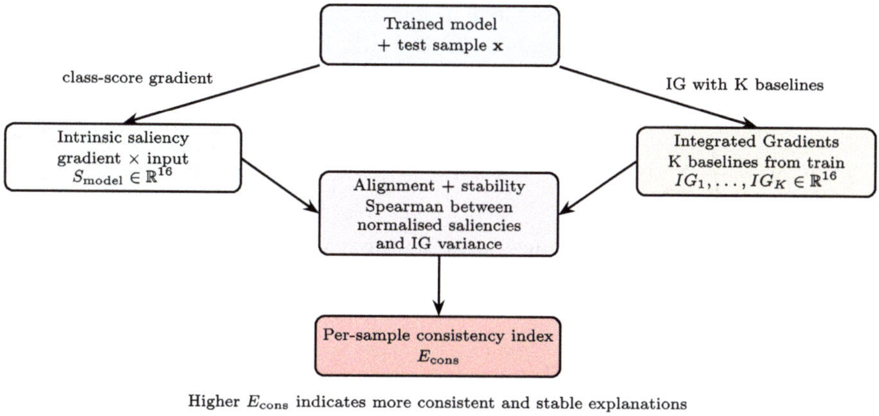

Fig. 2. Explainability workflow. For each test sample, intrinsic saliency (gradient × input) and Integrated Gradients with multiple baselines are combined into the per-sample consistency index E_{cons}.

We combine gradient × input and IG because they provide complementary views of explainability. The first reflects the model's internal sensitivity, while the second is a standard post-hoc explainer with clear attribution semantics. By comparing S_{model} with IG and measuring IG variability across multiple baselines, we define a consistency index that captures both cross-method agreement and post-hoc stability [16]. To identify which component drives the observed effects, we perform two ablations: an alignment-only version with $s(x) = 1$, so that $E_{cons}(x) = a(x)$, and a stability-only version with $a(x) = 1$, so that $E_{cons}(x) = s(x)$. We also test the sensitivity of the method by varying the number of IG baselines $K \in \{1, 3, 5, 10\}$ and the variance penalty $\alpha \in \{0.1, 1, 5\}$ while keeping the supervised model fixed. We define E_{cons} as a per-sample score that combines intrinsic saliency and IG. For a test sample $\mathbf{x} \in \mathbb{R}^{16}$, we consider an intrinsic attribution vector $S_{model}(\mathbf{x}) \in \mathbb{R}^d$ and K IG attribution vectors $IG_1(\mathbf{x}), \ldots, IG_K(\mathbf{x}) \in \mathbb{R}^d$, with $d = 16$. We first summarise the IG explanations by their component-wise mean:

$$\bar{\mathbf{s}}_{IG}(\mathbf{x}) = \frac{1}{K} \sum_{k=1}^{K} IG_k(\mathbf{x}) \in \mathbb{R}^d. \tag{1}$$

We then normalise the two attribution profiles with the ℓ_1 norm:

$$\tilde{\mathbf{s}}_{model}(\mathbf{x}) = \frac{S_{model}(\mathbf{x})}{\|S_{model}(\mathbf{x})\|_1 + \varepsilon}, \qquad \tilde{\mathbf{s}}_{IG}(\mathbf{x}) = \frac{\bar{\mathbf{s}}_{IG}(\mathbf{x})}{\|\bar{\mathbf{s}}_{IG}(\mathbf{x})\|_1 + \varepsilon}. \tag{2}$$

This removes scale effects and allows a direct comparison of relative feature importance [16].

The first component of E_{cons} measures alignment between the two explanations. We compute the Spearman rank correlation between the normalised vectors and then rescale it to $[0, 1]$:

$$\rho(\mathbf{x}) = \text{Spearman}\left(\tilde{\mathbf{s}}_{model}(\mathbf{x}), \tilde{\mathbf{s}}_{IG}(\mathbf{x})\right), \qquad a(\mathbf{x}) = \frac{\rho(\mathbf{x}) + 1}{2}. \tag{3}$$

Higher $a(\mathbf{x})$ means better agreement in feature ranking.

The second component measures IG stability across baselines. For each feature $j \in \{1, \ldots, d\}$, we compute the mean and empirical variance across the K runs:

$$\mu_j(\mathbf{x}) = \frac{1}{K}\sum_{k=1}^{K} IG_{k,j}(\mathbf{x}), \qquad \mathrm{Var}_j(\mathbf{x}) = \frac{1}{K-1}\sum_{k=1}^{K}\left(IG_{k,j}(\mathbf{x}) - \mu_j(\mathbf{x})\right)^2. \tag{4}$$

We then average these variances over features and convert the result into a stability factor:

$$v(\mathbf{x}) = \frac{1}{d}\sum_{j=1}^{d} \mathrm{Var}_j(\mathbf{x}), \qquad s(\mathbf{x}) = \frac{1}{1+\alpha\, v(\mathbf{x})}. \tag{5}$$

Here, $\alpha > 0$ controls the penalty strength. In the main experiments, we use $\alpha = 1.0$.

Finally, we define the consistency index as

$$E_{\text{cons}}(\mathbf{x}) = a(\mathbf{x}) \cdot s(\mathbf{x}) = \frac{\rho(\mathbf{x})+1}{2} \cdot \frac{1}{1+\alpha\, v(\mathbf{x})}. \tag{6}$$

High $E_{\text{cons}}(\mathbf{x})$ values indicate explanations that are both well aligned and stable across IG baselines.

For a fixed trained model, computing $E_{\text{cons}}(x)$ requires one gradient$\times$input pass and K IG runs. The per-sample cost therefore grows linearly with K and with the feature dimension d (here $d = 16$). In our main setting ($K = 5$), this extra cost is modest and suitable for post-hoc reliability analysis. The two components of the score are illustrated in Fig. 3. The left panel shows the alignment term $a(\mathbf{x})$, which increases linearly from 0 to 1 as the Spearman correlation $\rho(\mathbf{x})$ goes from -1 to 1. The right panel shows the stability term $s(\mathbf{x})$, which decreases as the IG variance $v(\mathbf{x})$ increases. Thus, high alignment and low variance lead to larger E_{cons} values.

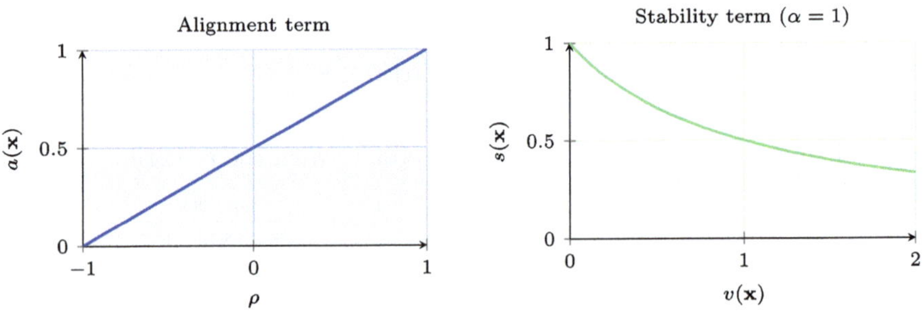

Fig. 3. Illustration of the two components of the consistency index. Left: alignment score $a(\mathbf{x})$ as a rescaled version of the Spearman correlation $\rho(\mathbf{x})$. Right: stability factor $s(\mathbf{x})$ as a decreasing function of the IG variance $v(\mathbf{x})$ for $\alpha = 1$.

For each analysed test sample $\mathbf{x}_i$ with true label y_i, predicted class $\hat{y}_i$, and predicted probability vector $\mathbf{p}i$, we compute $E\text{cons}(\mathbf{x}_i)$ and define the associated prediction confidence and correctness indicator as follows:

$$c_i = \max_c p_{i,c}, \qquad z_i = \begin{cases} 1, & \text{if } \hat{y}_i = y_i, \\ 0, & \text{otherwise.} \end{cases} \tag{7}$$

Here, c_i measures confidence in the predicted class, while z_i indicates whether the prediction is correct.

Let $\mathcal{T}$ denote the set of the N analysed samples. To examine how confidence and accuracy vary with explanation consistency, we sort $\mathcal{T}$ by $E_{\text{cons}}(\mathbf{x}_i)$ and form quantile-based groups. The low-consistency group contains the bottom 30% of samples, and the high-consistency group the top 30%. This provides clear separation while retaining enough samples for stable estimates. Nearby cut-offs, such as 25%/75% and 20%/80%, gave the same qualitative trends.

For any group $\mathcal{G} \subseteq \mathcal{T}$, we report three aggregate quantities: mean consistency, mean confidence, and local accuracy:

$$\overline{E}_{\text{cons}}(\mathcal{G}) = \frac{1}{|\mathcal{G}|} \sum_{\mathbf{x}_i \in \mathcal{G}} E_{\text{cons}}(\mathbf{x}_i), \qquad \bar{c}(\mathcal{G}) = \frac{1}{|\mathcal{G}|} \sum_{\mathbf{x}_i \in \mathcal{G}} c_i, \qquad \text{Acc}(\mathcal{G}) = \frac{1}{|\mathcal{G}|} \sum_{\mathbf{x}_i \in \mathcal{G}} z_i. \tag{8}$$

These quantities summarise the typical consistency level, average confidence, and fraction of correct predictions in the group. Comparing the low- and high-consistency groups allows us to test whether higher E_{cons} is associated with more confident and more reliable predictions [17].

The consistency index is applied to all three neural architectures considered in this study: MLP, 1D CNN, and LSTM. All models are trained on the same stratified split, with the same standardised inputs, optimisation settings, and explainability hyperparameters (K IG baselines and α), enabling direct comparison across architectures. The MLP is treated as the main reference because it achieves the highest test accuracy on BEED. For this model, we provide the most detailed XAI analysis, including score distributions, scatter plots, low/high consistency comparisons, and reliability-style curves. The 1D CNN and LSTM serve as complementary baselines, for which we report the same summary quantities: test accuracy, mean E_{cons}, and group-wise confidence and local accuracy. This allows us to assess whether the main conclusions about E_{cons} extend beyond a single architecture. Figure 4 summarises the role of the three models in the study.

3 Results

We first compare the three architectures on the full BEED test set. All models are trained on the same 6,400 training samples and evaluated on the same 1,600 test samples, using the same preprocessing and optimisation pipeline. The MLP achieves the best test accuracy, 91.00%, while the 1D CNN and LSTM reach 69.56% and 71.50%, respectively. This suggests that, for the compact BEED representation with 16 features per segment, the MLP is better suited to the

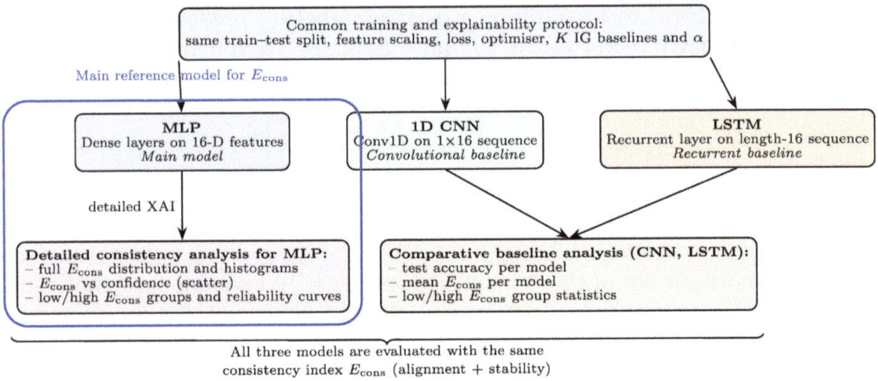

Fig. 4. Schematic comparison of the three architectures used in the consistency analysis. All models share the same training and explainability protocol. The MLP is analysed in detail, while the 1D CNN and LSTM serve as complementary baselines under the same E_{cons} framework.

data than sequence-based models. We therefore use the MLP as the main model in the detailed explainability analysis, while the CNN and LSTM are kept as baselines.

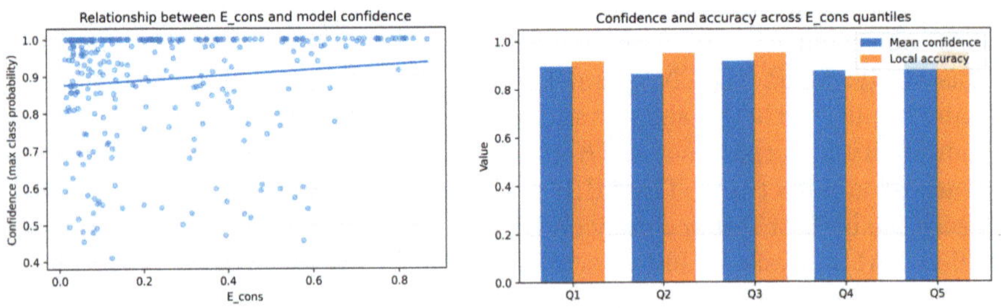

(a) E_{cons} versus prediction confidence for the MLP on $N = 300$ test samples.

(b) Mean confidence and local accuracy across E_{cons} quintiles for the MLP.

Fig. 5. MLP results at sample and group level.

We then analyse the MLP in more detail using the $N = 300$ test samples selected for the consistency study. For each sample, we consider the triplet $(E_{\text{cons}}(\mathbf{x}_i), c_i, z_i)$, where $E_{\text{cons}}(\mathbf{x}_i)$ is the consistency score, c_i is the prediction confidence, and z_i indicates whether the prediction is correct. Figure 5a shows that most samples lie in a high-confidence region, while their consistency scores still span a wide range. This means that confidence and consistency are related, but not identical. To make this trend clearer, we sort the samples by E_{cons} and divide them into five equal groups. Figure 5b shows the mean confidence and local accuracy for each quintile. Confidence stays high in all groups, but local accuracy increases slightly from low- to high-consistency samples. The highest-consistency group gives the best accuracy. This suggests that E_{cons} adds useful information beyond confidence alone.

Table 1. Summary of reliability results for low- and high-E_{cons} groups in the MLP on $N = 300$ test samples.

(a) Consistency, confidence, and local accuracy

Group	$\overline{E}_{\text{cons}}$	$\overline{c}$	Acc
Low E_{cons} (bottom 30%)	0.0455	0.8875	0.9111
High E_{cons} (top 30%)	0.5531	0.9041	0.9222

(b) Consistency, confidence, and local accuracy

Metric	Low-E_{cons}	High-E_{cons}
ECE ↓	0.0212	0.0157
Brier ↓	0.1649	0.0921

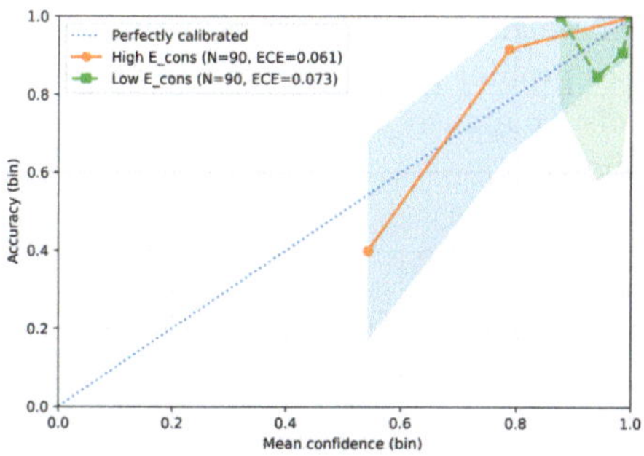

Fig. 6. Reliability curves for low- and high-consistency MLP samples. Each curve shows mean confidence versus empirical accuracy per confidence bin, with the dashed diagonal indicating perfect calibration. High-E_{cons} predictions are generally closer to the diagonal, suggesting better calibration and reliability.

Table 2. Cross-architecture comparison of classification performance and consistency statistics on the BEED dataset. E_{cons} is computed on $N = 300$ test samples for each model.

Model	Acc.	$\mathbb{E}[E_{\text{cons}}]$	Acc_{low}	Acc_{high}
MLP	0.9100	0.2535	0.9111	0.9222
1D CNN	0.6956	0.4463	0.7222	0.7444
LSTM	0.7150	0.5192	0.6444	0.8222

We also compare two broader groups: the bottom 30% and the top 30% of E_{cons} values. Table 1a shows that the high-consistency group is slightly more confident and more accurate. This agrees with the quintile analysis and supports the use of E_{cons} as a per-sample trust signal. Table 1b shows the same trend for calibration. The high-E_{cons} group has lower ECE and a much lower Brier score. Figure 6 is consistent with this result: the high-E_{cons} curve stays closer to the diagonal, especially in the medium-to-high confidence range. This suggests that samples with more consistent explanations are not only slightly more accurate, but also better calibrated.

We next compare E_{cons} across architectures. For the 1D CNN and LSTM, we use the same $N = 300$ test samples and the same explainability settings. Table 2 reports test accuracy, mean E_{cons}, and local accuracy in the low- and high-consistency groups. The MLP achieves the best test accuracy, but not the highest mean E_{cons}: the CNN and LSTM have larger average consistency scores. This shows that E_{cons} is not a simple proxy for accuracy and that its absolute scale depends on the architecture, so it is best interpreted within each model. The within-model trend is consistent. In all three architectures, the high-E_{cons} group attains equal or higher local accuracy than the low-E_{cons} group. The gain is small for the MLP and CNN, but larger for the LSTM. This suggests that higher consistency is associated with more reliable predictions across fully connected, convolutional, and recurrent models.

To illustrate what E_{cons} captures, we inspect two MLP test samples: one from the top decile and one from the bottom decile of E_{cons}. For each sample, we compare the ℓ_1-normalised intrinsic saliency $S_{\text{model}}(\mathbf{x})$ with the mean IG attribution $\bar{\mathbf{s}}_{\text{IG}}(\mathbf{x})$ over the 16 input features. Figure 7 shows that the high-consistency example has very similar attribution patterns under the two methods, with the same small set of features receiving similar weights. The low-consistency example shows clearer disagreement and a noisier profile. These examples support the quantitative results: high E_{cons} corresponds to aligned and stable explanations, whereas low E_{cons} reflects less coherent ones.

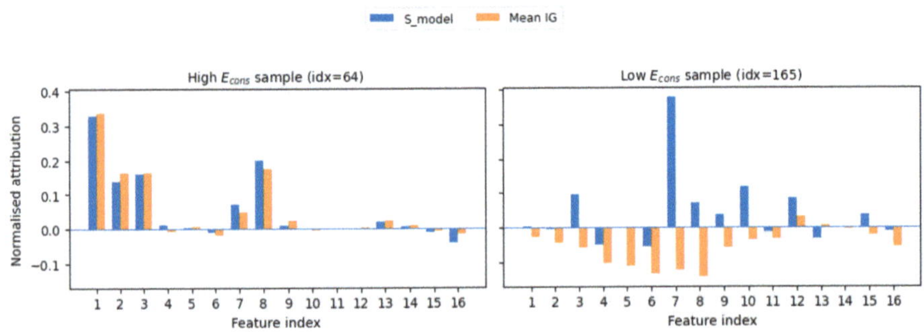

Fig. 7. Qualitative examples of explanation consistency for the MLP model on the BEED dataset. Each panel shows ℓ_1-normalised feature attributions for the intrinsic saliency S_{model} (gradient $\times$ input) and for the mean IG profile over the 16-dimensional input. The left panel displays a high-consistency example (large E_{cons}), where both attribution methods highlight largely the same subset of features with similar relative importance. The right panel shows a low-consistency example (small E_{cons}), where the two attribution profiles emphasise different features and exhibit a noisier pattern.

We also test whether E_{cons} can be used for selective prediction. For the same $N_{\text{examples}} = 300$ MLP test samples, we sort predictions in descending order of E_{cons} and keep only the top fraction of samples. We then compute the mean consistency, mean confidence, and local accuracy at coverage levels of 100%, 80%, 60%, 40%, and 20%. As a baseline, we repeat the same procedure using prediction confidence c_i instead of E_{cons}.

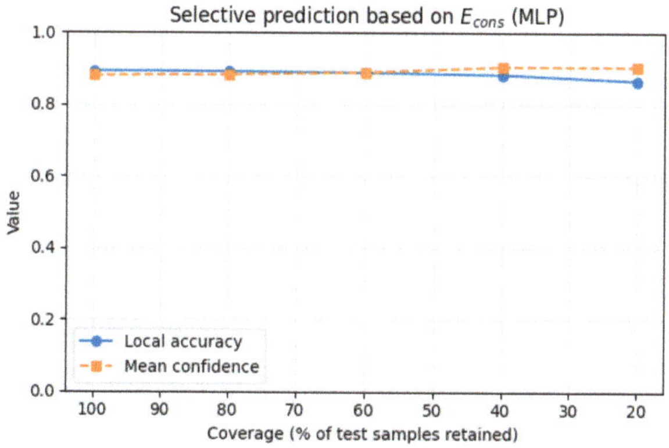

Fig. 8. Selective prediction analysis for the MLP based on E_{cons}. The plot shows local accuracy and mean confidence as functions of coverage when only the top-ranked test samples by E_{cons} are retained. Lower coverage corresponds to stricter consistency thresholds and yields subsets with higher average consistency, local accuracy, and mean confidence.

Figure 8 shows that, as coverage decreases, the retained subset has higher mean E_{cons}, higher local accuracy, and higher mean confidence. The effect is clearest at low coverage, where only the most consistent predictions are kept. This supports the use of E_{cons} as a practical signal for selecting more reliable predictions. In the risk–coverage analysis, confidence gives a lower AURC on this in-distribution test set. Still, tuning (K, α) improves the selective behaviour of E_{cons}, and E_{cons} adds information that confidence alone does not provide, especially about explanatory support and calibration [18]. This makes it useful as a complementary reliability signal, not a replacement for confidence.

4 Conclusion

This work investigated explanation consistency as a reliability signal for neural EEG-based epilepsy classifiers. Using the BEED dataset, we trained an MLP, a 1D CNN, and an LSTM under the same protocol, and introduced E_{cons}, a per-sample index that combines agreement between gradient×input saliency and Integrated Gradients with the stability of IG across multiple baselines. Among the tested models, the MLP achieved the highest test accuracy. This should be interpreted in the context of the input representation: the BEED dataset provides a compact tabular description with only 16 pre-computed features per segment, which is naturally well suited to an MLP. Thus, the results do not suggest that CNNs or LSTMs are less suitable for EEG in general, but only that the MLP is the best match for this specific feature-based setting.

The main contribution of the study is the proposal of E_{cons} as a simple quantitative measure of explanation quality. The results show that E_{cons} is distinct from accuracy: its average value varies across architectures, and a higher mean consistency does not necessarily imply better classification performance. However, within each model, samples with higher E_{cons} tend to be more reliable. For the MLP, these samples show slightly better local accuracy, better calibration, and more coherent qualitative explanations. Selective prediction analysis further shows that retaining the most consistent cases leads to subsets with improved accuracy and confidence.

The findings suggest that E_{cons} is a useful complement to standard performance metrics. It is lightweight, easy to compute after training, and can serve as an additional post-hoc reliability signal for auditing, triage, and decision support. Although it does not solve trust on its own, it provides useful information about whether a prediction is supported by aligned and stable explanations. This makes it a practical step toward more transparent and clinically usable EEG-based decision systems.

Acknowledgments. This research is partially supported by the project HRIA, Smart Growth, Digitization and Financial Instruments Program, 2021–2027, no. 351416.

References

1. Najmusseher, P., Azar, A.T.: Impact of multi-domain features for EEG based epileptic seizures classification. In: Proceedings of the 10th International Conference on Advanced Intelligent Systems and Informatics 2024: Volume 2, vol. 220, p. 317. Springer, Heidelberg (2024)
2. Najmusseher, B.P.N., Janardhan, D.C.: Automated epileptic seizure classification using adaptive fast Fourier transform with non-uniform sampling and improved deep belief network. Int. J. Intell. Eng. Inf. **12**(4), 460–512 (2024)
3. Paneru, B.: Interpretable machine learning for epileptic seizure detection on the beed using lime with an ensemble network. In: medRxiv 2025–09 (2025)
4. Najmusseher, P., Azar, A.T., Kamal, N.A., Alzahrani, A.: Exploration and analysis of seizure spikes through spectral domain. In: Business Intelligence and Information Technology: Proceedings of BIIT 2023, vol. 394, p. 99 (2024)
5. Zhou, M., et al.: Epileptic seizure detection based on EEG signals and CNN. Front. Neuroinform. **12**, 95 (2018)
6. Beed, P.S., Salmen, B., Schmitz, D.: Gluk2-mediated excitability within the superficial layers of the entorhinal cortex. PLoS ONE **4**(5), e5576 (2009)
7. Rajput, V.: Beed: Bangalore EEG epilepsy dataset, https://www.kaggle.com/datasets/varunrajput/beed-bangalore-eeg-epilepsy-dataset. kaggle dataset. Accessed Dec 2025
8. Banu, P.N., Azar, A.T., Kamal, N.A., et al.: Explainable artificial intelligence in epilepsy management: Unveiling the model interpretability. In: Explainable AI in Healthcare Imaging for Medical Diagnoses, pp. 175–211. Elsevier (2025)
9. Vijyalakshmi, K., Rajakannu, A., Kamarudden, M., Ramachandran, K., AV, S.R.: Intelligent fault diagnosis using deep learning algorithms: a comparative analysis of MLP, CNN, RNN and LSTM (2024)
10. Shoji, T., Yoshida, N., Tanaka, T.: Automated detection of abnormalities from an EEG recording of epilepsy patients with a compact convolutional neural network. Biomed. Signal Process. Control **70**, 103013 (2021)

11. Tsoulos, I.G., Charilogis, V., Tsalikakis, D.: Improving the performance of constructed neural networks with a pre-train phase. Symmetry **17**(9), 1567 (2025)
12. Costi, F., Covaci, E., Roman, O., Onchis, D.: Neurofuzzxai: a hybrid artificial intelligence framework for epileptic seizure detection. Appl. Med. Inf. **47** (2025)
13. Farooqui, N.A., Mishra, A.K., Mehra, R.: Concatenated deep features with modified LSTM for enhanced crop disease classification. Int. J. Intell. Rob. Appl. **7**(3), 510–534 (2023)
14. Saha, T.S., Mukherjee, J., Kar, M., Chakrabarti, A., Guha, K.: A LLM and explainable AI assistive model for improvement of user's query in a chatbot for women cancer awareness for Indian lay people. In: International Conference on Data Management, Analytics & Innovation, pp. 211–226. Springer, Heidelberg (2025). https://doi.org/10.1007/978-981-96-5860-2_13
15. Becattini, F., et al.: Viscounth: a large-scale multilingual visual question answering dataset for cultural heritage. ACM Trans. Multimed. Comput. Commun. Appl. **19**(6), 1–20 (2023)
16. Ruwanpathirana, G.P., Thilakarathna, S., Baduge, S.K.: Interpretation and understanding of asphalt crack detection deep learning models using integrated gradient (IG) maps. Case Stud. Constr. Mater. **22**, e04613 (2025)
17. Altman, N., Krzywinski, M.: Predicting with confidence and tolerance. Nat. Methods **15**(11), 843–845 (2018)
18. Qi, Z., Khorram, S., Li, F.: Visualizing deep networks by optimizing with integrated gradients. In: AAAI, vol. 34, pp. 11890–11898 (2020)

Verification of Synchronization of LLM Agents in the Integrated Model of Distributed Systems Using Petri Net

Denny B. Czejdo[1], Wiktor B. Daszczuk[2](✉), and Waldemar Grabski[2]

[1] Department of Mathematics and Computer Science, Fayetteville State University, Fayetteville, NC, USA
bczejdo@uncfsu.edu

[2] Institute of Computer Science, Warsaw University of Technology, Nowowiejska Str. 15/19, 00-665 Warsaw, Poland
{wiktor.daszczuk,waldemar.grabski}@pw.edu.pl, wbd@ii.pw.edu.pl

Abstract. The Integrated Model of Distributed Systems (IMDS) provides a formal basis for describing asynchronous distributed systems by separating server states from agent-generated messages. This distinction captures both resource ownership and communication causality, enabling a precise representation of distributed behavior without global synchronization. In this paper, we present IMDS together with its interpretation as a Petri net, which preserves the operational semantics of actions and exposes concurrency, locality, and dependencies in a structural form. The Petri-net interpretation supports reachability analysis, deadlock detection, and termination reasoning using existing verification tools. We further show that IMDS naturally extends to contemporary LLM-based multi-agent workflows, whose message-driven interactions exhibit similar patterns of autonomy, nondeterminism, and potential deadlocks. Modeling such agents in IMDS enables systematic analysis of coordination failures and reasoning loops that are difficult to detect in procedural agent frameworks. The results demonstrate that IMDS, combined with its Petri net semantics, forms a unified and rigorous foundation for analyzing both classical distributed systems and modern multi-agent AI architectures.

Keywords: Integrated model of distributed systems · IMDS · Petri net interpretation · Asynchronous distributed systems · Multi-agent systems · LLM-based agents · Communication duality · Formal verification · Deadlock analysis

1 Introduction

Distributed systems form the foundation of modern computing, ranging from traditional microservice architectures and cloud platforms to the emerging paradigm of Large Language Model (LLM) based Multi-Agent Systems [1]. While traditional distributed components rely on programmed logic, autonomous AI agents introduce a new layer of complexity: they operate with significant autonomy, non-deterministic decision-making, and asynchronous communication patterns that are difficult to predict.

W. Zamojski et al. (Eds.): DepCoS-RELCOMEX 2026, LNNS 2014, pp. 68–79, 2026.
https://doi.org/10.1007/978-3-032-27927-9_7

Whether managing cloud resources or coordinating a swarm of AI agents (e.g., *Coder* and *Reviewer* agents), these systems share defining characteristics: the lack of a global clock, autonomous components, and inherently asynchronous communication [1]. In the context of Generative AI, these characteristics create severe risks. A system of interacting LLM agents can easily fall into "hallucinated loops" (infinite conversation cycles), "token exhaustion" (resource deadlock), or conflicting state updates [1]. Therefore, properties such as deadlock freedom, proper resource usage, and distributed termination must be addressed during early design stages using precise formal models.

Classical formalisms such as CSP [2] and CCS [3] provide strong foundations but often rely on synchronous interactions, which are ill-suited to the latency and asynchrony of LLM agent communication. Conversely, notations such as UML or BPMN lack the execution semantics required to rigorously verify that a swarm of agents will actually complete their tasks.

The *Integrated Model of Distributed Systems* [4, 5] offers a unique solution to these challenges. By explicitly separating server states (representing context windows, tools, or memory) from *agent-associated messages* (prompts and API calls), IMDS provides a framework that mirrors the architecture of modern Agentic AI. The separation enables *communication duality* [5], allowing the system to be analyzed from the perspective of the resources (e.g., "Is the database locked?") or the agents (e.g., "Did the Reviewer agent finish its task?").

This paper presents IMDS and its systematic Petri net interpretation [5] as a method for verifying these complex, asynchronous interactions. We demonstrate that established verification techniques—such as reachability analysis and deadlock detection—can be applied to ensure the reliability of autonomous multi-agent systems.

2 Related Work

The general conceptual framework underlying this work is rooted in foundational studies on autonomous agent systems and their coordination mechanisms. A comprehensive perspective on the rise, capabilities, and limitations of Large Language Model–based agents is presented in [1], which highlights their potential as autonomous reasoning entities interacting through structured communication. A detailed architectural view of agentic systems, including key components such as memory or context windows and tool invocation mechanisms, is provided in the extensive survey [6]. These architectural elements align closely with the separation of server states and agent-associated messages in IMDS.

Recent research has investigated how reasoning and acting can be combined within language models to support agent autonomy. In particular, [7] analyzes the integration of planning and execution in LLM-based agents, revealing challenges related to coordination, causal dependencies, and non-terminating behaviors. Multi-agent conversational frameworks are further explored in [3], where agent roles such as Planner, Coder, and Critic are introduced as interacting personas. These role-based architectures implicitly assume structured control flow and message causality, properties that are explicitly captured by agent processes in IMDS.

Synchronization and coordination remain critical challenges in multi-agent LLM systems. Meta-programming approaches for collaborative agent frameworks are discussed in [8], where Standard Operating Procedures (SOPs) are encoded into prompts to enforce strict execution order and synchronization. This concept is closely related to shared-state coordination mechanisms and resembles the notion of server-centric control captured by IMDS server processes. Complementary work on synchronization through role-playing strategies is presented in [9], emphasizing structured interaction patterns to reduce nondeterministic behavior.

The effectiveness of sequential barriers in preventing erroneous agent behavior is demonstrated in [6], which shows that explicitly enforced execution ordering can significantly improve reliability in multi-agent systems.

Overall, while existing agentic frameworks rely primarily on procedural or prompt-based coordination mechanisms, they lack a formally defined operational semantics. IMDS, together with its Petri net interpretation [5], complements this body of work by providing a rigorous formal foundation for analyzing control flow, synchronization, deadlocks, and termination in both classical distributed systems and modern LLM-based multi-agent architectures.

3 Overview of the IMDS Formalism

The Integrated Model of Distributed Systems (IMDS) is a formal framework designed to describe asynchronous distributed systems using a small and orthogonal set of modeling concepts. The model distinguishes two fundamental carriers of computation: server states, representing localized resources with internal values, and messages associated with migrating agents, representing distributed computations that request for services (performed by servers). This separation allows IMDS to unify message-passing and resource-oriented interpretations within a single semantic structure.

An IMDS specification is defined over four finite domains: a set of servers S, a set of possible state values V, a set of services R offered by servers, and a set of agents A representing distributed computations. Each server $s \in S$ has exactly one state at any moment, represented by a pair $(s,v) \in S \times V$. Messages are represented as triples $(a,s,r) \in A \times S \times R$, indicating that agent a requests service r from server s.

A configuration of an IMDS system consists of the set of current server states together with the set of all pending messages (exactly one state for a server, at most one message for an agent). This configuration captures the complete global state of the distributed system while maintaining a clear separation of local server resources and communication requests. Execution proceeds by applying actions, which are the only atomic events in IMDS.

Each action consumes exactly one server state and one matching message and produces a new server state and a new message. Formally, an action is expressed as

$$((S, V), (a,s,r)) \rightarrow ((S, V'), (a, s', r')) \tag{1}$$

where the output message may target the same or a different server. This structure enforces action locality: all decisions are made using only the local state of a server and the message addressed to it. No action may inspect or modify the state of another server directly, and no global synchronization is permitted. Note that the input and output messages concern the same agent.

A defining feature of IMDS is communication duality. The same model may be interpreted from two equivalent viewpoints. In the server view, servers are passive entities that react to incoming requests and update their states accordingly. In the agent view, agents are active processes that migrate between servers by issuing messages. Both interpretations describe the same behavior and share identical semantics, differing only in the projection of computation.

The operational semantics of IMDS is defined as a labeled transition system (LTS) whose states correspond to configurations and whose transitions correspond to enabled actions. An action is enabled whenever its required server state and message are present in the current configuration. The execution proceeds through arbitrary interleavings of independent actions. This semantics supports formal reasoning about reachability, deadlock situations, and termination conditions.

IMDS explicitly supports reasoning about distributed correctness properties. Communication deadlock (server deadlock) occurs when messages cannot be consumed by any server state (at the moment and forever). Resource deadlock (agent deadlock) arises when a message of an agent cannot be consumed by any action (at the moment and forever). Thus, communication deadlock occurs between servers, while resource deadlock occurs between agents. Agent termination is modeled by the absence of further agent messages. Distributed termination concerns a set of agents. These properties form the foundation for the Petri net interpretation presented in the next section [5].

4 Petri Net Interpretation

Petri nets [10] provide a well-established mathematical framework for modeling concurrency, synchronization, and resource flows in distributed systems. One of the key strengths of IMDS is that its semantics can be systematically translated into a *Petri net that is behaviorally equivalent* to the original specification. This section presents the construction, intuition, and implications of this mapping.

The IMDS $\rightarrow$ Petri net interpretation [5] preserves all essential properties of the system: locality of actions, asynchronous communication, communication duality, and the independence of concurrently enabled operations. As a result, Petri net–based analysis—including structural verification, reachability examination, deadlock detection, and invariant computation—can be naturally applied to IMDS models.

4.1 Basic Mapping Principles

The Petri net associated with an IMDS specification contains two fundamental classes of places: *state places*, corresponding to server state pairs $(s,v) \in S \times V$, and *message places*, corresponding to messages $(a,s,r) \in A \times S \times R$. Each IMDS *action* becomes a *transition* in the Petri net. The input and output arcs reflect exactly the server state and message consumed and produced by the action.

Formally, the IMDS action $((s,v), (a,s,r)) \rightarrow ((s,v'), (a,s',r'))$ is mapped to a Petri net transition t with input places: (s,v) and (a,s,r) and output places: (s,v') and (a,s',r').

This mapping ensures a one-to-one correspondence between IMDS atomic steps and Petri net firings.

4.2 Stepwise Refinement Construction

Although the formal mapping is straightforward, an important contribution of IMDS research is a stepwise refinement procedure that constructs the Petri net in stages. This method clarifies how IMDS separates concerns and introduces successive modeling elements.

Step 1—Unified System Place (Fig. 1a). The construction begins with a single place representing the entire system state. While not operationally useful, this abstraction introduces the idea that each IMDS action corresponds to a single Petri net transition.

Step 2—Server vs. Message Decomposition (Fig. 1b,c). The unified place is split into: a state place for server states, and a message place for messages. This is the first structural separation directly corresponding to the dual carriers of IMDS semantics.

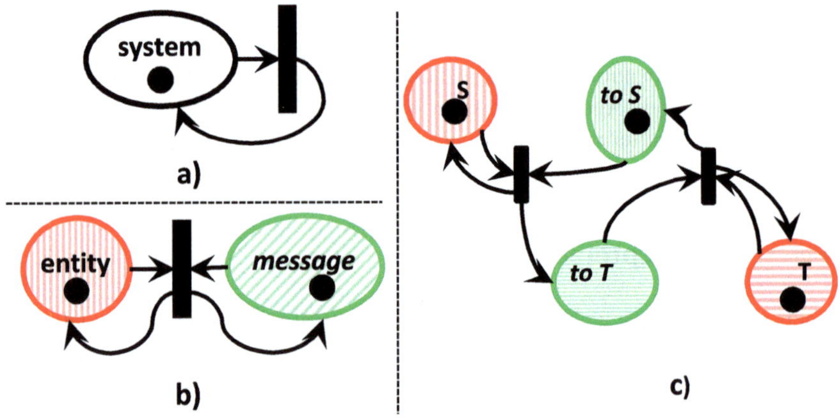

Fig. 1. The model (**a**) a system (**b**) working system with action; (**c**) working system with messages and two actions.

Step 3—Multiple Servers (Fig. 2a). The model introduces individual state places for each server $s \in S$. This reveals the *locality of execution*: transitions for one server do not directly manipulate the state of another.

Step 4—Agents (Fig. 2a). Message places are further refined by introducing agent identifiers. Each message becomes a pair: (a,s), representing an agent's presence at a given server.

Step 5—Services (Fig. 2b). To capture services offered by servers, messages are extended to triples: (a,s,r), indicating that agent a requests service r from server s.

Step 6—Server State Values (Fig. 2b). State places are expanded to pairs (s,v), to reflect internal data values. Transitions now encode both control flow and state manipulation.

Step 7—Terminating Transitions. An optional final refinement introduces transitions that do not produce any output message, modeling agent termination. This supports the detection of distributed termination and allows the modeling of finite computations. These refinement steps collectively result in a fully structured Petri net that accurately mirrors IMDS semantics. An example of a terminating action would be the absence of an arc leading to a (B,U,J) message in Fig. 2b. The absence of a continuation message terminates the agent.

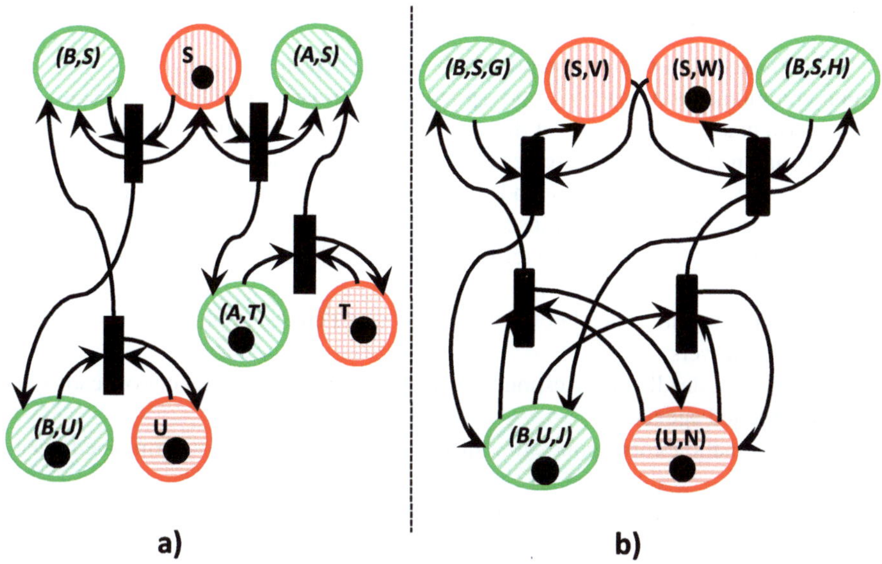

Fig. 2. (**a**) introducing servers S, T, U, and agents A, B; (**b**) introducing server state values V, W, N, and services G, H, J.

In this article, we use a version of IMDS without the ability to create new servers and agents in action. This is due to the requirement that the system subject to model checking be finite.

4.3 Structural Properties

The resulting Petri net exhibits key structural characteristics:

- *Bipartite structure*: All transitions have exactly one state place and one message place as inputs and outputs, reflecting IMDS's atomic action model.
- *1-safeness*: Each server state place contains exactly one token at all times; thus, race-free access is guaranteed by construction.
- *Natural concurrency*: Independent IMDS actions map to transitions without shared input places, and therefore can fire concurrently in the Petri net.
- *Modularity*: Each server and each agent induces well-defined subnetworks, facilitating analysis.

4.4 Behavioral Equivalence

The central semantic result is that the Petri net (PN) reachable markings graph is *isomorphic* to the IMDS labeled transition system:

$$\mathrm{Reach}_{\mathrm{PN}} \cong \mathrm{Reach}_{\mathrm{IMDS}} \tag{2}$$

This equivalence ensures that:

- all IMDS execution paths are representable in the Petri net,
- total communication deadlock corresponds to terminal marking with present agent tokens (in places of messages of those agents) of all non-terminated agents (only the servers having those tokens are deadlocked, others are idle),
- total resource deadlock corresponds to terminal marking with present agent tokens of all non-terminated agents; those agents are deadlocked,
- total termination corresponds to terminal marking with tokens in places of all agents absent,
- partial communication deadlock corresponds to a Petri net siphon losing tokens of deadlocked servers,
- partial resource deadlock corresponds to a Petri net siphon losing tokens of deadlocked non-terminated agents,
- partial termination corresponds to a marking in which tokens of terminated agents are absent,
- state-space exploration tools can be used directly to analyze IMDS models.

5 Processes in IMDS

Processes in IMDS arise semantically from the execution of actions rather than being defined syntactically. Since each action consumes a server state and a message and produces their successors, two complementary views of system behavior naturally emerge: server processes and agent processes. These projections describe the same global execution but emphasize different structural aspects of the model.

Server processes emphasize locality and state evolution. A server maintains exactly one state token, which ensures that its behavior forms a single sequential thread regardless of global concurrency. This perspective describes how servers react to incoming messages, how their internal states change, and where resource-related conflicts or local deadlocks may appear. Because all decisions are strictly local, server processes reveal the dependencies that arise when multiple agents compete for the same resource. The server view is shown in Fig. 3a.

Agent processes emphasize control flow and message causality. An agent's behavior is defined by the sequence of messages it produces across servers, revealing computation logic, dependencies, and potential deadlocks or loops. In LLM-based multi-agent systems, this shows how agents trigger tasks, rely on responses, and coordinate via messaging. The agent view is shown in Fig. 3b.

The two views are fully consistent: server behavior constrains how agents can progress, while agent behavior determines how servers evolve in response to requests.

This duality is preserved in the Petri-net interpretation of IMDS [5], where server processes correspond to transitions connected by state places, and agent processes correspond to transitions linked through message places. Together, they provide a comprehensive representation of both resource-level and communication-level behavior in distributed systems.

In modern LLM-based multi-agent workflows, server processes naturally correspond to shared tools, APIs, or functions accessed by autonomous agents. The server-oriented perspective precisely captures contention and consistency issues arising when multiple agents rely on the same shared components.

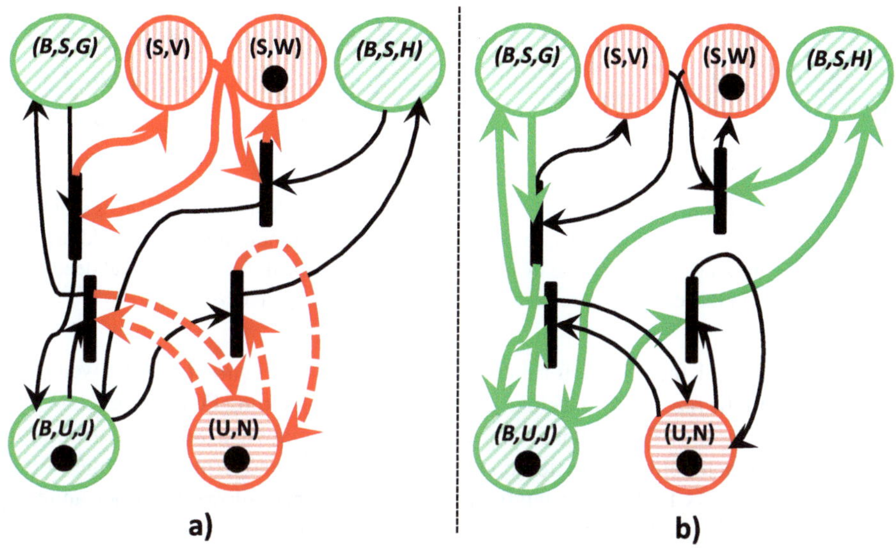

Fig. 3. (**a**) server view of an example system (red arrows–server processes–2 servers *S*, *U*) (**b**) agent view (green arrows–agent process *B*).

6 Semantics and Verification

The semantics of an IMDS specification is defined as the set of all configurations reachable through the execution of enabled actions. Each configuration consists of current server states and pending messages. The operational behavior of the system can be formally represented as a labeled transition system (LTS), in which states correspond to configurations and transitions correspond to individual actions. An action is enabled when both its required server state and associated message are present, and its execution replaces them with a new state and a new message. Multiple actions may be enabled simultaneously, and system behavior is defined by arbitrary interleavings of independent actions.

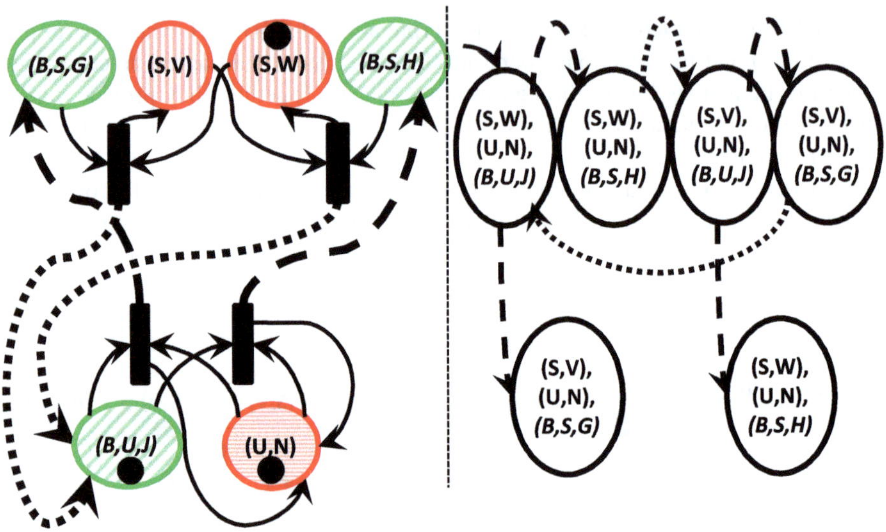

Fig. 4. (**a**) system from Fig. 3, transition in the servers shown near their states; messages sent from *U* to *S* shown as dashed bold arrows, responses (also messages) showed as dotted bold arrows; (**b**) reachable markings graph; a node includes the names of marked places. Two total deadlocks were identified (in this case, they are both server deadlocks and agent deadlocks).

A key result of the IMDS formalism is the *behavioral equivalence* between the IMDS LTS and the reachable markings graph of the corresponding Petri net. Each IMDS configuration maps to a Petri-net marking, and each action maps to a transition firing. As a consequence, reachability properties, concurrency relations, and terminal configurations are preserved across both representations. This equivalence allows Petri-net analysis techniques to be applied directly to IMDS models without semantic adaptation.

IMDS supports explicit reasoning about different classes of deadlocks. Communication deadlocks occur when messages remain pending, but a server state is not available to consume them, while resource deadlocks arise from messages that cannot be processed (due to cyclic dependencies among servers that prevent further progress or other situations). In the Petri net, these situations correspond to markings with no enabled transitions or to structurally blocked subnetworks (siphons). IMDS also supports analysis of distributed termination, which occurs when agents complete execution and no further messages are produced.

The formal semantics of IMDS has been implemented in tools such as Dedan [4], enabling automated reachability analysis, deadlock detection, and Petri net–based structural analysis (Fig. 4). This demonstrates that IMDS is not only a theoretical formalism but also a practical framework for verifying asynchronous distributed systems.

7 Verification of LLM-Based Multi-Agent Systems

The recent rise of "Agentic AI" frameworks (such as *AutoGen*, *CrewAI*, or *LangGraph*) has necessitated rigorous methods for synchronizing autonomous agents. In these systems, agents act as independent entities that exchange natural language messages to solve complex tasks. IMDS is particularly well-suited for modeling these workflows because its fundamental structure—autonomy of decisions and locality of actions [11]—directly maps to the behavior of LLM agents.

7.1 Mapping LLM Agents to IMDS

To verify a multi-agent workflow, we map the components of an AI swarm to the IMDS tuples (s, v) and (a, s, r) described in Sect. 2:

- *Agents* (A): Represent the active AI personas (e.g., a *Planner*, *Coder*, or *Critic*) [12]. In IMDS, these agents migrate between contexts by issuing messages [4].
- *Server States* $(S \times V)$: Represent the local "context window" or "memory" of an agent, as well as shared tools (e.g., a file system or vector database) [13]. Crucially, only the owner of the context can update it, enforcing the *locality of actions* [4].
- *Messages* (M): Represent the prompts or structured outputs passed between agents. An IMDS message (a, s, r) indicates that agent a is requesting a specific function r (e.g., ReviewCode) from the context of server s [4].

7.2 Modeling Synchronization and Handoffs

Synchronization in LLM systems often involves a "handoff," where one agent yields control to another (e.g., a *Coder* passes a script to a *Tester*; they are both agents starting from their own servers). In standard programming, this is a function call; in asynchronous AI systems, it is a message exchange that can fail.

Consider a scenario with two agents, *Producer* and *Consumer*, synchronizing over a shared buffer (Server s). In IMDS, this is modeled not by global variables, but by the availability of a specific server state.

- The *Producer* agent sends a message to Server s to store data. This action is only enabled if Server s is in a *Ready* state [7].
- The action updates Server s to a *Full* state and sends the *Producer* agent to a wait state.
- The *Consumer* agent sends a request to retrieve data. This action is only enabled if Server s is in the *Full* state.
- If the *Consumer* requests data while the server is *Ready* (empty), the action is disabled. The message remains pending, creating a natural, non-blocking wait typical of asynchronous systems [14].

7.3 Verifying Distributed Termination in AI Swarms

A critical failure mode in LLM-based multi-agent systems is "infinite conversation loop," where agents politely thank each other endlessly or debate a plan without executing it. This corresponds to the IMDS concept of no *Distributed Termination* (but some actions are executed) [11].

By translating the agent workflow into the Petri net interpretation (as detailed in Sect. 3, [5]), we can mathematically verify if the system reaches a terminal state.

- *Termination*: A successful workflow results in a Petri net marking where all agent tokens are absent (consumed by terminating transitions) [5].

- *Communication Deadlock:* If the *Coder* agent sends a message *Verify* from the *Coder* server to the *Tester* agent on the *Tester* server, but the *Tester* is stuck in a state where it cannot accept *Verify* messages, the system enters a communication deadlock [11]. In the Petri net, this is detected as a reachable marking where the agent token remains in a message place permanently.
- *Resource Deadlock*: If two agents cyclically wait for each other's tools (e.g., Agent *A* holds the *WriteFile* tool while waiting for *ReadFile*, and Agent *B* holds *ReadFile* while waiting for *WriteFile*), IMDS identifies this as a resource deadlock [11].

This verification capability allows developers to prove that an autonomous agent swarm *will* eventually stop and produce a result, rather than simply hoping the probabilistic nature of the LLM resolves the interaction.

8 Discussion and Conclusions

This paper presents an updated IMDS model with a Petri net interpretation, showing its applicability to both classical asynchronous systems and LLM-based multi-agent workflows. By separating server states from agent messages, IMDS clearly represents resource ownership, causality, and component autonomy.

The semantics-preserving mapping to Petri nets enables standard analyses—such as reachability, deadlock detection, and termination checking—supporting the study of common issues in AI agents, including communication deadlocks, resource conflicts, and infinite loops.

The examples in the paper illustrate that multi-agent LLM workflows—often modeled informally in current frameworks—can be expressed precisely within IMDS. Their correctness properties can then be verified using existing Petri-net tools without modifying the underlying semantics. This provides a formal link between high-level agent orchestration and rigorous verification.

Despite the inherent risk of state explosion, IMDS offers a structurally simple and conceptually clean model that supports verification of both classical distributed systems and modern autonomous AI agents. Future work includes extending the approach toward timed or probabilistic variants, improving tool scalability, and integrating IMDS-based verification directly into agentic AI development frameworks.

References

1. Guo, T., Chen, X., Wang, Y., Chang, R., Pei, S., Chawla, N. V., Wiest, O., Zhang, X.: Large language model based multi-agents: a survey of progress and challenges. In: 33rd International Joint Conference on Artificial Intelligence, Jeju, Korea, 3–9 Aug 2024, pp. 8048–8057 (2024). https://doi.org/10.24963/ijcai.2024/890
2. Hoare, C.A.R.: Communicating sequential processes. Commun. ACM **21**, 666–677 (1978). https://doi.org/10.1145/359576.359585
3. Milner, R.: Communication and Concurrency. Cambridge University Press (2009)
4. Daszczuk, W.B.: Integrated model of distributed systems. Springer Nature, Cham, Switzerland (2020). https://doi.org/10.1007/978-3-030-12835-7

5. Chrobot, S., Daszczuk, W.B.: Communication Dualism in distributed systems with petri net interpretation. Theor. Appl. Informatics. **18**, 261–278 (2006)
6. Qian, C., Liu, W., Liu, H., Chen, N., Dang, Y., Li, J., Yang, C., Chen, W., Su, Y., Cong, X., Xu, J., Li, D., Liu, Z., Sun, M.: ChatDev: Communicative agents for software development. In: Proceedings of the 62nd Annual Meeting of the Association for Computational Linguistics, vol 1, pp. 15174–15186. ACL, Stroudsburg, PA (2024). https://doi.org/10.18653/v1/2024.acl-long.810
7. Tanenbaum, A.S., Van Steen, M.: Distributed systems: Principles and Paradigms. Pearson (2017)
8. Yao, S., Zhao, J., Yu, D., Du, N., Shafran, I., Narasimhan, K., Cao, Y.: ReAct: synergizing reasoning and acting in language models. In: 11th International Conference on Learning Representations, ICLR 2023, 1–5 May 2023, pp. 1–16 (2023)
9. Li, G., Hammoud, H.A.A.K., Itani, H., Khizbullin, D., Ghanem, B.: CAMEL: communicative agents for "mind" exploration of large language model society. In: NIPS '23: 37th International Conference on Neural Information Processing Systems, pp. 51991–52008 (2023). https://doi.org/10.48550/arXiv.2303.17760
10. Reisig, W.: Understanding Petri Nets. Springer Berlin Heidelberg, Berlin, Heidelberg (2013). https://doi.org/10.1007/978-3-642-33278-4
11. Daszczuk, W.B.: Threefold analysis of distributed systems: IMDS, Petri Net and distributed automata DA3. In: Federated Conference on Computer Science and Information Systems, FEDCSIS'17, Prague, Czech Republic, 3–6 Sept. 2017, pp. 377–386. IEEE Comput. Soc. Press (2017). https://doi.org/10.15439/2017F32
12. Wu, Q., Bansal, G., Zhang, J., Wu, Y., Li, B., Zhu, E., Jiang, L., Zhang, X., Zhang, S., Liu, J., Awadallah, A.H., White, R.W., Burger, D., Wang, C.: AutoGen: Enabling Next-Gen LLM Applications via Multi-Agent Conversation (2024)
13. Xi, Z., Chen, W., Guo, X., He, W., Ding, Y., Hong, B., Zhang, M., Wang, J., Jin, S., Zhou, E., Zheng, R., Fan, X., Wang, X., Xiong, L., Zhou, Y., Wang, W., Jiang, C., Zou, Y., Liu, X., Yin, Z., Dou, S., Weng, R., Cheng, W., Zhang, Q., Qin, W., Zheng, Y., Qiu, X., Huang, X., Gui, T.: The rise and potential of large language model based agents: a survey (2023). https://doi.org/10.48550/arXiv.2309.07864
14. Lamport, L.: Time, clocks, and the ordering of events in a distributed system. Commun. ACM **21**, 558–565 (1978). https://doi.org/10.1145/359545.359563

Radio Communication Systems for Special Military Applications

Tomasz Dlugosz, Wiktor Zietara(✉), and Łukasz Bak

DSW Univeristy of Lower Silesia, Strzegomska Street 55, Wroclaw, Poland
{tomasz.dlugosz,wiktor.zietara}@dsw.edu.pl

Abstract. The article analyzes the application of radio communication systems and networks for special purposes in military operations, using the Israel–Hamas conflict as a case study. It demonstrates how modern military telecommunications—including Software-Defined Radio (SDR) systems, Mobile Ad hoc Networks (MANET), Unmanned Aerial Vehicles (UAVs), and integrated C4ISR systems—affect combat tactics, command processes, and the development of information superiority. Particular emphasis is placed on the integration of communications with predictive analytics, artificial intelligence, and the concept of network-centric warfare. The article also discusses challenges related to cybersecurity and electronic warfare, as well as future development perspectives for military communication systems in the context of 6G technologies and quantum communication. Israel is presented as an example of a state effectively adapting modern telecommunication technologies to the realities of asymmetric warfare.

Keywords: Military telecommunications · Radio communication systems · Asymmetric warfare · MANET networks · C4ISR · cybersecurity

1 Introduction

The transformation of the modern battlefield is increasingly determined by the dynamic development of information technologies, telecommunications, and data processing systems [1, 2]. The ongoing digitalization of the operational environment means that information—its acquisition, transmission, analysis, and protection—has become one of the key resources determining the effectiveness of military operations. Consequently, modern armed forces are compelled to implement new organizational and technical models, particularly in operational and strategic communications, which enable the conduct of operations under conditions of high dynamics and uncertainty [2].

In the realities of asymmetric and hybrid conflicts, where the adversary often operates below the threshold of conventional warfare and the front line becomes blurred, reliable and secure communications constitute not only a command-support tool but one of the fundamental determinants of operational superiority. Modern military telecommunications, based on advanced radio systems, mobile ad hoc networks, and network-centric architectures, enable real-time synchronization of combat operations, integration of distributed sensors, and the conduct of reconnaissance, intelligence, and electronic warfare activities [1, 3].

W. Zamojski et al. (Eds.): DepCoS-RELCOMEX 2026, LNNS 2014, pp. 80–89, 2026.
https://doi.org/10.1007/978-3-032-27927-9_8

A particular challenge for military communication systems is operations conducted in urban environments, characterized by high infrastructure density, electromagnetic interference, and the presence of civilian populations. Under such conditions, radio communication systems must demonstrate high resistance to interference, topological flexibility, and the ability to autonomously reconfigure in the event of partial infrastructure loss. These very characteristics determine their suitability for special operations and asymmetric activities.

The Israeli– Palestinian conflict, and in particular the cyclical clashes between Israel and Hamas in the Gaza Strip, constitutes one of the most representative contemporary case studies in the use of advanced radio communication systems. The nature of these operations—encompassing urban combat, special forces operations, intensive use of unmanned aerial platforms, and parallel activities in cyberspace and the information domain—has forced the Israel Defense Forces (IDF) to implement innovative communication solutions at the tactical, operational, and strategic levels [2, 4].

For many years, Israel has been developing the concept of integrated command and control systems (C4ISR), in which radio communications form the backbone of the entire battlefield information architecture. The integration of SDR systems, MANET networks, unmanned platforms, and predictive analytics tools enables the IDF to achieve so-called temporal superiority, understood as the ability to detect threats, analyze situations, and make decisions faster than the adversary [3].

The aim of this article is to analyze the role of radio communication systems and networks designed for special military applications in the context of contemporary armed conflicts, using the Israel–Hamas conflict as an example. The study focuses on the technical, operational, and strategic aspects of the use of military communications, with particular emphasis on their impact on decision-making processes, combat effectiveness, and the development of information superiority. The analysis is based on recent research reports by the RAND Corporation, the Center for Strategic and International Studies (CSIS), and the Brookings Institution, as well as on academic literature concerning modern armed conflicts and information warfare.

2 Abbreviations and Definitions

This study uses a number of abbreviations and specialized terms characteristic of military, telecommunications, and information security terminology. Their unambiguous definition is necessary to ensure precision of the discussion and correct interpretation of the analyzed issues.

AI (Artificial Intelligence)—artificial intelligence

A set of methods and algorithms that enable information systems to autonomously process data, recognize patterns, perform machine learning, and support or automate decision-making processes. In military applications, AI is used, among others, for the analysis of intelligence data, management of information flows, and prediction of adversary behavior.

C2 (Command and Control)—command and control

A basic system for managing military operations, encompassing decision-making, planning, issuing orders, and monitoring their execution. The effectiveness of the C2 system largely depends on the quality and reliability of communication means.

C4ISR (Command, Control, Communications, Computers, Intelligence, Surveillance and Reconnaissance)

An integrated command architecture encompassing command, control, communications, information systems, intelligence, surveillance, and reconnaissance. C4ISR constitutes the foundation of modern network-centric warfare systems, enabling the creation of a common operational picture.

HUMINT (Human Intelligence)—human intelligence

The acquisition of information through human sources such as agents, informants, interrogations, or direct observation. HUMINT remains an important complement to data obtained from technical systems.

IDF (Israel Defense Forces) – Israel Defense Forces

The armed forces of the State of Israel, responsible for conducting land, air, naval, cyber, and information-domain operations.

MANET (Mobile Ad hoc Network)—mobile ad hoc network

A self-configuring wireless network that does not require fixed infrastructure. Network nodes simultaneously perform transmitting and relay functions, enabling dynamic reconfiguration of connections under combat conditions.

OSINT (Open-Source Intelligence)—open-source intelligence

The acquisition and analysis of publicly available information, such as media content, internet resources, reports, or scientific publications. OSINT plays a growing role in information conflicts and psychological operations.

SDR (Software-Defined Radio)—software-defined radio

A radio communication system in which key functions, such as modulation, coding, and spectrum management, are implemented in software. SDR enables rapid adaptation of operating parameters to be changing operational conditions.

SIGINT (Signals Intelligence)—signals intelligence

The acquisition of information through the interception and analysis of electromagnetic signals, including radio communications, radar emissions, and adversary data transmissions.

UAV (Unmanned Aerial Vehicle)—unmanned aerial vehicle

An uncrewed aerial platform used for observation, reconnaissance, signal relay, and offensive operations. UAVs constitute an important element of contemporary C4ISR systems.

VHF/UHF (Very High Frequency/Ultra High Frequency)

Radio frequency bands are commonly used in tactical military communications, characterized by good propagation in urban environments but susceptibility to interference.

6G (Sixth Generation of Wireless Technology)—sixth generation of wireless technology

An emerging generation of wireless communication systems aimed at achieving ultra-low latency, high throughput, and increased resistance to interference and external manipulation, including military applications.

3 Contemporary Military Telecommunications

3.1 System Resilience and Operating Environment

Contemporary military telecommunications constitute one of the key components of the national security system and the foundation for the effective functioning of armed forces in the environment of modern armed conflicts [3, 5]. They comprise an integrated set of technologies, procedures, and organizational structures whose purpose is to ensure continuous, secure, and interference-resistant information exchange between all levels of command—from the strategic to the tactical level. Unlike civilian systems, military telecommunication systems are designed to operate under conditions of deliberate adversary interference, infrastructure degradation, and high operational dynamics [3].

One of the fundamental characteristics of modern military telecommunications is their resistance to electromagnetic interference and attempts to intercept transmissions. This is achieved with advanced encryption methods, frequency-hopping techniques, adaptive modulation, and multichannel transmission. These solutions make it possible to maintain communications even under conditions of intensive electronic warfare, which has now become an integral element of armed conflict [6].

3.2 MANET and SDR as Core Enablers

Another important aspect is the topological flexibility of communication systems. Contemporary military operations are increasingly conducted by dispersed, mobile, and autonomous subunits, which necessitates moving away from centralized communication models. In this context, MANET-type networks play a key role, as they enable the dynamic creation and reconfiguration of connections without the need for permanent telecommunications infrastructure. Each network element can serve both as a user and as a relay node, which significantly increases the resilience of the entire system to damage and combat losses [7].

An important step in the development of military telecommunications is the use of Software-Defined Radio (SDR) technology. These solutions enable rapid adjustment of radio operating parameters to current operational conditions, the type of mission being conducted, and the characteristics of the adversary. Through the implementation of software-based functions, it is possible to integrate various communication standards, frequency bands, and operating modes within a single device, thereby increasing interoperability between units and allies [5].

3.3 C4ISR, Common Operational Picture, and Information Superiority

Military telecommunications are not limited solely to voice and data transmission; they constitute an integral element of command-and-control system architectures. Within the C4ISR concept, communication systems enable the integration of data originating from reconnaissance sensors, unmanned platforms, radar systems, and intelligence sources such as SIGINT and HUMINT. The result is the creation of a Common Operational Picture, which forms the basis for real-time decision-making [2, 3].

From a military perspective, telecommunications serve both defensive and offensive functions. In addition to ensuring continuity of command and coordination of one's own forces, they enable operations aimed at disrupting adversary systems, intercepting and analyzing enemy transmissions, and conducting disinformation activities. In this sense, radio communication systems for special applications should be viewed as an element of a broader information and electromagnetic warfare architecture, in which information superiority becomes as important as kinetic superiority.

The development of contemporary military telecommunications is closely linked to the ongoing network-centric transformation of the battlefield. Each combat unit, sensor, or unmanned platform becomes a node in a data-exchange network, enabling decentralization of the decision-making process and shortening of the observe–orient–decide–act (OODA) cycle. Under conditions of asymmetric conflicts and high temporal dynamics, this capability constitutes one of the key factors determining the operational effectiveness of armed forces.

4 Tactical Use of Communication Systems in IDF Operations

4.1 Urban Operational Environment

The tactical use of communication systems in the operations of the Israel Defense Forces (IDF) constitutes one of the key factors determining operational effectiveness under conditions of asymmetric conflict [2, 4]. The specific characteristics of the operational environment in the Gaza Strip—including dense urban development, extensive underground infrastructure, and a high level of threat dynamics—necessitate the use of communication solutions with a high degree of reliability, flexibility, and resistance to interference and interception.

In urban combat conditions, IDF communication systems perform a role that goes far beyond the traditional transmission of orders. They constitute an integral element of the command process at the tactical level, enabling ongoing synchronization of actions by small, dispersed subunits, rapid response to a changing situation, and minimization of the risk of both military and civilian casualties. Of particular importance here is the ability to transmit data in real time, such as imagery from unmanned aerial platforms, positional information, or threat warnings.

4.2 SDR-MANET Integration and Elbit E-LynX

One of the pillars of IDF tactical communications is the integration of SDR radios with mobile MANET networks. Such a solution makes it possible to create self-configuring, damage-resistant communication networks that retain functionality even in the event of the loss of some nodes. An example of the implementation of this concept is the Elbit E-LynX system, used by Israeli ground units. The technical parameters of an example SDR radio station used in this system are presented in Table 1, which allows the tactical analysis to be placed in an empirical context.

Table 1. Example parameters of a tactical SDR system (Elbit E-LynX)

Parameter	Value
Maximum data throughput	up to 12 Mb/s
Operational range	up to 50 km
RF output power	0.5 W
Encryption	AES-256
Resilience / networking techniques	frequency hopping, MIMO, QoS, IP mesh

Source Author's own elaboration based on manufacturer data and technical analyses [5].

This system enables the simultaneous transmission of voice, data, and imagery, dynamic spectrum management, and the generation of a tactical situational picture available to commanders at the platoon and company levels.

4.3 UAV-Supported Communications and Silent Modes

An important element of the tactical application of communication systems is their close integration with unmanned aerial platforms (UAVs). Drones used by the IDF perform not only reconnaissance functions but also serve as mobile signal relays, extending the range of MANET networks in urban and subterranean environments. This makes it possible to maintain communications with subunits operating in buildings, tunnels, or areas with limited radio-wave propagation.

A practical confirmation of the effectiveness of these solutions was Operation "Shield and Arrow," conducted in 2023, during which Israeli special forces carried out parallel combat operations in multiple sectors of the Gaza Strip. The communication systems employed enabled the synchronization of ground maneuvers with precision artillery and air support, based on data transmitted in real time. A particularly important role was played by low-emission communication modes (silent communication), using packet-based transmission and adaptive frequency management, which reduced the likelihood of detection and targeting of units by the adversary.

4.4 Predictive Analytics and Electronic Warfare at the Tactical Level

At the tactical level, the IDF also makes intensive use of communication systems to support intelligence and analytical activities. Data collected from radio networks, battlefield sensors, and unmanned platforms are integrated and analyzed by predictive analytics systems, often supported by artificial intelligence algorithms. Analysis of the adversary's communication patterns makes it possible to identify potential concentration points of Hamas forces, predict directions of attack, and optimize the deployment of one's own subunits [4, 8].

IDF tactical communication systems also play a key role in electronic warfare operations. These include both the protection of one's own transmissions against jamming and interception, as well as active measures against the adversary's communication systems. Jamming, deception, and interception of Hamas signals constitute an important element of combat support and the enhancement of information superiority at the tactical level.

As a result, the communication systems used by the IDF at the tactical level should be viewed as multifunctional tools integrating command, reconnaissance, maneuver, and fire support. Their effectiveness stems not only from technological advancement but also from a doctrinal approach that treats information as a key resource of the battlefield.

5 Strategic Significance and Integration with Command Systems

5.1 C4ISR Architecture and Multidomain Integration

The significance of military telecommunications at the strategic level extends far beyond support for ongoing combat operations and encompasses the entire spectrum of processes related to planning, command, and coordination of armed operations at the national scale [2, 3]. In modern conflicts characterized by multidomain operations (land, air, sea, cyberspace, and the information domain), communication systems constitute a key element enabling the integration of activities conducted simultaneously across different operational environments.

The central component of this process is integrated command and control systems of the C4ISR class, which combine command, communications, data processing, intelligence, surveillance, and reconnaissance functions into a coherent information architecture. These systems enable the collection, fusion, and distribution of data from multiple sources—from reconnaissance satellites and ground-based radars, through unmanned platforms, to information obtained via SIGINT, HUMINT, and OSINT. The result of this integration is the creation of a common operational picture, which forms the basis for decision-making at the strategic and operational levels [2, 3].

In the case of the Israel Defense Forces, the integration of communication systems with the command architecture is systemic and doctrinal in nature. For many years, the IDF has been developing a command model based on maximizing the shortening of the decision-making cycle and decentralizing decision authority, while simultaneously maintaining coherence of actions at the strategic level. Military telecommunications serve in this model as the informational backbone, enabling rapid data flow between headquarters, operational commands, and executing units.

5.2 Temporal Superiority and Decision Acceleration

A key element of Israeli strategic doctrine is the concept of so-called temporal superiority, understood as the ability to detect threats, analyze situations, and make operational decisions faster than the adversary. The integration of communication systems with advanced analytical tools, including artificial intelligence systems, allows the IDF to automate parts of the decision-making process and prioritize targets in near-real time [3].

An important aspect of the strategic integration of command systems is the capability to conduct multi-level and multi-phase operations, in which tactical actions are directly linked to strategic objectives. Thanks to the network-centric C4ISR architecture, information obtained at the lowest levels of command can be almost immediately used in operational and strategic planning processes, increasing the flexibility and adaptability of the entire command system.

5.3 Network-Centric Warfare and Information Influence

Israel also intensively develops the concept of network-centric warfare, in which every unit, combat platform, or sensor constitutes a node in a data-exchange network. Such an architecture enables not only increased precision and effectiveness of operations but also partial decentralization of the decision-making process, allowing autonomous decisions to be made at lower levels of command. Under conditions of a rapidly changing operational situation, this capability constitutes a significant factor of strategic advantage [2].

At the strategic level, IDF communication systems also serve as instruments of information and psychological influence. The integration of communication channels with information analysis systems enables the conduct of coordinated information operations, encompassing both defensive activities (protection of one's own information space) and offensive actions (shaping the adversary's perception and influencing public opinion). In this way, military telecommunications become an element of a broader national security strategy.

In summary, the integration of communication systems with command systems at the strategic level determines a state's ability to conduct modern military operations in a coordinated, flexible, and interference-resistant manner. The example of Israel demonstrates that an effective C4ISR architecture, supported by advanced military telecommunications, constitutes one of the foundations of strategic superiority in contemporary conflicts.

5.4 From Theoretical Framework to Empirical Verification

The theoretical considerations presented in this article demonstrate that contemporary military telecommunications constitute far more than a technical support function for command and control. They represent a foundational element of modern warfare, shaping the structure of decision-making processes, the tempo of operations, and the ability to achieve and maintain information superiority. The integration of Software-Defined Radio, Mobile Ad hoc Networks, and C4ISR architectures reflects a broader transformation of the battlefield into a network-centric, information-driven operational environment.

While the theoretical framework outlined above provides a systematic understanding of the principles, technologies, and doctrines underpinning modern military communications, its full significance can only be assessed through empirical analysis. The practical effectiveness of these systems depends on their performance under real operational conditions, particularly in asymmetric conflicts characterized by urban terrain, electromagnetic contestation, and intensive activity in cyberspace.

For this reason, the analytical focus is continued in a complementary study. Part II of this research examines the practical application of radio communication systems for special military purposes using the Israel–Hamas conflict as a case study, allowing the theoretical assumptions discussed herein to be confronted with operational reality. This empirical perspective enables an assessment of how advanced communication technologies are employed at tactical and strategic levels, how they contribute to information superiority, and how they evolve in response to emerging cyber and electronic warfare threats.

6 Conclusion

This paper has examined the role of special-purpose radio communication systems in contemporary military operations, with particular emphasis on the Israel–Hamas conflict as a representative case of asymmetric warfare. The analysis demonstrates that modern military telecommunications—especially SDR-based systems, MANET networks, UAV-integrated platforms, and C4ISR architectures—are no longer merely supportive tools but constitute a fundamental component of operational and strategic effectiveness.

The study highlights that the integration of communication systems with artificial intelligence and predictive analytics significantly enhances decision-making processes, situational awareness, and operational tempo. The Israeli Defence Forces (IDF) provide a compelling example of how network-centric warfare can be effectively implemented through highly adaptive, resilient, and data-driven communication infrastructures.

At the same time, the research underscores that the increasing reliance on advanced communication technologies introduces new vulnerabilities, particularly in the domains of cybersecurity and electronic warfare. The electromagnetic spectrum and cyber domain have become critical battlegrounds, where information superiority is contested alongside traditional kinetic operations.

Looking forward, the evolution of military telecommunications will be shaped by the convergence of AI-driven network management, 6G communication technologies, and quantum-secure transmission systems. These developments will further transform the battlefield into a multidimensional information environment, where speed, security, and reliability of data exchange will determine strategic advantage.

In conclusion, armed forces that successfully integrate advanced communication technologies with intelligent data processing and robust cybersecurity frameworks will gain a decisive edge in future conflicts. The case of Israel clearly demonstrates that investment in communication systems is not supplementary but foundational to modern military capability.

References

1. Cohen, R., et al.: The future of warfare in 2030. RAND. (2020)
2. Cohen, R., et al.: From cast Lead to protective edge. RAND. (2017)

3. Madison, A., et al.: Machine learning for C2. ICMCIS (2024)
4. Byman, D., McCaleb, A.: Understanding Hamas's uses of information technology. CSIS (2023)
5. Black, J., et al.: Command and control in the future. RAND. (2024)
6. Jeliazov, P.M., Marinov, M.S.: Military radar systems. In: IEEE conference. (2004)
7. Harding, E.: Israeli intelligence analysis. CSIS. (2023)
8. Cybercrime, I.N.C.D.: Report. (2025)
9. Chang, Y., Keblis, M., Li, R., Iakovou, E., White, C.: Misinformation and disinformation in modern warfare. Oper. Res. (2022). https://doi.org/10.1287/opre.2022.2263
10. Cordey, S.: Israeli unit 8200 study. ETH Zurich (2019)
11. Dickson, J., Harding. E.: Quantum potential, CSIS (2024)
12. INCD.: Annu. Rep. (2025)
13. Stanescu, G.: Informational war. (2023)
14. Wirtschafter, V.: Information ecosystem. (2023)

Performance Analysis of the Plasma Generator System Used in Medical Devices

Pawel Dobrowolski[1,2](✉) and Michal Wisniewski[2]

[1] MIKROVOLT sp. z o.o., 3A-5 Kasztanowa Street, 53-125 Wroclaw, Poland
office@mikrovolt.pl

[2] Wroclaw University of Science and Technology, Wroclaw, Poland
{pawel.dobrowolski,wisniewski.michal}@pwr.edu.pl

Abstract. The development of medicine in the XXI century caused a rise in the number of surgical procedures performed. The scalpel remains an essential tool for any surgeon. It can be used to easily cut through soft tissues and remove tumors or cysts. Despite its simplicity, the scalpel has its flaw: the operating area bleeds heavily and is prone to infection. Alternatively, a surgical diathermy can be used, which utilizes current flow between the active and return electrodes through the patient's body. In this method, tissues are cut and blood vessels are instantly closed, stopping bleeding and reducing the risk of infection. In this article, a safe method of generating plasma arcs for medical purposes is shown. It allows tissue vaporization and constitutes an excellent alternative to the traditional scalpel. The proposed method is characterized by high precision – under 1 mm, which is affected, among others, by the generator supply voltage. The operating principle of the generator, based on capacitive feedback that prevents current flow through the patient's body, is described. Characteristics of the power transmitted to the operating area and the impact of generator frequency on its power were determined. Identified risks are described with minimization methods applied to them in the proposed device. The risk analysis for medical devices was performed according to *PN-EN ISO 14971:2020-05 Medical devices—Application of risk management to medical devices*. The results of the tests are summarized by choosing a safe and efficient working point of the proposed generator.

Keywords: Plasma generator · ISO 14971 Risk Management · Plasma in medicine

1 Introduction

Plasma is a fourth state of matter, defined as the energy flow between two areas of high potential difference in the ionized gas environment [3,4]. This conditions allow arc ignition, which converts electrical energy into thermal energy. Plasma in medicine has a wide range of uses. Low temperature plasma (which temperature does not exceed 40 °C [4]), generated in an environment of air or noble gasses –

W. Zamojski et al. (Eds.): DepCoS-RELCOMEX 2026, LNNS 2014, pp. 90–99, 2026.
https://doi.org/10.1007/978-3-032-27927-9_9

argon or helium, allows to avoid thermal damage of skin tissues in the operation area. It can be used to disinfect and detoxify skin and other tissues [1,3,5, 7,8]. Due to its cleaning and antiseptic properties [1], it supports the healing process of wounds [1,3,5,7,8], scars [3], and positively affects the treatment of atopic dermatitis [3]. In these treatments large area electrodes are used; for example, in paper [1] the author uses an electrode of 15 mm diameter. In addition to non invasive procedures, a correctly generated plasma beam can be used to cut and coagulate tissues [2,8]. To generate a plasma beam, an alternating current is used at a certain frequency without DC component [2,5,7]. In article [7] the author uses the generator output voltage in the range of $2 \div 6\,\mathrm{kV_{PP}}$, at 1 MHz frequency, in noble gas environment. Another article [2] proposes a plasma generator with $1.5 \div 50\mathrm{kV_{PP}}$ voltage range, frequency less than 300 kHz, and noble gas environment with flow less than 5 l/min. In this example, the depth of penetration of the tissue was between 0.1–0.8 mm or 0.3–1.5 mm depending on the operating mode of the device, with arc length of more than 5 mm.

Plasma beam generation methods described in [1–8] utilize a classical approach – for the generator to function correctly, two electrodes are required (active and return). This causes current flow through the patient's tissues, which may cause its degradation and burns at the location of contact between the patient's skin and the return electrode. The proposed solution for plasma beam generation is characterized by high precision and no current flow through the patient's body, due to utilization of capacitive feedback and lack of return electrode. Plasma beam ignition and its maintenance can be precisely controlled by the operator and occur at a strictly defined distance between the end of the active electrode and the operated tissue. Increasing the distance or touching the tissue with the active electrode causes the plasma beam to be immediately extinguished.

2 Plasma Generator System

The plasma beam generation circuit was designed using an astable generator with built in current limitation, which is implemented by pairs of resistors and transistors marked (R83, Q74) and (R84, Q75) on the schematic presented in Fig. 1. The main element of the generator circuit is transformer T1 with four windings – three low voltage on the primary side (NP1, NP2, FB) and one high voltage on the secondary (NS). The NP1 and NP2 windings have a common connection such that it is the end of the NP2 winding and the beginning of the NP1 winding. The secondary side NS has 1500 turns, while on the primary side FB has 3 turns, NP1 and NP2 have 6 turns each. This results in a transformer having a 1:125 turn ratio. The common point of the primary side windings is used to supply power to that side, while other ends of NP1 and NP2 are connected to collectors of NPN transistors (Q74, Q75). These transistors are controlled by the voltage induced in the FB winding. When the base-emitter voltage of one transistor, Q74 or Q75, exceeds 0.7V that transistor starts to conduct. The maximum current the transistor can conduct, collector current I_c is limited by the R83 resistor for transistor Q74 and by R84 for transistor Q75.

To generate sinusoidal signal, an LC generator is used, consisting of C80 capacitor and the inductance of the transformer windings NP1 and NP2. Its resonance frequency is determined by the following equation:

$$f_{LC} = \frac{1}{2\pi\sqrt{LC}} \tag{1}$$

In the high voltage side, a capacitor C81 was placed in order to remove DC component of the output signal and to limit the output current, which results in limited output power. One of its leads is connected directly to the active electrode. That capacitor's reactance (imaginary part of the impedance) is strictly connected to its capacity and frequency of the voltage source. It can be determined using the following equation:

$$X_C = \frac{1}{2\pi f C} \tag{2}$$

The second connection of the secondary winding is connected to the flat electrode and spiral-wound electrode, designed in such a way that they fit in the handle of the device while providing capacitive feedback. The placement of a flat electrode and spiral-wound electrode placement is subject to patent P.429541 registered by the Patent Office of the Republic of Poland.

The generator power supply is protected by a polymer resettable PTC fuse (F1) and a Zener diode (D1). This limits the power converted to heat by elements in case of damage to one of the components and also protects the generator against excessive input voltage. In case of exceeding the Zener voltage of diode D1, it starts to rapidly conduct current. This causes the current passing through the PTC fuse to exceed its maximum value, which in turn causes it to rapidly increase in resistance and limit the generator input voltage.

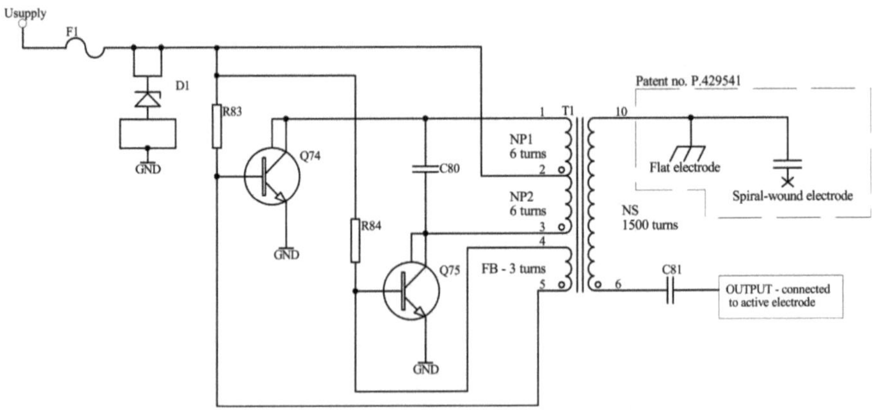

Fig. 1. Schematic of a plasma generator used in medicine.

The electronic circuit of the plasma generator, shown in the schematic in Fig. 1, is characterized by symmetric sinusoidal output voltage with zero DC component. Due to this, output voltage can be described by the following equation:

$$U_{wy} = Asin(\omega t) \quad (3)$$

where:
ω – pulsation, also defined as $\omega = 2\pi f$,
A – output voltage amplitude.

The working principle of the plasma beam generator is based on the resonance of the LC generator. Once the voltage is supplied to the generator input, it appears in the NP1 and NP2 windings of the transformer, and causes capacitor C80 to charge. At this point, transistors Q74 and Q75 are forward active. Due to the momentary undefined state, one of the transistors opens faster, which in turn causes the FB windings to induce voltage and close the second transistor. The first transistor also causes capacitor C80 to discharge, causing a voltage drop in the FB winding, cutting off the transistor, and reversing the polarity of the FB voltage. Again, capacitor C80 charges, causing the second transistor to start conducting. The cycle described previously starts to repeat itself until the generator voltage supply is turned off. The output power of the plasma generator is directly connected to its supply voltage, which affects the RMS voltage of the primary windings of the transformer. This relation can be described by the following equation:

$$V_{RMS_{NP}} = U_{SUPPLY} * (2.15 \pm 0.11) \quad (4)$$

During the operation of the generator, the current drawn from the source is limited by transistors Q74 and Q75. If its value is lower than defined by $I_c = \beta I_b$ it is not limited; however, once it exceeds that value, the collector-emitter voltage U_{CE} of transistors Q74 and Q75 increases until the maximum collector current I_C is reached. This in turn limits the output power of the plasma beam.

The described plasma beam generator utilizes capacitive feedback for energy transmission between its output and the patient's body. Due to its design, current does not flow through the patient's body to return electrode. In order to close the circuit, the generator utilizes the skin and tissues surface in proximity to the operating area, together with a flat electrode and a spiral-wound electrode in the device. Because the skin of the patient acts as a capacitor surface, its moisture level greatly affects the amount of energy transmitted. Moisture of the point of tissue vaporization also affects ignition and maintaining of the plasma arc as well as the transmitted energy. The less moisture in the operating area, the more concentrated the energy, leading to deeper and more precise tissue penetration.

The described plasma generator circuit is part of tests during the design process of a medical device. Therefore, during the conceptual, design and testing phases, it is necessary to perform a risk analysis according to *PN-EN ISO 14971:2020-05 Medical devices—Application of risk management to medical devices.* Its key elements are shown in Table 1.

Table 1. Selected elements of risk analysis according to *PN-EN ISO 14971:2020-05 Medical devices—Application of risk management to medical devices.*

Hazard	Result	Actions taken
Excessive output power	Excessive tissue vaporization/scarring after dermatological procedures	Series capacitor in high voltage circuit limiting maximal output current/Resistors polarizing and limiting base current controlling primary circuit
Accidental contact with the active electrode by the patient or operator	No arc ignition or its extinguishing/accidental puncture with an electrode	Utilization of capacitive feedback to ignite and maintain arc
Short circuit or open circuit in the secondary winding	No output voltage in the active electrode	Control of the transistors by voltage induced on primary circuit – idle state of the generator
Secondary winding breakdown (high-voltage side)	No output voltage in the active electrode	High frequency, low energy output voltage, not matched to capacitive feedback
Short circuit in the primary windings	No output voltage in the active electrode	Base current limit by resistors R83 and R84/No induced voltage on FB resulting in cutting off transistors Q74, Q75
Short circuit in auxiliary windings (FB)	No output voltage in the active electrode	Cutting off transistors Q74, Q75/PTC fuse tripping

3 Test Setup

The described plasma generator circuit can be configured, therefore in order to determine its characteristics – maximal power transmitted to generate plasma beam – impact of the following was tested:

- supply voltage U_{SUPPLY} in the range 3.5 VDC - 6.0 VDC;
- capacitor C80 capacity: 10nF, 22nF, 33nF, 47nF, 56nF, 100nF;
- resistance of the R83 i R84 resistors: R83 = R84: $3k\Omega$, $1k\Omega$;
- capacitor C81 capacity: 22pF, 100pF, 150pF, 220pF, 330pF, 470pF.

In the measurement circuit shown in Fig. 2, the following was measured:

- generator supply voltage using the RIGOL DM3068 digital multimeter;
- voltage drop across the measuring resistor $R_{POM} = 180\ mOhm \pm 1\%$ using the RIGOL MSO5354 oscilloscope in differential measurement mode. The measurement window was set at 6 s, while the measuring ranges were manually adjusted using the *Cursor* function. The built in algorithm calculated the voltage drop $V_{RMS_{R_{POM}}}$ according to Eq. (5);

– generated sinusoidal waveform on the capacitor C80 using the SIGLENT SDS 1022DL oscilloscope. The voltage $V_{RMS_{NP}}$ of winding NP = (NP1+NP2) according to Eq. (5) and the generator frequency were calculated by an internal numerical algorithm;

$$V_{RMS} = \sqrt{\frac{\sum_{i=1}^{n} x_i^2}{n}} \tag{5}$$

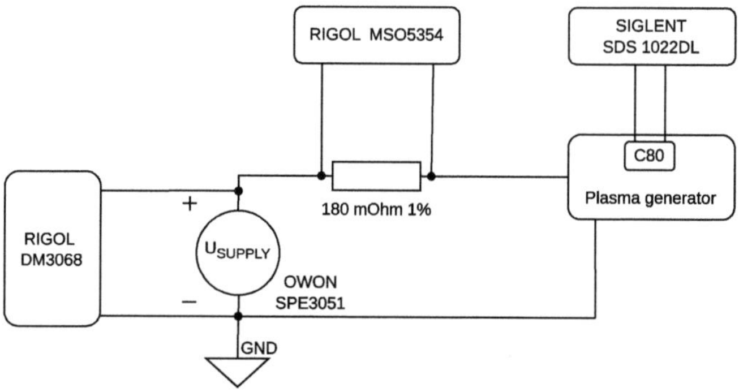

Fig. 2. Schematic diagram of the plasma generator measurement system.

In order to conduct the tests, animal tissue consisting of muscle and fat tissues was used. Constant environmental conditions of the workspace were maintained – stable temperature and humidity were ensured, monitored by a thermo-hygrometer TESTO 174H. The repeatability of the placement of the treatment electrode in relation to the generator was ensured by an aluminum head mounted on a plastic mounting element. The generator components were placed on a designed PCB board. The circuit was powered by an external DC power supply OWON SPE3051 with a maximum power of 150 W.

Due to the working principle of the plasma generator, characteristics, dynamics, and plasma beam instability, no access was possible for measuring equipment on the high-voltage side (NS windings). Therefore, the measurements were performed on the low voltage side. In order to determine the maximum power necessary to generate a plasma beam, the following was assumed:

– the voltage drop $V_{RMS_{RPOM}}$ across R_{POM} according to Eq. (5) reflects the amount of constant power drawn from the power source;
– the voltage drop $V_{RMS_{RPOM}}$ across R_{POM} in idle state (without beam generation) reflects the amount of power that was considered lost power P_{LOSS} (heat losses, transformer efficiency);
– the voltage drop $V_{RMS_{RPOM}}$ across R_{POM} during plasma beam generation minus power losses P_{LOSS} reflects the amount of power P_{WORK}, which is used to generate the plasma beam.

4 Test Results

Measurements of the plasma beam generator allowed power P_{WORK} characteristics to be generated. These characteristics were generated for a constant supply voltage U_{SUPPLY} and a constant resistance of the resistors R83 and R84. The values of capacitor C80 are placed on the OX axis and those of capacitor C81 on the OY axis. In order to better illustrate the characteristics, the measurement points were marked in red and cubic spline interpolation was performed to determine the remaining points of the characteristics.

Analyzing the characteristics obtained (Fig. 3) the following was concluded:

- resistors R83 and R84 affect the amount of power used in plasma beam generation. The higher the resistance, the shallower the penetration of tissue in the area of operation;
- capacitor C80 is responsible for the generator's operating frequency. It does not depend on the values of resistors R83 and R84 or capacitor C81, which confirmed consistency with the design assumptions;
- an increase in the supply voltage U_{SUPPLY} increases the amount of power transferred to plasma beam generation, resulting in deeper penetration of the tissue in the area of operation;
- as the operating frequency of the generator increases, the power P_{WORK} transmitted to the plasma beam generation increases;
- as the generator frequency increases, the plasma arc becomes more stable.

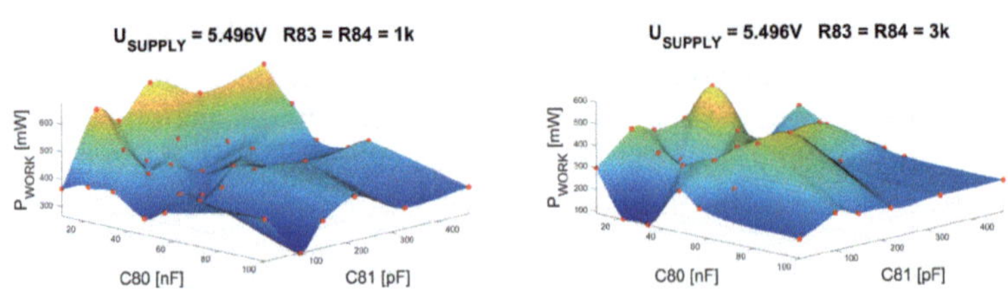

Fig. 3. Characteristics of the amount of power drawn from the source for plasma beam generation.

The design of the generator also affects the amount of energy transmitted to generate a plasma beam. In the described circuit, one of the limitations is the inductance of the transformer windings and the core used, which affects the maximum work frequency of the generator. The higher the operating frequency of the system, the better the energy transfer and the greater the beam stability.

By analyzing the characteristics obtained, points were identified that were characterized by an increase in the amount of power drawn from the source to generate the plasma arc, resulting in deeper penetration of tissue – Table 2.

Table 2. Potentially optimal points P_{WORK}.

U_{SUPPLY}	C80	C81	R83	R84	P_{WORK}
5.496 V	22 nF	220 pF	3 k	3 k	604.95 mW
5.496 V	10 nF	220 pF	1 k	1 k	677.72 mW
5.496 V	10 nF	470 pF	1 k	1 k	650.17 mW
5.996 V	47 nF	220 pF	3 k	3 k	734,65 mW
5.996 V	22 nF	220 pF	3 k	3 k	669,71 mW
5.996 V	10 nF	220 pF	1 k	1 k	1118.94 mW
5.996 V	22 nF	470 pF	1 k	1 k	788.39 mW

Based on the data in Table 2, it was observed that for the parameters C80 = 10 nF, C81 = 220 pF, R83 = R84 = 1 kΩ a significant increase (65%) in power is present. The generator at this point was stable; however, a further increase in voltage U_{SUPPLY} would cause damage to the NP windings of the transformer.

During testing, measurements were performed for the following supply voltages: U_{SUPPLY} = 3.496 V, 4.497 V, 5.496 V, 5.996 V. The analysis revealed that in most cases the voltage U_{SUPPLY} = 3.496 V was too low to cause plasma beam ignition. In case of supply voltage U_{SUPPLY} = 4.497 V the ignition appeared more frequently, however, it was characterized by its instability. For U_{SUPPLY} = {5.496 V; 5.996 V} ignition and stability were acceptable. Taking into account the transformer ratio, it can be concluded that stable plasma arc ignition for the presented method is possible for an output voltage induced on the NS winding of approximately $U_{RMS_{HV}}$ = 1400 V (Table 3).

Table 3. Output voltage of the generator $U_{RMS_{HV}}$.

$\mathbf{U_{SUPPLY}}$	3.496 V	4.497 V	5.496 V	5.996 V
$\mathbf{V_{RMS}}$	7.6V	9.6 V	11.8 V	12.8 V
$\mathbf{U_{RMS_{HV}}}$	950 V	1200 V	1475 V	1600 V

When comparing power P_{WORK} with the operating frequency of the plasma beam generator system, an increase in power P_{WORK} was observed for frequencies of 132 kHz and 142 kHz, see Fig. 4. Due to technological limits, no further

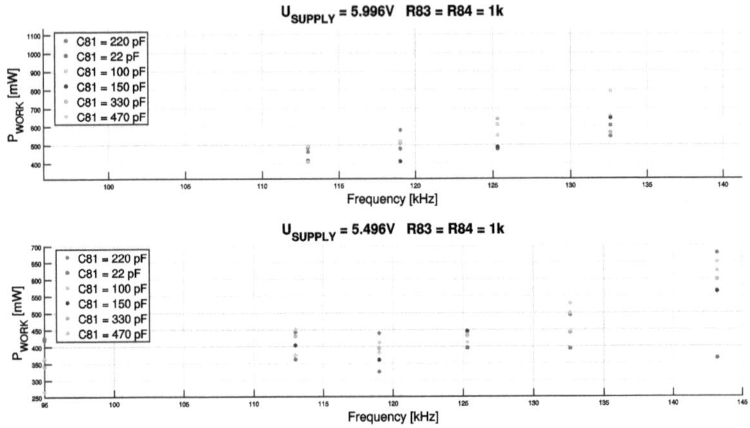

Fig. 4. P_{WORK} as a function of the operating frequency of the generator system.

tests were possible. This observation leaves room for further research into systems capable of operating at frequencies higher than 142 kHz.

For the configurations tested C80 and C81 with R83 and R84 equal to $3k\Omega$, the random excitation of the generator occurred with a frequency of 645 kHz for C80 $=$ 10 nF and C81 $= \{220\,\text{pF}, 330\,\text{pF}, 470\,\text{pF}\}$. This phenomenon did not occur when the resistors R83 and R84 had resistance of $1k\Omega$.

5 Summary

The proposed circuit is the result of R&D work that has led to the definition of the architecture of the plasma beam generator system, which does not require a return electrode. This minimizes current flow through the patient's body to the values allowed by the standards, i.e., less than $10\,\mu\text{A}$ (according to PN-EN 60601-1:2011 with collateral standards). The system's ability to transfer energy depends mainly on the type and properties of the transformer, including its inductance for frequencies above 100 kHz, maximum power, and voltage V_{RMS} i $U_{RMS_{HV}}$.

The developed plasma beam generator circuit requires components with appropriate parameters for its correct operation, i.e. capacitor C80 determining the operating frequency of the generator, capacitor C81 limiting the maximum output power and blocking the DC component of the output voltage, and resistors R83 and R84 limiting the maximum output power by limiting the current flowing through the primary winding. As a result of the tests, the characteristics of the power consumed from the source for plasma arc generation were generated (Fig. 3), the efficient and stable operating point was determined, i.e. C80 $=$ 10 nF, C81 $=$ 220 pF, R83 $=$ R84 $=$ $1\,\text{k}\Omega$. For the selected point, the power consumed from the source for plasma arc generation was 1118.94 mW.

The presented plasma beam generator circuit meets the normative requirements for use in a medical device. Risk analysis performed according to *PN-EN*

ISO 14971:2020-05 Medical devices—Application of risk management to medical devices which fragments were included in Table 1 shows that the system is safe for both the patient and the operator.

Further research and analysis of the results concerning the effectiveness of the proposed plasma beam generator will be the subject of subsequent studies and scientific publications in the field of medical sciences (recovery time, tissue scarring).

Acknowledgments. This study was founded by MIKROVOLT sp. z o.o. and presents the results of R&D work under project no. RPDS.01.02.01-02-0001/20 entitled "Development and implementation of an innovative treatment device using cold plasma technology by MIKROVOLT sp. z o.o."

References

1. Woedtke, T., Metelmann, H.-R., Weltmann, K.-D.: Clinical plasma medicine: state and perspectives of in vivo application of cold atmospheric plasma. Contrib. Plasma Phys. **54**(2), 104–117 (2014). https://doi.org/10.1002/ctpp.201310068
2. Ly, L., et al.: A new cold plasma jet: performance evaluation of cold plasma, hybrid plasma and argon plasma coagulation. Plasma **2018**(1), 189–200 (2018). https://doi.org/10.3390/plasma1010017
3. Bernhardt, T., Semmler, M.L., Schäfer, M., Bekeschus, S., Emmert, S., Boeckmann, L.: Plasma medicine: applications of cold atmospheric pressure plasma in dermatology. Oxid. Med. Cell. Longev. **2019**(1), 1–10 (2019). https://doi.org/10.1155/2019/3873928
4. Woedtke, T., Weltmann, K.-D.: Plasma medicine–current state of research and medical application. Plasma Phys. Controlled Fusion **59**(2017), 1–11 (2017). https://doi.org/10.1088/0741-3335/59/1/014031
5. Laroussi, M.: Cold plasma in medicine and healthcare: the new frontier in low temperature plasma applications. Front. Phys. **8**(74), 1–7 (2020). https://doi.org/10.3389/fphy.2020.00074
6. Tabares, F.L., Junkar, I.: Cold plasma systems and their application in surface treatments for medicine. Molecules **26**(1903), 1–16 (2021). https://doi.org/10.3390/molecules26071903
7. Tanaka, H., et al.: State of the art in medical applications using non-thermal atmospheric pressure plasma. Rev. Mod. Plasma Phys. **1**(3), 1–89 (2017). https://doi.org/10.1007/s41614-017-0004-3
8. Martusevich, A.K., Surovegina, A.V., Bocharin, I.V., Nazarov, V.V., Minenko, I.A., Artamonov, M.Y.: Cold argon athmospheric plasma for biomedicine: biological effects, applications and possibilities. Antioxidants **11**(1262), 1–15 (2022). https://doi.org/10.3390/antiox11071262

A Scalable PSO–MC Hybrid Metaheuristic for Multiprocessor Task Scheduling

Dariusz Dorota(✉)

Cracow University of Technology, st. Warszawska 24, 31-155 Cracow, Poland
dariusz.dorota@pk.edu.pl

Abstract. This paper proposes a structurally integrated hybrid metaheuristic for multiprocessor task scheduling that combines Particle Swarm Optimization (PSO) with a Modified Muntz–Coffman (MC) constructive procedure. The PSO-MC algorithm aims at minimizing makespan while preserving high load uniformity across processors. Unlike conventional hybrids, where local search is loosely coupled with the global optimizer, the MC component in PSO-MC operates as a problem-specific correction mechanism applied at each iteration. This ensures feasibility with respect to precedence and resource constraints while intensifying exploitation in promising regions of the search space.

Experimental evaluation was performed for task sets from 10 to 10^6 jobs and processor configurations of 4, 8, 16, and 32 units. The results show that PSO-MC consistently outperforms PSO, SA, TS, MPA, ACO, and deterministic MC in terms of makespan, while maintaining near-ideal load uniformity (LUI $\approx$ 1.0) and near-linear parallel scalability. Statistical analysis confirms the significance of the improvements.

Keywords: Multiprocessor Scheduling · Hybrid Metaheuristics · Particle Swarm Optimization (PSO) · Load Balancing · Precedence-Constrained Scheduling · Cyber-Physical Systems

1 Introduction

Embedded systems and modern cyber-physical systems are characterized by growing computational complexity, high data-processing dynamics, and strict temporal determinism requirements. In such environments, synchronously executed multiprocessor tasks play a crucial role in meeting reliability, functional, and architectural constraints. In classical embedded systems, they support fault tolerance, jitter reduction, stable response times, and efficient utilization of heterogeneous platforms integrating CPUs, DSPs, and hardware accelerators. They also enable computational redundancy, precise real-time synchronization, and effective processing of sensory data streams.

In modern domains such as Digital Twins, the Internet of Vehicles (IoV), the Internet of Robots (IoR), and integrated Cyber-Physical Systems (CPS), synchronous multiprocessor execution becomes even more critical. These systems operate on high-throughput streaming data in dynamic environments where

W. Zamojski et al. (Eds.): DepCoS-RELCOMEX 2026, LNNS 2014, pp. 100–110, 2026.
https://doi.org/10.1007/978-3-032-27927-9_10

millisecond-level delays may directly impact safety. Deterministic multiprocessor scheduling enables reliable data fusion, state estimation, motion planning, multi-agent coordination, and computationally intensive perception tasks.

From both perspectives, synchronous multiprocessor tasks constitute a fundamental design element of contemporary computing architectures, ensuring predictable timing, resilience to disturbances, efficient heterogeneous mapping, and support for advanced simulation and control components.

These challenges motivate the development of specialized scheduling strategies. To address them, this work proposes the PSO-MC hybrid algorithm, combining the global exploration capability of Particle Swarm Optimization [8] with a problem-oriented MC heuristic [6], tailored for deterministic and tightly constrained embedded and CPS environments.

2 Problem Formulation

A deterministic representation of system parameters provides a practical and analytically convenient approach for studying scheduling methods in real-time multiprocessor environments and embedded systems. Although non-deterministic models may more accurately capture complex system behavior, a deterministic formulation is appropriate at the stage of defining and analyzing heuristics. In this work, we introduce the PSO-MC hybrid algorithm, integrating global search performed by Particle Swarm Optimization with the local MC heuristic. This combination enables effective exploration of the solution space for the scheduling of independent multiprocessor tasks while ensuring synchronous execution.

To formalize the considered problem, we define the task set

$$T = \{T_1, T_2, \ldots, T_n\},$$

and the processor set

$$M = \{M_1, M_2, \ldots, M_m\}.$$

Each task T_i consists of a single, indivisible operation executed synchronously on $a_i \geq 1$ processors for a duration p_i. When preemption is allowed, a task is composed of a sequence of x_i operations:

$$O_i = \{o_{i,1}, o_{i,2}, \ldots, o_{i,x_i}\}, \qquad \sum_{k=1}^{x_i} |o_{i,k}| = p_i,$$

where each operation requires exactly a_i processors.

The Multi-Processor Ratio (MPR, a_i) characterizes multiprocessor tasks:

- $a_i = 1$ – single-processor tasks,
- $a_i = k$ – each task requires exactly k processors,
- $a_i \leq k$ – tasks require between 1 and k processors.

In embedded systems, the most common cases are $a_i = 2$ and $a_i = 3$, resulting from redundant architectures in which reliability is increased through synchronous execution on multiple processors. The following feasibility constraint must hold:

$$\forall t: \quad \sum_{T_i \ | \ t \in [s_i, f_i)} a_i \leq m.$$

We use the three-field notation $\alpha \mid \beta \mid \gamma$ [14], extended according to [9] by incorporating the MPR parameter into the β field. The instances analyzed in this work belong to the class:

$$P \mid \circ,\, a_i = k,\, a_i \leq k,\, pmtn,\, prec \mid C_{\max},$$

where:

- *pmtn* – preemptive execution,
- *prec* – precedence constraints,
- $C_{\max}$ – minimized makespan.

System representation significantly influences heuristic effectiveness and the exploration of the solution space [8]. In this study, Task Graphs (TG) are employed as a standard modeling approach for independent and partially dependent tasks. Experimental instances are generated using a modified TGFF-based procedure [5], resulting in the formulation:

$$P \mid MPR \leq k \mid C_{\max}.$$

Traditional deterministic scheduling assumes that tasks execute on a single processor, even in multiprocessor environments. In contrast, synchronous execution on multiple processors significantly enhances system reliability. Dual redundant execution may lead to ambiguous results, whereas triple execution enables majority voting. These mechanisms are widely applied in high-reliability embedded systems, shared-memory architectures, VLSI testing [2,10], and fully parallel hardware systems [15]. Synchronization of task execution is defined as:

$$\forall T_i,\ \forall p, k \in P_i: \quad start_{i,p} = start_{i,k},$$

and

$$f_i = s_i + p_i.$$

The problem

$$P \mid MPR \leq 3 \mid C_{\max}$$

is strongly analogous to the Strip Packing (Cutting Stock) problem, where allocating tasks to processors corresponds to the two-dimensional placement of rectangles:

$$T_i \rightarrow \text{rect}(\text{width} = p_i,\ \text{height} = a_i),$$

with the lower bound:

$$C_{\max} \geq \frac{1}{m} \sum_i a_i p_i.$$

This complexity, combined with data-flow constraints and the requirement of synchronous resource allocation, motivates the use of metaheuristic methods. The proposed PSO-MC algorithm combines: A) PSO – global exploration of the permutation-based scheduling space, B) MC (Modified Muntz–Coffman) – local optimization through rapid elimination of processor conflicts, precedence violations, and synchronization inconsistencies.

The scheduling of dependent tasks on multiprocessor systems can be formulated as the NP-hard problem $P|prec|C_{\max}$, while removing precedence constraints reduces it to the independent scheduling case $P||C_{\max}$, which represents a special case of the general problem. Therefore, evaluating the proposed algorithm on independent tasks allows the intrinsic optimization capability of the hybrid PSO–MC method to be analyzed without interference from precedence constraints, providing a controlled benchmark for assessing its scheduling performance.

This combination is particularly effective for deterministic workloads typical of embedded systems, where tasks have fixed processor demands, limited preemption, and precisely defined temporal dependencies. The formalized problem provides the basis for subsequent analysis of the quality and performance of the PSO-MC algorithm compared with classical heuristics and modern metaheuristics.

3 Motivation of Hybrid Algorithm Introduction

Research on hybrid metaheuristics has become a central direction in multiprocessor scheduling optimization. These problems—characterized by precedence constraints, limited resources, and large instance sizes—challenge global approaches such as PSO [17], evolutionary methods [21,22], or swarm-based algorithms (e.g., GWO [20]). Although effective in exploration, these methods often fail to preserve schedule feasibility at each iteration. Conversely, local search techniques typical of memetic algorithms [21,22], while precise, incur high computational cost and limited scalability.

Memetic algorithms introduced by Moscato [21] and further analyzed by Neri et al. [22] integrate global search with intensive local refinement. However, for large-scale scheduling, the computational overhead of local search becomes prohibitive. Similarly, PSO hybrids incorporating structural heuristics such as tabu search or repair operators [16] improve solution quality at the expense of increased complexity. Despite its simplicity and parallelism [17], standard PSO does not guarantee preservation of precedence constraints in multiprocessor scheduling.

Talbi [24] identifies a key limitation of existing hybrids: the lack of deterministic, computationally efficient structural operators suitable for iterative integration within metaheuristics. Most approaches rely on neighborhood-based local search, limiting scalability. Surveys of swarm-based algorithms, including GWO [20], confirm that these methods do not inherently enforce schedule validity under precedence constraints.

The PSO–MC Hybrid addresses this gap by combining global PSO exploration with the deterministic Muntz–Coffman (MC) scheduler. MC is a discrete-time multiprocessor scheduling algorithm that executes tasks in parallel using a multi-component priority system (w_i, a_i, h_i). Its complexity is dominated by precedence graph processing $O(n + |E|)$ and the linear execution loop $O(\sum p_i)$. Crucially, MC eliminates processor overloads and precedence violations deterministically, without neighborhood exploration.

Embedding MC at every PSO iteration and for each particle ensures low-cost structural correction, stabilizes the population, and significantly reduces infeasible solutions. This enables scalability to instances up to 10^6 tasks—a behavior not observed in classical PSO+LS hybrids [16] nor in broader metaheuristic frameworks [24]. Moreover, the PSO–MC architecture is inherently parallel, consistent with findings on parallel evolutionary computation [3].

Therefore, PSO–MC represents a structurally integrated hybrid that overcomes limitations of conventional PSO+LS approaches and enables efficient optimization of large-scale, precedence-constrained multiprocessor scheduling problems.

4 Description of the Approach

The system specification was based on the TGFF task graph model [5]. We consider a special case where $E = \emptyset$, which implies the absence of precedence constraints (*prec*) in the β field. This approach required introducing appropriate notation for multiprocessor tasks ($a_i \geq 1$). The adopted specification was used in this study for task sets ranging from 10 to 1,000,000 tasks. Each generated instance was then scheduled using one of the three considered algorithms. The obtained results were collected and compared in the subsequent section.

Below, we present the pseudocode of the PSO–MC algorithm, based on previous formulations of MC [6] and PSO [8]. All algorithms were parallelized to accelerate the computation of the makespan.

Tasks in the system can be represented as follows:

$$
\begin{aligned}
T = \{&\{T_0, 2, 13\}, \{T_1, 3, 35\}, \{T_2, 1, 21\}, \{T_3, 1, 28\}, \{T_4, 2, 40\}, \{T_5, 4, 5\}, \\
&\{T_6, 2, 23\}, \{T_7, 3, 10\}, \{T_8, 1, 9\}, \{T_9, 2, 30\}\}
\end{aligned}
\tag{1}
$$

In (1), each element of the set represents a task in the form $\{T_i, a_i, p_i\}$, where:

- T_i – task identifier,
- a_i – number of processors required for task execution,
- p_i – processing time (in time units).

Algorithm 1 PSO with Permutation Difference Operator and MC Correction

Randomly initialize particle positions $X_i(0)$ and velocities $V_i(0)$ for $i = 1 \dots N$.
Compute objective values $F(X_i(0))$ and set $pBest_i = X_i(0)$; $gBest = \arg\min_i F(X_i(0))$.
for $t = 1 \dots T_{\max}$ **do**
  for each particle i **do**
    Update velocity:

$$V_i(t) = \omega V_i(t-1) \oplus c_1 r_1 (pBest_i - X_i(t-1)) \oplus c_2 r_2 (gBest - X_i(t-1)),$$

where $\oplus$ denotes the permutation difference operator.
    Update position:

$$X_i(t) = X_i(t-1) \oplus V_i(t).$$

    Perform local correction:

$$X_i'(t) = MC(X_i(t)),$$

eliminating resource and precedence conflicts.
    Evaluate the objective function $F(X_i'(t))$.
    if $F(X_i'(t)) < F(pBest_i)$ **then**
      $pBest_i = X_i'(t)$
    end if
    if $F(X_i'(t)) < F(gBest)$ **then**
      $gBest = X_i'(t)$
    end if
  end for
end for
Return $gBest$ as the final solution.

The proposed algorithm is a permutation-based variant of Particle Swarm Optimization designed for scheduling problems in which solutions are represented as task permutations. Particle positions and velocities are updated using a permutation difference operator that replaces classical vector arithmetic and enables directional movement in permutation space. The velocity of each particle is computed as a combination of inertia, cognitive, and social components, corresponding respectively to movement toward the previous direction, the particle's individual best solution, and the global best solution. The new position is obtained by composing the current permutation with its velocity, which may produce infeasible schedules. To eliminate resource conflicts and precedence violations, a local MC procedure is applied as a repair operator. The MC correction transforms the resulting permutation into a feasible schedule without exploring the neighborhood space. After repair, each particle updates its *pBest*, and the global *gBest* is updated if an improved solution is found. Iterations continue until a time or iteration limit is reached. The algorithm ultimately returns the best discovered permutation as the optimal or near-optimal solution.

According to the current state of knowledge, deterministic scheduling of multiprocessor tasks has been studied in [4,9]. For the general case, heuristic algorithms were proposed in [6,11], while under uncertain or special conditions additional solutions are presented in [4,10,23].

5 Computational Experiments

This chapter describes the conducted computer experiments and provides the obtained results. An exemplary benchmark set has been considered, shown as specification 1.

Each instantiation is a mix of independent tasks with various $a_i \in \{1, 2, 3, 4\}$ specified as the "processors" in "task profile". Further details can be found in [9]. We can calculate the priority of a task as $w_i = a_i * p_i$. For each instance a set of independent tasks was generated. The benchmark set contains 8 instances with sizes $n \in \{10, 50, 100, 500, 1000, 10000, 100000, 1000000\}$. For each of the algorithms, the same instances specifying the system were used to compare not only the numerical values of the determined makespan, but also efficiency, speedup and LUI (Load Unevenness Index) of the algorithm. The experimental results are shown in Table 1 (PSO-MC algorithm) and in Table 2 (MC, TS, PSO, SA, ACO algorithms) according to previous papers [7,8]. In these tables, some abbreviations are introduced, which mean the following: A) Id - instantiation of system specification, B) Tasks - number of tasks in instantiation, C) MS - makespan, D) LUI - Load Unevenness Index, E) S - speedup correlated to 4-processor architecture, F) E - efficiency. However some of this factors where closely correlated with goal function, another must by calculated using some equations:

$$S_p = \frac{T_4}{T_p} \qquad E_p = \frac{S_p}{p/4} \tag{2}$$

p
where:

- T_4 – makespan obtained for 4 processors (reference configuration),
- T_p – makespan obtained for p processors,
- S_p – speedup relative to the 4-processor configuration,
- E_p – parallel efficiency with respect to ideal scaling from 4 processors.

6 Discussion of the Results

This article proposes a hybrid metaheuristic algorithm for scheduling independent multiprocessor tasks that combines Particle Swarm Optimization (PSO) as the global exploration mechanism with the Modified Muntz–Coffman (MC) procedure acting as a local exploitation operator. The proposed approach was compared with previously studied deterministic and metaheuristic scheduling algorithms (TS, PSO, SA, MPA, MC, and ACO), selected based on methods analyzed in the author's earlier studies.

Table 1. Scalability analysis of PSO-MC: makespan, LUI (4 CPU), speedup (S) and efficiency (E)

Tasks	4 CPU		8 CPU			16 CPU			32 CPU		
	MS	LUI	MS	S	E	MS	S	E	MS	S	E
10	56	1.00000	40	1.40	0.70	40	1.40	0.35	40	1.40	0.18
50	315	1.00000	159	1.98	0.99	82	3.84	0.96	50	6.30	0.79
100	635	1.00291	319	1.99	1.00	162	3.92	0.98	85	7.47	0.93
500	3408	1.00236	1705	2.00	1.00	855	3.99	1.00	430	7.93	0.99
1000	6932	1.00036	3467	2.00	1.00	1735	3.99	1.00	871	7.96	1.00
10000	68451	1.00012	34226	2.00	1.00	17115	4.00	1.00	8560	8.00	1.00
100000	687876	1.00001	343939	2.00	1.00	171971	4.00	1.00	85988	8.00	1.00
1000000	6872276	1.00000	3436139	2.00	1.00	1718071	4.00	1.00	859038	8.00	1.00

Table 2. Makespan (MS) and Load Uniformity Index (LUI) comparison for TS, PSO, SA, MPA, ACO and MC algorithms

ID	Tasks	TS		PSO		SA		MPA		ACO		MC	
		MS	LUI	MS	LUI	MS	LUI	MS	LUI	MS	LUI	MS	LUI
1	10	58	1.09	80	1.33	55	1.08	56	1.07	62	1.12	66	12
2	50	374	1.01	422	1.03	370	1.01	388	1.02	395	1.02	423	18
3	100	817	1.00	966	1.01	816	1.00	868	1.01	890	1.01	942	20
4	500	4187	1.00	4746	1.00	4168	1.00	4565	1.00	4689	1.00	2239	45
5	1000	8549	1.00	9789	1.00	8545	1.00	9396	1.00	9523	1.00	9696	43
6	10000	85641	1.00	97117	1.00	85470	1.00	96329	1.00	96842	1.00	97253	45
7	100000	859321	1.00	978481	1.00	859189	1.00	975309	1.00	977112	1.00	977840	47
8	1000000	8588773	1.00	9776667	1.00	8586536	1.00	9769111	1.00	9772450	1.00	9777551	9

The PSO–MC hybrid demonstrates consistent quantitative superiority across all evaluated problem scales. Comparative results indicate that PSO–MC systematically achieves the lowest makespan values. The observed improvement ranges from 25% to 42% relative to classical metaheuristic algorithms, and this advantage remains stable for instances ranging from 50 to 1,000,000 tasks. Such scale-invariant behavior confirms the robustness and scalability of the proposed hybrid optimization architecture.

In addition to makespan reduction, the Load Uniformity Index (LUI) was analyzed to evaluate processor load balancing for four processors. The average percentage improvement of PSO–MC across benchmark instances (10–100,000 tasks) equals 1.24% over TS, 4.02% over PSO, 1.24% over SA, and 1.27% over MPA, indicating slightly improved load balancing compared with classical metaheuristic approaches.

The most significant effect is observed in comparison with the deterministic MC algorithm, where PSO–MC reduces load imbalance by approximately 96% on average. This result indicates that MC alone produces substantial processor load disparity, whereas its integration within the PSO framework effectively eliminates this imbalance.

7 Conclusion

The selection of comparative algorithms spans distinct optimization paradigms: the deterministic MC scheduler, memory-based local search (TS) [13], probabilistic optimization (SA) [18], population-based PSO [17], and the bio-inspired Marine Predators Algorithm (MPA) [12]. This benchmark enables evaluation of the proposed method across constructive, local, stochastic, and modern population-based optimization strategies.

Hybrid metaheuristics constitute a dominant direction in combinatorial optimization [24]. Combining global exploration with local refinement improves convergence stability and solution quality [1,19]. However, in most hybrid approaches the local component acts as a generic refinement or repair mechanism, without structural adaptation to the scheduling model.

PSO–MC addresses this limitation by embedding the deterministic Modified Muntz–Coffman scheduler directly within the population-based search. The MC module is not a generic local operator but a problem-specific constructive scheduler designed for multiprocessor task allocation. The integration is structural rather than sequential. At each PSO iteration, candidate permutations are deterministically transformed using MC, restricting the search space to feasible and resource-consistent schedules. PSO provides diversification, whereas MC ensures deterministic exploitation and systematic load redistribution.

This architecture yields: (i) balanced exploration and exploitation, (ii) reduced premature convergence compared with standalone PSO, (iii) mitigation of deterministic greediness, and (iv) near-optimal load balancing (LUI ≈ 1) while minimizing makespan. A comparative analysis relative to the deterministic MC scheduler shows that TS and SA reduce the makespan by approximately 12% on average, while MPA and ACO achieve smaller improvements of about 3–4% and 2–3%, respectively, and the classical PSO provides almost no improvement. In contrast, the proposed PSO–MC hybrid reduces the makespan by approximately 29–30% on average relative to MC (excluding the anomalous instance for 500 tasks).

In the conducted experiments, the algorithms TS, PSO, SA, MPA, ACO, and PSO–MC achieved LUI values close to 1.0, indicating highly balanced processor utilization, whereas the MC algorithm produced significantly higher LUI values, revealing substantial load imbalance between processors. The results confirm that PSO–MC significantly improves load uniformity relative to MC while maintaining performance comparable to or better than advanced metaheuristics without degrading load balance. Therefore, PSO–MC should be viewed as a structurally integrated hybrid tailored specifically for multiprocessor scheduling rather than a simple PSO variant with local search.

References

1. Alba, E., Dorronsoro, B.: Hybrid cellular genetic algorithms for multiprocessor scheduling. Futur. Gener. Comput. Syst. **21**(5), 718–728 (2005)
2. Błażewicz, J., Liu, Z.: Scheduling multiprocessor tasks with chain constraints. Eur. J. Oper. Res. **94**(2), 231–241 (1996). https://doi.org/10.1016/0377-2217(96)00126-9
3. Cantú-Paz, E.: Efficient and Accurate Parallel Genetic Algorithms. Springer (2000)
4. Dereniowski, D., Kubiak, W.: Shared processor scheduling of multiprocessor jobs. Eur. J. Oper. Res. **282**(2), 464–477 (2020). https://doi.org/10.1016/j.ejor.2019.09.033
5. Dick, R.P., Rhodes, D.L., Wolf, W.: TGFF: task graphs for free. In: Proceedings of the Sixth International Workshop on Hardware/Software Codesign (CODES/CASHE'98), pp. 97–101. IEEE (1998). https://doi.org/10.1109/HSC.1998.666245
6. Dorota, D.: Dual-processor tasks scheduling using modified Muntz-Coffman algorithm. In: Zamojski, W., Mazurkiewicz, J., Sugier, J., Walkowiak, T., Kacprzyk, J. (eds.) DepCoS-RELCOMEX 2018. AISC, vol. 761, pp. 151–159. Springer, Cham (2019). https://doi.org/10.1007/978-3-319-91446-6_15
7. Dorota, D.: A comparative study of heuristic and metaheuristic approaches to multiprocessor task scheduling. In: International Conference on Intelligent Systems in Production Engineering and Maintenance, pp. 255–268. Springer (2025)
8. Dorota, D.: Multiprocessor task scheduling-comparison of some heuristics and metaheuristics. In: International Conference on Dependability and Complex Systems, pp. 51–60. Springer (2025)
9. Dorota, D.P.: Szeregowanie zadan wieloprocesorowych w warunkach niepewnosci. Ph.D. thesis, Politechnika Wroclawska (2023)
10. Drozdowski, M.: On the complexity of multiprocessor task scheduling. Bull. Pol. Acad. Sci. Tech. Sci. **43**(3) (1995)
11. Drozdowski, M.: New applications of the Muntz and Coffman algorithm. J. Sched. **4**(4), 209–223 (2001). https://doi.org/10.1002/jos.76
12. Faramarzi, A., Heidarinejad, M., Stephens, B., Mirjalili, S.: Marine predators algorithm: a nature-inspired metaheuristic. Expert Syst. Appl. **152**, 113377 (2020)
13. Glover, F.: Future paths for integer programming and links to artificial intelligence. Comput. Oper. Res. **13**(5), 533–549 (1986)
14. Graham, R.L., Lawler, E.L., Lenstra, J.K., Kan, A.R.: Optimization and approximation in deterministic sequencing and scheduling: a survey. Ann. Discrete Math. **5**, 287–326 (1979). https://doi.org/10.1016/S0167-5060(08)70356-X
15. Ibrahim, H., Salih, M.H.: Design and implementation of embedded true parallelism jammer system using FPGA-soc for low design complexity. ARPN J. Eng. Appl. Sci. **13**(24), 9410–9420 (2018)
16. Kao, Y., Zahara, K.: A hybrid particle swarm optimization for multimodal functions. Appl. Soft Comput. **8**(4), 1718–1724 (2008)
17. Kennedy, J., Eberhart, R.: Particle swarm optimization. In: Proceedings of the IEEE International Conference on Neural Networks, pp. 1942–1948 (1995)
18. Kirkpatrick, S., Gelatt, C.D., Vecchi, M.P.: Optimization by simulated annealing. Science **220**(4598), 671–680 (1983)
19. Li, X., Zhang, J., Gen, M.: A hybrid PSO algorithm for multiprocessor scheduling. Appl. Soft Comput. **12**(9), 2631–2641 (2012)
20. Mirjalili, S.: Grey wolf optimizer. Adv. Eng. Softw. **69**, 46–61 (2014)

21. Moscato, P.: On evolution, search, optimization, genetic algorithms and martial arts: toward memetic algorithms. Technical report, 826, Caltech Concurrent Computation Program (1989)
22. Neri, F., Cotta, C., Moscato, P.: Handbook of Memetic Algorithms. Springer (2012)
23. Spychalski, P., Arendt, R.: Mechanizm szeregowania procesów w systemie wieloagentowym wspomagającym projektowanie układów sterowania. Zeszyty Naukowe Wydziału Elektrotechniki i Automatyki Politechniki Gdańskiej, pp. 135–136 (2017)
24. Talbi, E.G.: Metaheuristics: From Design to Implementation. Wiley (2009)

Optimization of Deep Neural Networks for Improving the Effectiveness of Cancer Treatment

Amelia Draga[1] and Konrad Pempera[2(✉)]

[1] Faculty of Medicine, Wrocław Medical University, Wrocław, Poland
amelia.draga@student.umw.edu.pl

[2] Department of Control and Quantum Computing, Faculty of Information and Communication Technology, Wrocław University of Science and Technology, Wrocław, Poland
konrad.pempera@pwr.edu.pl

Abstract. Deep Neural Networks (DNNs) are transforming oncological care, yet their clinical deployment is hindered by challenges such as overfitting, data scarcity, and high computational costs. This paper provides a comprehensive review of optimization strategies—ranging from architectural enhancements to specialized algorithmic tuning—aimed at improving diagnostic precision. To confirm the findings derived from the literature review and provide a practical benchmark under resource-constrained conditions, a comparative study was conducted using the BreakHis breast cancer dataset. The work evaluates the impact of gradient-based optimizers, regularization techniques, and Transfer Learning under a rigorous patient-level split. Experimental results demonstrate that a baseline lightweight architecture trained with standard stochastic gradient descent achieves a robust 84.94% accuracy, providing a highly stable and effective alternative compared to volatile optimizers like Adam. However, Transfer Learning with fine-tuned DenseNet-121 proved superior, yielding a realistic clinical accuracy of 86.94% by successfully eliminating data leakage. The study concludes that while deep, pre-trained models offer the best precision, properly optimized lightweight architectures provide a viable, efficient alternative for resource-constrained clinical environments.

Keywords: Deep Learning · Cancer Diagnosis · Optimization · Transfer Learning · CNN · Medical Imaging

1 Introduction

Cancer remains a leading global health challenge, accounting for nearly 10 million deaths annually [24,25]. Despite therapeutic advances, survival rates are hindered by late detection and the biological complexity of tumors. Genetic and epigenetic heterogeneity, as described in clonal evolution models, often leads to

W. Zamojski et al. (Eds.): DepCoS-RELCOMEX 2026, LNNS 2014, pp. 111–120, 2026.
https://doi.org/10.1007/978-3-032-27927-9_11

treatment-resistant subpopulations, constituting a significant barrier to effective therapy [5,11,12,20].

Effective management of common malignancies, such as lung or breast cancer, requires analyzing multidimensional biomedical data to navigate complex care stages—from surgery to systemic treatment [2,4,15]. Consequently, Artificial Intelligence (AI) is increasingly utilized to personalize therapy, supporting clinicians in precise dosage planning and minimizing side effects based on genomic profiling [1,9,14].

Deep Neural Networks (DNNs) have become a cornerstone of this analysis due to their capacity for nonlinear modeling and feature extraction [8,18]. They demonstrate high efficiency in diagnosis and the integration of multi-omics data, enabling the discovery of novel biomarkers and significantly improving immunotherapy response predictions [2,3,10,26].

However, clinical implementation faces challenges such as overfitting, lack of interpretability, data scarcity, and high hardware requirements [5,16,22]. To address these, diverse optimization methods are employed, ranging from architectural modifications to algorithmic tuning. The primary objective of this paper is to bridge the gap between theoretical optimization strategies and their practical application. Beyond a comprehensive review of state-of-the-art techniques, we present an empirical comparative study using the BreakHis breast cancer dataset. By evaluating specific gradient-based optimizers, regularization techniques, and Transfer Learning, we seek to identify the optimal trade-off between computational efficiency and diagnostic precision in resource-constrained medical environments [8,14,23].

2 Literature Review

Recently, dynamic advances in Deep Learning (DL) have exerted a fundamental impact on oncological research, leading to significant improvements in diagnostic accuracy, treatment strategy planning, and clinical outcome prediction. Among the wide spectrum of Machine Learning (ML) techniques, Deep Neural Networks (DNNs) have gained particular and well-deserved attention. These models have proven their exceptional capability to process complex, multidimensional medical data and to extract essential, often latent, patterns that are crucial for both diagnostic and predictive purposes [2,4]. Currently, the primary area of development in this field is DNN optimization, which is evolving in a multifaceted manner, encompassing diverse network architectures and specific clinical applications. This section provides an overview of key scientific achievements in this domain, with a particular emphasis on optimization methods and strategies that have measurably increased the efficiency of AI systems in tasks related to the fight against cancer.

2.1 Methods Overview

In the context of breast cancer diagnostics, the development of an evolving deep convolutional neural network, which is a hybrid of a pre-trained ZFNet,

Extreme Learning Machine (ELM), and an improved chimp optimization algorithm (SWChOA), merits particular attention [19]. A key innovation by the authors was the replacement of the fully connected layers of ZFNet with ELM, which facilitated a drastic reduction in training time while maintaining high-quality feature extraction from the images. To maximize performance, the ELM parameters were precisely fine-tuned using the SWChOA algorithm. The final model (ZFNet-SWChOA-ELM) achieved impressive metrics on the studied dataset: 96.89% accuracy, 96.73% precision, 94.98% specificity, and 96.87% sensitivity, effectively validating the efficacy of combining hybrid models with metaheuristic optimization in mammographic analysis.

Expanding on hybrid strategies, Awotunde et al. [3] proposed an intelligent healthcare framework for leukemia diagnosis, utilizing Internet of Medical Things (IoMT) devices to collect physiological data. Crucially, a novel hyperparameter optimization algorithm, HORD, was introduced to effectively fine-tune CNN parameters within a multidimensional search space. The combination of PSO with the HORD algorithm yielded a CNN classifier with exceptional accuracy reaching 99.9%, highlighting the immense potential of integrating advanced metaheuristics with neural network architecture engineering.

An analysis of the 20 reviewed methodological studies indicates a dominance of specific, publicly available datasets. The BreakHis dataset and The Cancer Genome Atlas (TCGA) database were the most frequently utilized. BreakHis, containing histopathological images of breast cancer, served as the foundation for training and validating CNN-based classification models [6,9]. Conversely, the TCGA database, offering a rich array of clinical, genomic, and imaging data, enabled researchers to develop advanced multimodal models integrating diverse types of biological information [2,7].

2.2 Application Overview

Deep Neural Networks (DNNs) have revolutionized oncology, particularly in diagnostics, which remains the primary focus of optimization research (Fig. 1a). While models are successfully applied across widespread imaging modalities (CT, MRI, histopathology) for detecting breast, lung, and skin cancers [1,12,20], applications supporting prognosis and treatment planning are expanding.

Analysis of input data (Fig. 1b) confirms the dominance of medical imaging, driven by the visual processing capabilities of CNNs. However, multi-omics approaches—integrating imaging with genomic (RNA-seq, mutations) and clinical data—are increasingly adopted to capture complementary biological signals and improve predictive outcomes [2,7].

Significant advancements have also been recorded in the domain of Computer-Aided Drug Design (CADD) and response prediction. Recent research highlights models such as Deep-Affinity and PADME, which utilize deep learning to predict drug-target interactions by analyzing chemical structures and protein sequences. These models streamline the pharmacokinetic assessment process, estimating properties like ADMET with high precision. A notable divergence from general-purpose architectures is P-NET, a biologically informed

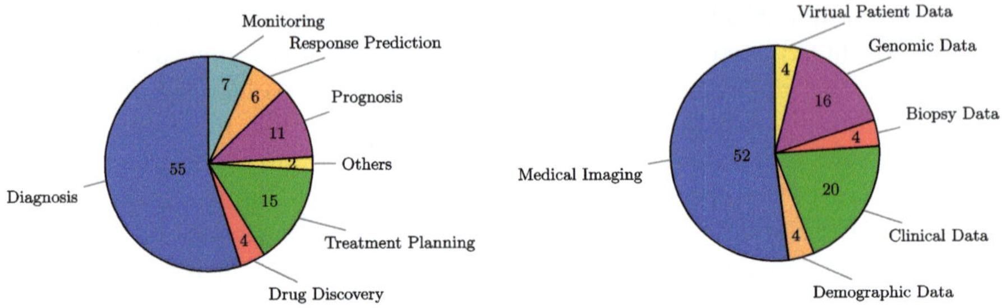

(a) Distribution of treatment stages [%] (b) Distribution of data types [%]

Fig. 1. Statistical overview of reviewed studies: (a) distribution of treatment stages, (b) distribution of utilized data types.

deep neural network designed for prostate cancer stratification. Unlike standard "black-box" models, P-NET embeds prior biological knowledge—specifically gene pathways—directly into its architecture. This design not only reduces computational complexity but also enhances interpretability, offering mechanistic insights into disease progression and treatment resistance at a molecular level [7,12].

Recent studies demonstrate the efficacy of specialized optimizations. For instance, thermography-based CNNs utilizing metaheuristic tuning (LFR-COA) and multi-scale architectures for bone marrow analysis have achieved accuracies exceeding 99% [9,14]. Table 1 details these high-performing models. Beyond diagnostics, the field is evolving towards multimodal frameworks, incorporating NLP for electronic health records and reinforcement learning on virtual patient data to personalize therapy and predict drug responses [4,10,23].

3 Methods

This section presents the methodological foundations underlying the optimization of Deep Neural Networks (DNNs) in oncological applications. We begin with a synthesis of representative approaches from recent literature, placing particular emphasis on key architectures and optimization strategies implemented to enhance diagnostic and predictive performance. Subsequently, we present a structured overview of the optimization techniques most commonly employed in this field, categorizing them into main classes.

3.1 Common Optimization Techniques in Deep Neural Networks for Oncology

Neural Networks (DNNs) applied in oncology face unique challenges, including data scarcity, class imbalance, the risk of overfitting, and the necessity of ensuring interpretability within clinical settings. To address these issues, researchers employ a wide array of optimization strategies, which have been categorized into

Table 1. Classification Accuracy and Optimization Types of Selected Deep Learning Models

Source	Model	Cancer Type	Optimization Type	ACC (%)
[9]	LFR-COA-DenseNet121-BC	Breast cancer	Metaheuristic, Transfer Learning	99.97
[3]	CNN + HORD + PSO	Leukemia	Metaheuristic, Parameter Tuning	99.90
[14]	AMCNN	Leukemia	Architectural (attention, multiscale)	99.70
[9]	CNN + ICOTL	Breast cancer	Transfer Learning, Optimization Layers	99.30
[6]	HCNN (Inception)	Breast cancer	Architectural, Gradient-based	99.05
[17]	SiCNN (FL)	Brain tumors	Federated Learning, Lightweight CNN	98.78
[21]	Paige Prostate	Prostate cancer	Transfer Learning, Domain tuning	98.30
[8]	MobileNet + SAdagrad	Colorectal cancer	Optimizer modification	98.00
[13]	BgDL (ResNet-18)	Gastric cancer	Attention-based	97.00

six main groups. Analysis of the data presented in Fig. 2 reveals an interesting trend: the most frequently employed group consists of "Specialized Methods" (16 instances), encompassing advanced techniques such as Explainable AI (XAI) and Federated Learning (FL), indicating a growing need for transparency and data privacy. Hyperparameter tuning (13 studies) and architectural enhancements (12 studies) follow closely.

Table 2 details these categories, providing specific examples of algorithms. Among gradient-based optimizers (used in 10 studies), SGD, Adam, and their variants such as SAdagrad predominate. Regarding overfitting reduction (9 studies), Dropout and Batch Normalization techniques are pivotal. Data strategies include, among others, SMOTE for addressing class imbalance and Transfer Learning. Conversely, architectural enhancements often utilize biologically inspired approaches (e.g., P-NET) or evolutionary algorithms for automated network design.

Crucially, these optimization categories rarely function in isolation. Synergistic combinations—such as coupling architectural enhancements with hyperparameter tuning and data augmentation—are essential to combat oncological data scarcity. Furthermore, integrating XAI with standard optimizers marks a shift toward trustworthy, clinically deployable AI. This co-optimization is vital for overcoming the inherent heterogeneity of cancer histopathology and genomics.

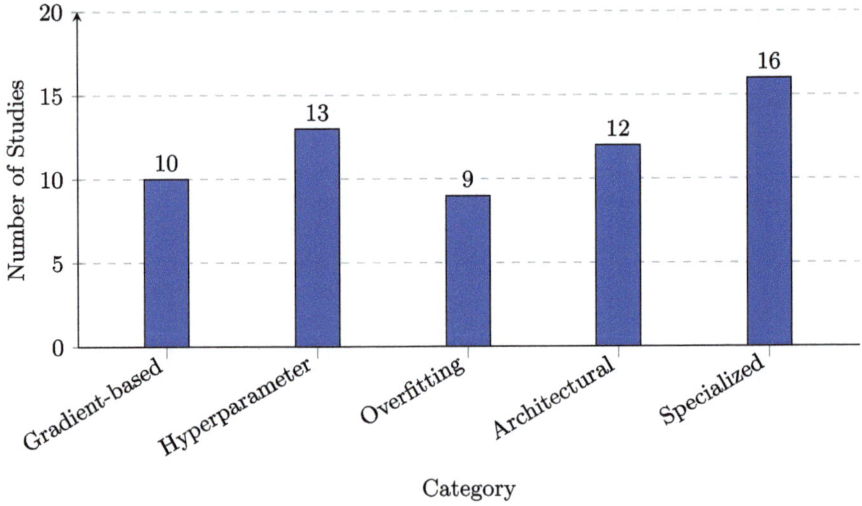

Fig. 2. Number of reviewed studies utilizing categories of optimization technique.

4 Experimental Evaluation

To empirically assess optimization strategies, we conducted a comparative study on the BreakHis dataset using a custom 3-layer CNN (baseline) and DenseNet-121. The objective was to evaluate the trade-offs between predictive performance (Accuracy, AUC) and computational efficiency. Accuracy measures the proportion of correct diagnoses, while AUC evaluates the model's ability to distinguish between benign and malignant classes across various thresholds. Simultaneously, computational efficiency is assessed by recording total training and validation runtimes, ensuring the models are viable for real-time clinical use.

4.1 Setup and Methodology

Experiments were implemented in PyTorch on an NVIDIA GPU using 3-fold stratified cross-validation. The BreakHis dataset was used in a binary classification setting, where histopathological images were categorized into two classes: benign and malignant. To prevent magnification-level data leakage, the cross-validation split was performed strictly at the patient level. Confining all images from a single patient exclusively to one fold prevents the memorization of patient-specific artifacts, ensuring a robust and realistic clinical evaluation.

Input images were resized to 128×128. The baseline is a lightweight custom CNN featuring three convolutional blocks (16, 32, 64 filters; 3×3 kernels; padding 1; ReLU; 2×2 Max Pooling). A flattening layer connects to two fully connected layers (128 and 2 neurons) for binary classification. This simplified structure provides a clear benchmark for evaluating advanced optimization strategies.

We evaluated modifications across four categories: (1) **Optimizers:** Adam, Adagrad vs. baseline Stochastic Gradient Descent; (2) **Architecture:** Adding

Table 2. Summary of Optimization Techniques in Deep Learning for Oncology

Category	Description	Examples
Gradient-based optimizers	Update network weights via gradient descent algorithms	SGD, NAG, Adam, Adagrad, SAdagrad
Hyperparameter tuning	Search for optimal settings like learning rate, number of layers	Grid Search, Random Search, PSO, Bayesian Optimization
Overfitting reduction	Improve generalization by reducing overfitting risk	Dropout, Batch Normalization, L1/L2 Regularization, Early Stopping
Data strategies	Enhance training through data preprocessing and augmentation	Transfer Learning, Data Augmentation, PCA, SMOTE, Autoencoders
Architectural enhancements	Modify model structure or use biologically inspired design	P-NET, BioNet, CNN-GA, Multi-task Learning, Evolutionary Design
Specialized methods	Support privacy, explainability, or complex decision processes	Federated Learning (FL), Deep Reinforcement Learning (DRL), XAI

a 4th convolutional layer; (3) **Transfer Learning (TL):** DenseNet-121 (Fine-Tuning). Training was set to 10 epochs (batch size 32, learning rate 0.01) for convergence evaluation in a resource-aware scenario.

4.2 Results and Discussion

Table 3 summarizes the key findings and standard deviations across the 3-fold CV. The baseline 3-layer CNN achieved 84.94% ± 2.33% accuracy and 0.8340 ± 0.0343 AUC. **Adagrad** provided stable learning (81.83% ± 2.34% accuracy, 0.8530 ± 0.0467 AUC), whereas Adam exhibited significant instability under default hyperparameters, leading to suboptimal convergence (68.85% ± 9.70% accuracy, 0.5000 ± 0.0000 AUC). This suggests that for small, heterogeneous datasets like BreakHis with patient-level splits, Adam's adaptive momentum may require more extensive tuning compared to the inherent stability of Adagrad or SGD. Applying **Dropout** yielded 83.99% ± 2.90% accuracy and 0.8397 ± 0.0388 AUC. The increased variance reflects the natural heterogeneity of patient-level splitting. Similarly, adding a **4th convolutional layer** reduced accuracy to 83.71% ± 3.10% and yielded 0.8372 ± 0.0481 AUC, showing that merely deepening a custom network on limited medical data exacerbates overfitting. Furthermore, while the deep DenseNet-121 experienced a drastic performance drop when shifting from standard image-level splits to rigorous patient-level splits (revealing prior data leakage exploitation), the 3-layer CNN remained consistent. This proves that lightweight architectures' constrained parameter spaces prevent the memorization of patient-specific artifacts. Thus, shallow networks learn generalizable features rather than overfitting, making them suitable for clinical settings with prevalent data heterogeneity.

Table 3. Comparison of Optimization Strategies (10 Epochs, 3-Fold CV): Accuracy, AUC, and Total Time

Method	Accuracy (%)	AUC	Time (s)
Baseline (CNN 3-layer)	84.94% ± 2.33	0.8340 ± 0.0343	10275
CNN 4-layer	83.71% ± 3.10	0.8372 ± 0.0481	10524
CNN + Adam	68.85% ± 9.70	0.5000 ± 0.0000	10325
CNN + Adagrad	81.83% ± 2.34	0.8530 ± 0.0467	10302
CNN + Dropout	83.99% ± 2.90	0.8397 ± 0.0388	10430
DenseNet-121 (Fine-Tune)	**86.94% ± 3.75**	**0.9153 ± 0.0085**	36799

Transfer Learning via DenseNet-121 fine-tuning achieved the highest predictive accuracy: 86.94% ± 3.75% with an AUC of 0.9153 ± 0.0085. Crucially, implementing a rigorous patient-level cross-validation split eliminated the data leakage at the magnification-level that frequently inflates BreakHis results. This normalization from near-perfect literature benchmarks (99%) to a realistic 86.94% confirms that without strict patient grouping, deep models merely memorize patient-specific artifacts rather than true malignant features. However, this precision comes at a substantial computational cost (36799 s), taking over three times longer than lightweight CNN variants. This highlights our core finding: while deep pre-trained networks maximize metrics under clinical validation, optimized lightweight architectures provide a resource-efficient alternative.

Future Work: Subsequent research will address current limitations by implementing dynamic learning rate schedules and early stopping to optimize convergence. We plan to integrate metaheuristic algorithms, such as SWChOA or PSO, for automated hyperparameter tuning. To bridge the gap between accuracy and clinical trust, future studies will incorporate Explainable AI (XAI) frameworks like Grad-CAM or SHAP.

These tools will enable pathologists to verify histological features through visual decision-making analysis. Finally, we aim to validate these frameworks on multi-center datasets to assess generalization across diverse cancer subtypes.

5 Conclusion

This study analyzed Deep Neural Network optimization strategies for oncological diagnosis. Combining a literature review with empirical experiments on the BreakHis dataset, we demonstrated that optimization choices are as critical as model architecture.

Evaluated under a strict patient-level cross-validation framework to prevent magnification-level data leakage, our results provide a realistic clinical benchmark. Consequently, performance metrics normalized compared to often-inflated literature standards. Transfer Learning via fine-tuned DenseNet-121 achieved the highest predictive power (86.94% accuracy), confirming the value of pre-learned features for complex medical imaging. However, for resource-constrained scenarios, lightweight models utilizing Dropout regularization proved highly effective, achieving 83.99% accuracy. This configuration significantly outperformed unstable variants like the Adam optimizer, requiring only a fraction of the computational resources.

Ultimately, while deep pre-trained models maximize precision, clinical deployment requires balancing accuracy, rigorous validation, and computational efficiency. Integrating robust, lightweight architectures with appropriate regularization and Explainable AI (XAI) is essential to foster the trust required for real-world medical applications.

References

1. Alsharif, F.: Artificial intelligence in oncology: applications, challenges and future frontiers. Faculty of Nursing, King Abdulaziz University (2024). Part: Drug discovery and development
2. Amanzholova, A., Coskun, A.: Enhancing cancer stage prediction through hybrid deep neural networks: a comparative study. Front. Big Data **7**, 1359703 (2024)
3. Awotunde, J.B., et al.: An enhanced hyper-parameter optimization of a convolutional neural network model for leukemia cancer diagnosis in a smart healthcare system. Sensors **22**(24), 9689 (2022). https://doi.org/10.3390/s22249689
4. Azadinejad, H., Rad, M.F., Shariftabrizi, A., Rahmim, A., Abdollahi, H.: Optimizing cancer treatment: exploring the role of ai in radioimmunotherapy. Diagnostics **15**(3), 397 (2025)
5. Bohmrah, M.K., Kaur, H.: Advanced hybridization and optimization of DNNs for medical imaging: a survey on disease detection techniques. Artif. Intell. Rev. **58**, 122 (2025). https://doi.org/10.1007/s10462-024-11049-x
6. Burcak, K.C., Baykan, O.K., Uguz, H.: A new deep convolutional neural network model for classifying breast cancer histopathological images and the hyperparameter optimisation of the proposed model. J. Supercomput. **77**, 973–989 (2021). https://doi.org/10.1007/s11227-020-03321-y
7. Elmarakeby, H.A., et al.: Biologically informed deep neural network for prostate cancer discovery. Nat. Med. (2021)
8. Elshamy, R., Abu-Elnasr, O., Elhoseny, M., Elmougy, S.: Enhancing colorectal cancer histology diagnosis using modified deep neural networks optimizer. Multimedia Tools Appl. (2024). https://doi.org/10.1007/s11042-023-16766-3
9. Emam, M.M., Houssein, E.H., Samee, N.A., Alohali, M.A., Hosney, M.E.: Breast cancer diagnosis using optimized deep convolutional neural network based on transfer learning technique and improved coati optimization algorithm. Biomed. Sig. Process. Control (2024). Abstract, introduction and methodology
10. Gallagher, K., et al.: Mathematical model-driven deep learning enables personalized adaptive therapy. Cancer Res. **7**, 9–10 (2024)

11. Hajim, W.I., Zainudin, S., Daud, K.M., Alheeti, K.: Optimized models and deep learning methods for drug response prediction in cancer treatments: a review. Paper Review, pp. 27–29 (2024)
12. Jiang, X., Hu, Z., Wang, S., Zhang, Y.: Deep learning for medical image-based cancer diagnosis. Cancers, 37–40 (2023)
13. Jiang, Y., Zhang, Z., Wang, W., et al.: Biology-guided deep learning predicts prognosis and cancer immunotherapy response. Nat. Commun. **14**, 5135 (2023). https://doi.org/10.1038/s41467-023-40890-x
14. Kandasamy, V., Simic, V., Bacanin, N., Pamucar, D.: Optimized deep learning networks for accurate identification of cancer cells in bone marrow. Sci. Rep. (2025). Sec. 3.1-3.4, sec. 4.1-4.5
15. Li, C., et al.: Deep neural network provides personalized treatment recommendations for de novo metastatic breast cancer patients. J. Cancer **15**(20), 6668–6685 (2024). https://doi.org/10.7150/jca.101293
16. Olumuyiwa, B.I., Han, T.A., Shamszaman, Z.U.: Enhancing cancer diagnosis with explainable & trustworthy deep learning models (2024)
17. Onaizah, A.N., Xia, Y., Hussain, K.: FL-SiCNN: an improved brain tumor diagnosis using Siamese convolutional neural network in a peer-to-peer federated learning approach. Alex. Eng. J. **114**, 1–11 (2025). https://doi.org/10.1016/j.aej.2024.11.063
18. Pfeffer, M.A., Ling, S.H.: Evolving optimised convolutional neural networks for lung cancer classification (2022). Sec. 1-4
19. Qian, L., Bai, J., Huang, Y., Zeebaree, D.Q., Saffari, A., Zebari, D.A.: Breast cancer diagnosis using evolving deep convolutional neural network based on hybrid extreme learning machine technique and improved chimp optimization algorithm. Biomed. Signal Process. Control **87**, 105492 (2023)
20. Reis, T.C.: Deep learning in oncology: transforming cancer diagnosis, prognosis, and treatment. Ph.D. thesis, Universidade de Lisboa, Faculdade de Medicina, Lisboa, Portugal (2024). Sec. 2.1-2.2
21. Riaz, I.B., Harmon, S., Chen, Z., Naqvi, S.A.A., Cheng, L.: Applications of artificial intelligence in prostate cancer care: a path to enhanced efficiency and outcomes. Mayo Clin. Proc. Digit. Health **2**(2), 182–192 (2024). https://doi.org/10.1016/j.mcpdig.2024.03.004
22. Sriraman, H., Badarudeen, S., Vats, S., Balasubramanian, P.: A systematic review of real-time deep learning methods for image-based cancer diagnostics. Multidiscip Healthc (2024)
23. Wang, J., et al.: The clinical application of artificial intelligence in cancer precision treatment. Exp. Hematol. Oncol. **14** (2025). Part: Intraoperative guidance
24. World Health Organization: Cancer fact sheet. https://www.who.int/news-room/fact-sheets/detail/cancer. Accessed 12 Dec 2025
25. World Health Organization: Global cancer burden growing, amidst mounting need for services (2024). https://www.who.int/news/item/01-02-2024-global-cancer-burden-growing--amidst-mounting-need-for-services. Accessed 14 May 2025
26. Xiang, Q., et al.: Quantum classical hybrid convolutional neural networks for breast cancer diagnosis. Sci. Rep. (2024)

Hybrid Quantum Image Recognition on Limited-Qubit Hardware

Olaf Duda[1], Mateusz A. Kucharski[2(✉)], and Radosław Idzikowski[2]

[1] Faculty of Science, University of Copenhagen, Universitetsparken 5, 2100 Copenhagen, Denmark
olaf.duda@op.pl

[2] Department of Control and Quantum Computing, Wrocław University of Science and Technology, Wrocław, Poland
{mateusz.kucharski,radoslaw.idzikowski}@pwr.edu.pl

Abstract. This paper addresses the problem of image recognition using a hybrid quantum neural network combined with a classical autoencoder. The proposed approach is evaluated using the well-known MNIST dataset, consisting of images of handwritten digits. In the first stage, a classical autoencoder is trained to compress the input images into a low-dimensional representation consisting of three features while minimizing reconstruction loss. The reduced feature vector is then processed by a quantum neural network. The hybrid model is trained on a classical computer, whereas its evaluation is performed on a real quantum computer, namely the Odra 5. A key contribution of this work is the identification and analysis of practical software tools and frameworks that enable effective cooperation with real quantum hardware in quantum machine learning workflows. The study also highlights the impact of design choices on training efficiency and model performance.

Keywords: Quantum computing · Hybrid approach · Autoencoders · QCNN

1 Introduction

Since the turn of the millennium, quantum computing has transitioned from a theoretical concept based on quantum physics to prototype machines built in science labs to a commercially available product in recent years. While still viable only to research institutions and burdened by many problems, like noise and limited size, this breakthrough allowed researchers to tackle different problems, and not simply focus on the computers themselves. One such topic is quantum machine learning (QML). As classical machine learning and artificial intelligence are constantly pushing new boundaries and have essentially become integral part of our daily lives, the question arose: will it be possible to translate these algorithms into the quantum realm?

The researchers have already provided an answer, as they almost immediately started developing solutions for quantum neural networks (QNNs). It was since

W. Zamojski et al. (Eds.): DepCoS-RELCOMEX 2026, LNNS 2014, pp. 121–130, 2026.
https://doi.org/10.1007/978-3-032-27927-9_12

utilized in many areas, such as healthcare [16], physics [4], chemistry [6], as well as more narrow fields such as acoustics [14]. Unfortunately, the problems of quantum computing in general, also translate to QML. The most impactful one is the limited number of qubits. Both on real quantum hardware and possible to simulate on classical computers. At the present the largest computers are available at IBM through their remote services, but due to being located in United States using them to train QNNs from Europe is not viable. European IQM machines, while much closer, currently provide only up to 54 qubits with plans to build 150 qubit machines. In Poland at Wrocław University of Science and Technology, where the authors are from, there was recently installed a 5 qubit IQM Spark class machine called Odra 5. The number of qubits is directly connected to the number of numerical features possible to be processed. This means, that the possible input of a QNN deployed on Odra 5 quantum computer can only have at most 5 numerical values. Simply increasing the number of qubits is not as viable as it may seem either. As most quantum processor architectures do not have the fully connected qubits, thus needing complex transpilation when deploying the quantum circuit onto the machine. This means that the problems possible to solve using fully quantum QNNs is limited at the moment.

Due to the limited availability of qubits, most researchers started exploring hybrid options [2]. This approach allows to conduct initial calculations and pre-processing including data heavy inputs with classical layers and after narrowing information to the most important kernel to process final layers with a quantum algorithm. A common type of neural networks utilizing this concept are convolutional neural networks (CNNs), which process matrices, such as pixels of an image, narrowing the kernel to the point of being possible to be processed by a quantum computer. In this paper, we propose a ML framework supported by the use of Autoencoders to prepare input for a quantum machine. The use of Autoencoders is well known in the field of classical machine learning, and at the same time it is a clean and simple tool to implement for an already existing project. However, this method is not yet researched as much in the field of QML. Even though, ideally it can prepare an input vector to perfectly fit the quantum machine.

2 Literature Review

Since the emergence of quantum computing and rise of its availability, the topics of AI and machine learning started to be both utilized alongside [5] as well as implemented using quantum computing. Because of the current limitations of quantum hardware, there are not that many QNNs fully realized on a quantum computer and even those that try to map all features onto a quantum circuit utilize an optimization loop performed on a classical computer [13]. As the most common input mapping method is to assign a single numerical value to a single qubit via a rotation angle, and the qubit numbers of available gate-based quantum computers is limited, it is hard to implement a QNN that will be able to process high-feature data. As mentioned before, at the moment, the largest quantum computers available are the 127-qubit IBM machines. Therefore, the largest possible images to be processed would be 11×11 pixels. This also does not take into account the transpilation needs and is only theoretically possible [12].

Works on QNNs that need to process images often use a hybrid network, typically in conjunction with a classical CNN. There are several possible ways to approach this methodology, of which the most prevalent one is to utilize classical layers at the beginning of the model, where the size is the largest, and then add a small quantum layer further, when the neuron structure does not need as many inputs [9]. Another approach is to utilize a quantum kernel for the convolutions, as it needs only to search limited area of an image [7].

The Autoencoder-based approach does not seem to be a topic well explored in the area of quantum machine learning, thus sparking interest with the authors and being an inspiration for this work. However, quantum autoencoders themselves appear to be a topic that is actively being explored by researchers [15] and may have great potential for future quantum networks [3].

3 Dataset and Tools

The MNIST (Modified National Institute of Standards and Technology) dataset is a well-established benchmark in the fields of machine learning and computer vision, widely used for the development and evaluation of image classification algorithms [10]. It comprises a total of 70,000 grayscale images of handwritten digits (0–9), each image standardized to a fixed resolution of 28×28 pixels and annotated with the corresponding numeric label; 60,000 of these samples form the training set, while an additional 10,000 are reserved for testing and performance assessment. Originating from a re-processing of the National Institute of Standards and Technology's Special Database collections, the original binary digit scans were normalized, anti-aliased to grayscale, and centered within the image frame, resulting in a dataset that is both easily accessible and computationally tractable for algorithmic benchmarking. Due to its relative simplicity, balanced class distribution, and ease of use, MNIST has become a standard initial evaluation task for supervised learning, particularly in neural network research, and serves as a foundational reference point for comparing novel models and methods in handwritten digit recognition.

To create the model a decision was made to use PyTorch over TensorFlow. Even though, TensorFlow has a more mature pipeline, Pytorch has better results over it, both in speed and quality – especially, when used with smaller dataset [1]. Moreover, as a younger framework, it seems to offer a higher support and ease of integration for newer technologies like quantum machine learning. On the other hand the disadvantage of PyTorch is its age. Since it is a younger library, and the technology of quantum machines is very new, the documentation of many tools in this field is lacking. This is likely due to very fast development of these tools caused by very recent emergence of technology availability.

4 Methodology

The main concern in any kind of quantum algorithm in the Noisy Intermediate Quantum Computing (NISQ) era, is to create an environment in which the quantum machine, produces effective calculations. The current state of quantum machines forces users, to reduce as many sources of noise as possible before running any type of quantum task. There are a plethora of methods being researched for the purpose of maximizing the quality of our results. Whether it is improving physical state of the quantum machine, reducing a Hamiltonian of the circuit into smaller parts, minimizing depth of the circuit or post/pre-processing of the data. In this article, the team explored different frameworks of reducing images to 3 or 4 features in a way, that would lose the smallest amount of information. At the same time creating the best input for the quantum circuit.

4.1 Quantum Convolutional Neural Networks

One of the most popular approaches – the Quantum Convolutional Neural Network (QCNN) [11] – is to use quantum layers enwrapped by the classical CNN model. In the context of quantum machine learning methods, this one is well known and well documented. Although, not many sources tend to implement it on an actual quantum hardware.

The architecture can be divided into 4 main parts, first one being the initial CNN layers. In the field of neural networks it is currently possible to reduce the data points into smaller sizes and shapes. This kind of reduction also evolves and corrects itself over time with each epoch. This is a perfect tool, to prepare the data for a quantum machine. Then, there is a need to input the data, onto the quantum circuit, with an encoding layer. In this research the ZZFeatureMap was used to map the correlated features of each data point, onto the entangled qubits. Although, due to later mentioned hardware limitations, a more specified encoding layer might have been useful. Assuming U_{ZZ} represents ZZ-gate layer enveloped by an entanglement scheme. For some data vector $\overrightarrow{x}$ of size N, the ZZFeatureMap prepares the following state:

$$\overrightarrow{x} \xrightarrow{ZZFM} |\psi(\overrightarrow{x})\rangle = \left(U_{ZZ}(\overrightarrow{x}) \left(\bigotimes_{k=1}^{N} P(\overrightarrow{x_k}) \right) H^{\otimes N} \right) |0\rangle \tag{1}$$

Finally, it is needed to design the convolutional layer equivalent. Quite often, for this step the RealAmplitude ansatz is used. The disadvantage of this architecture is the fact, that this one uses the CNOT gate to entangle qubits and in the gate set of the Odra 5 computer the only entangling option is the CZ gate. Therefore to minimize additional gates, which are being created – as well as take into account the topology of qubits for Odra 5 – during the transpilation process, it is possible to design an ansatz specifically for that quantum machine. Its topology can be seen in Fig. 1. Due to time constraints, the exploration and testing of different types of ansatz compositions was decided to be left for future work. Consequently, the RealAmplitude ansatz was selected for the current study.

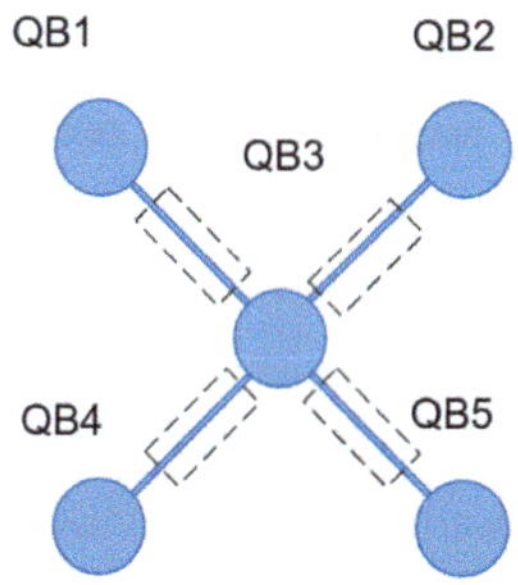

Fig. 1. 5-qubit arrangement of physical qubits on the Odra 5 quantum computer. Img. source: https://enccs.github.io/qas2023/notebooks/intro-to-helmi-qiskit

The final step is to calculate the $\langle Z \rangle = \langle \psi | Z | \psi \rangle$ – expectation value of the Pauli-Z for each remaining qubit. It can be found by measuring our circuits final states multiple times and approximating the expectation value based on our finite set of measurements. This expectation value provides a real number between -1 and 1, which can be used to determine probability of the qubit being in the state $|0\rangle$ or $|1\rangle$. The task of properly estimating the expectation value, without too much computation is a complicated problem. Especially, since the limitation of the type of observables possible to be measured and by the number of measurements that can be run [8].

When using the Pytorch library for QNNs, there exist two popular classes for the expectation value approximation. The EstimatorQNN and SimulatorQNN, which define how to use the measurements of the quantum circuit in the classical layers of the machine learning. The Estimators are estimating the expectation value of the state by looking at the circuit and assuming what the expectation value of density matrix should be for this specific arrangement of quantum gates and observable measurements. Where in a Simulator every shot of the experiment is being run and measured (on real backend or the simulator), based on which the distribution of every state-vector is used to calculate the approximation of the final state.

From a practical standpoint, the team came to a conclusion that when using these Pytorch tools, the SamplerQNN calculates more legitimate results when the program is being run on a simulator. However, the Estimator is much faster, since it doesn't have to run every shot for every data point. It is also easier to control and interpret the answers since we get only the measrument of the specific observable. While, executing the same piece of code on an actual quantum machine, EstimatorQNN was also much more convenient to work with. There are not many differences a person needs to apply, to swap between those two tools. Therefore, it is advised to compare both the methods in a controlled and simple environment. One needs to remember that, a choice of a wrong gradient type for the EstimatorQNN can massively slow down process of training the model.

During training of each type of the model – both QCNN and QCNN supported by Autoencoders – it was decided to compare the epochs with binary cross entropy with logits loss function. Defined as:

$$l(x, y) = l_1, ..., l_N{}^T, \quad (2)$$

where

$$l_i = -w_i[y_i \cdot log_n + (1 - y_i) \cdot log(1 - p_i)] \quad \forall i \in N. \quad (3)$$

This function is commonly used for binary classification problem, which was ideal for this task.

4.2 QCNN with Autoencoders

While QCNN is fairly recognizable, there is a limited number of sources, where authors try to enhance their quantum model with autoencoders [17]. The idea behind that framework is to prepare the data for the quantum circuit with a classically trained autoencoder. Ideally, one could tune the autoencoder model using the final state of the quantum circuit and the returned results, a topic the authors plan to tackle in future research. An autoencoder is a neural network divided into two main parts. An encoder, which prepares an encoding of data. And a decoder that decodes this encoding back into our input. During the training process the output data is then compared to the original and based on the difference between the two a optimization algorithm tunes the encoder.

When preparing the data for the quantum machine with an autoencoder, a data point $x \in X, X = \mathbb{R}^n$ is reduced into q features, s.t. q is the number of qubits on our machine and $q < n$. Assuming, that $z \in Z, Z = \mathbb{R}^q$; θ, ϕ are variables that tune our encoder E_θ and decoder D_ϕ and $L(x, y)$ is a loss function. Then the training process works by finding $z = E_\theta(x)$ and $x' = D_\phi(z)$, then calculating loss value $L(x, x')$ and applying changes to the variables of E_θ and D_ϕ, based on that value. The choice of the loss function varies based on the final task of the encoded data, but quite often the Mean Squared Error (MSE) loss is used [17].

The training phase of the proposed model reveals several advantages of the framework. The autoencoder is designed to learn a compact representation of the input data that is compatible with the quantum circuit; consequently, it can be trained once on the entire dataset. Although the quantum component of the framework addresses a k-class classification problem, retraining the autoencoder is not required when modifying or extending the classification task. Moreover, when new data become available, the existing autoencoder can be adapted via mixed fine-tuning that incorporates both the original and newly samples. Finally, the variational learning process is largely shifted to the quantum component, which has the potential to yield a significant computational acceleration.

After the autoencoder model is created, it can be used at every training step of the QML model to encode a data point into features used by the quantum circuit. Any quantum circuit that works on q number of qubits can be used. In this case the same q-layer as in the QCNN was used. Additionally, a single linear layer was also added for the model after the q-layer, so that the model might slightly tune the interpretation of the quantum output.

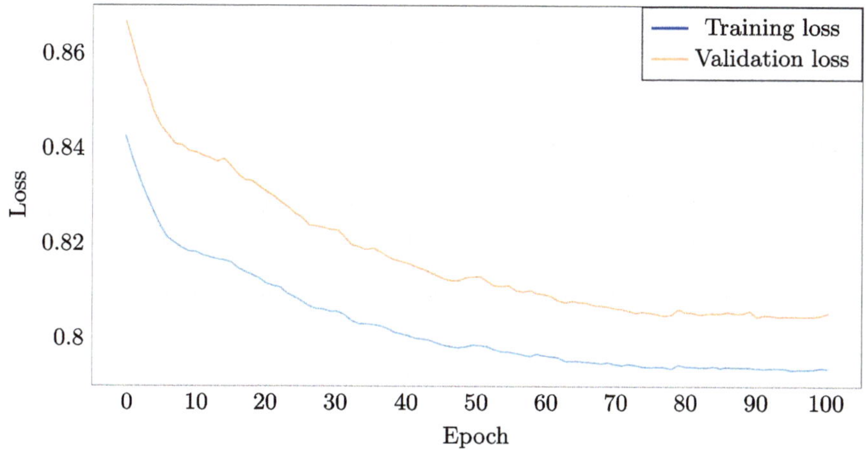

Fig. 2. The loss values over each epoch of training the QCNN model on MNIST dataset.

5 Experiments and Results

Different variations of the model were extensively tested – on both simulators and real quantum backend. An analysis was conducted to identify the variables, tools, and methods with the greatest impact on accuracy. The time authors had for this project was, unfortunately, not enough to fully test all of the possibilities.

After training the autoencoder to encode the MNIST dataset, a work on the quantum model was started. Here, the set of layers and the quantum circuit were very similar to the previously utilized version. Testing different loss functions and

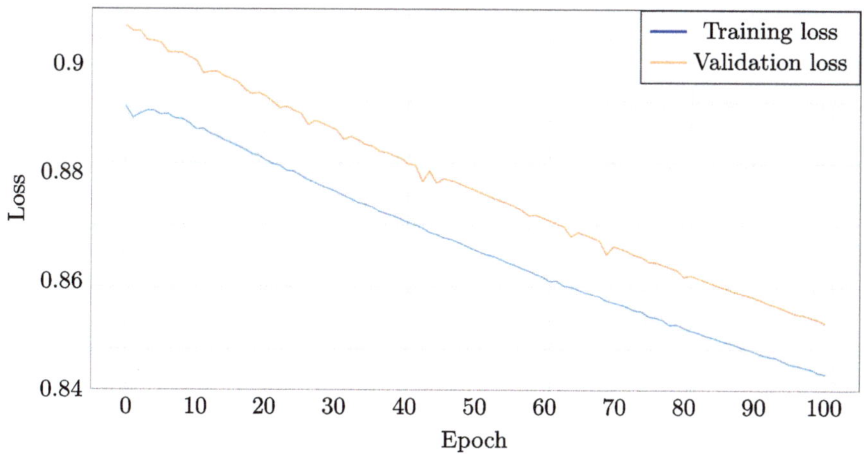

Fig. 3. The loss values over each epoch of training the AEQCNN model on MNIST dataset.

parameters was conducted for both classical layers and EstimatorQNN. It can be easily seen how the model performed for the MNIST data can get on Figs. 2 and 3. At the same time it can as easily be seen, that the loss values are not satisfactory. Those are the final versions for this dataset that was used and ran on Odra 5 quantum computer.

Until now the main measurement that has been analyzed in the paper, was the loss value. Although, throughout the training of ML models it is needed to measure the accuracy of how well the trained model is predicting.

When testing the model it was noticed, that for the most part it estimates a single answer. That is also related to the fact that the loss value delta is smaller than desired. However, when testing the same model on a highly reduced dataset – of only 10 images per class – on a simulator, it was discovered that there is always a breaking point when the model starts estimating other answers. As it is presented on Fig. 4 this breaking point begins at around 0.4 ± 0.05 loss value. It varied based on different types of variables, number and type of images in the class and the type of loss function that was used. It is also understandable, why the breaking point would be present at that value. Since, for the final answer there are only classes 0 and 1, the estimation of 0.38 and 0.43 is the same when it comes to the final answer. Therefore, the breaking point should begin at values ~0.5 of the loss value – depending on the loss function it might be closer to 0.3.

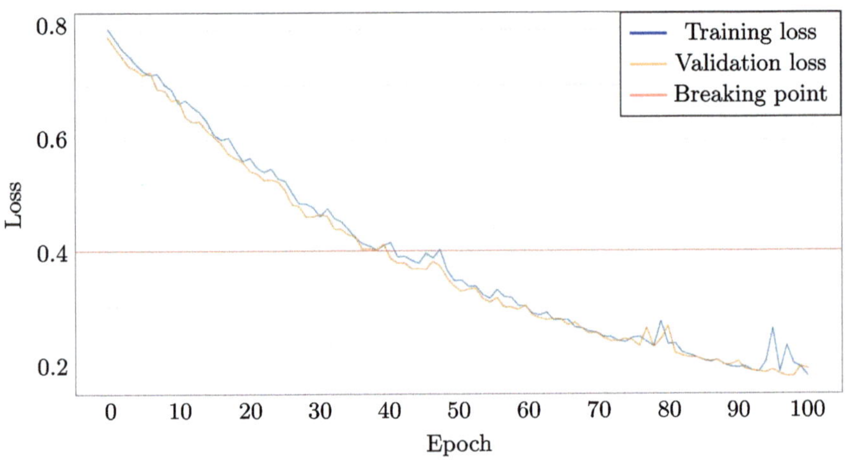

Fig. 4. The loss values over each epoch of training the QCNN model on reduced MNIST dataset (only 10 images per class). This version of the model was tested on a simulated backend without any noise. The cross section between the graph and the red line was the breaking point, for which the model started estimating different answers than the original guess.

When examining adjustable tools and parameters, the study identified potential in using more than one observable. This approach resulted in a vector of measurements being passed to the classical component of the model, which was expected to facilitate easier correlation and adaptation by the classical layers. However, experimental results showed that this modification had little to no impact on overall model performance, while significantly increasing the training time. Consequently, further exploration of this approach was discontinued.

6 Conclusions

One of the aims of this work was to analyze what tools are available, when writing a QML project. As well as, to compare the results of using an autoencoder in QML in comparison to standard QNN. While comparing tools used in this field, the two most impactful choices one could make early on when writing such a program – with Pytorch and Qiskit libraries – was the choice between using an EstimatorQNN and SimulatorQNN. The former changes the way of using the quantum measurements. The training when using Sampler and Estimator can give much different set of results based on implementation of the rest of the model. At the same time, the fact of time complexity should not be ignored. When training a model with a Sampler on a set of 10 images per class, on the other hand it took around 1 h to train the model for a 100 epochs on a simulator. The same training with an Estimator took only around 15 min. It could be argued that Sampler returns more trustworthy results, when it comes to simulating a real quantum backend, but that is the case mostly, because Estimator class is not well presented by PyTorch. It lacks proper documentation, which would explain what happens in such a framework with more details.

The result of around 0.8 loss value might not be a full success however, both graphs on Fig. 2 and 3 appear as if there might be some further progress if trained for a longer period of time. Especially, in the case of the AEQCNN. The team believes, that if the training time of an autoencoder would be longer, the results should give greater accuracy. It can already be seen that the training delta is 3 times greater than for QCNN, although the results start with a higher loss value, which implies a better training evolution. With added time and additional adjustments to the classical layers and the ansatz this framework might have great potential. An additional small change, that had a massive impact on the output of AEQCNN, was a use of at least a single classical layer after the quantum part of the framework. Even, if the layer was a very simple linear layer, it had a massive impact on the training loss graph.

References

1. Paszke, A., et al.: Pytorch: an imperative style, high-performance deep learning library. In: NIPS Papers (2019)
2. Alam, M., Ghosh, S.: DeepQMLP: a scalable quantum-classical hybrid deep neural network architecture for classification. In: 2022 35th International Conference on VLSI Design and 2022 21st International Conference on Embedded Systems (VLSID), pp. 275–280. IEEE (2022)
3. Asaoka, H., Kudo, K.: Quantum autoencoders for image classification. Quantum Mach. Intell. **7**(2), 71 (2025)

4. Bauer, C.W., et al.: Quantum simulation for high-energy physics. PRX Quantum **4**(2), 027001 (2023)
5. Bożejko, W., Trotskyi, S., Uchroński, M., Wodecki, M.: Optimizing two-machine scheduling in flexible manufacturing systems using autonomous ai and quantum computing. Neurocomputing, 132067 (2025)
6. Cao, Y., et al.: Quantum chemistry in the age of quantum computing. Chem. Rev. **119**(19), 10856–10915 (2019)
7. Chang, D.T.: Parameterized quantum circuits with quantum kernels for machine learning: a hybrid quantum-classical approach. arXiv preprint arXiv:2209.14449 (2022)
8. Crawford, O., van Straaten, B., Wang, D., Parks, T., Campbell, E., Brierley, S.: Efficient quantum measurement of Pauli operators in the presence of finite sampling error. Quantum **5**, 385 (2021)
9. De Luca, G.: A survey of NISQ era hybrid quantum-classical machine learning research. J. Artif. Intell. Technol. **2**(1), 9–15 (2022)
10. Deng, L.: The MNIST database of handwritten digit images for machine learning research. IEEE Signal Process. Mag. **29**(6), 141–142 (2012)
11. Fan, F., Shi, Y., Guggemos, T., Zhu, X.X.: Hybrid quantum-classical convolutional neural network model for image classification. IEEE Trans. Neural Netw. Learn. Syst. (2023)
12. Gnatowski, A., Niżyński, T.: Quantum circuits compilation for superconducting quantum processing units: a systematic mapping study. In: International Conference on Intelligent Systems in Production Engineering and Maintenance, pp. 269–284. Springer (2025)
13. Jędruszek, M., Idzikowski, R., Kucharski, M.A.: Experimental analysis of quantum neural networks on Odra 5 quantum computer. In: Proceedings of the 21st International Conference on Dependability of Computer Systems. Submitted for publication
14. Kucharski, M.A.: A survey on quantum machine learning in speech acoustics. In: 2025 IEEE 25th International Symposium on Computational Intelligence and Informatics (CINTI), pp. 235–240 (2025). https://doi.org/10.1109/CINTI67731.2025.11311733
15. Ngairangbam, V.S., Spannowsky, M., Takeuchi, M.: Anomaly detection in high-energy physics using a quantum autoencoder. Phys. Rev. D **105**(9), 095004 (2022)
16. Rani, S., Pareek, P.K., Kaur, J., Chauhan, M., Bhambri, P.: Quantum machine learning in healthcare: developments and challenges. In: 2023 IEEE International Conference on Integrated Circuits and Communication Systems (ICICACS), pp. 1–7. IEEE (2023)
17. Sakhnenko, A., O'Meara, C., Ghosh, K.J., Mendl, C.B., Cortiana, G., Bernabé-Moreno, J.: Hybrid classical-quantum autoencoder for anomaly detection. Quantum Mach. Intell. **4**(2), 27 (2022)

Solving Capacitated Vehicle Routing Problem in a Distributed Computing System

Andrzej Gnatowski(✉) and Konrad Pempera

Department of Control and Quantum Computing, Faculty of Information and Communication Technology, Wrocław University of Science and Technology, Wrocław, Poland

andrzej.gnatowski@pwr.edu.pl, konrad.pempera@pwr.edu.pl

Abstract. This paper examines resource allocation strategies in a distributed computing system for solving Capacitated Vehicle Routing Problem (CVRP) instances using Branch and Bound algorithms. Tasks arrive online in a stochastic manner, and the system orchestrates their execution through two strategies: cluster-exclusive latency (`CEL`), which dedicates all resources to minimize latency for a single task, and throughput node-instance (`TNI`), which processes multiple tasks concurrently to maximize throughput. The approach integrates task orchestration with sequential and distributed solvers. Empirical evaluations via simulations under varying loads demonstrate that the `CEL` strategy offers superior responsiveness in low-traffic conditions but exhibits rapid saturation, whereas the `TNI` strategy provides stability across a broader range of workloads. The study addresses the latency–throughput trade-off and outlines directions for adaptive orchestration in future research.

Keywords: Capacitated Vehicle Routing Problem · Distributed Computing · Branch and Bound · Latency vs Throughput · Ray Framework

1 Introduction

In the context of Industry 4.0 and the expanding e-commerce sector, logistics operations are under increasing pressure to achieve real-time optimization [1]. Contemporary platforms for last-mile delivery handle a continuous influx of transport requests that require grouping into optimal routes to minimize fleet operational costs and ensure adherence to delivery schedules [16]. One of the fundamental combinatorial problems in this domain is the Capacitated Vehicle Routing Problem (CVRP), classified as NP-hard [3,15]. Within this framework, CVRP instances arise during system runtime and necessitate timely resolution. Due to the large computational volume, the application of parallel and distributed computing techniques becomes a necessity.

W. Zamojski et al. (Eds.): DepCoS-RELCOMEX 2026, LNNS 2014, pp. 131–141, 2026.
https://doi.org/10.1007/978-3-032-27927-9_13

While heuristics and metaheuristics are commonly used for fast responses and large instances [10], exact methods remain preferred for guaranteeing optimality or benchmarking approximations [14]. Exact methods like Branch and Bound (B&B) support parallelization through search tree decomposition [18]. Theoretically, exploring disjoint subtrees on separate units yields linear speedups [9]; however, irregular tree structures cause load imbalance, and efficient pruning requires timely global bound propagation, creating communication trade-offs in distributed settings [7].

This study examines a distributed computing system for solving CVRP instances (tasks) via B&B algorithms. Tasks arrive online, with unknown timing and specifics but known distribution. The system orchestrates task execution using two strategies: assigning all nodes exclusively to one request for minimal latency (*Cluster-Exclusive Latency*, `CEL`) or processing multiple requests concurrently for maximal throughput (*Throughput Node-Instance*, `TNI`). We evaluate the performance trade-offs between these approaches. This paper provides the following contributions:

- Development of a distributed computing system for solving CVRP instances that combines high-level task orchestration with low-level Branch and Bound.
- Empirical evaluation of the two resource-allocation strategies based on simulations under stochastic request arrival patterns.
- Demonstration and discussion of the performance trade-offs between the strategies, including recommendations regarding the operational regimes in which each approach is most appropriate.

The remainder of the paper is organized as follows. Section 2 reviews related literature. Section 3 describes the considered problem and Sect. 4—solving algorithms. Section 5 reports the results of the computational experiments. Section 6 summarizes the findings and indicates directions for future research.

2 Literature Review

The Capacitated Vehicle Routing Problem (CVRP) constitutes one of the foundations of combinatorial optimization and logistics [17]. Due to its membership in the class of NP-hard problems, the CVRP remains a computational challenge despite decades of intensive research [19]. Although heuristics and metaheuristics dominate industrial applications [11], exact methods such as Branch and Cut algorithms are essential for determining reference solutions [5].

With the reaching of physical limits in processor frequency scaling, parallel computing allows for further scaling [4]. Gendron and Crainic [7], in their comprehensive review, classify parallel B&B algorithms based on the degree of synchronization and the knowledge management structure (centralized vs. distributed). The foundation of parallel Depth-First Search (DFS) algorithms is the ability to independently explore subtrees rooted at different nodes, which theoretically allows for the minimization of communication between processors. However, in practice, the structure of search trees is highly irregular and difficult to estimate a priori, which causes static task partitioning to lead to significant load imbalance and necessitates the use of dynamic control strategies [8].

Online scheduling is a problem in which jobs arrive sequentially and must be assigned to machines without knowledge of future tasks [6]. Günther et al. [12] introduced the concept of competitive-ratio approximation schemes, which allow the design of online strategies whose performance is nearly optimal for a given problem. This approach is based on an abstract characterization of online algorithms as decision maps and a drastic simplification of input instances. By reducing the infinite set of strategies to a finite number of significant equivalence classes, it becomes possible to compute the optimal competitive ratio with arbitrary precision $1 + \epsilon$. Another recent approach is the Random Order Model (ROM) proposed by Albers et al. [2], where jobs are presented in a random permutation. This model allows the identification of stable features of the input sequence. The strategy maintains a specific machine load profile by selecting among three candidate machines and analyzing the largest jobs processed so far. This method achieves a ratio of 1.8478, representing the first breakthrough in online algorithm performance for this problem in nearly two decades.

Novak et al. [13] tackle centrifuge batching by optimizing Patient Turnaround Time (TAT), defined as the duration from sample collection to result reporting. Their strategy focuses on reducing the 0.95 quantile of TAT to mitigate outliers that negatively impact the quality of care and patient length of stay. The researchers demonstrate that online policies utilizing distributional knowledge can achieve results nearly identical to the theoretical upper bounds established by perfect-knowledge offline algorithms.

3 Problem Statement

The problem addressed in this study can be decomposed into two interrelated levels: (1) task orchestration and (2) solving instances of the Capacitated Vehicle Routing Problem (CVRP). In this section, we define both levels.

3.1 CVRP

The Capacitated Vehicle Routing Problem (CVRP) aims to find a set of minimum-cost vehicle routes starting and ending at a depot, such that all customer demands are served. Each customer must be visited exactly once, while the total demand on each route cannot exceed the vehicle capacity. This problem naturally models real-world distribution and delivery planning.

It is assumed that there are N service locations in the set $\mathcal{N} = \{1, 2, \ldots, N\}$, extended by the depot node 0 (start and end point for each vehicle), forming $\mathcal{D} = \{0\} \cup \mathcal{N}$. Each location is assigned a pair of coordinates, which are used to compute the travel cost between any two points. These costs are stored in the distance matrix $d_{i,j}$, where $d_{i,j}$ represents the cost of traveling from node i to node j, for all $i, j \in \mathcal{D}$. A set of vehicles $\mathcal{V}$ is considered, with each vehicle $v \in \mathcal{V}$ subject to a capacity limit of Q. It is further assumed that $x_{v,i,j}$ is a binary decision variable equal to 1 if vehicle $v \in \mathcal{V}$ travels directly from node i to node j, and 0 otherwise. Moreover, $z_{v,i}$ is a binary decision variable that takes the value 1 if location $i \in \mathcal{N}$ is served by vehicle v, and 0 otherwise. Finally, y_i denotes the demand associated with location $i \in \mathcal{N}$, which must be satisfied by the vehicle assigned to that location.

The objective function minimizes the total travel cost incurred by all vehicles:

$$\min \sum_{v\in\mathcal{V}} \sum_{i\in\mathcal{D}} \sum_{j\in\mathcal{D}} x_{v,i,j} d_{i,j}, \tag{1}$$

where $d_{i,j}$ denotes the travel cost between locations i and j, and $x_{v,i,j}$ is a binary decision variable indicating whether vehicle v travels directly from i to j.

The objective function (1) is minimized subject to the following constraints:

$$\sum_{v\in\mathcal{V}} \sum_{j\in\mathcal{D}\setminus i} x_{v,i,j} = 1, \qquad \forall i \in \mathcal{N} \tag{2}$$

$$\sum_{j\in\mathcal{D}\setminus i} x_{v,i,j} - \sum_{j\in\mathcal{D}\setminus i} x_{v,j,i} = 0, \qquad \forall v \in \mathcal{V}, i \in \mathcal{N} \tag{3}$$

$$z_{v,i} = \sum_{j\in\mathcal{D}\setminus i} x_{v,j,i}, \qquad \forall v \in \mathcal{V}, i \in \mathcal{N} \tag{4}$$

$$\sum_{i\in\mathcal{N}} x_{v,0,i} = \sum_{i\in\mathcal{N}} z_{v,i}, \qquad \forall v \in \mathcal{V} \tag{5}$$

$$\sum_{i\in\mathcal{N}} x_{v,i,0} = \sum_{i\in\mathcal{N}} z_{v,i}, \qquad \forall v \in \mathcal{V} \tag{6}$$

$$x_{v,i,i} = 0, \qquad \forall v \in \mathcal{V},\ \forall i \in \mathcal{D} \tag{7}$$

$$\sum_{i\in\mathcal{N}} z_{v,i} y_i \le Q, \tag{8}$$

$$x_{v,i,j}, z_{v,i} \in \{0,1\}, \qquad \forall v \in \mathcal{V}, i, j \in \mathcal{D} \tag{9}$$

$$p_{v,i}, q_{v,i}, w_{v,i}, t_{v,i} \ge 0. \qquad \forall v \in \mathcal{V}, i \in \mathcal{D} \tag{10}$$

Constraint (2) guarantees that every customer node is visited exactly once by a single vehicle. Constraint (3) imposes flow conservation, meaning that whenever a vehicle enters a node, it must also leave it. Constraints (5) and (6) ensure that each active route begins and terminates at the depot. Constraint (4) links the visit indicator variable with the routing variables. Constraint (7) prevents trivial solutions by disallowing arcs from a node to itself, thereby eliminating self-loops that would otherwise satisfy flow conditions without actual travel. Finally, constraint (8) restricts the total customer demand assigned to each vehicle. It ensures that the sum of demands served by vehicle v remains within the vehicle capacity limit Q, thus maintaining the feasibility of the load distribution.

3.2 Task Orchestration

The system under consideration is an online optimization task orchestration framework. Over a defined time horizon, computational tasks become available for processing. These tasks arrive stochastically, with inter-arrival times following a specific distribution, which may be unknown a priori. Each task consists of solving a computational problem. Processing occurs on available computing nodes, each equipped with multiple processor cores. A task can be assigned to any non-empty subset of processor cores, and its processing time depends on the selected solving algorithm, the assigned set of cores, concurrent tasks in the system, and the inherent characteristics of the task instance. The objective is to assign tasks to processor cores and select solving algorithms to optimize selected quality-of-service (QoS) metrics.

In this paper, we model a typical operating day of a dynamic delivery platform, such as those used in quick-commerce or ride-hailing services. The system operates from 08:00 to 18:00, spanning 10 h. Transport requests, formulated as CVRP instances, arrive stochastically throughout this period. The solution specifies which task to process at any given time and under which strategy or algorithm variant. Here, we consider a single algorithm—Branch and Bound (B&B), detailed in Sect. 4—and two resource allocation strategies, also described in Sect. 4. Tasks not completed by 18:00 are classified as *unfinished*.

Given that the B&B algorithm guaranties solution optimality, the primary QoS metric is the Turnaround Time (TAT), defined as:

$$\mathrm{TAT} = T_{\text{wait}} + T_{\text{compute}}, \tag{11}$$

where T_{wait} is the time a request spends in the queue awaiting resource allocation, and T_{compute} is the solver execution time. Since TAT is computed per task, we aggregate it using the following surrogates: the 95th percentile (P_{95}) of TAT, computed over the entire day (D95) and per hour (H95), as well as the daily number of unfinished tasks. The P_{95} metric provides a conservative measure of tail latency, commonly used in performance analysis to mitigate the influence of extreme outliers [13].

4 Algorithms

This section describes the algorithms employed in the study. First, a sequential Branch and Bound algorithm for solving CVRP instances on a single processor core is presented. Next, a distributed Branch and Bound algorithm that utilizes multiple cores, potentially distributed across computing nodes, is outlined. Finally, the orchestration strategies applied at a higher level are detailed.

4.1 Sequential Branch and Bound

A classical Branch and Bound (B&B) solver was implemented, where each CVRP instance is solved independently on a single logical thread. The solver recursively explores all feasible routes while maintaining the current best solution. The search strategy is based on depth-first search (DFS), meaning that the algorithm extends the current partial route as deep as possible before backtracking.

For each partial route, a lower bound is computed by summing the current route cost with the minimal distances from each unvisited city to any other city. The pruning function evaluates the bound by adding, for each unvisited node, the cost of its minimum outgoing edge. If the resulting bound exceeds the current best (upper bound) solution, the branch is cut off. Any branch whose lower bound exceeds the current best solution is pruned, ensuring that subtrees incapable of yielding an improved solution are not explored.

4.2 Distributed Branch and Bound

Parallelization is achieved through decomposition by fixing the first two cities visited after the depot (city-pair decomposition), generating $O(N^2)$ independent subproblems that are dispatched across the available workers (processor cores). Each worker processes at most one subproblem at a time; thus, not all subproblems may be assigned immediately. A queue of subproblems is maintained, and upon completion of a subproblem, the next one from the queue is assigned. The subproblems in the queue are not sorted in any particular order.

Active synchronization and pruning are incorporated to enhance efficiency. A global upper bound is maintained centrally, and local upper bounds are periodically refreshed to prevent redundant exploration of dominated subtrees (synchronization intervals: every 100 expanded nodes or 200 ms).

4.3 Orchestration Strategies

The orchestration system determines, at any given time, which task to process and in what manner. The following decision rules are assumed: if computing resources are available, the task with the longest waiting time T_{wait} is selected for processing. The processing follows one of two strategies: cluster-exclusive latency strategy (`CEL`) or throughput node-instance strategy (`TNI`). The `CEL` strategy assigns the task to all worker nodes using the distributed B&B algorithm, occupying the entire cluster's computing power for the duration of the CVRP solution. In contrast, the `TNI` strategy assigns the task to a single worker (processor core) using the sequential B&B algorithm, blocking only that worker for the solving duration. It is assumed that the orchestration system employs either the `CEL` or `TNI` strategy consistently throughout the considered time span.

5 Computational Experiments

To validate the theoretical assumptions regarding the latency-throughput trade-off of the considered orchestration strategies, we designed a simulation environment that mimics the operational reality of a modern logistics platform.

5.1 Hardware

The experimental cluster consisted of a 8 worker nodes managed by a dedicated coordinator node. Each worker node was equipped with a processor with 4 physical cores, 8 logical threads, 3.6 GHz max boost and approximately 5.7 GiB of RAM. The nodes operated on a Linux environment. The coordinator node was responsible solely for the Ray Global Control Store (GCS) and task scheduling, ensuring that computational results were not skewed by orchestration overhead.

5.2 Test Scenarios

The goal of the test is to check the performance of the proposed orchestration strategies given a specific scheduling algorithm. Therefore, the CVRP instance sizes were adjusted to match the usability range of the B&B algorithm, $N \in [11, 16]$. To reflect the diversity of the jobs in real systems, we divided them into three categories: *easy* ($N \in [11, 12]$), 50% of the workload, *medium* ($N \in [13, 14]$), 32% of the workload, *hard* ($N \in [15, 16]$): 18% of the workload.

Here, we assume a problem instance is a single day of operations. This assumption is valid since a lot of businesses work between X and Y, like couriers. We assume the scheduling jobs flow with uniform intensity throughout the day, characterized by Tasks Per Hour (TPH). We considered TPH from 10 to 170 and randomly generated 30 instances for each TPH. Due to a low throughput of `CEL` we did not test it for TPH > 35.

All simulations were executed on the hardware infrastructure described in Sect. 5.1, with the simulation speed accelerated (1 real minute ≈ 1 simulation hour). We logged timestamps for arrival, start, and completion. Task generation ceased at 17:00, with a hard simulation cutoff at 18:00. Any requests pending at the cutoff were marked as `UNFINISHED`.

Two tail-latency metrics were computed to capture system congestion. First, D95 is defined as the daily 95th percentile of the total system time, calculated for each day and then averaged over 30 independent instances for a given load. Analogously, H95 is defined as the hourly 95th percentile, computed separately for each hour of operation, assuming a fixed load and averaged across test instances. In the experiments, we assumed TPH = 30. In addition, the daily number of unfinished tasks (NUT) was measured.

5.3 Results and Discussion

The primary objective of the computational experiments was to empirically verify the trade-off between minimizing single-task latency and maximizing total system throughput.

Daily TAT. First, to study the robustness of orchestration, we computed the mean daily 95th percentiles of TAT, D95; for different numbers of tasks per hour (TPH). The results are shown in Fig. 1. The two proposed approaches exhibit fundamentally different behaviors in response to increasing workload.

In the low-TPH zone, the `CEL` strategy demonstrates superior responsiveness. For $TPH = 10$, the D95 is approximately 1000 s, compared to roughly 4400 s for the `TNI` strategy. This confirms that when the system is not saturated, dedicating the entire cluster's computing power to a single request significantly reduces the TAT, offering a clear advantage for high-priority, time-sensitive tasks. A critical inflection point is observed at $TPH \approx 30$. The serialization of tasks in the `CEL` model causes a rapid accumulation of the queue and increases TAT to almost 8000 s at just $TPH = 35$.

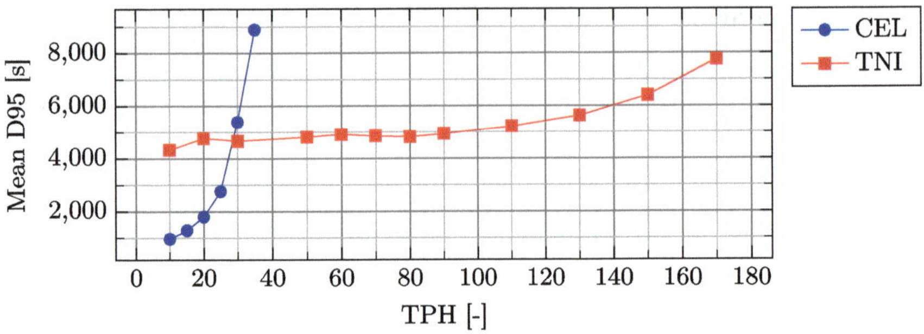

Fig. 1. Impact of task arrival rate (TPH) on mean daily 95th percentile of TAT (D95).

Conversely, the TNI strategy exhibits high stability across a wide range of workloads. Despite a higher initial baseline latency (due to solving instances on single threads), the curve remains relatively flat up to TPH $\approx$ 130. This indicates that the strategy effectively utilizes the available hardware concurrency, maintaining predictable service levels even under heavier traffic. Significant congestion is only observed when the arrival rate exceeds 150 TPH, which is nearly 5$\times$ the saturation point of CEL.

Hourly Dynamics. To investigate the hourly dynamics of the system, we selected a boundary scenario with TPH = 30. Each day, tasks were grouped by their arrival hour; then, for each hour, the 95th percentile of TAT (H95) was computed. Results are shown in Fig. 2.

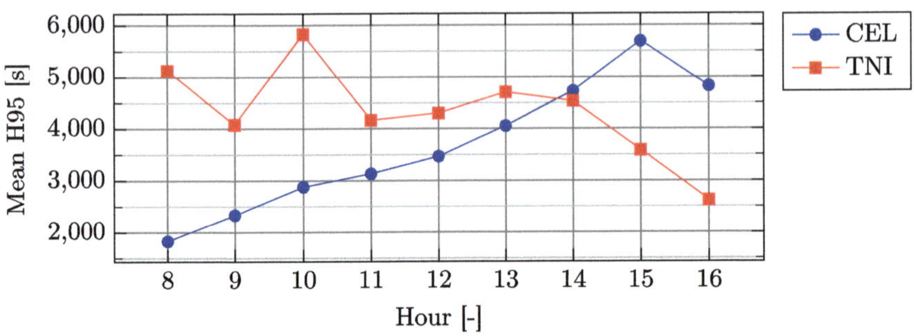

Fig. 2. Mean hourly 95th percentile (H95) of TAT and 30 TPH.

CEL exhibits a steep, nearly linear growth in system time, indicating rapid saturation. Since this strategy processes tasks sequentially, the accumulation of computationally expensive tasks creates a bottleneck, causing wait times to compound continuously for all subsequent requests. The apparent drop in latency observed for 16:00–17:00 is influenced by the experimental setup, where new task arrivals ceased at 16:00. Moreover, tasks stuck in the growing backlog were not completed before the hard cutoff and were thus excluded from the average.

`TNI`, despite higher absolute values in the middle of the day, exhibits a smoother trajectory. The distributed architecture ensures that difficult tasks are isolated on individual nodes, preventing them from blocking the global throughput. Similarly to `CEL`, a drop in H95 is observed near the end of the day.

Unfinished Tasks. A key indicator of system adequacy is the number of unfinished tasks per day (NUT). This metric was measured for simulated days characterized by varying tasks per hour (TPH) values; the relationship is illustrated in Fig. 3. The `CEL` strategy yields NUT $= 0$ for TPH ≤ 25, but saturates rapidly for higher TPH, which is consistent with observations from prior experiments. In contrast, the `TNI` strategy exhibits a substantially higher saturation threshold at TPH ≈ 150; however, even at lower TPH, it provides no assurance of NUT $= 0$, because computationally demanding tasks arriving near the end of the day may exceed the time available for resolution on a single node.

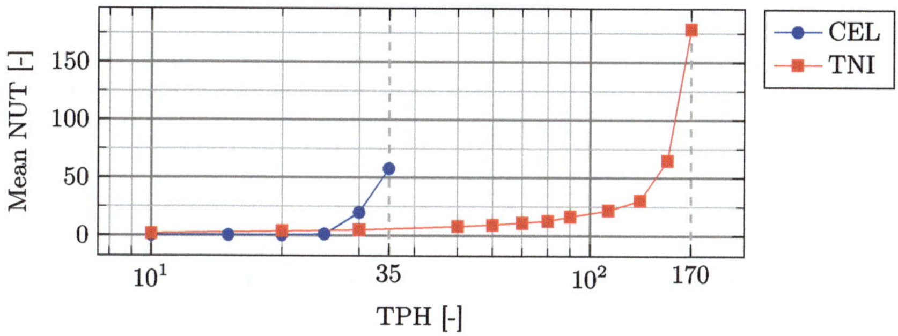

Fig. 3. Mean number of unfinished tasks (NUT) with relation to TPH.

6 Conclusions and Future Work

This paper presents an empirical analysis of resource allocation strategies for solving the Capacitated Vehicle Routing Problem (CVRP) in a distributed, online orchestration system. We evaluated two distinct strategies: cluster-exclusive (`CEL`) and node-instance (`TNI`) under dynamic load conditions.

The `CEL` is superior in low-load regimes, offering lower latency for individual requests. However, it is extremely fragile; exceeding the saturation point leads to a fast accumulation of the backlog. The `TNI` strategy demonstrates a more gradual degradation of the quality of service metrics with the increase in load. By isolating tasks on separate processing units, it prevents computationally expensive hard instances from starving the entire system; thus, it performs well under much higher and diverse loads.

Future research will focus on building more realistic orchestration models. We will explore task priorities (as in [13]), more resource allocation strategies, and more diverse solving algorithms—in practice, some tasks may be solved heuristically, while others must not. Finally, we plan to shift towards dynamic orchestration strategies, adapting to the current stress of the system and the backlog of tasks.

References

1. Agnusdei, G., Gnoni, M., Sgarbossa, F., Govindann, K.: Challenges and perspectives of the industry 4.0 technologies within the last-mile and first-mile reverse logistics: a systematic literature review **45**, 100896 (2022)
2. Albers, S., Janke, M.: Scheduling in the random-order model. Algorithmica **83**(9), 2803–2832 (2021)
3. Ardon, L.: Reinforcement learning to solve NP-hard problems: an application to the CVRP (2022). https://doi.org/10.48550/arXiv.2201.05393
4. Asanovic, K., et al.: The landscape of parallel computing research: a view from Berkeley. Technical report, Technical Report UCB/EECS-2006-183, EECS Department, University of California, Berkeley (2006)
5. Claudia, N.B., Speranza, M.G.: Branch-and-cut algorithms for the split delivery vehicle routing problem. Eur. J. Oper. Res. **238**(3), 685–698 (2014)
6. Dwibedy, D., Mohanty, R.: Online scheduling with makespan minimization: state of the art results, research challenges and open problems (2020). https://doi.org/10.48550/arXiv.2001.04698
7. Gendron, B., Crainic, T.G.: Parallel branch-and-bound algorithms: survey and synthesis. Oper. Res. **42**(6), 1042–1066 (1994)
8. Grama, A., Kumar, V.: State of the art in parallel search techniques for discrete optimization problems. IEEE Trans. Knowl. Data Eng. **11**(1), 28–35 (1999)
9. Jansen, J., Sijstermans, F.: Parallel branch-and-bound algorithms. Futur. Gener. Comput. Syst. **4**(4), 271–279 (1989)
10. Kır, S., Yazgan, H.R., Tüncel, E.: A novel heuristic algorithm for capacitated vehicle routing problem. J. Ind. Eng. Int. **13**(3), 323–330 (2017)
11. Liu, X., Chen, Y.-L., Por, L.Y., Ku, C.S.: A systematic literature review of vehicle routing problems with time windows. Sustainability **15**(15) (2023)
12. Lübbecke, E., Maurer, O., Megow, N., Wiese, A.: A new approach to online scheduling: approximating the optimal competitive ratio. ACM Trans. (2016)
13. Novak, A., Gnatowski, A., Sucha, P.: Urgent samples in clinical laboratories: stochastic batching to minimize patient turnaround time (2025). https://doi.org/10.1007/s10729-026-09756-8
14. Pempera, K., Jaroszczuk, M.: Analysis of the impact of time window lengths on maintenance vehicle routing problem efficiency. In: Advances in Dependable Systems and Networks, pp. 181–190. Springer, Cham (2025)
15. Sajid, M., et al.: A novel algorithm for capacitated vehicle routing problem for smart cities. Symmetry **13**(10) (2021)
16. Shuaibu, A.S., Mahmoud, A.S., Sheltami, T.R.: A review of last-mile delivery optimization: strategies, technologies, drone integration, and future trends. Drones **9**(3) (2025)
17. Toth, P., Vigo, D., Toth, P., Vigo, D.: Vehicle Routing: Problems, Methods, and Applications, 2nd edn., USA (2014)

18. Vandelli, M., Ferrari, F., Dragoni, D.: Parallel splitting method for large-scale quadratic programs (2025). https://doi.org/10.48550/arXiv.2503.16977
19. Vidal, T., Laporte, G., Matl, P.: A concise guide to existing and emerging vehicle routing problem variants. Eur. J. Oper. Res. (2020)

Quantum Circuit Compilation for IQM Spark

Andrzej Gnatowski[(✉)], Aleksander Biskup, Karolina Berniker, Marcel Owczarek, and Eryk Sikora

Wrocław University of Science and Technology, Wybrzeże Wyspiańskiego 27, 50-370 Wrocław, Poland
andrzej.gnatowski@pwr.edu.pl, aleksanderbiskup@icloud.com, eryk_sikora@o2.pl

Abstract. The crucial stage in efficient quantum compilation is the qubit routing process, which involves inserting SWAP operations to adapt quantum circuits to the restricted physical connectivity of a Quantum Processing Unit (QPU). Universal routing algorithms like SABRE are typically general-purpose and often overlook the specific architectural features of individual devices, which may lead to suboptimal results. Motivated by these considerations, we study a specific star topology used, for example, in IQM Spark. We provide problem properties that allow us to reduce the search space, and we illustrate their use in an exact algorithm. The proposed method performs an exhaustive search within a reduced solution space to identify the global minimum for the SWAP gate count. To ensure computational efficiency, the algorithm employs a pruning mechanism that discards decision paths found to be inferior to the currently known optimum. The performance of our approach was evaluated on a dataset of 280 instances. The conducted research demonstrates that the dedicated algorithm enables an observable reduction in SWAP overhead compared to standard compilation tools, thereby leading to a higher fidelity of computations on real hardware.

Keywords: Quantum circuit compilation · Transpilation · Exact algorithm · Routing · Sabre

1 Introduction

In the NISQ (Noisy Intermediate-Scale Quantum) era, quantum computing is becoming a reality. The primary method of formalizing a computational task for a Quantum Processing Unit (QPU) is a quantum circuit. The transition from an abstract circuit to its physical execution is complex, with compilation being one of the fundamental stages. Compilation is the process of transforming an abstract quantum circuit into executable instructions for a specific, imperfect QPU, while simultaneously minimizing the impact of those imperfections. Currently, one of the most widely adopted frameworks is Qiskit [4,6]. However,

W. Zamojski et al. (Eds.): DepCoS-RELCOMEX 2026, LNNS 2014, pp. 142–152, 2026.
https://doi.org/10.1007/978-3-032-27927-9_14

because it is designed to support a wide variety of quantum circuits, its methods are inherently generalized. This generality may potentially lead to a loss of efficiency ("no free lunch").

In this paper, we focus on routing as a vital and extensively researched stage of the compilation process, frequently utilizing the commonly used and efficient SABRE algorithm [3,4,7,18]. Nevertheless, it is a heuristic, which precludes the guaranty of reaching exact solutions. On the other hand, since NISQ-era circuits are necessarily limited in size due to high error rates, employing exact methods becomes computationally feasible. Consequently, this paper introduces an exact routing algorithm tailored to the 5-qubit IQM Spark architecture with it's star topology (i.e., all qubits are connected exclusively to the center qubit). This paper provides the following contributions:

- Formal model of routing and layout compilation stages, with an efficient permutation-based solution representation for star architectures.
- Exact routing algorithm using the introduced solution representation.
- Study on the performance bounds of SABRE algorithm for small instances on IQM Spark architecture.

The remainder of the paper is organized as follows. Section 2 reviews related work. Section 3 formulates the routing problem and introduces the permutation-based representation. Section 4 describes the proposed exact algorithm and its implementation. Section 5 presents the experimental setup and results. Section 6 concludes and discusses future directions.

2 Related Work

Since quantum compilation is crucial, it remains a highly active area of research [4]. Among prevalent methodologies, there are heuristic approaches, frequently developed as extensions of the SABRE [7] algorithm, which is widely recognized as the state-of-the-art baseline for quantum circuit mapping [11,15]. Notable examples include LightSABRE [18] and Route-Forcing [3], the latter of which is designed to handle large-scale systems. On the other hand, exact algorithms provide optimal solutions at the cost of worse scalability. For instance, the exact mapper implemented in the MQT QMAP framework [16] ensures optimality through constraint satisfaction. Similarly, OLSQ2 [10] ensures optimal circuit depth and SWAP count by utilizing a succinct SMT formulation.

The prominence of SABRE-based heuristics in the recent literature does not imply that they are flawless. Despite their efficiency and status as a standard, a critical analysis reveals inherent deficiencies in the original design. These limitations can be grouped into four primary categories, each targeting a specific barrier to efficient quantum synthesis.

Firstly, researchers address the shallow search horizon inherent in greedy heuristics. SABRE's localized approach often prevents it from reaching an optimal solution in complex circuits. Zhou et al. [17] address this by employing Monte Carlo Tree Search (MCTS) to deepen the exploration of potential SWAP sequences. Similarly, Li et al. [9] introduced the FIDLS algorithm, which analyzes combinations of up to k swaps, resulting in a nearly twofold reduction in gate overhead. Furthermore, Ping et al. [15] proposed a multilevel framework (ML-SABRE) that utilizes a coarsening strategy to prevent the algorithm from becoming trapped in local optima during large-scale mapping.

The second area concerns the integration of hardware properties. While the original SABRE algorithm treats hardware connections as uniform, Niu et al. [13] and Murali et al. [12] demonstrate that the incorporation of real-time noise variability drastically improves program success rates. These extensions, such as Variation-Aware SABRE [2], replace static distance metrics with dynamic calibration data to optimize throughput while minimizing error rates.

Another trend in research focuses on temporal aspects and scheduling. Li et al. [8] prove that ignoring single-qubit gate execution times leads to unnecessary resource blocking. Their SQGM solution introduces buffering and progress indicators to synchronize SWAP insertions with actual qubit inactivity. In addition, Cheng et al. [1] presented the Duostra framework, which utilizes an "occupied time" function for more precise management of circuit latency.

Furthermore, considering structural flexibility, Liu et al. [11] identify isolated compilation stages as a source of lost optimization. Their PAM algorithm utilizes permutation-aware synthesis to enable flexible input-output mappings. Complementing this, Itoko et al. [5] leverage gate commutation rules to break away from SABRE's rigid fixed-layer formulation.

3 Problem Formulation

We focus on the layout and routing stages of quantum circuit compilation: assigning logical qubits to physical qubits and inserting SWAP gates to meet QPU connectivity constraints. We consider the 5-qubit IQM Spark QPU with a star topology (with center qubit 3 connected to qubits 1, 2, 4, and 5); however, the approach extends to star architectures with more qubits. We also assume that qubits are identical.

The input is a quantum circuit as a DAG of 2-qubit gates[1]. Gates must execute in an order that respects the DAG, and only one 2-qubit gate can be executed at a time due to the shared center qubit. Each gate requires its logical qubits to be mapped to connected physical qubits. Layouts (logical-to-physical mappings) change between gates via SWAPs on connected pairs. The goal is to find layouts and a gate order that minimizes the total number of SWAPs.

3.1 Mathematical Formulation

Let $Q_p = \{1, 2, 3, 4, 5\}$ be the physical qubits and $Q_l = \{1, 2, 3, 4, 5\}$ the logical qubits. The QPU connectivity is $P = \{\{1, 3\}, \{2, 3\}, \{3, 4\}, \{3, 5\}\}$. At any time, each physical qubit participates in at most one gate, implying at most one 2-qubit gate executes simultaneously. The circuit is a DAG $C = (G, E)$, where

[1] 1-qubit gates can always be executed; k-qubit gates for $k \geq 3$ are decomposed earlier.

$G = \{1, 2, \ldots, n\}$ is the set of n 2-qubit gates. For gate $i \in G$, let $g_i^1, g_i^2 \in Q_l$ be its logical qubits ($g_i^1 \neq g_i^2$). The edge set E contains (i, j) iff gate j must execute directly after i and they share a logical qubit (data dependency).

A solution is a permutation $\pi : G \to G$ (gate execution order) that is a linear extension of the partial order induced by E (i.e., if $(i, j) \in E$, then $\pi^{-1}(i) < \pi^{-1}(j)$), and a sequence of bijective layouts $L_{\pi(1)}, \ldots, L_{\pi(n)} : Q_l \to Q_p$. For each $k = 1, \ldots, n$, let $i = \pi(k)$. Gate i executes under L_i iff $\{L_i(g_i^1), L_i(g_i^2)\} \in P$. The initial layout $L_{\pi(1)}$ is arbitrary. For $k = 1, \ldots, n-1$, transition from $L_{\pi(k)}$ to $L_{\pi(k+1)}$ via a sequence of SWAPs, each on a pair $\{a, b\} \in P$. A SWAP on $\{a, b\}$ updates the layout to L' where $L'(q) = b$ if $L(q) = a$, $L'(q) = a$ if $L(q) = b$, and $L'(q) = L(q)$ otherwise (for $q \in Q_l$). The problem is to find π and $\{L_{\pi(k)}\}_{k=1}^n$ minimizing the total number of SWAPs across all transitions.

3.2 Permutational Solution Representation

A key limitation of the introduced solution representation is its limited usability for metaheuristics and similar optimization methods. Most solutions are infeasible, and many feasible ones yield identical objective values. This results in a sparse and redundant search space. We therefore propose a simpler alternative representation that is far more amenable to local search, genetic algorithms, and other iterative heuristic techniques.

Remark 1. For any layout L and any 2-qubit gate on logical qubits $u, v \in Q_l$ ($u \neq v$), there exists a sequence of at most one SWAP leading to a layout L' such that $\{L'(u), L'(v)\} \in P$.

Proof. Let c be the center physical qubit. If $L(u) = c$ or $L(v) = c$, then $\{L(u), L(v)\} \in P$, since all peripheral qubits connect to c. In this case, no SWAP is needed ($L' = L$). Otherwise, both $L(u)$ and $L(v)$ are peripheral. Let $w \in Q_l$ such that $L(w) = c$ and $w \neq u, v$. Perform a SWAP on physical pair $\{c, L(u)\} \in P$. This yields L' where $L'(u) = c$, $L'(w) = L(u)$, and $L'(q) = L(q)$ for $q \neq u, w$. Thus, $\{L'(u), L'(v)\} = \{c, L(v)\} \in P$.

Following Remark 1, it is easy to observe that the number of SWAPs required to transition between layouts depends only on whether the center logical qubit changes. Since qubits are identical and peripheral positions do not affect gate executability, we can track only the logical qubit at the center for each layout. Thus, a solution simplifies to a gate order π and a sequence $L^c = (c_{\pi(1)}, \ldots, c_{\pi(n)})$, where $c_i \in \{g_i^1, g_i^2\}$ is the logical qubit at the center during gate i's execution. Now, the objective function of a feasible solution can be given by

$$f(\pi, L^c) = \sum_{k=1}^{n-1} \mathbb{1}_{c_{\pi(k)} \neq c_{\pi(k+1)}}, \tag{1}$$

which counts the number of center changes, each requiring one SWAP. The problem is to minimize $f(\pi, L^c)$ over feasible π and L^c.

Now, another step to simplify the solution representation is to formulate a method of computing optimal L^c for any feasible π. A direct enumeration could be done in $\Theta(2^n)$ time on a RAM (Random Access Memory) machine, which is computationally prohibitive. However, there exists a much more efficient method. Let $c_{\pi(1)}$ be given. Then, any following center is chosen in the following way: if the center can remain unchanged, it is unchanged. If it must be changed, a lookahead method is used: that is, both options are explored, and the option with more consecutive gates to be performed without a change in center is chosen. Formally, for $k = 1, \ldots, n-1$, let $U = \{g^1_{\pi(k+1)}, g^2_{\pi(k+1)}\}$. Define the gate streak length $s(q) = \max\{m \geq 1 : \forall j = 0, \ldots, m-1, q \in \{g^1_{\pi(k+1+j)}, g^2_{\pi(k+1+j)}\}\}$ for $q \in Q_l$. Then,

$$c_{\pi(k+1)} = \begin{cases} c_{\pi(k)} & \text{if } c_{\pi(k)} \in U, \\ \arg\max_{q \in U} s(q) & \text{otherwise.} \end{cases} \tag{2}$$

It is easy to observe that such a procedure yields, for a given π, an optimal L^c.

Remark 2. For any feasible π, an optimal L^c (minimizing $f(\pi, L^c)$), as well as the objective function value, can be computed in $\Theta(n)$ time on a RAM machine.

Proof. Trivial: compute L^c using the above procedure (2), then evaluate (1).

This leads to an efficient representation of the solution as just π, with the constraints being partial ordering satisfaction; and the objective function computable in $\Theta(n)$ time on RAM machine.

4 Algorithm

The proposed solving algorithm relies directly on the permutational solution representation introduced in Sect. 3. Its idea can be summarized in Alg. 1. The method exhaustively enumerates all valid gate orders while efficiently computing the minimal SWAPs for each, ensuring an optimal solution for small circuits.

Algorithm 1 Routing algorithm

1: Load the quantum circuit from OpenQASM.
2: Simplify the circuit.
3: Extract the DAG $C = (G, E)$ from the simplified circuit.
4: For C, generate all feasible topological orders Π.
5: **for** $\pi \in \Pi$ **do**
6:     Compute a corresponding optimal L^c using Eq. (2).
7:     Evaluate the objective function $f(\pi, L^c)$ using Eq. (1).
8: **end for**
9: Select the π and corresponding L^c with the minimum f.

The implementation of the proposed algorithm is integrated into the Qiskit framework as a custom transpiler pass. To balance flexibility with performance, a hybrid architecture is employed: the management layer (Python) handles data orchestration (Lines 1–4 and 9), while the computational core (C++) performs the computationally intensive objective function evaluation (Lines 5–8).

The process begins with circuit ingestion and structure optimization (1–3). The input OpenQASM circuit is loaded and converted into a Directed Acyclic Graph (DAG) representation using Qiskit. An important component of the implementation is the preprocessing stage outlined in Line 2. To reduce the search space for the subsequent routing steps, the DAG is analyzed to aggregate sequential gates acting on the same qubits. This operation preserves unitary equivalence while significantly reducing the depth of the circuit. The reduction algorithm iterates through two-qubit gate nodes in topological order. If both incoming edges originate from the same predecessor node (excluding input/output nodes), the gates are candidates for aggregation. This procedure is repeated iteratively until the circuit structure stabilizes. Minimizing the number of input gates directly translates to fewer permutations generated in Line 4.

Following preprocessing, the Python layer generates valid topological orders (Π) and converts them into a lightweight data structure for transfer. The execution then transitions to the C++ module via the `pybind11`[2] interface to minimize data transfer overhead. The core loop (Lines 5–8) is executed natively, where the cost function f and the optimal layout L^c are computed for each permutation. Finally, the metrics are returned to the management layer, which identifies the optimal permutation (Line 9).

5 Experiments

The experiments aim to validate the proposed algorithm and determine the instance sizes it can solve in an acceptable time. Additionally, we investigate whether the SABRE algorithm yields suboptimal results even for small instances, as our approach provides optimal solutions.

The section is organized as follows. First, the instance generator used to create the benchmarks is described. Then, the experimental setup is outlined, including the benchmarks and evaluation procedure. Finally, the experimental results are presented and discussed.

5.1 Instance Generator

Standard benchmark suites (e.g., QAOA, QFT) may exhibit structural regularities that could lead to easier or harder than expected problem instances. To address this, we developed a custom generator that produces instances specifically designed to stress-test the routing phase on the target star architecture.

The generator constructs circuits iteratively, layer by layer, based on a user-defined depth. Routing difficulty is controlled by the interaction scope parameter:

[2] https://github.com/pybind/pybind11.

1. *Native*: restricts interactions to physically connected pairs. Used primarily for pipeline validation (sanity check); easiest.
2. *Uniform*: samples qubit pairs uniformly; average.
3. *Adversarial*: samples from qubits without a direct connection; hard.

The final steps utilize the density parameter to control parallelism by determining the probability of inserting a second, disjoint two-qubit gate within the same layer (i.e. sampled from the remaining qubits). This approach enables the creation of a benchmark suite with tunable complexity.

5.2 Experimental Setup

The evaluation was performed on a dataset of 280 synthetic circuit instances generated according to the procedure defined in Subsect. 5.1. We fixed the circuit width to 5 and restricted the gate set to CZ (native to IQM Spark). The benchmark suite spans four depth levels: $\{8, 12, 16, 20\}$. For the `uniform` and `adversarial` types, the density parameter was in $\{0.0, 0.5, 1.0\}$. For the `native` type, density was restricted to 0.0, as all physical connections share the central qubit. Each configuration was replicated with 10 unique random seeds, resulting in a total of 280 instances (7 configs $\times$ 4 depths $\times$ 10 seeds). We benchmarked the proposed exact algorithm against Qiskit's SABRE. SABRE was configured with the `decay` heuristic and executed with 20 independent trials per instance; the minimum SWAP count was recorded. The primary performance metrics were the number of inserted SWAP gates and the total compilation time.

5.3 Results

This subsection presents the empirical findings from the evaluation. First, we analyze the computational scalability (runtime) of the proposed algorithm compared to the heuristic baseline. Subsequently, we assess the routing quality (SWAP overhead), discussing the specific impact of circuit depth, density, and interaction locality on performance.

For `native` instances, the exact algorithm outperforms SABRE performance-wise by approximately two orders of magnitude, achieving an execution time $\approx 0.2ms$ compared to SABRE execution time $\approx 10ms$. Both algorithms solved the instances optimally.

Computation times of both tested algorithms are shown in Fig. 1 (`uniform` instances) and in Fig. 2 (`adversarial` instances). A steep increase in execution time is evident for the proposed algorithm, especially for `uniform` instances characterized by a heavy-tailed distribution attributable to "wide" DAGs (high parallelism). These structures may admit more valid topological orderings (possibly $\Omega(2^n)$), forcing the exact algorithm to traverse a large search space to guaranty optimality. However, the structures of DAGs may also lead to a low number of feasible solutions, resulting in a wide range of execution times, e.g. from milliseconds to hundreds of seconds for a depth of 20 and `uniform`. On

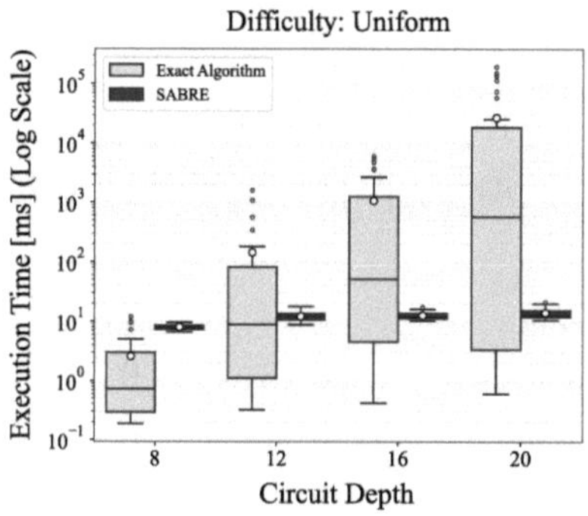

Fig. 1. Computational time for `uniform` instances

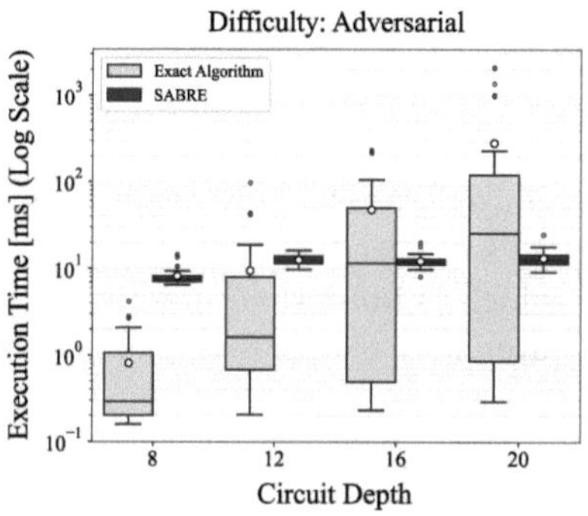

Fig. 2. Computational time for `adversarial` instances

the other hand, the execution time of SABRE remained mostly unaffected by instance parameters.

Table 1 provides a summary of the routing performance of the proposed exact algorithm and the SABRE baseline across the tested instances. The column "SABRE Non-optimal" indicates the percentage of instances where the exact algorithm outperformed the heuristic. "SABRE Gap" shows aggregated statistics on the difference between the SABRE's and optimal solutions (i.e., mean, median, min, max). "Optimal SWAPs" reports the number of SWAP gates in the optimal solutions (mean, median, min, max).

The `uniform` instances proved unexpectedly challenging for the heuristic, with the proposed algorithm achieving reductions of up to 4 SWAP gates and a visibly lower distribution of SWAP counts compared to the SABRE baseline. The exact algorithm outperforms SABRE in 70% of cases at depth 16 and 56.7% at depth 20. In the `adversarial` category, which is designed to be more challenging, the exact algorithm outperforms SABRE in up to 40% of cases at depth 20. In general, as circuit depth increases from 8 to 20, both the gap to optimality and the percentage of non-optimal SABRE solutions rise. This suggests that an exact algorithm could be preferred over SABRE for small—but relevant—instances.

Table 1. Routing performance comparison, excluding trivial `native` instances

Difficulty	Depth	SABRE non-optimal [%]	SABRE Gap Mean	Med	Min	Max	Optimal SWAPs Mean	Med	Min	Max
`adversarial`	8	20.0%	0.20	0.0	0	1	4.83	4.0	2	8
	12	10.0%	0.13	0.0	0	2	6.83	6.5	3	12
	16	20.0%	0.23	0.0	0	2	8.77	9.0	4	13
	20	40.0%	0.57	0.0	0	2	10.23	11.0	5	16
`uniform`	8	20.0%	0.20	0.0	0	1	4.17	4.0	2	7
	12	43.3%	0.57	0.0	0	2	5.67	6.0	3	10
	16	70.0%	1.10	1.0	0	4	7.10	7.5	3	12
	20	56.7%	1.10	1.0	0	4	8.83	9.0	5	13

While the considered instances are relatively small, they are of practical size in the NISQ era. The 16 SWAPs required for some depth 20 instances translate into 48 native 2-qubit gates, which is borderline practical for ~99% 2-qubit fidelity on the IQM Spark hardware.

6 Conclusions

This paper presents an exact routing algorithm for the IQM Spark 5-qubit star architecture based on a permutation representation of gate orders (π) with greedy lookahead for optimal center-qubit sequences (L^c). The approach illustrates the efficiency of the proposed permutational formulation (with $\Theta(n)$ objective function evaluation) and serves as a benchmark for heuristic quality.

Experiments on relatively small, 280 synthetic instances showed that the exact method achieves acceptable runtimes for some instances but scales exponentially for deeper DAGs due to the number of feasible permutations. SABRE was near-optimal in the shallowest instances but suboptimal in over 50% of the larger ones, with mean gaps of ~1.1 SWAPs (max 4). Thus, for small instances, a dedicated exact algorithm is preferable to SABRE.

Future work includes generalizing to arbitrary star architectures (proving the at-most-one-SWAP property) and applying the π-based representation in metaheuristics (e.g., local search, genetic algorithms) for larger instances. Since many metaheuristics perform well on permutation encodings and benefit from linear-time objective evaluation, this is a promising direction. A natural next step is to develop a custom neighborhood that preserves feasibility or provides objective guaranties, similar to block properties in scheduling [14].

References

1. Cheng, C.Y., Yang, C.Y., Kuo, Y.H., Wang, R.C., Cheng, H.C., Huang, C.Y.R.: Robust Qubit Mapping Algorithm via Double-Source Optimal Routing on Large Quantum Circuits. ACM Trans. Quantum Comput. **5**(3), 1–26 (2024). https://doi.org/10.1145/3680291
2. Das, P., Tannu, S.S., Nair, P.J., Qureshi, M.: A case for multi-programming quantum computers. In: Proceedings of the 52nd Annual IEEE/ACM International Symposium on Microarchitecture, pp. 291–303. ACM, Columbus OH USA (Oct 2019). https://doi.org/10.1145/3352460.3358287
3. Escofet, P., Gonzalvo, A., Alarcón, E., Almudéver, C.G., Abadal, S.: Route-forcing: scalable quantum circuit mapping for scalable quantum computing architectures. In: 2024 IEEE International Conference on Quantum Computing and Engineering (QCE), vol. 01, pp. 909–920 (2024). https://doi.org/10.1109/QCE60285.2024.00110
4. Gnatowski, A., Niżyński, T.: Quantum circuits compilation for superconducting quantum processing units: a systematic mapping study. In: Proceedings of the Fifth International Conference on Intelligent Systems in Production Engineering and Maintenance ISPEM 2025, pp. 269–284. Springer Nature Switzerland, Cham (2026). https://doi.org/10.1007/978-3-032-14163-7_19
5. Itoko, T., Raymond, R., Imamichi, T., Matsuo, A., Cross, A.W.: Quantum circuit compilers using gate commutation rules. In: Proceedings of the 24th Asia and South Pacific Design Automation Conference, pp. 191–196. ACM, Tokyo Japan (Jan 2019). https://doi.org/10.1145/3287624.3287701
6. Javadi-Abhari, A., et al.: Quantum computing with qiskit (2024). https://doi.org/10.48550/arXiv.2405.08810
7. Li, G., Ding, Y., Xie, Y.: Tackling the qubit mapping problem for nisq-era quantum devices. In: Proceedings of the Twenty-Fourth International Conference on Architectural Support for Programming Languages and Operating Systems, ASPLOS 2019, pp. 1001–1014. Association for Computing Machinery, New York (2019). https://doi.org/10.1145/3297858.3304023
8. Li, S., Nguyen, K.D., Clare, Z., Feng, Y.: Single-qubit gates matter for optimising quantum circuit depth in qubit mapping. In: 2023 IEEE/ACM International Conference on Computer Aided Design (ICCAD), pp. 1–9. IEEE, San Francisco, CA, USA (Oct 2023). https://doi.org/10.1109/ICCAD57390.2023.10323863
9. Li, S., Zhou, X., Feng, Y.: Qubit Mapping Based on Subgraph Isomorphism and Filtered Depth-Limited Search (Sep 2021). https://doi.org/10.48550/arXiv.2004.07138
10. Lin, W.H., Kimko, J., Tan, B., Bjørner, N., Cong, J.: Scalable optimal layout synthesis for nisq quantum processors. In: 2023 60th ACM/IEEE Design Automation Conference (DAC), pp. 1–6 (2023). https://doi.org/10.1109/DAC56929.2023.10247760
11. Liu, J., Younis, E., Weiden, M., Hovland, P., Kubiatowicz, J., Iancu, C.: Tackling the qubit mapping problem with permutation-aware synthesis. In: 2023 IEEE International Conference on Quantum Computing and Engineering (QCE), pp. 745–756. IEEE, Bellevue, WA, USA (Sep 2023). https://doi.org/10.1109/QCE57702.2023.00090
12. Murali, P., Baker, J.M., Abhari, A.J., Chong, F.T., Martonosi, M.: Noise-Adaptive Compiler Mappings for Noisy Intermediate-Scale Quantum Computers (Jan 2019). https://doi.org/10.48550/arXiv.1901.11054
13. Niu, S., Suau, A., Staffelbach, G., Todri-Sanial, A.: A Hardware-Aware Heuristic for the Qubit Mapping Problem in the NISQ Era. IEEE Trans. Quantum Eng. **1**, 1–14 (2020). https://doi.org/10.1109/TQE.2020.3026544

14. Nowicki, E., Smutnicki, C.: A fast tabu search algorithm for the permutation flow-shop problem. Eur. J. Oper. Res. **91**(1), 160–175 (1996). https://doi.org/10.1016/0377-2217(95)00037-2
15. Ping, S., Sathishkumar, N., Lin, W.H., Wang, H., Cong, J.: A High-Performance Multilevel Framework for Quantum Layout Synthesis (May 2025). https://doi.org/10.48550/arXiv.2505.24169
16. Wille, R., Burgholzer, L.: Mqt qmap: efficient quantum circuit mapping. In: Proceedings of the 2023 International Symposium on Physical Design, ISPD 2023, pp. 198–204. ACM (Mar 2023). https://doi.org/10.1145/3569052.3578928
17. Zhou, X., Feng, Y., Li, S.: Quantum Circuit Transformation: A Monte Carlo Tree Search Framework (Jan 2022). https://doi.org/10.48550/arXiv.2008.09331
18. Zou, H., Treinish, M., Hartman, K., Ivrii, A., Lishman, J.: LightSABRE: A lightweight and enhanced sabre algorithm (2024). https://doi.org/10.48550/arXiv.2409.08368

Comparative Evaluation of Text Detection and Extraction Methods for Use in Comics

Hedi Hassan and Marek Woda(✉)

Wrocław University of Science and Technology, Faculty of Information and Communication Technology, Wrocław 50-370, Poland
hedi.hassan28@gmail.com, marek.woda@pwr.edu.pl

Abstract. The precision of digital archives and accessibility tools for graphic narratives depends on the successful identification and extraction of text, but traditional OCR systems fail to read their complex visual structures. The combination of nonstandard page organization, vertical writing, and decorative font styles in comics and manga makes traditional OCR systems unable to recognize their content. This study evaluates three OCR engines, PaddleOCR, EasyOCR, and AzureOCR, on the Japanese Manga109 benchmark data set. Performance was assessed using six metrics, including F1-score, Character Error Rate (CER), and processing time. The cloud-based AzureOCR system produced the most stable results among all tested engines, as evidenced by an F1-score of 0.277 and a CER of 49.1%. The open-source PaddleOCR and EasyOCR systems performed poorly when dealing with vertical text and decorative elements. The results indicate that general-purpose OCR systems remain inadequate for complex comic layouts, highlighting the need for layout-aware and domain-adapted models.

Keywords: Text Recognition · Comic Books · Comic Image Processing

1 Introduction

Comics and manga combine visual storytelling with complex layout structures, including irregular panel arrangements, vertical text, decorative typography, and multilingual content. These characteristics violate the structural assumptions of traditional optical character recognition (OCR) systems, which are typically trained on well-structured horizontally aligned documents [8,19]. As digital manga archives continue to expand, accurate and scalable OCR solutions are required for preservation, translation, and accessibility purposes [6].

Current OCR models do not recognize comics effectively because they handle complex visual and structural elements in these documents [3]. This study evaluates three widely used OCR systems, PaddleOCR, EasyOCR, and Azure-OCR on the Manga109 benchmark dataset. Performance is assessed using six metrics: Precision, Recall, F1-score, Character Error Rate (CER), Intersection

W. Zamojski et al. (Eds.): DepCoS-RELCOMEX 2026, LNNS 2014, pp. 153–162, 2026.
https://doi.org/10.1007/978-3-032-27927-9_15

over Union (IoU), and processing time [1,15]. In addition to quantitative benchmarking, qualitative error analysis is conducted to identify systematic failure patterns in comic-specific text recognition.

2 Related Work

The detection and recognition of text in comics remains challenging due to irregular layouts, decorative fonts, vertical scripts, and integration of text within illustrations. Early rule-based approaches, such as connected component analysis and speech balloon detection [9,18], proved insufficient for such complex documents. Traditional OCR systems designed for structured text documents also struggle in this domain [13]. Although deep learning methods like CRAFT, YOLO, and Faster R-CNN have improved scene text detection [5,17], they remain limited when applied to manga layouts. The unique combination of decorative fonts and vertically arranged text and illustration integration in manga comics creates major obstacles for both detection and recognition tasks [8,14]. The evaluation of OCR models depends on three main datasets which include Manga109 [15], eBDtheque [7], and COMICS [10] yet these datasets fail to represent the complete range of global comic styles and languages [12,14].

Current research leaves several problems unaddressed; specifically, models struggle to understand text-artwork interactions and identify vertical or multilingual scripts. The current models lack ability to understand how text elements interact with artwork elements and they struggle to identify text written in vertical scripts and multiple languages [2,21].

The development of real-time OCR applications has not kept pace with growing requirements for scalable translation and preservation and accessibility systems [20].

3 Methodology

3.1 Dataset and Preprocessing

OCR performance was evaluated on the Manga109 dataset, a widely used benchmark for Japanese manga [1,15]. It contains 109 books with diverse artistic styles, page layouts, vertical text, non-standard fonts, and speech bubbles integrated within illustrations, all of which present significant challenges for text detection and recognition. For this study, 1,053 pages from the first 11 manga titles were selected, representing the Manga109 collection. OCR models were evaluated in their default configurations without fine-tuning, so results reflect out-of-the-box performance on manga data, including bounding box coordinates $(x_{min}, y_{min}, x_{max}, y_{max})$ for ground-truth comparison. XML annotations were converted to plain text to enable direct evaluation of OCR outputs. No manual editing or image preprocessing (e.g., denoising, binarization) was applied, preserving the full visual complexity of the pages. The evaluation focused on out-of-the-box OCR performance without any fine-tuning, model adaptation, or data augmentation. Consequently, the results provide insights into default model capabilities but may not generalize to the entire Manga109 collection, Western comics, or multilingual graphic narratives. Future research should include a broader dataset and investigate domain-specific fine-tuning and layout-aware adaptations to enhance OCR robustness and applicability.

3.2 OCR Models

The evaluation assessed three OCR systems because they represent common applications and demonstrate different technical approaches and support Japanese language functionality.

- **PaddleOCR** [4]: functions as an open-source modular framework which unites Differentiable Binarization (DB) for detection with CRNN or SVTR for recognition. The system operates with more than 80 languages and provides optimal performance in different visual environments.
- **EasyOCR** [11]: A lightweight PyTorch-based library designed for fast deployment. The system enables users to select from more than eighty languages through an easy setup procedure.
- **AzureOCR** [16]: The commercial cloud-based OCR solution from Microsoft Azure provides high-accuracy text recognition for printed and handwritten documents in 100+ languages through API integration.

All models were evaluated using their default pretrained configurations to reflect real-world, out-of-the-box usability. The local execution of PaddleOCR and EasyOCR took place on a GPU-enabled system (NVIDIA RTX 3060) but AzureOCR operated through REST API access.

Limitations: The default pretrained configurations enable fair model comparison yet they might not demonstrate the complete capabilities of these models when working with manga data. Effective recognition of comics requires training strategies specifically adapted to vertically arranged and artistically stylized text.

3.3 Evaluation Pipeline

The development of a single evaluation system aimed to maintain uniform results between all OCR systems. The pipeline consisted of the following stages.

1. **Image Loading:** Manga page images were loaded and processed in their original resolution and color format.
2. **Model Inference:** The models performed text detection and recognition tasks on the same image collections through controlled testing environments.
3. **Output Formatting:** The system combined detected text instances with their bounding boxes into a single format which used text<TAB>xmin,ymin,xmax,ymax to match the ground truth data.

4. **Prediction Matching:** OCR predictions were matched to ground-truth using an IoU threshold of 0.3, with successful matches compared at the character level.

This modular pipeline can benchmark other OCR systems or comic datasets, following the full workflow in Fig. 1.

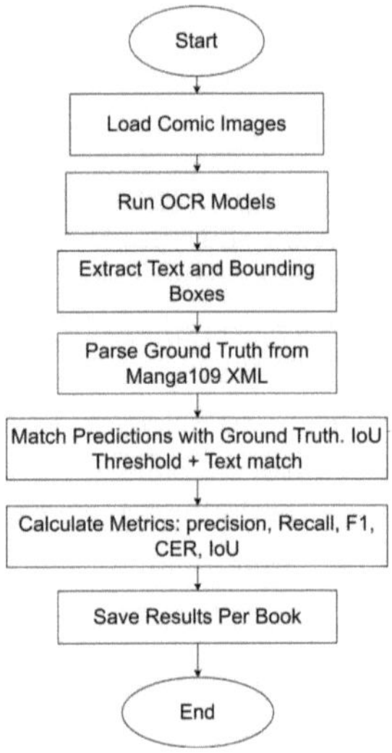

Fig. 1. Flowchart of the OCR evaluation pipeline.

3.4 Evaluation Metrics

Model performance was assessed using six metrics:

- **Precision, Recall, F1-Score:** Measures the quality of text detection based on overlap and accuracy of predicted bounding boxes.
- **Character Error Rate (CER):** Captures text recognition accuracy via normalized Levenshtein distance.
- **Average Intersection over Union (IoU):** Assesses spatial accuracy of bounding box predictions.
- **Processing Time per Image:** Evaluates runtime performance and suitability for real-time or batch deployment.

3.5 Implementation Environment

Experiments were implemented in Python 3.9. The system used PaddleOCR and EasyOCR through local GPU acceleration by running PyTorch and OpenCV together with their necessary libraries. The AzureOCR service operated through Microsoft Computer Vision API under a student-tier plan which restricted API requests to 10 calls per second.

4 Experimental Results and Discussion

This section presents the evaluation results of AzureOCR, EasyOCR, and PaddleOCR on 1,053 manga pages from the Manga109 dataset. The evaluation employed six metrics: Precision, Recall, F1-score, Character Error Rate (CER), Average Intersection over Union (IoU), and processing time per image.

4.1 Quantitative Evaluation

As shown in Table 1, AzureOCR achieved the highest F1-score 0.277 and lowest CER 49.1%, demonstrating the best balance between detection and recognition accuracy. EasyOCR performed moderately with a CER of 52.1% and lower IoU (0.452), while PaddleOCR scored the lowest across all metrics, confirming its difficulty with vertical and stylized text. Although its Character Error Rate (CER) is high, it provides a lower-bound reference for performance on this exceptionally complex document domain. EasyOCR achieved average results, showing a decent CER of **52.1%**, indicating its recognition module performs moderately well when text is successfully detected. However, its lower Intersection over Union (IoU) score of 0.452 suggests that while it recognizes characters, its bounding box predictions are less spatially accurate than AzureOCR's. Conversely, PaddleOCR produced the worst performance across all evaluation criteria, scoring an F1-score of just **0.047**. This confirms its difficulty in handling the non-standard structures of manga text.

Table 1. Performance Metrics of OCR Models on Manga109

Model	Precision	Recall	F1	CER (%)	IoU	Time (s)
AzureOCR	0.185	0.545	0.277	49.1	0.513	1.902
EasyOCR	0.142	0.490	0.219	52.1	0.452	1.059
PaddleOCR	0.044	0.050	0.047	83.5	0.231	0.172

Figure 2 visualizes the comparison across the main evaluation metrics. AzureOCR consistently outperformed the others, though it had the longest processing time due to its cloud-based architecture.

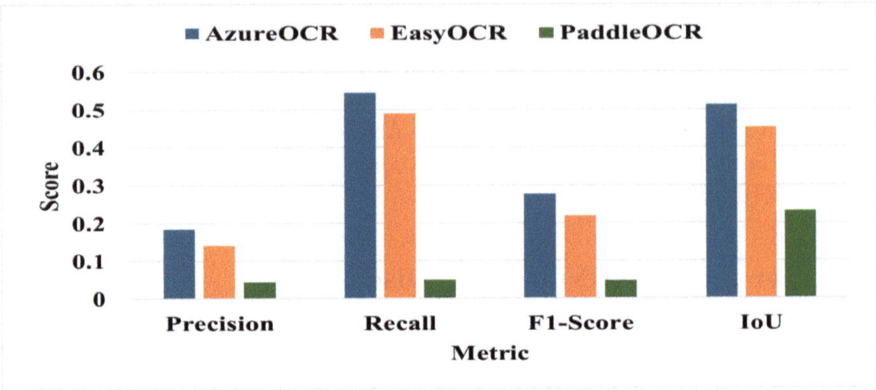

Fig. 2. Performance comparison of OCR models across evaluation metrics.

4.2 Qualitative Analysis and Common OCR Failure Patterns

Qualitative evaluation supports the quantitative results. AzureOCR performed best in layout-aware detection, handling dense text and vertical Japanese script, but with the longest processing time due to its cloud-based architecture. EasyOCR performed moderately but often misaligned boxes or split lines near panel borders. PaddleOCR had the most severe failures, missing stylized or vertical text (Fig. 3), reflected in its low F1-score (0.047) and IoU (0.231). Even AzureOCR has limitations, emphasizing the need for comic-specific OCR.

Common Failure Patterns

1. **Vertical Text Detection:** PaddleOCR struggles with vertical Japanese script because its DB detection algorithm is optimized for horizontal text, leading to broken or misaligned boxes.
2. **Stylized Font Recognition:** All models, including AzureOCR, often fail to detect highly decorative or occluded text. Recognition networks trained on clean digital fonts cannot reliably map these glyphs to Unicode.
3. **Layout Segmentation Errors:** Bounding boxes overlapping decorative elements (speed lines, shadows) cause partial or incorrect character recognition due to multi-stage pipeline noise.

General-purpose OCR models require layout-aware and language-adaptive improvements to achieve reliable recognition for manga and other graphic narratives.

Figure 4 shows a side-by-side comparison of OCR outputs, which provides visual validation for the results in Table 1. AzureOCR delivers the most precise text detection results because it detects both vertical and dense text areas with excellent layout analysis performance. EasyOCR shows average performance but it frequently fails to detect text and it splits extended lines and places bounding boxes in incorrect positions. PaddleOCR produced the sparsest and most fragmented output, failing to detect heavily stylized or decorative text elements.

Fig. 3. Failure case from PaddleOCR on Manga109 dataset: missed vertical speech balloon and misaligned detection boxes [1].

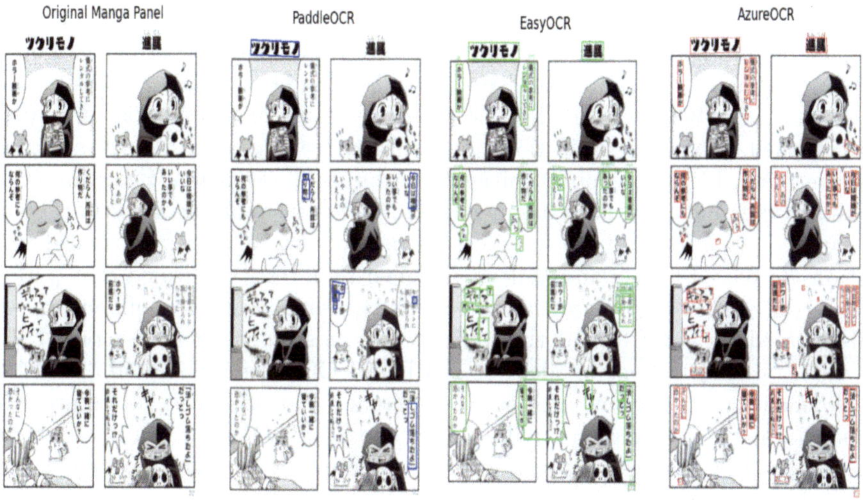

Fig. 4. Side-by-side OCR results on Manga109 dataset: AzureOCR is dense and accurate; EasyOCR is moderate; PaddleOCR struggles with vertical and stylized text [1].

Table 2 shows complete OCR results for five example text samples. AzureOCR successfully identified all text strings which proved its ability to handle both vertical and stylized written content. The two models EasyOCR and PaddleOCR failed to detect two different instances but PaddleOCR correctly identified two instances from the total. The models demonstrated varying levels of recognition precision in their operations.

Table 2. Visual comparison of OCR results for five strings in a single manga panel using PaddleOCR, EasyOCR, and AzureOCR

Text	AzureOCR	EasyOCR	PaddleOCR
ツクリモノ	✓	✓	✓
展	✓	×	×
ホラ映画か	✓	✓	×
式の参考に	✓	×	×
今日は嫌が	✓	✓	✓

4.3 Performance Insights

AzureOCR achieved the highest recall and IoU, accurately detecting text and maintaining layout, though at the cost of longer cloud-based processing times. EasyOCR performed moderately, with acceptable F1-score and CER, but misaligned bounding boxes reduced spatial accuracy. PaddleOCR performed poorly across all metrics, struggling with vertical and stylized text. While this evaluation used out-of-the-box OCR models, no domain-specific fine-tuning or adaptation was applied. Observed limitations, such as misreading vertical text and decorative fonts, highlight opportunities for future improvements. Fine-tuning on comic-specific datasets, data augmentation, or layout-aware architectures could enhance recognition and address current methodological gaps. Statistical validation via paired t-tests and Wilcoxon signed-rank tests confirmed that performance differences were significant ($p < 0.001$). Per-book 95% confidence intervals showed AzureOCR had the highest and most consistent F1-score (0.277 ± 0.037), EasyOCR was moderate, and PaddleOCR remained weak. Practical deployment considerations include cloud latency and API costs for AzureOCR versus local GPU execution for EasyOCR and PaddleOCR, which is faster and free of per-call charges but less accurate. These factors inform system choice depending on operational constraints.

5 Conclusion and Future Work

This study presents a systematic diagnostic evaluation of three widely used OCR systems, AzureOCR, EasyOCR, and PaddleOCR on 1,053 pages from the Manga109 dataset. AzureOCR achieved the strongest overall performance, particularly in detecting vertical and densely arranged Japanese text. However, all evaluated systems exhibited substantial limitations when processing stylized typography and complex layouts.

The findings confirm that general-purpose OCR solutions remain insufficient for manga digitization without domain adaptation. Future research will focus on manga-specific fine-tuning, layout-aware detection models, multilingual transferability, and statistical validation of performance differences. While this study focuses on a comparative evaluation of existing OCR systems, it does not introduce new architectures or perform domain-specific fine-tuning. The results highlight the performance gap of general-purpose OCR engines on manga content, providing a clear motivation for future work in model adaptation, transfer learning, and fine-tuning with vertically oriented and stylized text. Incorporating data augmentation or specialized pretraining could further improve recognition accuracy and robustness for comics and other graphic narratives.

5.1 Practical Applications and Accessibility Benefits

The visual complexity of comics makes it difficult for readers who are visually impaired and language learners to understand, but new OCR technology enables better integration of screen readers and audio narration systems for enhanced reading accessibility. The technology enables real-world applications through mobile apps which translate manga content by using AzureOCR to extract text for machine translation. The system allows educational institutions to perform immediate translation through text simplification and real-time overlay translation. The advanced OCR technology enables digital libraries to preserve comics through their precise indexing systems. As digital comics grow, layout-aware OCR systems will be essential for broader access and cultural preservation.

Acknowledgments. The authors would like to thank the Faculty of Information and Communication Technology at Wrocław University of Science and Technology for their support. The conference fees were kindly waived by the Institute.

Disclosure of Interests. The authors have no competing interests to declare that are relevant to the content of this article.

References

1. Aizawa, K., et al.: Building a manga dataset "manga109" with annotations for multimedia applications. IEEE Multimedia **27**, 8–18 (2020). https://doi.org/10.1109/MMUL.2020.2987895
2. Bhunia, S., Sharma, R., Singh, K.: Text detection and extraction in comics: challenges and advances. IEEE Trans. Image Process. **30**, 215–228 (2021)
3. Chen, Y.C., Jhala, A.: A computational model of comprehension in manga style visual narratives. In: Proceedings of the 43rd Annual Meeting of the Cognitive Science Society, vol. 43, pp. 1153–1158. Cognitive Science Society (2021)
4. Du, Y., Chen, M., Cui, C., Zhang, R., Wang, W., Li, H.: Pp-ocr: a practical ultra lightweight ocr system. Preprint arXiv:2009.09941, arXiv (2020), arXiv preprint
5. Dubray, D., Laubrock, J.: Deep cnn-based speech balloon detection and segmentation for comic books. In: Proceedings of the 5th IAPR International Conference on Document Analysis and Recognition (ICDAR) Workshops, pp. 1237–1243. IEEE, Sydney, Australia (2019)
6. Dutta, A., Biswas, S., Das, A.K.: Emocomicnet: a multi-task model for comic emotion recognition. Pattern Recogn. **150**, 110261 (2024)

7. Guérin, C., et al.: ebdthèque: a representative database of comics. In: Proceedings of the 12th International Conference on Document Analysis and Recognition (ICDAR), pp. 1145–1149. IEEE, Washington, DC, USA (2013)
8. Hartel, R., Dunst, A.: An ocr pipeline and semantic text analysis for comics. In: Bimbo, A.D., et al. (eds.) Pattern Recognition. ICPR International Workshops and Challenges. LNCS, vol. 12663, pp. 213–222. Springer International Publishing, Cham (2021). https://doi.org/10.1007/978-3-030-68780-9_19
9. Ho, A.K.N., Burie, J.C., Ogier, J.M.: Panel and speech balloon extraction from comic books. In: Proceedings of the 10th IAPR International Workshop on Document Analysis Systems (DAS), pp. 424–428. IEEE (2012). https://doi.org/10.1109/DAS.2012.66
10. Iyyer, M., Manjunatha, V., Guha, A., Vyas, Y., Boyd-Graber, J.L., III, H.D., Davis, L.S.: The amazing mysteries of the gutter: drawing inferences between panels in comic book narratives. In: Proceedings of the IEEE Conference on Computer Vision and Pattern Recognition (CVPR), pp. 6478–6487. IEEE, Honolulu, HI, USA (2017)
11. Jaided AI: Easyocr: Ready-to-use ocr with 80+ supported languages (2020). https://github.com/JaidedAI/EasyOCR
12. Jiang, T., Kise, K.: Multilingual ocr for comics: challenges and solutions. In: Proceedings of the IEEE Conference on Computer Vision, pp. 456–468. IEEE (2023)
13. Jung, K., Kim, K.I., Jain, A.K.: Text information extraction in images and video: a survey. Pattern Recogn. **37**(5), 977–997 (2004)
14. Laubrock, J., Dunst, A.: Computational approaches to comics analysis. Top. Cogn. Sci. **12**(1), 274–310 (2020)
15. Matsui, Y., et al.: Sketch-based manga retrieval using manga109 dataset. Multimedia Tools Appli. **76**(20), 21811–21838 (2016). https://doi.org/10.1007/s11042-016-4020-z
16. Microsoft Corporation: Azure cognitive services – optical character recognition (ocr) (2021). https://azure.microsoft.com/en-us/services/cognitive-services/computer-vision/
17. Kompatsiaris, I., Huet, B., Mezaris, V., Gurrin, C., Cheng, W.-H., Vrochidis, S. (eds.): MMM 2019. LNCS, vol. 11296. Springer, Cham (2019). https://doi.org/10.1007/978-3-030-05716-9
18. Rigaud, C., Tsopze, N., Burie, J.-C., Ogier, J.-M.: Robust frame and text extraction from comic books. In: Kwon, Y.-B., Ogier, J.-M. (eds.) GREC 2011. LNCS, vol. 7423, pp. 129–138. Springer, Heidelberg (2013). https://doi.org/10.1007/978-3-642-36824-0_13
19. Soykan, G., Yuret, D., Sezgin, T.M.: A comprehensive gold standard and benchmark for comics text detection and recognition. In: Mouchère, H., Zhu, A. (eds.) Document Analysis and Recognition - ICDAR 2024 Workshops, pp. 168–197. Lecture Notes in Computer Science, Springer Nature Switzerland, Cham (2024)
20. Wu, S., Wang, J., Ma, W., Jin, L.: Precise detection of Chinese characters in historical documents with deep reinforcement learning. Pattern Recog. **107**, 107503 (202https://doi.org/10.1016/j.patcog.2020.107503, https://www.sciencedirect.com/science/article/pii/S003132032030306X
21. Yang, Y.T., Chu, W.T.: Manga text detection with manga-specific data augmentation and its applications on emotion analysis. In: MultiMedia Modeling: 29th International Conference, MMM 2023, Bergen, Norway, 9–12 January 2023, Proceedings, Part II. LNCS, vol. 13742, pp. 29–40. Springer, Cham (2023https://doi.org/10.1007/978-3-031-27077-2_3

Hammering Test for Adhesive-Applied Tile Wall Using Deep Learning

Atsushi Ito[1,1](✉), Yuto Kurosawa[1], Masako Saito[2], and Masafumi Koike[2]

[1] Chuo University, Hachioji, Tokyo, Japan
{atc.00s,a23.6nxw}@g.chuo-u.ac.jp
[2] Utsunomiya University, Tochigi, Japan
m_saito@cc.utsunomiya-u.ac.jp, koike@is.utsunomiya-u.ac.jp

Abstract. In Japan, the aging of public structures and buildings is becoming a significant social problem. There are several methods for inspecting them, including electromagnetic waves or infrared cameras; the most widely used is the hammer test. Hammering tests are widely used for building inspection, but their accuracy depends heavily on the inspector's skill. In this study, we investigate AI-based hammering inspection for adhesive-applied tile walls, which are more difficult to evaluate due to reduced acoustic contrast. Through cross-user experiments, we reveal that hammering sound classification is strongly influenced by the inspector, introducing a distribution shift problem. To address this issue, we propose a multi-user fine-tuning strategy. The results demonstrate that fine-tuning using data from multiple inspectors significantly improves model generalization. In addition, the proposed method enables not only binary classification based on the 60% adhesion criterion defined in JASS19, but also estimation of adhesion levels with a resolution of 20%, which is useful for practical inspection. These results indicate that AI-assisted hammering inspection for adhesive-applied tile walls is feasible and that handling operator dependency is essential for real-world deployment.

Keywords: Hammering test · Non-destructive testing · Adhesive-applied tile wall · Deep learning · Fine-tuning

1 Introduction

In recent years, the aging of social infrastructure and buildings has emerged as a significant concern. So, it is essential to perform regular assessments and maintenance to ensure the safety and functionality of them. Inspection of infrastructure and buildings involves a broad range of maintenance and diagnostic activities. For example, in the case of large-scale bridges or high-rise buildings, there has been a shift toward using robots and drones as alternatives to conventional methods [1].

Although a variety of non-destructive testing techniques are available for identifying internal defects in materials such as concrete—including infrared thermography [2] and electromagnetic wave testing [3]—the hammering test remains one of the most widely adopted methods. This approach evaluates internal structural conditions by interpreting

W. Zamojski et al. (Eds.): DepCoS-RELCOMEX 2026, LNNS 2014, pp. 163–174, 2026.
https://doi.org/10.1007/978-3-032-27927-9_16

the acoustic response generated when a surface is tapped or rubbed with a hammer or inspection rod. The basic principle of the hammering test is shown in Fig. 1. If there is a cavity inside, we can hear an echo after the impact.

The hammering test is primarily used for low-rise residential buildings. In Japan alone, approximately 6.4 million apartment units across 120,000 buildings were recorded as of 2022 [4], and these units require regular inspection. Many residential buildings are decorated with tiles. However, the hammering test relies on the inspector's skill and experience. However, the declining number of experienced inspectors—mainly due to an aging workforce—is a critical issue. Therefore, it is necessary to support the shortage of skilled inspectors with the aid of AI. Our previous research introduced a deep learning–based framework for hammering inspections to assist less-experienced personnel in accurately classifying acoustic signals during inspections of mortar-applied tile walls [5].

There are two types of tile walls: mortared-applied tiles and adhesive-applied tiles. Usually, the leaping area of mortar-applied tiles is broad and produces clearer sound anomalies; on the other hand, in adhesive-applied tile walls, the reaping is localized. The adhesive material also suppresses both sound and vibration, making it harder to detect problems during the hammering test. Building on our earlier study of applying neural networks to hammering tests of concrete structures, this study aims to develop a robust AI system capable of accurately interpreting acoustic data from hammering tests conducted on adhesive-applied tile surfaces.

The main contributions of this study are as follows.

1. **Identification of operator dependency in hammering sound classification.** Through cross-user experiments, we show that hammering sound data contains strong inspector-dependent characteristics, which cause a distribution shift problem when models are applied to different inspectors.
2. **Proposal of a multi-user fine-tuning strategy to improve model generalization.** We demonstrate that fine-tuning with data from multiple inspectors can significantly improve model performance and enhance cross-user robustness.
3. **Practical adhesion condition estimation for building inspection.** In addition to binary classification based on the 60% adhesion criterion defined in JASS19, the proposed method evaluates adhesion levels at 20% resolution, enabling a more detailed assessment of tile detachment conditions.

2 Related Works

2.1 Testing of Adhesive-Applied Tile Wall

Adhesive methods are commonly used to fix tiles to wall surfaces, and evaluating the long-term durability and bonding performance of these methods remains a key area of investigation. Prior studies have addressed various aspects of tile adhesion, particularly for exterior wall applications. For instance, a study [6] reported that stresses at the adhesive interface can, in certain conditions, exceed the adhesive's bonding capacity, leading to eventual tile detachment. Similarly, research presented in [7] demonstrated that tile systems using adhesives tend to exhibit a higher relative maximum amplitude in the acoustic response zone when compared with those installed using conventional mortar methods. This fact suggests that adhesive characteristics have a significant influence on the structural integrity and longevity of tile applications. Moreover, as noted in [8], the presence of a cavity beneath a tile can generate two distinct acoustic frequency peaks, resulting in a characteristically dull sound upon inspection.

Additionally, [9] highlights the challenge of acoustically detecting minor delaminated or floating sections beneath tiled surfaces. While these insights underscore the importance of adhesive performance and acoustic evaluation in tile installation quality control, a notable research gap remains in applying neural network-based techniques for such assessments. This study aims to explore the feasibility and effectiveness of introducing neural networks to enhance the detection and evaluation of tile adhesion issues.

2.2 Hammering Test

Several investigations have examined hammering-based inspection techniques supported by information and communication technology (ICT). For example, studies such as [10] and [11] propose sophisticated methods for evaluating concrete conditions without relying on traditional hammering. Despite their technical advantages, these systems typically require high-cost equipment, which poses practical limitations to their widespread adoption. Consequently, although such techniques hold potential as alternatives to conventional inspection practices, the associated financial burden remains a key barrier to broader implementation. Furthermore, there is a noticeable lack of research specifically addressing tiled wall systems. While a study outlined in [12] explores acoustic hammering inspection approaches, its focus does not extend to walls finished with tiles. In contrast, our previous research [13] focused on applying neural networks to analyze hammering sounds from mortar-fixed tile walls. The insights and methodologies developed in that study have directly informed the present work.

2.3 Deep Learning for Acoustic and Vibration-Based Inspection

In recent years, deep learning has been widely applied to structural health monitoring (SHM) and non-destructive testing using acoustic and vibration signals. Cha et al. [14] demonstrated that deep learning models can automatically extract damage-related features from structural response data, enabling effective damage detection without manual feature engineering. Similarly, Azimi et al. [15] provided a comprehensive review of data-driven SHM approaches, highlighting the growing importance of machine learning techniques for infrastructure inspection.

Vibration-based damage detection methods have also been extensively studied, as summarized by Avci et al. [16], where various signal processing and machine learning techniques are used to identify structural defects. These studies indicate that data-driven approaches are promising for detecting subtle changes in structural conditions.

In addition, recent research has emphasized the importance of model reliability and robustness in practical applications. For example, Ahmed et al. [17] discussed the impact of distribution shift on structural damage detection models and highlighted the need for robust learning strategies when the data distribution changes.

In this context, this study focuses on hammer sound analysis for adhesive-applied tile walls and investigates the effect of operator-dependent distribution shifts. Furthermore, we explore a fine-tuning approach using multi-user data to improve model generalization (Fig. 1).

Fig. 1. The mechanism of the hammering test

3 Outline of the Research

3.1 The Difference Between This Study and Our Previous Work

In Fig. 2, we summarize the difference between this study and our previous works. We performed AI-based hammering tests on concrete and tile walls with mortar-fixed tiles in 2024, and achieved good results.

In this paper, we focus on the research for real use conditions as follows;

(1) Identification of operator dependency
(2) Multi-user fine-tuning strategy to improve model generalization.
(3) Practical adhesion condition estimation.

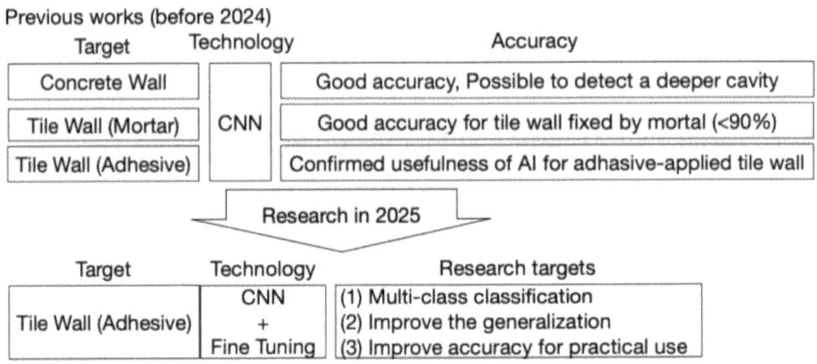

Fig. 2. Research evolution and new challenges addressed in this study

3.2 The Target of This Research

This study aims to verify whether an AI model can be used in a real situation, especially whether a general AI model can work for many inspectors.

This research also aims to develop an AI model to determine whether a tile's adhesion ratio is above or below 60%.

The Architectural Institute of Japan (AIJ) publishes the JASS (Japanese Architectural Standard Specification), a standard specification for building construction in Japan. On page 163 of JASS19 [18], it is stated that at least 60% of the back of a tile should be bonded with adhesive.

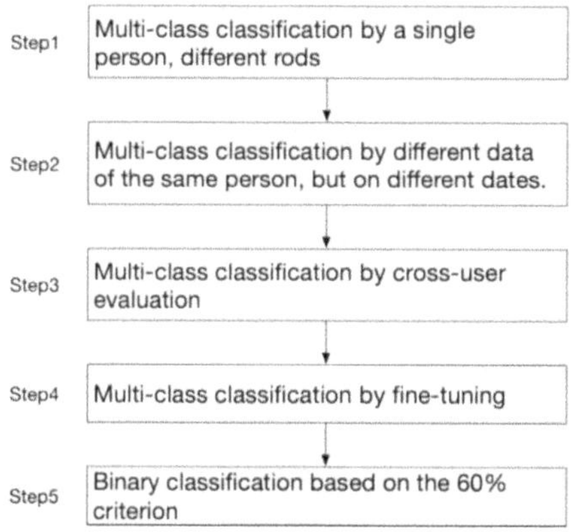

Fig. 3. Design of this research

Figure 3 explains the five steps to achieve the above research targets.

3.3 Test Specimen

We made six test specimens with varying adhesive rates, ranging from 100% to 0%. The size is 320 × 320 × 35 mm with 18 tiles on its face. Every tile has a size of 95 × 45 mm and is fixed by organic adhesives, specifically modified silicon resin adhesive.

3.4 Inspection Hammers

We used three inspection hammers as shown in (Fig. 4).

- Standard rod with a small iron ball (17φ)
- Large rod with a large iron ball (21φ)
- Rolling head rod (20φ).

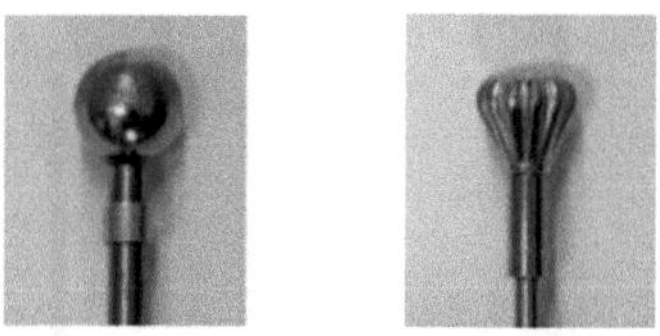

(1)Ball head (2)Rolling head (Pumpkin head)

Fig. 4. The heads of the inspection rod

4 Experiments

4.1 Sound Data Processing

We rubbed the surface of the test specimens with one of the hammering rods and collected sound data using a digital camera (Sony ZV-E10A) or an iPhone. First, record video and then extract sound data. The recording duration is between 60 s and 90 s. Then, extract features using the Short-Time Fourier Transform (STFT). We used the STFT from "scipy.signal". We applied 75% overlap and four different segment lengths, 92.8, 46.4, 23.2 and 11.6 ms, to find the most appropriate length. We used three examinees, ex1, ex2, and ex3, in the following part.

4.2 Learning Model Development

By using the Keras-tuner to select the most accurate model based on a combination of a number of hierarchies, learning rate, drop rate, etc.

- Number of hierarchies 1–4
- Number of epochs 7 or 20
- Dropout 01. or 0.5
- Learning rate 1e-4 or 1e-2
- Number of Dense parameters from 32 to 256 (32 steps).

4.3 Experiment-1 (Multi-Class Classification)

Use three types of inspection rods to confirm that a learning model created using data acquired by rubbing six different specimens with the standard rod can be recognized in the same way with the large rod and pumpkin ball rod. The ex1 performed this test. The result is in Table 1.

The orange shading indicates an incorrect result and is classified in a different group (err1); the blue shading is also incorrect.

However, the result falls within the same category (err2). If the STFT length was 23.2 ms, there is only one err2, and the other cases were correct. ***It is evident that a learning model does not depend on the type of hammering rod***.

Table 1. Result of experiment-1

STFT length	Rod	Type of specimen (adhesive rate)					
		100%	*80%*	*60%*	*40%*	*20%*	*0%*
92.8msec	Large	0 %	80 %	0 %	80 %	20 %	0 %
	Pumpkin	100 %	80 %	60 %	40 %	20 %	0 %
46.4msec	Large	60 %	80 %	60 %	20 %	20 %	0 %
	Pumpkin	100 %	80 %	60 %	40 %	20 %	0 %
23.2msec	Large	100 %	80 %	60 %	20 %	20 %	0 %
	Pumpkin	100 %	80 %	60 %	40 %	20 %	0 %
11.6msec	Large	80 %	80 %	40 %	40 %	20 %	0 %
	Pumpkin	0 %	100 %	60 %	40 %	20 %	0 %

4.4 Experiment-2 (Multi-Class Classification by Different Data of the Same Person, but Different Dates)

The same person (ex2) collected data on different dates and used the results to compare the accuracy and universality of a learning model using the standard hammering rod (in Table 2). Therefore, based on this result, *a learning model can work as a personal model.*

Table 2. Result of experiment-2

STFT length	data	Type of specimen (adhesive rate)					
		100%	*80%*	*60%*	*40%*	*20%*	*0%*
92.8msec	ex-2-1	100 %	80 %	60 %	40 %	20 %	0 %
	ex-2-2	100 %	80 %	60 %	40 %	20 %	0 %
46.4msec	ex-2-1	100 %	80 %	60 %	40 %	20 %	0 %
	ex-2-2	100 %	80 %	60 %	40 %	20 %	0 %
23.2msec	ex-2-1	100 %	80 %	60 %	40 %	20 %	0 %
	ex-2-2	100 %	80 %	60 %	40 %	20 %	0 %
11.6msec	ex-2-1	100 %	80 %	60 %	40 %	20 %	0 %
	ex-2-2	100 %	80 %	60 %	40 %	20 %	0 %

4.5 Experiment-3 (Multi-Class Classification by Cross-User Evaluation)

We tested whether the learning model built from the sound data by ex1 could work when we input the data from ex2. The result is in Table 3.

The results reveal that hammer sound classification is strongly dependent on the inspector. A model trained using data from one inspector fails to generalize to another inspector. This indicates a distribution shift caused by operator differences. It is not easy to reuse a learning model by different people.

Table 3. Result of experiment- 3

STFT length	Type of specimen (adhesive rate)					
	100%	*80%*	*60%*	*40%*	*20%*	*0%*
92.8msec	0 %	0 %	20 %	20 %	20 %	0 %
46.4msec	0 %	0 %	0 %	20 %	20 %	0 %
23.2msec	0 %	0 %	0 %	0 %	0 %	0 %
11.6msec	0 %	0 %	0 %	0 %	0 %	0 %

4.6 Experiment-4 (Multi-Class Classification by Fine-Tuning)

We tried to use fine-tuning. Based on a learning model trained on ex2's data, we fine-tuned it on ex1's data. We froze all layers except the last two layers and performed fine-tuning.

Initial fine-tuning using only data from a different inspector did not improve performance, indicating that naive fine-tuning is insufficient to address operator dependency. The accuracy remained poor even after 1,000 epochs, as shown below.

accuracy: 0.6278, val_accuracy: 0.0699.

However, when fine-tuning was performed using data from multiple inspectors, the model achieved significantly improved generalization, as shown in Table 4.

We tried fine-tuning using both ex1's data and ex2's data on the learning model trained on ex2's data. The result is displayed in Table 4.

Table 4. Result of experiment-4

STFT length	**data**	**Type of specimen (adhesive rate)**					
		100%	***80%***	***60%***	***40%***	***20%***	***0%***
92.8 msec	ex2-1	100 %	80 %	60 %	40 %	20 %	0 %
	ex2-2	100 %	80 %	60 %	20 %	20 %	0 %
	ex1(pumpkin)	100 %	80 %	60 %	40 %	20 %	0 %
	ex-3	100 %	80 %	60 %	40 %	20 %	0 %
46.4 msec	ex2-1	100 %	80 %	60 %	20 %	20 %	0 %
	ex2-2	100 %	80 %	60 %	40 %	20 %	0 %
	ex1(pumpkin)	100 %	80 %	60 %	40 %	20 %	0 %
	ex-3	100 %	80 %	60 %	20 %	20 %	0 %
23.2 msec	ex2-1	100 %	80 %	60 %	20 %	20 %	0 %
	ex2-2	100 %	80 %	60 %	40 %	20 %	0 %
	ex1(pumpkin)	100 %	80 %	60 %	40 %	20 %	0 %
	ex-3	100 %	80 %	60 %	20 %	20 %	0 %
11.6 msec	ex2-1	100 %	80 %	20 %	20 %	20 %	0 %
	ex2-2	100 %	80 %	60 %	40 %	20 %	0 %
	ex1(pumpkin)	0 %	80 %	60 %	60 %	20 %	0 %
	ex-3	100 %	80 %	20 %	20 %	20 %	0 %

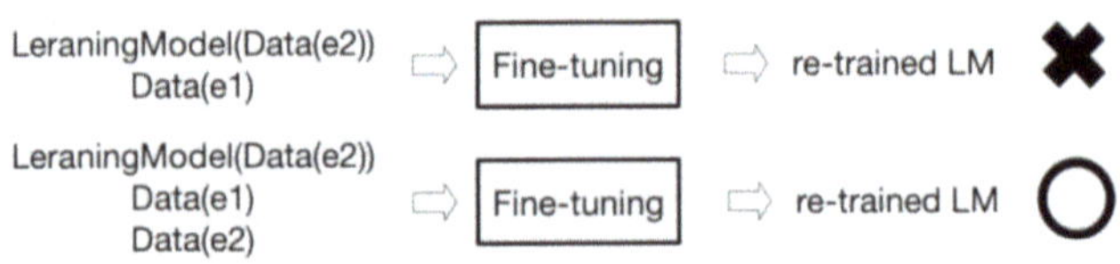

Fig. 5. Fine-tuning in Experiment-4

In the case of STFT, the data length is 92.8 ms, and there was only one mistake (but not so serious: err2). Additionally, in the cases of 46.4 and 23.2 ms, there were two errors in each instance. Therefore, we believe *the learning model developed through fine-tuning, including the data used to develop the base learning model, exhibits good universality. Also, this model is not dependent on the STFT data length.* The difference between these two trials is shown in Fig. 5.

4.7 Experiment-5 (Binary Classification Based on the 60% Criterion)

Finally, we would like to summarize this result as a binary classification based on the 60% criterion. Table 5 shows the result. We find that the personalized learning model and the fine-tuned learning model perform well on the hammering test.

Table 5. Result of experiment-5

Case	STFT length	Type of specimen (adhesive rate)	
		Over 60%	*Under 60%*
Multi-class classification	92.8msec	67 %	67 %
	46.4msec	83 %	83 %
	23.2msec	100 %	100 %
	11.6msec	33 %	100 %
Multi-class classification by different data of the same person, but different dates	92.8msec	100 %	100 %
	46.4msec	100 %	100 %
	23.2msec	100 %	100 %
	11.6msec	100 %	100 %
Multi-class classification by cross-user evaluation	92.8msec	0 %	67 %
	46.4msec	0 %	67 %
	23.2msec	0 %	33 %
	11.6msec	0 %	33 %
Multi-class classification by fine-tuning	92.8msec	100 %	92 %
	46.4msec	100 %	83 %
	23.2msec	100 %	83 %
	11.6msec	75 %	75 %

5 Discussions

The experimental results indicate that hammering sound data are strongly influenced by the inspector, which can be interpreted as a distribution shift problem in machine learning. In practical deployment, it is important to detect such shifts and ensure reliable model behavior.

Although this study focuses on feasibility evaluation, future work should consider model-monitoring techniques, such as uncertainty estimation and distribution-shift detection, to improve reliability in real inspection scenarios.

Based on the experiments, the following ideas for the practical use were obtained.

(1) If a construction company has N inspectors, it is possible to construct a highly accurate model by conducting fine-tuning using data from M (2≦M≦N) of those inspectors.
(2) Each individual could build their training model using a sample for testing. One possible approach would be for each person to own a testing device or to install the learning model on an SD card or similar device.

6 Conclusion

This study investigated AI-based hammering inspection for adhesive-applied tile walls, which are more difficult to evaluate due to reduced acoustic contrast caused by adhesive materials.

The experimental results revealed that hammering sound classification is strongly influenced by the inspector, introducing an operator-dependent distribution shift problem. This finding indicates that models trained on data from a single inspector cannot be directly applied to other inspectors.

To address this issue, we proposed a multi-user fine-tuning strategy. The results demonstrated that fine-tuning using data from multiple inspectors significantly improves model generalization and enables more robust inspection performance.

In addition, the proposed method enables not only binary classification based on the 60% adhesion criterion defined in JASS19, but also estimation of adhesion levels with a resolution of 20%, which is valuable for practical inspection.

These results suggest that AI-assisted hammering inspection for adhesive-applied tile walls is feasible, and that handling operator dependency is a key requirement for real-world deployment.

Further study issues are as follows.

(1) A collection of much hammering inspection sound data using many examinees to develop a more valuable and universal learning model.
(2) Performing trials using real tile wall buildings to check the accuracy of the developed learning model.

Acknowledgement. The authors thank Mr. Nishimura of Fujiyogyo Co., Ltd. For developing test specimens. This research is supported by JSPS Kakenhi (24K14987, 23K28338, 23H03649, 22K12598, 20H05702) and Tokyo Metropolitan Small and Medium Enterprise Support Center.

References

1. Terra Drone. Inspection services. https://terra-drone.net/service/inspection [Accessed: 2026–03–20]
2. Ministry of Land Infrastructure Transport and Tourism. Guidelines for exterior wall surveys by infrared surveys (including infrared surveys by unmanned aerial vehicles) under the periodic reporting system. https://www.mlit.go.jp/jutakukentiku/build/content/001474154.pdf [Accessed: 2026–03–20]
3. Just-Corporation. Non-destructive testing by electromagnetic waves. https://www.just-ltd.co.jp/service/existing/concreteexploration/electrowave/ [Accessed: 2026–03–20]
4. Condominium management companies association. Survey Report, 2024. https://www.kanrikyo.or.jp/report/doukou_r04.html (in Japanese) [Accessed: 2026–03–20]
5. General Construction Association of Fukuoka. Seminar on construction problems seen in tile peeling and problem-solving from a legal point of view. http://fukukenkyo.org/slide-koga.pdf [Accessed: 2026–03–20]
6. Soeda, T., Mikami, T.: Basic study into a diagnostic system for exterior tile debonding. AIJ J. Technol. Des. **81**, 1779–1787 (2016). Accessed: 2026–03–20. https://doi.org/10.3130/aijs.81.1779
7. Yokoi, M.: A study on diagnosis method using hammering test. J. Osaka Sangyo Univ. Nat. Sci., **131**, 59–69 (2021). https://osu.repo.nii.ac.jp/record/2353/files/059-069.pdf [Accessed: 2026–03–20]
8. Makatayama, M., Tanano, H., Nemoto, K.: Research on the diagnostic accuracy of the tile finish external wall by various measuring methods. AIJ J. Technol. Des., **21**, 919–924 (2015). Accessed: 2026–03–20. https://doi.org/10.3130/aijt.21.919
9. Kasahara, L., Fujii, J., Yamashita, H.: Research on the diagnostic accuracy of the tile finish external wall by various measuring methods. AIJ J. Technol. Des., **21**, 919–924 (2015). Accessed: 2026–03–20. https://doi.org/10.3130/aijt.21.919
10. Kurita, K., Oyado, M., Tanaka, H., Tottori, S.: Active infrared thermographic inspection technique for elevated concrete structures using remote heating system. Infrared Phys. Technol., 208–213 (2009). Accessed: 2026–03–20, https://doi.org/10.1016/j.infrared.2009.07.010
11. Akamatsu, R., Sugimoto, T., Utagawa, N., Katakura, K.: Proposal of non contact inspection method for concrete structures using high-power directional sound source and scanning laser doppler vibrometer. Jpn. J. Appl. Phys., **52**. Accessed: 2026–03–20. https://doi.org/10.7567/JJAP.52.07HC12
12. Ministry of Land Infrastructure, Transport and Tourism.: Periodic report of the result of investigation of tiles on exterior walls. https://www.mlit.go.jp/jutakukentiku/build/jutakukentiku_house_tk_000161.html [Accessed: 2026–03–20]
13. Ministry of Land Infrastructure, Transport and Tourism.: Study on rationalization of periodic survey methods for wet exterior walls. https://www.mlit.go.jp/common/001129988.pdf [Accessed: 2026–03–20]
14. Cha, S., Choi, T., Büyüköztürk, O.: Deep learning-based damage detection using convolutional neural networks. J. Struct. Eng. (2017). https://doi.org/10.1080/15732479.2016.1257060[Accessed:2026-03-20]
15. Azimi, M., Eslamlou, A., Pekcan, M.: Data-driven structural health monitoring and damage detection through deep learning: State-of-the-art review. Sensors (2020). https://doi.org/10.3390/s20102778[Accessed:2026-03-20]

16. Avci, O., Abdeljaber, M., Kiranyaz, S., et al.: A review of vibration-based damage detection in civil structures: From traditional methods to machine learning and deep learning applications. Mech. Syst. Signal Process. (2021). https://doi.org/10.1016/j.ymssp.2020.107077[Accessed: 2026-03-20]
17. Ahmed, M., Sharma, A., et al.: Unveiling out-of-distribution data for reliable structural damage assessment. Mech. Syst. Signal Process. (2023). https://doi.org/10.1016/j.ymssp.2023.110325[Accessed:2026-03-20]
18. JASS19 Ceramic Tiling.: (2022). ISBN 978–4–8189–1554–1 [Accessed: 2026–03–20]

A New Method for Finding Local Extrema Using a Quantum Annealer

Michał Jaroszczuk(✉)

Department of Control and Quantum Computing, Faculty of Information and Communication Technology, Wrocław University of Science and Technology, Wrocław, Poland
michal.jaroszczuk@pwr.edu.pl

Abstract. This work introduces a new method for detecting local extrema using a quantum annealer. A literature review is conducted to demonstrate its potential applications and the current state of knowledge and methods in this field, focusing on quantum annealers. Next, a MILP model is formulated that achieves 100% accuracy in detecting a given number of local minima or maxima. Based on it, the QUBO model is presented, which achieves an correctness of 92% when run on the D-Wave Leap solver. Experiments conducted on real test data in the field of sound processing and gold price analysis prove the high effectiveness and universality of the presented approach.

Keywords: quantum computing · D-Wave · local search · quantum annealing · QUBO Model · Gurobi solver · multiple local extrema

1 Introduction

Quantum computing is currently a rapidly developing field in response to the limitations of traditional computers. This is made possible by the use of qubits, which, unlike bits, instead of representing a high or low state, possess a full range of intermediate states, resulting in significantly higher performance [13]. Since the first quantum computer was presented by D-Wave in 2007 [4], numerous machines have been created using various technologies to take advantage of quantum phenomena. However, despite dynamic development, we are still in the era of NISQ (Noisy Intermediate-Scale Quantum) computers, which constantly cannot surpass the performance and stability of traditional machines, partly due to difficulties in constructing devices that replicate quantum phenomena. Achieving a system that allows for reliable calculations and eliminates quantum decoherence remains a significant challenge. Many technologies exist that allow for the acquisition of quantum states, such as computers based on superconductors, photons, ion traps, laser technologies, and quantum annealers [9,13]. Due to their specific properties, it is important to note that general-purpose quantum computers do not currently exist; instead, they are dedicated to specific applications. Moreover, at this point it is not known which technology will prove to

W. Zamojski et al. (Eds.): DepCoS-RELCOMEX 2026, LNNS 2014, pp. 175–185, 2026.
https://doi.org/10.1007/978-3-032-27927-9_17

be leading in the future [9]. In the field of discrete optimization the most commonly used type of quantum computer is the quantum annealer (QA), for which the objective function must be represented using the Quadratic Unconstrained Binary Optimization (QUBO) models, which in contrast to Mixed Integer Linear Programming (MILP) models, forces the addition of model constraints to the objective function [6]. These models are run using solvers, and their output is a global extremum solution. However, it is often necessary to find multiple local minima or maxima, depending on the problem posed, which is not the intended use of these solvers. There are already tools that allow searching the solution space of a function and finding multiple local minima, such as `islocalmin` / `islocalmax` function in Matlab [2] or `optimize` in SciPy [14], but yet there is no information about their equivalents for quantum technologies.

The aim of this work is to elaborate QUBO model for finding multiple local extrema of a discrete function, as its desired application is to find one global extremum and avoid stucking in the local ones. Similar problems can probably be solved much faster and more accurately using classical methods, but the aim of this work is not to beat classical solutions but to show the feasibility of the presented problem using quantum technologies. The model is described along with its variants, which allow for finding a given number of minima or maxima. A MILP equivalent for the quantum model is also presented, ready to be run using classical solvers. The presented approach is compared with the current literature to examine the methods used in other similar works. Experiments using a D-Wave Leap hybrid quantum solver are conducted for various types of problems and datasets. The results are compared with classical methods to assess the effectiveness of the proposed approach.

2 Literature Review

The goal of this literature review is to answer the question of whether implementing a quantum model capable of finding multiple local extrema is possible, based on the literature found. Next, it is examined whether similar works exist in the literature, focusing on multiple local optimization of a discrete function. Finally, potential applications for the created model is presented.

For quantum annealer, discrete optimization problems can be mapped to Hamiltonian, which is later transformed using quantum fluctuations to the Ising Hamiltonian representing the minimum energy configuration of the system [7]. Since formulating the Hamiltonian of the problem is not an easy task, it is possible to represent it using the QUBO model, which can then be easily transformed into the Ising model run on the QPU [5]. The most difficult task in QUBO formulation is the inclusion of constraints in the objective function, as described by Glover in [6]. It must be noted that QA is intended to return a single global minimum of the problem, which is however not guaranteed. This forces the examination of the current state of research in this field to explore approaches used to find multiple local minima as it is not a standard application of QA.

One of the work presented by Shaydulin et al. [16] consider multistart optimization approach with Quantum Approximate Optimization Algorithm

(QAOA) to find a global minimum of the funtion. Despite the main goal of the authors, such approach may be interesting in a given problem where there is a need to find the local optimas of the function. However, as the authors conclude, the main difficulty is to find optimal parameters, as starting point, and it is not possible to run algorithm on real data instances due to the NISQ era. Similar approach is presented in [3] where the dataset is divided into several scopes and the problem of finding the optimum is solved for them. In [8], the results returned by the quantum annealer, which are not global extremes but may constitute potential local minima, were compared with the results of Simulated Annealing (SA). The results showed that some local valleys are omitted by QA and, with a sufficiently long computation time, SA outperforms QA.

Many problems require finding many local extrema instead of one global one, an example of which is the work presented by Obuchowski et al. [10], where detecting local peaks is necessary to reveal a machine bearing or gearbox failure. Information on the frequency of local peaks is used to create a spectrogram that can be used to predict the occurrence of a machine failure. Another practical example is given by Sezan [15], where a local minimas called peaks are detected to specify the gray levels that are selected as quantization levels to reduce image data. The detection of multiple local optima for energy flow control is analyzed in [18], where the authors propose a mathematical model for energy balancing in power systems. Solving the problem of finding multiple local minima can also be useful in image analysis, as shown in the paper [1].

Summarizing the analysis of the selected works, it can be concluded that there is a significant need for efficient algorithms for detecting local extrema, which can be applied in many areas of industry and business. Although the main function of quantum annealers is to find the global minimum energy configuration for a given problem, it seems possible to formulate the model so that it can detect multiple local extrema - both minima and maxima. However, there are no works that use a QUBO model for this purpose, and moreover, run it on a real dataset.

3 Model Formulation

While considering the mathematical model for the posed problem, attempts were made to find inspiration in other fields of discrete optimization that could be used to formulate it. Since the author's primary research area is the optimization of vehicle routing problems (VRP), an attempt was made to translate the models developed for this purpose, for example in [11], into the problem of finding local extrema. It can be seen that the arguments of a discrete function can be reduced to the locations that the vehicle must visit with given function value, and the extrema can be reduced to the vehicles themselves. Each vehicle must find exactly one location that earns the highest reward for visiting, assuming that a single location can be visited by a maximum of one vehicle. Rules provided by the field of mathematical optimization are used to calculate the reward itself.

3.1 Input Data

The first step in formulating a model is defining the input data on which the model operates. Since the goal of creating a model is to find the local extremas of a discrete function, it is necessary to define this function as:

$$f : X \to Y. \tag{1}$$

Therefore, the two lists X, Y of input data are given, each of size s. Then a list of points $\mathcal{P}$ is defined, in which the found local extrema are stored. The list consist of m elements $p \in \mathcal{P}$. As in order to find a local extremum, the difference between the currently considered point and the neighboring points are calculated, an additional list $\mathcal{Z} = X \setminus \{x_1, x_s\}$ is introduced, which does not contain the first and the last element of list X. Finally, in order to store the found solutions, a binary decision variable $x_{i,p}$ is defined, which indicates whether the element with index $i \in \mathcal{Z}$ has been assigned to a specific place $p \in \mathcal{P}$ in the extremum list.

3.2 MILP Model

The goal function for the presented problem can then be specified as follows:

$$\max \sum_{i \in \mathcal{Z}} \sum_{p \in \mathcal{P}} x_{i,p} r_i. \tag{2}$$

Presented goal function has to be optimized under the following constraints:

$$\sum_{i \in \mathcal{Z}} x_{i,p} = 1 \qquad \forall p \in \mathcal{P}, \tag{3}$$

$$\sum_{p \in \mathcal{P}} x_{i,p} \leq 1 \qquad \forall i \in \mathcal{Z}, \tag{4}$$

$$r_i = (Y[i] - Y[i-1]) \cdot (Y[i] - Y[i+1]) \qquad \forall i \in \mathcal{Z}, \tag{5}$$

$$\sum_{p \in \mathcal{P}} x_{i,p} \cdot (Y[i] - Y[i+1]) \geq 0 \qquad \forall i \in \mathcal{Z}, \tag{6}$$

$$r_i \in (-10000, 10000) \qquad \forall i \in \mathcal{Z}, \tag{7}$$

$$x_{i,p} \in \{0, 1\} \qquad \forall i \in \mathcal{Z}, p \in \mathcal{P}. \tag{8}$$

Constraint (3) provides that each extremum point p is assigned to exactly one function argument. Since the task is to find many different local extrema, the (4) constraint is introduced, ensuring that one function argument can be assigned to at most one extremum, as the argument may not be an extremum point. The reward $r_i \in \mathbb{Z}$ for finding an extremum is calculated in (5) and is based on a simple mathematical observation. If a point is an extremum of a function, it means that the product of the difference in values between it and the points before and after it, is positive. Finally the range of decision variables is defined in (7) and (8), with the proviso that the range of the r variable should be adjusted to the problem being solved and the data set used.

The presented model is used to detect a given number of local maxima. To change the number of returned results, it is necessary to modify the constant m, which refers to the length of the list $\mathcal{P}$. To detect all local extrema, constraint (6) should be removed, whereas to detect local minima it is needed to modify it to the following form:

$$\sum_{p \in \mathcal{P}} x_{i,p} \cdot (Y[i] - Y[i+1]) \leq 0 \qquad \forall i \in \mathcal{Z}. \tag{9}$$

3.3 QUBO Model

To transform a MILP model into a QUBO model, it is necessary to reformulate the constraints so that they can be included in the objective function. This requires abandoning the inequality constraints and eliminating the additional decision variables introduced. This is typically accomplished by introducing a component of the objective function, multiplied by a large constants, to eliminate forbidden solutions or to identify specific properties of the problem being solved [6]. In this case, the objective function presented in Eq. (10) was separated into a principal component, and three additional constraints each multiplied by a different constant A_1, A_2, A_3, the value of which depends on the dataset used. Constant A_1 and A_3 allow to adapt the model to the problem being solved, whereas constant A_2 penalize selection of forbidden solutions. It should be noted, that unlike in the classical linear programming model, in the QUBO model, the objective function cannot be maximized due to the physical characteristics of the system, which strives for energy minimization of the Hamiltonian. This necessitates the inversion of the reward function so that it rewards outcomes that are not extremes of the function, as evidenced by the negative coefficient A_1.

$$\min K_a = -A_1 \cdot K_o + A_2 \cdot C_1 + A_3 \cdot C_2 + A_2 \cdot C_3, \tag{10}$$

The components of the objective function are described below. The main objective function (11) is a representation of (2), but additionally includes a reward calculation to eliminate the constraint (5).

$$K_o = \sum_{i \in \mathcal{Z}} \sum_{p \in \mathcal{P}} (x_{i,p} \cdot (Y[i] - Y[i-1]) \cdot (Y[i] - Y[i+1]))\,. \tag{11}$$

The constraints added to the model mirror those of the MILP model, which, however, had to be adapted due to the difficulties in formulating inequality constraints in QUBO. Constraint (12) ensures that each extremum is assigned to exactly one function argument, as this would then reset the sum to zero. Constraint (13) mirrors the MILP constraint (6), which determines whether local minima or maxima are considered. If local maxima are sought, the sign of the constant A_3 should be negative to favor solutions for which neighboring points have lower values. If the goal is to find local minima, the sign of the constant A_3 should be positive. Finally, constraint (14) forces the model to assign different function arguments to each extremum, preventing the same extremum from being returned multiple times, as in (4). If the solver were to assign a single function argument to multiple points p, then constraint C_3 would significantly increase the value of the objective function, as it is squared.

$$C_1 = \sum_{p \in \mathcal{P}} (\sum_{i \in \mathcal{Z}} x_{i,p} - 1)^2, \tag{12}$$

$$C_2 = \sum_{i \in \mathcal{Z}} \left((\sum_{p \in \mathcal{P}} x_{i,p}) \cdot ((Y[i] - Y[i-1]) + (Y[i] - Y[i+1])) \right), \tag{13}$$

$$C_3 = \sum_{i \in \mathcal{Z}} (\sum_{p \in \mathcal{P}} x_{i,p})^2. \tag{14}$$

4 Experiments

Having a defined mathematical model of the problem, its capabilities were tested whether it is truly effective in finding a given number of local extrema of different type depending on the given input data. The experiments were conducted to answer the following questions:

1. Is the model universal? Does it allow finding a given number of local extrema, both minima and maxima?
2. Whether the developed approach can be used in practical application?
3. What is the effectiveness of the presented MILP and QUBO models?

All the experiments were run using AMD Ryzen 3 7330U processor with 8 GB of RAM memory with Gurobi 13.0 linear solver and D-Wave Systems Leap hybrid quantum solver `hybrid_constrained_quadratic_model_version1p`. The sample with the lowest objective function value was used for testing. It is not possible to determine the QPU time consumed, what is an inherent feature of the Leap model. All the algorithms were written using `Python 3.12` programming language. Most of the tests used a modified dataset by David Rubiano, published on Kaggle in 2022 [12]. This dataset consists of .wav files containing recordings of the five syllables 'ha', 'he', 'hi', 'ho', and 'hu'. The dataset was processed using LPC (Linear Predictive Coding) to extract frequencies and their corresponding sound magnitudes. This resulted in a discrete function matching sound frequencies into corresponding magnitudes.

4.1 The Universality of the Model

First, it was tested whether the model could be successfully used to detect both local minima and maxima within a specified number of parameters. To verify this, the objective function and constraints were modified during the tests as described in Sects. 3.2 and 3.3. The results of the experiments are shown in Fig. 1 for Gurobi solver and in Fig. 2 for the Leap.

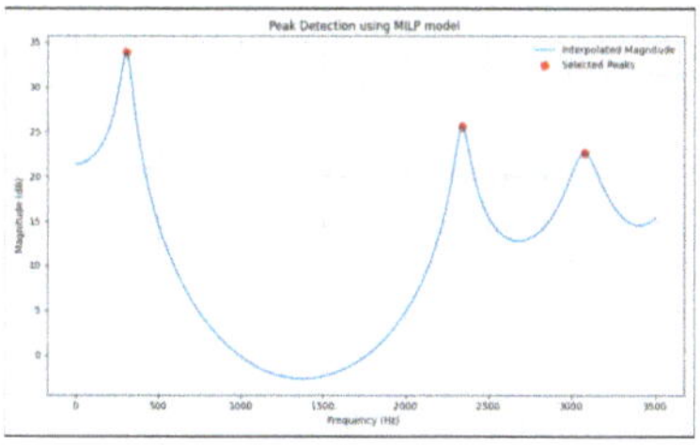

(a) Detection of 3 local maxima

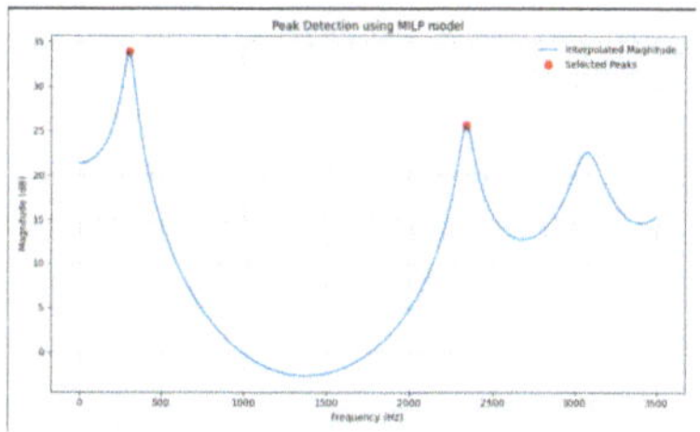

(b) Detection of 2 local maxima

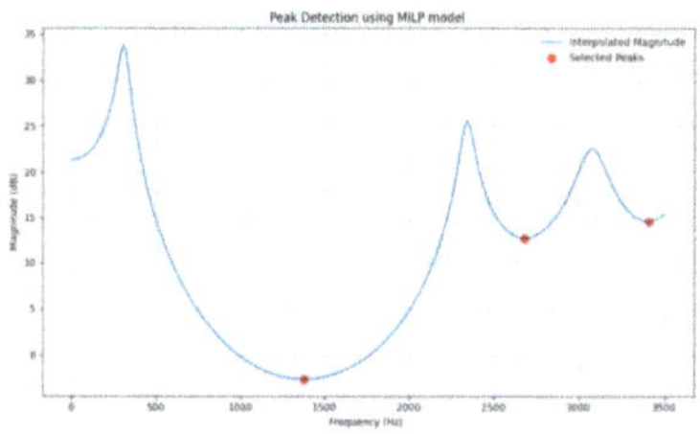

(c) Detection of 3 local minima

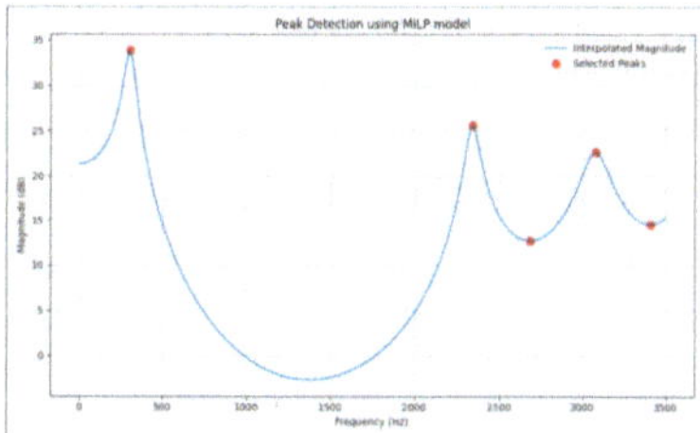

(d) Detecting 5 local extremes

Fig. 1. Checking the universality of the MILP model.

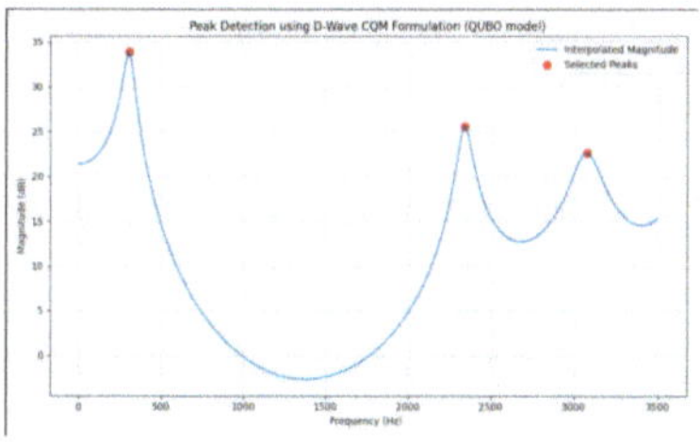

(a) Detection of 3 local maxima

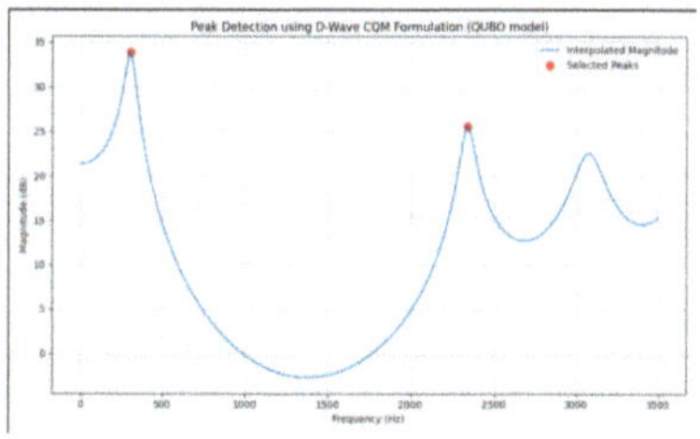

(b) Detection of 2 local maxima

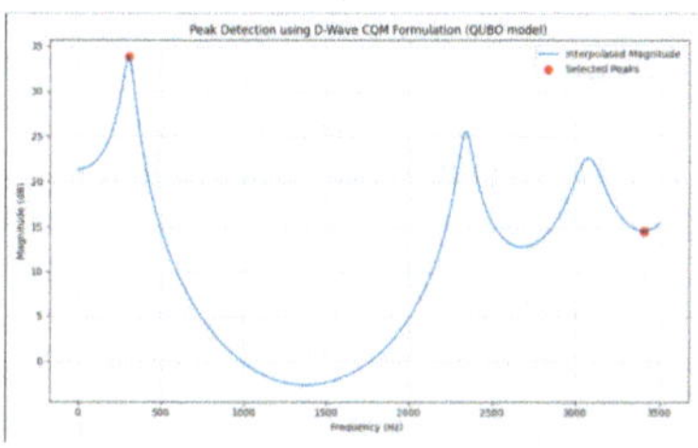

(c) Detection of 3 local minima

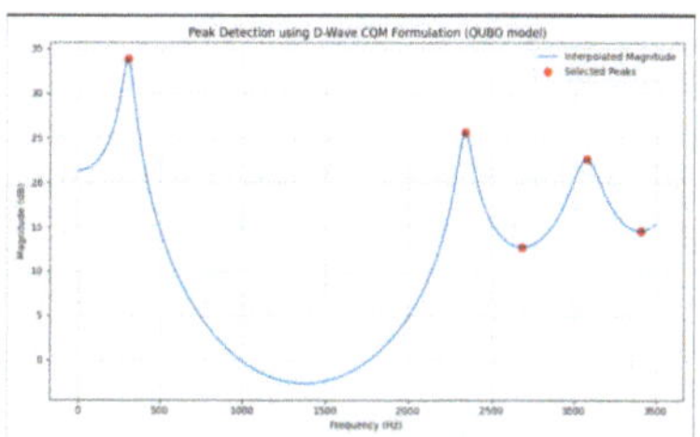

(d) Detecting 5 local extremes

Fig. 2. Checking the universality of the QUBO model.

The conclusion is that the model can be customized to detect different types and numbers of extremes. However, a problem arises if a larger number of extremes is specified than exists in the function. In such a situation, it may result in returning points surrounding the actual extremes, or the wrong types of extremes. It can be seen that both models favor extremes with sharp inflections, that is differences between the extreme and neighboring points. The Leap quantum solver fails with detection of minima for the same reason, that in presented example minima are flattened.

4.2 Practical Application on Different Datasets

The previous sections presented research on the models for detecting multiple local extrema, using the example of finding vowel formant features. However, this is a specific application, and it was decided to conduct research on one more example to prove that the presented model is suitable for many different applications. For the experiments the dataset with average yearly price of gold between 1969 - 2022 was taken [17]. The constants for QUBO were set to: $A_1 = 10$, $A_2 = 1000000$, $A_3 = 100$. The results obtained are shown in Fig. 3.

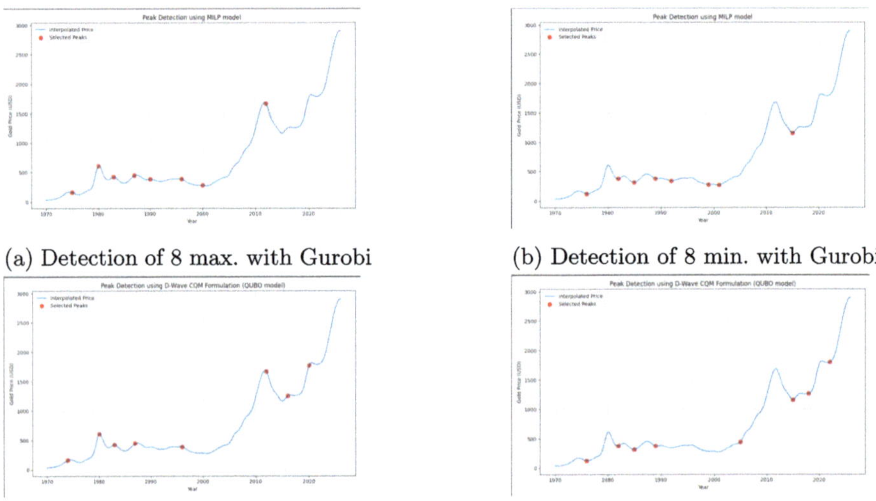

(a) Detection of 8 max. with Gurobi

(b) Detection of 8 min. with Gurobi

(c) Detection of 8 max. with Leap

(d) Detection of 8 min. with Leap

Fig. 3. Detection of extrema on gold price chart.

As can be seen in the graphs on Fig. 3, the Gurobi solver using the MILP model handled the task very well, achieving 100% accuracy even despite small differences between values for the given years. In contrast, the Leap solver based on the QUBO model detects extremes well but struggles to distinguish between minima and maxima and sometimes he took a common point as an extreme. This is most likely related to a problem with the implementation of the (13) constraint, for which a penalty factor would need to be adjusted, which, however, must be performed separately for each analyzed dataset. An interesting observation is that for this study the penalty constant had to be increased to the value $A_2 = 1000000$, as the gold price values on the y-axis were larger than the magnitude.

4.3 The Effectiveness of the Model

Knowing that we are currently in the NISQ era of quantum technology development, it was attempted to validate the results returned by the D-Wave quantum annealer. Additionally, it may have turned out that the MILP model contained some errors, so the results derived from it were verified against the results returned by Matlab functions `islocalmin` and `islocalmax`.

The results obtained for four different vowels, conducted on 20 files, which gives 60 local maxima, clearly showed that the classical approach using the MILP model achieved 100% accuracy in detecting local maxima compared to the reference method. The constants for QUBO were set to: $A_1 = 1000$, $A_2 = 10000$, $A_3 = 10$, as the values on the y-axis were not large. The quantum approach, on the other hand, showed 91.66% accuracy. Examples of errors returned by the quantum solver can be seen in Fig. 2c and mainly consisted of considering the same point twice or treating a local minimum as a maximum. This may be due to the noise inherent in current quantum solutions and insufficient tuning of the model to the problem by adjusting the constants. It seems that, a good way to improve these results would be to perform a larger number of samples and then consider the most frequently repeated results to minimize noise.

The computational time for different problem instance sizes was also checked, assuming detection of three local extrema ($m = 3$). It should be remembered that as the number of extrema to be detected increases, the computational time also increase, as it affects the number of decision variables in the given model. For a problem of size 1000 the computation time was 10 s, for the next tested sizes 1500 and 3000 it was 15 s and 64 s, respectively, while for 4000 it was as much as 114 s. It should be noted that with the limitations on the computation time, instances as large as 4000 samples may be a problem in conducting research.

5 Conclusions and Future Works

This paper presents a new approach to finding local extrema in a discrete function using a quantum annealer. This approach has been proven to yield satisfactory results using real-world test data and can be used to solve problems in practice. The presented MILP model performs well in detecting both local minima and maxima of a given number, achieving nearly 100% accuracy. In contrast, the QUBO model, run using the D-Wave Leap hybrid solver, achieves an accuracy of approximately 92% and performs well in detecting extrema with a high prominence above the surroundings. The errors that occur may be attributed to the early development of quantum technologies (NISQ) and the need to fine-tune the model to the problem. The results show that Leap solver may not give as good results as the linear solver despite the hybrid approach. Experiments have shown that the biggest challenge is the appropriate selection of penalty coefficients for the QUBO model, tailored to the problem being solved.

Future work on the model is planned to extend it to other practical applications and to more thoroughly investigate the obtained results. Since this work uses a hybrid quantum leap solver, it is planned to verify the model's performance using a native quantum approach and examine it for correctness of the returned results and maximum instance sizes. It is also planned to use the model in conjunction with models solving the VRP problem to implement two-stage optimization using a quantum annealer. Moreover, a detailed description of the process of selecting penalty coefficients for the problem being solved would be an interesting direction for the development of work and future research.

References

1. Al-Furaiji, O.J.M., Tuan, N.A., Yurevich, T.V.: A new approach of finding local extrema in grayscale images based on space-oriented masks. Eng. Lett. **28**, 586–593 (2020)
2. M. H. Center. Data preprocessing (2025). https://www.mathworks.com/help/matlab/preprocessing-data.html. Accessed: 31-01-2026
3. Chancellor, N.: Modernizing quantum annealing using local searches. New J. Phys. **19**, 023024 (2017)
4. Cohen, E., Tamir, B.: D-wave and predecessors: from simulated to quantum annealing. Int. J. Quan. Inf. **12**, 1430002 (2014)
5. Company, D.-W.: QUBOS and ISING models (2025). https://docs.dwavequantum.com/en/latest/quantum_research/qubo_ising.html. Accessed: 31-01-2026
6. Glover, F., Kochenberger, G., Hennig, R., Du, Y.: Quantum bridge analytics i: a tutorial on formulating and using QUBO models. Ann. Oper. Res. **314**, 141–183 (2022)
7. Hauke, P., Katzgraber, H.G., Lechner, W., Nishimori, H., Oliver, W.D.: Perspectives of quantum annealing: methods and implementations. Rep. Prog. Phys. **83**, 054401 (2020)
8. Koshka, Y., Novotny, M.A.: Comparison of D-wave quantum annealing and classical simulated annealing for local minima determination. IEEE J. Sel. Areas Inf. Theory, **1**, 515–525 (2020)
9. Ladd, T.D., Jelezko, F., Laflamme, R., Nakamura, Y., Monroe, C., O'Brien, J.L.: Quantum computers. Nature **464**, 45–53 (2010)
10. Obuchowski, J., Wyłomańska, A., Zimroz, R.: The local maxima method for enhancement of time–frequency map and its application to local damage detection in rotating machines. Mech. Syst. Signal Process. **46**, 389–405 (2014)
11. Pempera, K., Jaroszczuk, M.: Analysis of the impact of time window lengths on maintenance vehicle routing problem efficiency. Lecture Notes Networks Syst. **1427**, 2367–3370 (2025)
12. Rubiano, D.A.: Dataset of vowels (2022). https://www.kaggle.com/datasets/darubiano57/dataset-of-vowels. Accessed: 28-01-2026
13. N.S., Singh, H., A. U. N.: An extensive review on quantum computers. Adv. Eng. Softw. **174**, 103337 (2022)
14. SciPy. Optimization and root finding (2025). https://docs.scipy.org/doc/scipy/reference/optimize.html. Accessed: 31-01-2026
15. Sezan, M.I.: A peak detection algorithm and its application to histogram-based image data reduction. Comput. Vision Graphics Image Process. **49**, 36–51 (1990)
16. Shaydulin, R., Safro, I., Larson, J.: Multistart methods for quantum approximate optimization. arXiv (2019)

17. Yafeth, T.: Gold price historical data (1969–2022) (2022). https://www.kaggle.com/datasets/yafethtb/gold-price-historical-data-1969-2022. Accessed: 28-01-2026
18. Zhang, J., Han, Z.: Identifying multiple local optima for optimal power flow based on nonlinear dynamics. IEEE Access **8**, 129878–129888 (2020)

Experimental Analysis of Quantum Neural Networks on Odra 5 Quantum Computer

Małgorzata Jędruszek[1], Radosław Idzikowski[2](✉), and Mateusz A. Kucharski[2]

[1] Faculty of Information and Communication Technology, Wrocław University of Science and Technology, Wrocław, Poland
gosiajedruszek@wp.pl

[2] Department of Control and Quantum Computing, Wrocław University of Science and Technology, Wrocław, Poland
{radoslaw.idzikowski,mateusz.kucharski}@pwr.edu.pl

Abstract. This paper presents an in-depth experimental analysis of quantum neural networks executed on a real quantum computing hardware platform. The study is conducted using the well-known Iris dataset, which serves as a reference benchmark for classification tasks in quantum machine learning literature. Various optimization strategies for training parametrized quantum circuits are investigated, including both gradient-based and gradient-free optimizers. In addition to algorithmic aspects, the impact of quantum hardware conditions on computational performance is examined by evaluating the trained models at different times of day, accounting for hardware-state variability and daily calibration procedures. Furthermore, following preliminary experiments, a feature selection process is applied, reducing the original four input features to three to mitigate noise and improve the robustness and stability of the results. The analysis presented highlights the practical challenges and performance variability of quantum neural networks when deployed on real quantum hardware, underscoring the importance of optimizer choice, execution scheduling, hardware calibration, and feature selection in quantum machine learning experiments.

Keywords: Quantum Machine Learning · Optimizers · Feature Selection · Daily Calibration · Real-Hardware

1 Introduction

While still in its infancy, quantum computing technology has recently become a reality. With it came a lot of new algorithms and applications, ranging from optimization [3], cryptography [14], random number generation [6], and quantum machine learning (QML) – which will be the topic of this paper. There is still a noticeable shortage of quantum hardware, where most countries do not have even a single quantum computer. This is somewhat mitigated by remote access to services like IBM Quantum Platform, IQM Resonance or, in case of quantum

W. Zamojski et al. (Eds.): DepCoS-RELCOMEX 2026, LNNS 2014, pp. 186–195, 2026.
https://doi.org/10.1007/978-3-032-27927-9_18

annealers, D-Wave Leap. This solution however still has significant constraints – first is the fact that the computers can be accessed by a large number of scientists, creating a need for queuing and thus introducing delays, that will only grow as the interest in technology increases. The second constraint, that also contributes to the delay problem, is the fact that most quantum computers are stationed in North America, thus introducing delays for scientists stationed on other continents. This problem is also significant for training quantum neural networks (QNN), as they must repeatedly run the quantum circuit to adjust weights. The third constraint is simply monetary, which may be discouraging for scientist just starting their work in this field, as there is a risk of spending significant amounts without a guarantee of success. These problems noticeably discourage researchers from using real quantum hardware and steer towards simulating quantum circuits using classical computers [12].

Odra 5 is a quantum computer made by Finnish company IQM and located at Wrocław University of Science and Technology. It is a Spark-class machine that provides 5 qubits in superconducting, gate-based technology [16]. Although this size and type of quantum processors cannot meaningfully handle all quantum algorithms, such as VRP which requires a quantum annealer with thousands of qubits [9], it can be used for low-feature based QML.

This paper presents research of two main topics related to QML. The first is the impact of the optimization algorithms utilized for training the QNN on the final classification accuracy as well as the ease and speed of the training process. Since the optimization landscape of variational quantum algorithms (VQC) is often non-convex the choice of optimizer is crucial for the performance of QNN-based classifiers, as simple gradients like stochastic gradient descent (SGD) may get stuck in a suboptimal point. For experiments described Adam [11] and COBYLA [15] were chosen and compared with SGD as a reference.

The second is the impact of the time since the computer calibration on the accuracy of a trained QNN. Odra 5 is running continuously and needs to be calibrated in regular intervals to ensure correct readouts. At the moment it is being calibrated every 24 h using an algorithm provided by IQM, however it is seeing a low usage at the time of writing, due to limited access, and it is possible that the schedule would need to change if the demand for services rises. Because of this schedule and need for calibration, a hypothesis arose that results of quantum circuit execution would be the best right after the calibration, and the worst just before one.

2 Literature Review

Quantum Machine Learning (QML) is a relatively young research field, even within the broader domain of quantum computing. While foundational quantum algorithms—such as Grover's algorithm for database search—were introduced as early as 1996 [7], well before the realization of the first real quantum computers, the application of quantum computing to machine learning problems has gained significant attention only in recent years. With the rapid development of near-term quantum hardware and increased accessibility to quantum processors, a substantial increase in publications on QML can be observed. This includes not only algorithmic proposals and theoretical analyses but also survey papers aiming to systematize existing approaches, categorize QML methods, and identify open challenges in the field [13].

Several related studies in the literature compare different optimization strategies for training quantum machine learning models. In [19], various variants of the Adam optimizer—one of the most widely used optimization algorithms in classical machine learning—are investigated and compared against stochastic gradient descent (SGD)-based approaches. A similar comparative analysis is presented in [8], where different variants of SGD are examined, including experiments conducted on the same Iris dataset considered in this work. However, a major limitation of both studies is that the reported experiments are conducted exclusively using quantum circuit simulators.

Within the field of quantum machine learning, the most commonly analyzed datasets are feature-based, including benchmark problems such as heart disease classification [1] and cancer diagnosis [17]. More recently, an increasing number of studies have also explored image-based datasets, for example the LFW Face Database [2], reflecting growing interest in applying QML techniques to more complex data modalities. Regarding model architectures, quantum neural networks (QNNs) based on variational quantum circuits remain the dominant approach, particularly for small- and medium-sized datasets [18]. Notably, among the analyzed works, only a single study [10] reports experiments conducted on a real quantum computer of a comparable scale to the one considered in this paper (4-6 qubits). That work focuses on image classification using the MedMNIST dataset and employs quantum transformers in combination with QNNs.

The vast majority of the analyzed studies rely exclusively on quantum circuit simulators to evaluate the performance of quantum machine learning models. Consequently, the influence of real quantum hardware effects—such as noise, calibration drift, and temporal variability—is not addressed. By conducting a comprehensive experimental study on a real quantum computer and explicitly accounting for hardware-related factors, optimizer choice, execution scheduling, and feature selection, the present work aims to bridge this research gap and contribute to a more realistic assessment of quantum neural networks.

3 Dataset

Currently, quantum computers operate in the Noisy Intermediate-Scale Quantum (NISQ) era, in which their computational capabilities are still highly limited. Due to these constraints, increasing the probability of successfully solving a given task requires the use of shallow quantum circuits, and consequently, datasets with a small number of features. For this reason, the Iris dataset [4] constitutes a particularly suitable benchmark for analyzing the applicability of quantum machine learning methods.

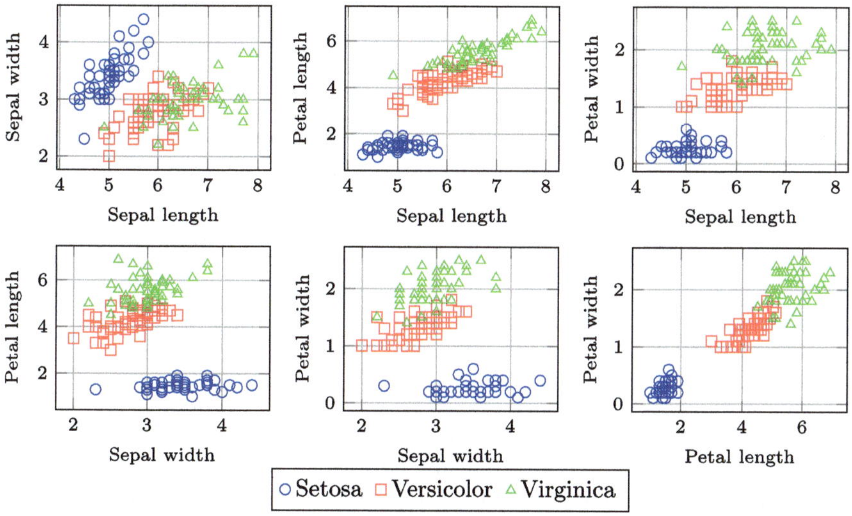

Fig. 1. Pairwise comparison of all features in the Iris dataset, illustrating the relationships and separability between feature pairs across.

3.1 Dataset Description

The dataset contains three classes (Setosa, Versicolor, and Virginica) and, most importantly, each sample is described by only four features: sepal length, sepal width, petal length, and petal width. This low-dimensional feature space enables a direct encoding of classical features into quantum states and their subsequent processing by a VQC, without the need for additional classical preprocessing layers or dimensionality reduction techniques such as autoencoders.

From a classification perspective, the Iris dataset is not particularly challenging, as the Setosa class is easily distinguishable from the remaining two classes. This property is clearly illustrated in Fig. 1, which presents pairwise comparisons of all features for each class (Setosa in blue, Versicolor in red, and Virginica in green). The blue markers are clearly separated from the samples belonging to the other two classes. As a result, the Iris dataset serves as a well-established and practical benchmark for experiments conducted on real quantum processing units (QPU).

3.2 Features Selection

A standard approach to QNN design is to construct quantum circuits according to the one-feature-per-qubit principle. However, variational ansätze typically require entangling operations between qubits. In the case of the Odra 5 QPU, five qubits are available, which theoretically exceeds the number of features in the Iris dataset by one. Nevertheless, the device provides only four native connections: a star-shaped topology in which a central qubit is connected to all remaining qubits, while no direct connections exist between the outer qubits.

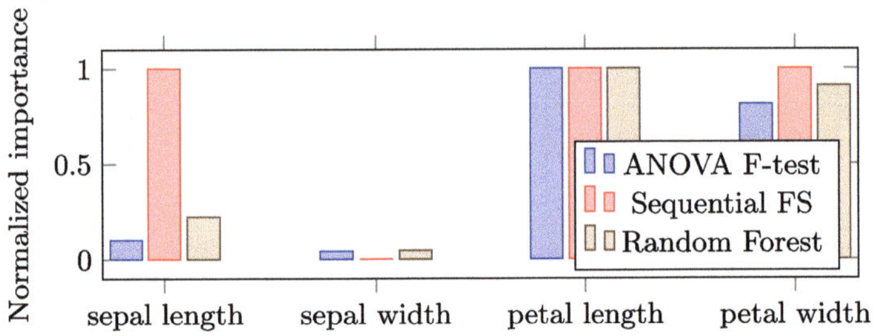

Fig. 2. Normalized feature importance scores for the four features of the Iris dataset, illustrating their relative relevance for the classification task.

Although a four-qubit circuit could in principle be executed, the transpilation process [5] would introduce a significant number of additional SWAP gates to compensate for the limited qubit connectivity. This substantially increases circuit depth and, consequently, the accumulated noise and error rates. For this reason, a three-qubit architecture—operating on only three features—constitutes a more favorable trade-off between expressivity and noise robustness.

Figure 2 presents the normalized results of the feature selection analysis obtained using three different methods: (a) ANOVA F-test, (b) Sequential Feature Selection (SFS), and (c) Random Forest feature importance. The analysis was conducted using the scikit-learn library. All three methods consistently indicate that petal length and petal width are the most informative features, while sepal width is identified as the least significant. Consequently, this feature is excluded from further analysis. Moreover, according to the Sequential Feature Selection method, sepal length is also considered a relevant feature. As a result, this feature is retained, and the final input set consists of three features: petal length, petal width, and sepal length.

4 Methodology

The quantum classification task considered in this work is based on training a variational quantum circuit (VQC) that constitutes the core of the quantum neural network (QNN). The training process aims at optimizing the parameters θ of parametrized quantum gates, such as R_Y rotations, which play a role analogous to neurons in classical neural networks. These parameters determine the transformation applied to the encoded input features and directly influence the classification output measurement probabilities. The optimization of the circuit parameters is performed using classical optimization algorithms, forming a hybrid quantum-classical learning loop.

4.1 Quantum Neural Network Architecture

The quantum circuit used as the QNN is composed of three main parts, that are analogous to a classical neural network model. The architecture can be observed in Fig. 3. The image is split vertically at the first barrier – this is simply for visible clarity and the circuit should be interpreted as one continuous entity. The first part, spanning from the beginning of the circuit on the left and indicated by qubit numbers until the first vertical barrier. It contains the ZZFeatureMap, which is the input layer of the network. It has a Hadamard gate enabling superposition, followed by rotation gates, where input data is encoded as angles. Finally, unlike a ZFeatureMap, it introduces initial entanglements needed for data with assumed correlation. After the first barrier appear the layer equivalents. They are constructed as the RealAmplitudes ansatz repetitions – in the image example there are 4 such layers. Increasing the number of variational layers enhances the expressive power of the model by introducing additional trainable parameters. However, in the context of NISQ-era devices, circuit depth directly impacts performance due to cumulative gate errors and decoherence effects. Each additional gate increases the probability of state corruption, which may degrade classification accuracy. In the final model, 15 repetitions of the RealAmplitudes ansatz were employed. This depth was determined experimentally as a compromise between sufficient representational capacity and acceptable noise-induced degradation. Beyond this point, performance gains from additional parameters were outweighed by the adverse effects of hardware noise. Finally, after the second barrier there is the measurement part, that collapses the qubits and reads out their values, which is then interpreted as output after repeating the circuit execution 1024 times. In Fig. 3 barriers are inserts added to visually distinguish circuit parts and are not used in real circuits.

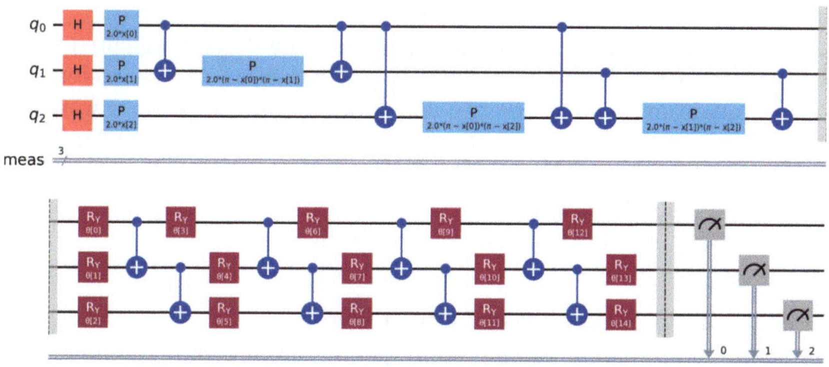

Fig. 3. The quantum circuit architecture with ZZFeatureMap and RealAmplitudes.

4.2 Optimizers for Training Process

To optimize the parameters of a QNN, we used three algorithms and compared their results. Gradient descent is a method for optimizing a differentiable function without constraints. In each iteration of the algorithm, it computes the gradients of the loss function and then chooses the next step where the gradient values are the smallest. The step size is determined by the learning rate α and gradient value. The stop conditions are either convergence at a local minimum or a number of iterations.

ADAM is an algorithm that uses first-order gradient to get to the global maximum. This optimizer keeps the results from a previous iteration to compute the next values. After calculating the gradient, the first- and second-order momentums are updated based on the new information and their values from the previous calculations. The next step is removing their biases, after which the learning rate is computed. It is then returned as the next step value, similar to how it worked in gradient descent.

COBYLA algorithm is a gradient-free method of optimizing a neural network's parameters. It is largely used for problems that aren't differentiable, making it hard to use algorithms based on gradient, such as ADAM. COBYLA works in a trust region, searching for the best option within it. The algorithm uses constraints, in this case set to $\pi \geq \theta_i \geq -\pi$. Because of its searching method, it is largely dependent on the beginning parameter values.

5 Experiments and Results

The experimental study is divided into two main stages. In the first stage, the focus is placed on comparing different optimization algorithms and analyzing their impact on the training quality and convergence behavior of the quantum neural network. In the second stage, the influence of the calibration state of the Odra 5 QPU on the model performance is investigated.

All models are initially trained using the Aer simulator augmented with the native hardware topology and a noise model reflecting the characteristics of the target quantum device. The training process is executed on a workstation equipped with an Intel® Core™ i7-13700K processor and 32 GB of RAM, running Ubuntu 24.04.2 LTS. The experiments are implemented in Python 3.12 using Qiskit 1.2.4, while the iqm-client 33.0.2 library is employed for communication with the QPU during real-hardware evaluation.

5.1 Experimental Analysis of Optimizers

All the optimizers were tested in the same way: first, 200 iterations using a noisy simulator and then 10 tests on Odra 5 using the best theta values computed during training. Each test has been conducted with the same set of classes and features chosen. The QNN has 15 real amplitude layers, a number based on previous tests on the Iris dataset.

For ADAM and gradient descent, the algorithms have been tested one time, with the second one taking slightly longer to finish while also finding the optimal parameters slower: 74 iterations for ADAM and 199 for gradient descent.

The COBYLA algorithm has been tested 10 times because of its dependency on beginning parameters and short execution time, yielding results ranging from

Table 1. Aggregated result from experimental analysis of optimizers.

Optimizer	Accuracy			Time[s]
	Average	Deviation	Max	
SGD	0.99	0.04	1.00	381
Adam	0.98	0.03	1.00	363
Cobyla	0.99	0.04	1.00	5

Table 2. Aggregated result from experimental analysis of the Cobyla optimizer.

Attempt	Accuracy		
	Average	Deviation	Max
1.	0.725	0.075	0.75
2.	0.870	0.080	0.95
3.	0.940	0.060	1.00
4.	0.625	0.075	0.70
5.	0.990	0.040	1.00
6.	0.705	0.095	0.80
7.	0.975	0.125	1.00
8.	0.750	0.100	0.80
9.	0.140	0.090	0.25
10.	0.615	0.085	0.70

an average evaluation accuracy on Odra 5 of 99% to as low as 15%, worse than randomly choosing classes. The comparison between all optimizers is shown in Table 1 and for 10 attempts of COBYLA in Table 2. The differences between average accuracies are minimal, with the same maximal and minimal values.

The average accuracy of COBYLA in 10 tries with 200 iterations in each of them is around 73%, each try taking approximately 5 s to compute.

5.2 Impact of Daily Calibration

The Odra 5 quantum computer is calibrated daily. To test its influence on the results of a QNN evaluation, we conducted hourly tests on the quantum computer. The parameters chosen for this test have been calculated using the ADAM optimizer in the previous test. Evaluations have been conducted 50 times every hour, each one of them containing 20 test samples. The results are shown in Fig. 4. We can see a jump in accuracy of about 1% between 5:00 and 6:00. A plausible explanation for this behavior is the completion of the daily calibration process, scheduled for approximately 4:00. Following calibration, the QPU may require additional time to stabilize before reaching optimal operational conditions. Once this stabilized state is achieved, the performance of the QPU remains relatively consistent for the majority of the day. Based on the

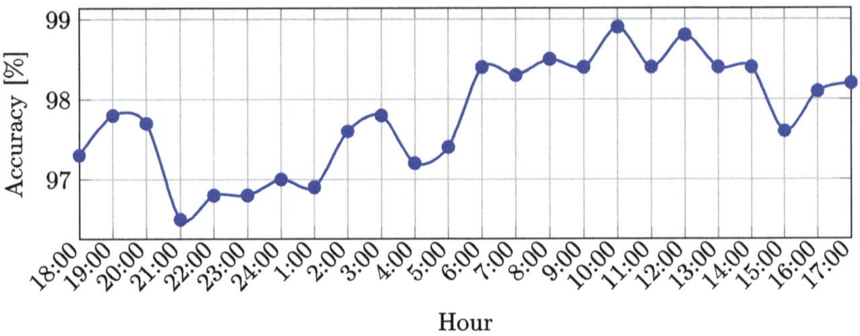

Fig. 4. Average accuracy measured at one-hour intervals on the Odra 5 QPU.

observed results, the most reliable time window for QNN evaluations appears to be between 6:00 and 17:00. Beyond this interval, a noticeable performance degradation was observed the previous day, indicating a decline in the quality of results for the tested quantum circuits.

6 Conclusions

While the COBYLA optimizer works the fastest out of the three, it is also the least reliable, making it necessary to test multiple times in order to get satisfying results. SGD and ADAM have comparable evaluation times, but ADAM finds the optimum faster. In a future version of this algorithm, we could add an early stop condition to shorten the execution time after finding parameters guaranteeing near 100% accuracy.

The hourly tests conducted on Odra 5 show that better accuracy can be achieved during the morning and early afternoon hours than later in the day. It shows that the daily calibration does have an impact, albeit later than expected, with the better accuracies measured after 5:00 AM instead of after 4:00 AM.

As future work, the investigation will be extended to other quantum circuit architectures, including designs explicitly tailored to the native topology and constraints of a given QPU. Furthermore, an important research direction is to adopt hybrid architectures that append classical neural network layers to the QNN, replacing the simple post-processing function currently used for output interpretation. In addition, more complex datasets can be explored by incorporating convolutional layers for feature extraction prior to quantum encoding, enabling the evaluation of quantum-classical models on higher-dimensional, structured data.

References

1. AL Ajmi, N.A., Shoaib, M.: Optimization strategies in quantum machine learning: a performance analysis. Appli. Sci. **15**(8), 4493 (2025)
2. ALRikabi, H., Aljazaery, I.A., Qateef, J.S., Alaidi, A.H.M., Roa'a, M.: Face patterns analysis and recognition system based on quantum neural network qnn. iJIM **16**(08), 35 (2022)
3. Bożejko, W., Trotskyi, S., Uchroński, M., Wodecki, M.: Optimizing two-machine scheduling in flexible manufacturing systems using autonomous ai and quantum computing. Neurocomputing, 132067 (2025)
4. Fisher, R.A.: Iris. UCI Machine Learning Repository (1936)
5. Gnatowski, A., Niżyński, T.: Quantum circuits compilation for superconducting quantum processing units: a systematic mapping study. In: International Conference on Intelligent Systems in Production Engineering and Maintenance, pp. 269–284. Springer (2025). https://doi.org/10.1007/978-3-032-14163-7_19
6. Gnatowski, A., Rudy, J., Niżyński, T., Święcicki, K.: True random number generators on iqm spark, arXiv preprint arXiv:2512.09862 (2025)
7. Grover, L.K.: A fast quantum mechanical algorithm for database search. In: Proceedings of the Twenty-eighth Annual ACM Symposium on Theory of Computing, pp. 212–219 (1996)
8. Huang, Y., Lei, H., Li, X.: An empirical study of optimizers for quantum machine learning. In: 2020 IEEE 6th International Conference on Computer and Communications (ICCC), pp. 1560–1566. IEEE (2020)
9. Jaroszczuk, M.: An introduction to quantum annealing and review of qubo models for vehicle routing problem. In: 2025 IEEE 25th International Symposium on Computational Intelligence and Informatics (CINTI), pp. 523–528. IEEE (2025)
10. Kerenidis, I., Mathur, N., Landman, J., Strahm, M., Li, Y.Y., et al.: Quantum vision transformers. Quantum **8**, 1265 (2024)
11. Kingma, D.P., Ba, J.: Adam: A method for stochastic optimization, arXiv preprint, arXiv:1412.6980 (2014)
12. Kucharski, M.A.: A survey on quantum machine learning in speech acoustics. In: 2025 IEEE 25th International Symposium on Computational Intelligence and Informatics (CINTI), pp. 235–240. IEEE (2025)
13. Li, Y., Tian, M., Liu, G., Peng, C., Jiao, L.: Quantum optimization and quantum learning: a survey. IEEE Access **8**, 23568–23593 (2020)
14. Pirandola, S., et al.: Advances in quantum cryptography. Adv. Optics Photonics **12**(4), 1012–1236 (2020)
15. Powell, M.J.: A direct search optimization method that models the objective and constraint functions by linear interpolation. In: Advances in Optimization and Numerical Analysis, pp. 51–67. Springer (1994). https://doi.org/10.1007/978-94-015-8330-5_4
16. Rönkkö, J., et al.: On-premises superconducting quantum computer for education and research. EPJ Quantum Technol. **11**(1), 32 (2024)
17. Sagingalieva, A., Kordzanganeh, M., Kenbayev, N., Kosichkina, D., Tomashuk, T., Melnikov, A.: Hybrid quantum neural network for drug response prediction. Cancers **15**(10), 2705 (2023)
18. Singh, U., Goldberg, A.Z., Heshami, K.: Coherent feed-forward quantum neural network. Quantum Mach. Intell. **6**(2), 89 (2024)
19. Vu, T.H., Le, V.T.D., Pham, H.L., Nakashima, Y.: Benchmarking variants of the adam optimizer for quantum machine learning applications. IEEE Open J. Comput. Soc. (2025)

Multi-Version AI Systems with Retraining: Availability Assessment Using Single- and Multi-Fragment Markov Models

Vyacheslav Kharchenko[1], Yuriy Ponochovnyi[2], Ievgen Babeshko[1,1(✉)], Oleksandr Ivasiuk[1,3], and Vitalii Mikhalevskyi[4]

[1] National Aerospace University KhAI, Kharkiv, Ukraine
{v.kharchenko,e.babeshko,o.ivasiuk}@csn.khai.edu
[2] Poltava State Agrarian University, Poltava, Ukraine
[3] Research and Production Company Radics, Kropyvnytskyi, Ukraine
[4] Khmelnytskyi National University, Khmelnytskyi, Ukraine

Abstract. This paper investigates the availability and reliability of multi-version artificial intelligence systems (AIS) with retraining, using single-fragment (SFM) and multi-fragment (MFM) Markov models. A classification of models is proposed according to the number of system versions and the structure used to represent retraining, taking into account key parameters that affect the rates of independent and common-cause failures, as well as scheduled and reactive retraining processes in a diversity-based architecture. The study develops Markov availability models for two-version AIS, detailing state-transition graphs, systems of Kolmogorov-Chapman differential equations, and numerical simulation methods. Two main models are considered: a single-fragment model with return to the initial state after retraining, and a multi-fragment model with adaptive improvement of parameters (reduced degradation rates) across consecutive fragments. Simulation results demonstrate the impact of retraining and diversity on system availability during long-term operation. A comparison of availability functions between the models highlights the advantages of the multi-fragment approach, providing key conclusions for improving the dependability of AIS in critical applications under dynamic and aggressive conditions.

Keywords: Multi-Version · AI Systems · Availability · Markov Models · MATLAB Simulation · Retraining

1 Introduction

1.1 Motivation and Related Works

The use of artificial intelligence (AI) in systems where the cost of failure is extremely high continues to grow despite certain limitations [1, 2]. Improving the reliability of artificial intelligence systems (AIS), together with certified solutions that ensure functional safety, facilitates their deployment in critical domains such as autonomous transportation, medicine, energy, and defense [3, 4]. However, AIS remain vulnerable due to risks

W. Zamojski et al. (Eds.): DepCoS-RELCOMEX 2026, LNNS 2014, pp. 196–207, 2026.
https://doi.org/10.1007/978-3-032-27927-9_19

of component failures, including physical hardware defects, software design faults, and AI model anomalies caused by data drift, limited training datasets, or cyberattacks [5, 6].

Especially hazardous are common-cause failures (CCF), in which the same defect manifests simultaneously across all system channels, or an attack exploits shared vulnerabilities, leading to catastrophic consequences [7]. To address these challenges, adaptation of the von Neumann paradigm (VNP) to AI-based systems has been proposed through the use of diversity and redundancy principles [8, 9]. Diversity in models, architectures, data, and training processes reduces the probability of CCF, while independent failures affecting only a single version can be mitigated through voting or decision-making mechanisms [10].

Traditional reliability assessment methods, such as fault tree analysis (FTA) and failure mode and effects analysis (FMEA), focus on static scenarios and do not account for the dynamic adaptation of AI through retraining [1, 2]. More advanced techniques, including Petri nets and reliability block diagrams (RBD), enable state-based modeling but lack sufficient detail for AIS with diversity and retraining processes [3]. Studies of multi-state systems (MSS) estimate parameters from sequences of observations; however, they typically ignore recovery and adaptation mechanisms, as well as AIS-specific characteristics [11].

In contrast, the Markov models proposed in this paper integrate single-fragment and multi-fragment models for single-version systems with retraining and extend them to multi-version AIS architectures. This approach makes it possible to capture dynamic transitions between states of operability, partial degradation, independent and common-cause failures, as well as scheduled and reactive retraining, thereby providing a more accurate assessment of availability under critical conditions [12–14].

1.2 Objectives and Methodology of Research

This paper considers a class of multi-version AIS with retraining, previously investigated by the authors in [8, 9]. Three strategies for implementing the diversity principle within the von Neumann paradigm (VNP) to ensure the dependability of AIS can be distinguished:

- diversity of datasets and methods for constructing AIS versions (channels) (e.g., different types of neural networks, as well as different development and implementation approaches);
- diversity of retraining procedures when problems are detected or additional datasets become available;
- diversity of health monitoring mechanisms for individual AIS versions (channels).

This motivates further development of the theory of multi-version dependable systems with respect to AIS [9]. A description of an AI system redundancy approach that partially implements the formulated principles is provided in [8, 13]. It is based on a multi-version system consisting of parallel-operating intelligent automata (versions); a data-subset analyzer; a converter that transforms analysis results into a control code for version selection while considering the operational status; a reconfiguration automaton that determines and controls system configuration; and retraining facilities under a federated organization. In [13], probabilistic models are presented for dependability assessment while accounting for model diversity and diversity of the software-hardware platform.

The objective of this study is to develop models that improve the trustworthiness (or adequacy) of availability assessment for multi-version AIS, taking into account independent and common-cause failures as well as retraining processes.

The research tasks are as follows: to provide a classification, describe assumptions, and define input parameters of the single-fragment (SFM) and multi-fragment (MFM) Markov models for single-version AI systems with retraining (Sect. 2); to present availability models, with emphasis on single- and multi-fragment models for single-version systems and their extension to a two-version architecture (Sect. 3); to discuss simulation results and provide conclusions regarding the impact of diversity and retraining on reliability and availability (Sect. 4); and to propose directions for further research and practical applications (Sect. 5).

Thus, the research methodology is based on combining the diversity principle for constructing dependable AIS, conceptual models of their information and technical states, and the apparatus of Markov processes for evaluating availability and dependability while accounting for the functional reliability parameters of AIS components. The study was conducted for generalized AIS structures; however, the parameter values were selected with reference to a diagnostic system for complex industrial equipment [15].

2 Classification, Parameterization, and Model Assumptions

To assess the availability of AIS in critical applications [14, 16, 17], it is necessary to use models that account for both AI model degradation and retraining processes. This section presents a classification of Markov models, their main assumptions, and the results of parameterization. The baseline case considered is single-version AI systems, which are subsequently extended to multi-version architectures with model and software-hardware diversity [6, 10].

2.1 AIS State Model

The AIS state model is based on an analysis of the hardware-software platform and the AI model performing control functions [8, 9]. The system state space includes operational states S_w, degradation and failure states S_f (including hardware, software, and AI-model-related failures), and maintenance states S_m associated with retraining:

$$S = S_w \cup S_f \cup S_m,$$

where $S_w = \{S_0, S_1\}$ – denote the states of full and partial operability, $S_f = \{S_2\}$– represents the inoperable state (critical failure or erroneous decisions), $S_m = \{S_3\}$– corresponds to the retraining state (Fig. 1).

Retraining is a key mechanism for adapting an AIS model to new operating conditions (data drift, environmental changes) or for eliminating detected errors. Retraining

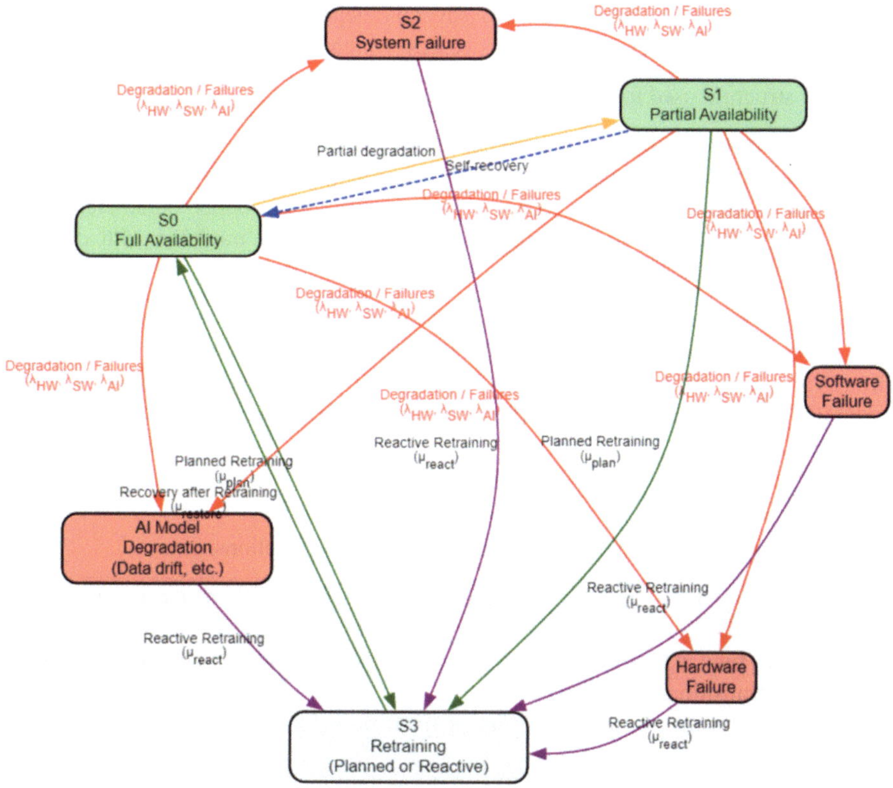

Fig. 1. State-transition graph of a single-version AIS

can be scheduled (periodic updates) or reactive (triggered when degradation or a failure is detected). Transitions between states are modeled by degradation rates λ_{rt} and retraining/recovery rates μ_{rt} [3, 5].

The proposed conceptual model describes a single-version AIS and serves as a basis for extension to multi-version architectures. In multi-version systems, several independent AI versions with diversity (architectures, data, training processes) operate in parallel, and their outputs are compared by a decision-making mechanism (a voter or decision maker). This enables tolerance to independent failures affecting only one version and reduces the risk of common-cause failures (CCF) due to the diversity principle [6, 10].

2.2 Classification of Markov Models

The Markov models are classified according to two main criteria:

1. By the number of system versions:

   – Single-version systems: a single AI model, which remains vulnerable to common-cause failures (CCF).

- Multi-version systems (N-version): several independent versions with diversity in architectures, data, and training processes, which reduces the risk of CCF [6, 10].

2. By the structure used to represent retraining:

   - Single-fragment models (SFM): AI model parameters are assumed to remain unchanged after retraining; the system returns to its initial state.
   - Multi-fragment models (MFM): retraining modifies model parameters (e.g., reduces the degradation rate); multiple fragments are modeled with different parameter sets [8, 9].

Single-fragment models are simpler to analyze but are less accurate in the long-term perspective. Multi-fragment models more adequately capture system adaptation through retraining, which is particularly important for multi-version architectures, where retraining may be coordinated or performed independently for each version [12].

2.3 Model Assumptions

The mathematical models are based on the following assumptions:

- the process is Markovian: the future state depends only on the current state and not on the past history;
- the sojourn time in each state and all state transitions follow exponential distributions (constant transition rates λ and μ);
- retraining is always successful or has a fixed probability of success D_p (in multi-fragment models);
- in single-version models, independent failures are represented by separate transition rates, whereas common-cause failures are captured by global transitions to the failure state;
- for multi-version systems (extensions), diversity is assumed to reduce the probability of CCF but not eliminate it completely; independent failures affect only a single version [6, 10].

These assumptions simplify computations but still enable an adequate evaluation of steady-state availability [11]. Model parameters are based on typical AIS operational scenarios (data drift, cyberattacks, and hardware faults) [3, 5].

The degradation rates λ decrease in subsequent fragments due to model adaptation, reflecting the effect of retraining. In multi-version extensions, additional parameters are introduced for independent failures (λ_{ind}) and common-cause failures (λ_{CCF}) [7, 10].

The proposed classification and parameterization provide a foundation for constructing availability models in the next section, enabling assessment of the impact of retraining and diversity on AIS dependability.

3 Markov's Availability Models for Multi-Version AISs with Retraining

3.1 Markov Single-Fragment Model of a Two-Channel Two-Version AIS

To improve the reliability and availability of AI-based systems in critical applications, a two-version (DMR – Dual Modular Redundancy) architecture based on the diversity principle is proposed [6, 10]. The system consists of two independent AI versions (Va and Vb), developed according to the same specification but with intentional diversity in model architectures, training data, and training processes. The outputs of the two versions are compared by a decision-making mechanism (a voter or decision maker). If the versions agree, the output is considered correct. In the case of disagreement, the system detects a potential error and may either switch to a degraded mode (using the output of a single version) or initiate reactive retraining.

The single-fragment Markov model (SFM) assumes that after retraining the system returns to the initial state with unchanged parameters (i.e., without improvement of degradation rates, in contrast to the multi-fragment model). The model accounts for two types of failures:

- independent failures, where a defect manifests in only one version;
- common-cause failures (CCF), where a defect affects both versions simultaneously.

The model state space includes five states:

S_0: Full availability—both versions are operational and their outputs are consistent.

S_1: Partial availability—one version is degraded (independent failure); the system operates using the healthy version.

S_2: Unavailability—a critical system failure (due to CCF or further degradation).

S_3: Scheduled retraining—coordinated updating of both versions.

S_4: Reactive retraining—triggered when a problem is detected (Fig. 2).

The system of Kolmogorov-Chapman differential equations for the state probabilities $P_i(t)$ is given by:

$$\frac{dP_0(t)}{dt} = -\left(2\lambda_{ind} + \lambda_{CCF} + \mu_{plan}\right)P_0(t) + \mu_{restore}[P_3(t) + P_4(t)]$$

$$\frac{dP_1(t)}{dt} = 2\lambda_{ind}P_0(t) - \left(\lambda_{deg} + \mu_{react1}\right)P_1(t)$$

$$\frac{dP_2(t)}{dt} = \lambda_{CCF}P_0(t) + \lambda_{deg}P_1(t) - \mu_{react2}P_2(t)$$

$$\frac{dP_3(t)}{dt} = \mu_{plan}P_0(t) - \mu_{restore}P_3(t)$$

$$\frac{dP_4(t)}{dt} = \mu_{react1}P_1(t) + \mu_{react2}P_2(t) - \mu_{restore}P_4(t)$$

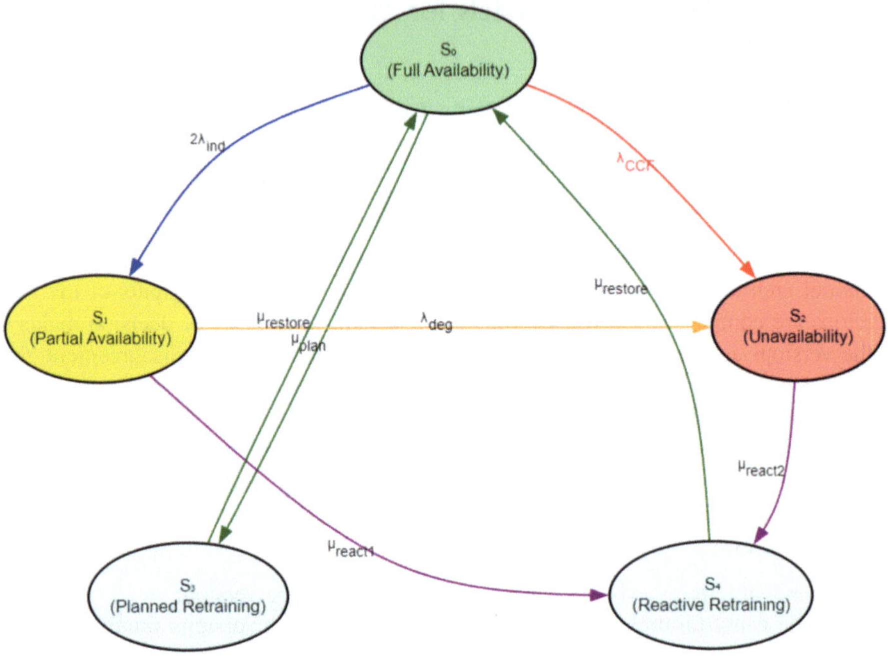

Fig. 2. State-transition graph of the single-fragment Markov model for a two-version AI system

The system availability function is defined as $A_g(t) = P_0(t) + P_1(t)$, since in both states the system remains operational (in S_1, operation is maintained using a single version).

Model parameters are adapted from typical AIS operational scenarios in critical environments [3, 7], taking into account the effect of diversity, which reduces the probability of common-cause failures and independent degradations compared with a single-version system (Table 1).

Table 1. Model parameters for the single-fragment two-version system

Parameter	Value (1/h)	Justification
λ_{ind}	0.007	Independent degradation rate of a single version (reduced due to diversity) [6, 10]
λ_{CCF}	0.001	Common-cause failure rate (low due to diversity) [7]
λ_{deg}	0.015	Further degradation in single-version operation
μ_{plan}	0.033	Scheduled retraining rate (periodicity ~1 per month)
μ_{react1}	0.1	Reactive retraining rate from the partial operability state
μ_{react2}	0.2	Reactive retraining rate from the critical failure state
$\mu_{restore}$	0.5	Average recovery rate after retraining (~2 h)

Simulation was performed in MATLAB using a numerical solution of the differential equation system (ode15s). Initial conditions: $P_0(0) = 1$, all other probabilities are set to zero. The simulation time was 200 h to reach steady-state operation (Fig. 3).

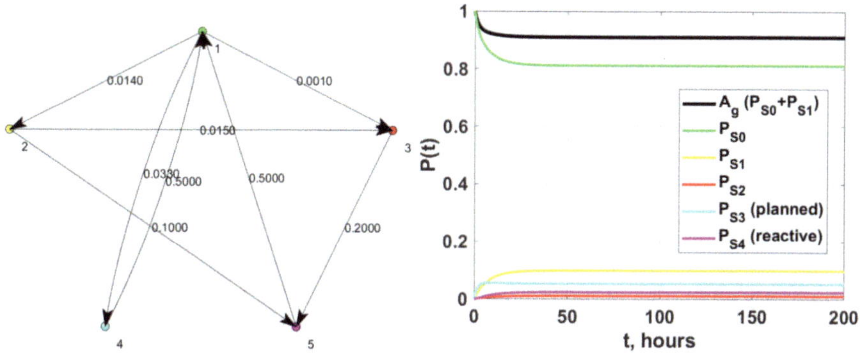

Fig. 3. Dynamics of state probabilities and availability coefficient for the single-fragment model of a two-version AIS

Simulation results show that the steady-state availability of the two-version system is approximately 0.945..0.950, which is significantly higher than that of a single-version system (~0.893) due to tolerance to independent failures and the reduced risk of common-cause failures (CCF) [6, 10]. Transient processes stabilize within the first 200–300 h of operation.

The proposed single-fragment model serves as a baseline for evaluating the availability of two-version AIS with retraining and demonstrates the benefits of the diversity principle in critical applications.

3.2 The Multi-Fragment Markov Model of Two-Channel Diverse AIS

The multi-fragment Markov model (MFM) is an extension of the single-fragment model and incorporates the adaptive effect of retraining: after successful or partially successful retraining, the system transitions to a new fragment with improved parameters (reduced degradation and common-cause failure rates) [8, 9]. This reflects the real adaptation process of AIS to new operating conditions (data drift, accumulated operational experience), enhancing the long-term availability of the system. In a two-version architecture, diversity is maintained within each fragment, and retraining can be coordinated for both versions.

The model consists of several consecutive fragments (in the example, three), each with the same five states as in the single-fragment model:

S_0: Full availability—both versions are consistent.

S_1: Partial availability—one version degraded due to an independent failure.

S_2: Unavailability—due to a common-cause failure (CCF) or further degradation.

S_3: Scheduled retraining.

S_4: Reactive retraining.

Transitions within a fragment are similar to those in the single-fragment model, but after retraining, the system returns to S_0 of the current fragment with probability D_p, or with probability 1-D_p it moves to S_0 of the next fragment with improved parameters (Fig. 4).

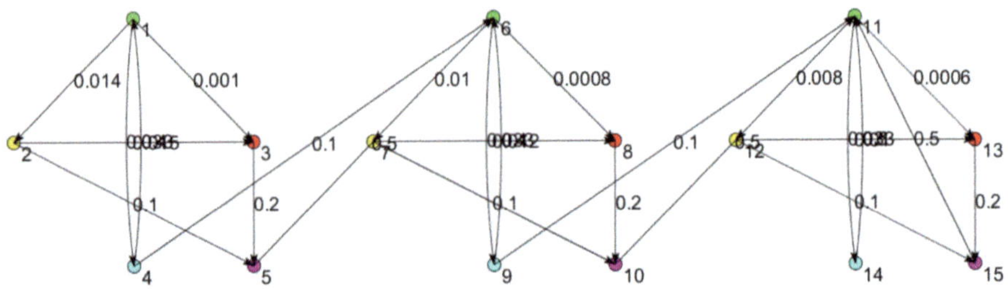

Fig. 4. State-transition graph of the multi-fragment Markov model for a two-version AI system

The system of differential equations is extended to cover all states across all fragments (15 states for three fragments). Inter-fragment transitions are introduced after the retraining states (S_3, S_4) with coefficients Dp and 1 − Dp.

Model parameters are specified for three fragments, where degradation rates decrease due to AIS model adaptation and accumulated diversity-related experience [6, 10] (Table 2).

The simulations were performed in MATLAB using the same functions for constructing the state graph and solving the system of equations. The initial condition places the system in S_0 of the first fragment. The simulation time was 1000 h (Fig. 5).

The results show that the steady-state availability of the multi-fragment model reaches 0.965.0.975, which is 2..3% higher than that of the single-fragment model (~0.945). This improvement is explained by the accumulation of adaptation: the system gradually transitions to fragments with lower failure rates. Transient processes are longer due to inter-fragment transitions; however, the long-term resilience of the system is significantly higher, especially under prolonged operation.

The multi-fragment model more adequately reflects the real adaptive behavior of two-version AIS with retraining and demonstrates the benefits of combining the diversity principle with reactive and scheduled model updates in critical applications [12].

4 Conclusion

The main contribution of the study is the development of an approach for modeling and evaluating the availability of multi-version AIS with retraining based on single-fragment (SFM) and multi-fragment (MFM) Markov models. The proposed models account for AI degradation levels, independent and common-cause failures, as well as scheduled and reactive retraining in a two-version architecture implementing the diversity principle.

This method supports decision-making aimed at improving the reliability and availability of AI-based systems in critical applications. The use of the multi-fragment model,

Table 2. Parameters of the multi-fragment model for a two-version AIS (three fragments)

Parameter	F 1	F 2	F 3	Justification
λ_{ind}	0.007	0.005	0.004	Reduction of independent degradations due to adaptation [6]
λ_{CCF}	0.001	0.0008	0.0006	Reduction of CCF due to accumulated experience and diversity [7, 10]
λ_{deg}	0.015	0.012	0.010	Reduced further degradation in single-version operation
μ_{plan}	0.033	0.033	0.033	Scheduled retraining (constant)
μ_{react1}	0.10	0.10	0.10	Reactive retraining from the partial state
μ_{react2}	0.20	0.20	0.20	Reactive retraining from the critical failure state
$\mu_{restore}$	0.50	0.50	0.50	Recovery within a fragment / in the final fragment
μ_{next}	0.50	0.50	–	Transition time to the next fragment
D_p	0.80	0.80	–	Probability of successful retraining without fragment transition [9]

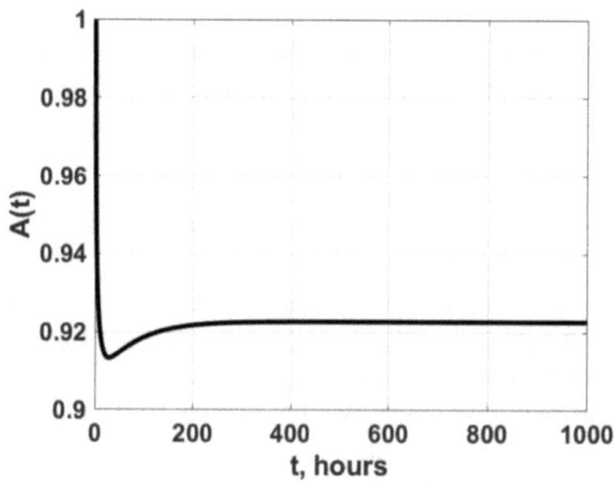

Fig. 5. Dynamics of the availability coefficient for the multi-fragment model of a two-version system

which captures the adaptive effect of retraining through a gradual reduction in degradation and common-cause failure rates, provides higher long-term availability compared with the single-fragment model, particularly under prolonged operation and a changing environment.

Simulation results show that the steady-state availability of the single-fragment model is approximately 0.945–0.950, whereas the multi-fragment model reaches 0.965.0.975, i.e., 2..3% higher due to accumulated adaptation across consecutive fragments. Although transient processes in the MFM are longer because of inter-fragment transitions, the long-term resilience of the system is substantially higher. This confirms the advantages of combining the diversity principle with retraining mechanisms over traditional single-version approaches.

The developed Markov models (SFM and MFM) are intended for practical use in critical AIS. For example, in autonomous vehicles [3, 4], the single-fragment model can represent a two-version control system tolerant to independent AI model degradation (e.g., due to data drift under different weather conditions), whereas the multi-fragment model reflects gradual adaptation through periodic retraining on accumulated operational data. Similarly, in medical diagnostic systems [4], the multi-fragment structure enables modeling of progressive accuracy improvement through reactive retraining after detecting discrepancies between versions, thereby ensuring high availability in long-term operation.

Future research may focus on applying the proposed method to improve AIS availability by integrating coordinated retraining strategies and extending multi-fragment models to architectures with a larger number of versions ($N > 2$). Another promising direction is the study of hybrid systems combining AI model degradation with hardware and software factors, as well as incorporating probabilistic effects of retraining success for a broader range of engineering domains.

References

1. Rahmani, A.M., Rezazadeh, B., Haghparast, M., Chang, W.-C., Ting, S.G.: Applications of artificial intelligence in the economy, including applications in stock trading, market analysis, and Risk Management. IEEE Access **11**, 80769–80793 (2023). https://doi.org/10.1109/access.2023.3300036
2. Mirzarazi, F., Danishvar, S., Mousavi, A.: The safety risks of AI-driven solutions in Autonomous Road Vehicles. World Electr. Veh. J. **15**(10), 438 (2024). https://doi.org/10.3390/wevj15100438
3. Moskalenko, V., Kharchenko, V., Moskalenko, A., Kuzikov, B.: Resilience and Resilient Systems of Artificial Intelligence: Taxonomy, models and methods. Algorithms **16**(3), 165 (2023). https://doi.org/10.3390/a16030165
4. Illiashenko, O., Kharchenko, V., Babeshko, I., Fesenko, H., Di Giandomenico, F.: Security-informed safety analysis of autonomous transport systems considering AI-powered cyberattacks and protection. Entropy **25**(8), 1123 (2023). https://doi.org/10.3390/e25081123
5. Liu, F., Zheng, H., Ma, S., Zhang, W., Liu, X., Chua, Y., Shi, L., Zhao, R.: Advancing brain-inspired computing with Hybrid Neural Networks. Natl. Sci. Rev., **11**(5), (2024). https://doi.org/10.1093/nsr/nwae066
6. Mendonça, J., Machida, F., Völp, M.: Enhancing the reliability of perception systems using N-version programming and rejuvenation. In: 2023 53rd Annual IEEE/IFIP international conference on Dependable Systems and Networks Workshops (DSN-W), pp 149–156 (2023). https://doi.org/10.1109/dsn-w58399.2023.00044
7. Wood, R.T., et al.: Diversity strategies for nuclear power plant instrumentation and control systems (2010). https://doi.org/10.2172/1000417

8. Kharchenko, V., Ponochovnyi, Y.: Markov's models of AI systems availability considering re-learning processes. ProfIT AI' **25**, 2025 (2025)
9. Kharchenko, V.: Conceptual foundations of dependable artificial intelligence systems. Rep. Natl. Acad. Sci. Ukr., (2), 11–23 (2025). https://doi.org/10.15407/dopovidi2025.02.011
10. Avizienis, A.: The N-version approach to fault-tolerant software. IEEE Trans. Softw. Eng., SE-11(12), 1491–1501 (1985). https://doi.org/10.1109/tse.1985.231893
11. Wotawa, F.: On the use of available testing methods for verification & validation of AI-based software and systems. In: CEUR workshop proceedings, 2808 (2021). https://ceur-ws.org/Vol-2808/Paper_29.pdf
12. Shayea, G.G., Zabil, M.H., Habeeb, M.A., Khaleel, Y.L., Albahri, A.S.: Strategies for protection against adversarial attacks in AI models: An in-depth review. J. Intell. Syst., 34(1), (2025). https://doi.org/10.1515/jisys-2024-0277
13. Kharchenko, V., Odarushchenko, O.: Trustworthy AI systems from untrustworthy components: Development von Neumann's paradigm using principle of diversity. In: Proceedings 4th International Workshop of IT-professionals on Artificial Intelligence (ProfIT AI 2024), September 25–27,2024, Cambridge, MA, USA, pp. 392–404. https://ceur-ws.org/Vol-3777/paper24.pdf
14. Mishchuk, V., Fesenko, H., Kharchenko, V.: Deep learning models for detection of explosive ordnance using autonomous robotic systems: trade-off between accuracy and real-time processing speed. Radioelectron. Comput. Syst. (4), 99–111. https://doi.org/10.32620/reks.2024.4.09
15. Kharchenko, V., Shchehlov, V., Ivasiuk, O., Morozova, O.: Digital twin-based lifecycle methodology for ensuring safety of NPP/SMR I&C systems. Technologies **14**(1), 46 (2026). https://doi.org/10.3390/technologies14010046
16. Perez-Cerrolaza, J., et al.: Artificial Intelligence for safety-critical systems in industrial and transportation domains: A survey. ACM Comput. Surv. **56**(7), 1–40 (2024). https://doi.org/10.1145/3626314
17. Babu, C.V., Simon, P.A.: Adaptive AI for dynamic cybersecurity systems. Adv Comput Intell Robot, 52–72. https://doi.org/10.4018/979-8-3693-0230-9.ch003

Evaluation of AI-Generated Image Quality for Selected Artistic Styles

Maciej Kopczynski(✉) and Jakub Ogar

Faculty of Computer Science, Bialystok University of Technology, Wiejska 45A, 15-351 Białystok, Poland
m.kopczynski@pb.edu.pl

Abstract. This work evaluates the quality of images generated by selected artificial intelligence models in the context of reproducing distinct artistic styles. The study includes two classical styles: van Gogh and cubism. Several widely used open-source generative models - Stable Diffusion 3.5, Flux.1 Dev and Qwen Image, are compared against reference datasets composed of original artworks representative of each style. The evaluation is conducted using a combination of quantitative, perceptual and semantic metrics, including KID, SSIM and CLIP. The results reveal model performance differences depending on different evaluation criterion.

Keywords: Artificial intelligence · Image generation · Generative models · Image quality metrics

1 Introduction

In recent years, generative artificial intelligence has grown rapidly, particularly in the domain of synthesizing images from textual descriptions. Text-to-image systems, once purely theoretical, now achieve a level of sophistication that enables the creation of photorealistic and artistically complex compositions often indistinguishable from human-made works. This progress raises a key question: how can the quality of such generated images be assessed objectively, especially in terms of their ability to accurately reproduce specific artistic styles?

Users typically expect that a prompt such as "a portrait of a woman in the style of van Gogh" will yield an image that not only appears coherent and aesthetically consistent, but also reflects the defining characteristics of the given style: its brushstroke textures, color palette, composition and recurring visual motifs, while preserving the semantic content specified in the prompt.

Such expectations motivate this study, which investigates whether modern AI models can faithfully capture essential attributes of selected artistic styles and compare their relative performance. The research focus goes beyond determining whether an image is "realistic" or "visually appealing" emphasizing instead the model's capacity to reconstruct stylistic features characteristic of a given artistic

W. Zamojski et al. (Eds.): DepCoS-RELCOMEX 2026, LNNS 2014, pp. 208–217, 2026.
https://doi.org/10.1007/978-3-032-27927-9_20

tradition. Addressing this requires quantitative evaluation methods that allow systematic comparison of generated images with datasets of authentic artworks.

Recent progress in generative image modeling has been driven primarily by diffusion-based text-to-image architectures [1–4]. Multimodal representation learning enabling semantic comparison between images and text was introduced in [5] and further developed in visionlanguage pretraining frameworks [6,7]. Some approaches to basic evaluation of generated images based on distribution-based metrics such as KID and FID [8,9], structural similarity measures [10,11,14] or perceptual similarity metrics based on deep features [12,13] exist.

This study employs three leading open-source generative models: Stable Diffusion 3.5, Flux.1 Dev and Qwen Image. Each model was tasked with generating images based on a unified set of prompts representing two styles: van Gogh and cubism. Corresponding collections of original artworks or photographs served as reference datasets. The generated images were evaluated using three complementary metrics: Kernel Inception Distance (KID) for global feature-space similarity [9], Structural Similarity Index Measure (SSIM) for local structural and textural alignment [10] and Contrastive LanguageImage Pre-training (CLIP) semantic scoring to assess the correspondence between textual prompts and visual output [5].

The paper is organized as follows. In Sect. 2 some information about the generative AI models, image quality metrics and reference datasets are provided. Also the dataset used during the research is described. The Sect. 3 focuses on a description of the system architecture and the implementation structure, while the Sect. 4 is devoted to the experimental results.

2 Images Generation and Quality Metrics

This section outlines the key concepts used in the study, including the generative AI models, image quality metrics and reference datasets necessary for the experimental evaluation.

2.1 Generative Models

Open-source generative models play an important role in research by providing transparency, modifiability and the ability to run computations locally. This study employs three leading models available under such licenses: Stable Diffusion 3.5, Flux.1 Dev and Qwen Image.

Stable Diffusion 3.5 developed by Stability AI, is a latent diffusion model (LDM) that performs image generation in latent space to reduce computational complexity [15]. The model uses multiple text encoders (CLIP-G/14, CLIP-L/14 and T5 XXL) to transform textual prompts into semantic conditioning vectors that guide the denoising process. Image synthesis begins from noise in latent space, which is iteratively refined using timestep conditioning until a coherent image representation is obtained.

Flux.1 Dev developed by Black Forest Labs, is a rectified-flow transformer-based generative model designed for high structural fidelity and prompt alignment [16]. The architecture combines Dual Stream Blocks, which process visual and textual features separately with cross-attention and Single Stream Blocks, where both modalities are merged into a unified representation. Image generation starts from latent noise conditioned on text embeddings (CLIP-L/14 and T5 XXL) and timestep information, followed by reconstruction into pixel space.

Qwen Image developed by Alibaba Cloud, is a multimodal text-to-image diffusion model integrating the Qwen2.5 VL visionlanguage architecture with transformer-based diffusion generation [17]. The model encodes textual prompts into conditioning vectors that guide latent diffusion through Multi-Modal Diffusion Transformer (MMDiT) blocks. After iterative denoising in latent space, the representation is decoded into the final image.

2.2 Image Quality Metrics

To objectively assess how well generative models reproduce specific artistic styles, quantitative image quality metrics are required. Each metric examines images differently—through statistical alignment, perceptual similarity, or semantic consistency.

KID (Kernel Inception Distance) is a distribution-based metric used to evaluate the quality and diversity of image sets generated by generative models. It was introduced in [9]. Metric is based on squared Maximum Mean Discrepancy (MMD), which eliminates the need to assume that feature distributions are Gaussian. It's defined as:

$$\mathrm{KID} = \mathrm{MMD}^2 = \mathbb{E}_{x,x' \sim p_r}[k(x,x')] + \mathbb{E}_{y,y' \sim p_g}[k(y,y')] - 2\,\mathbb{E}_{x \sim p_r,\, y \sim p_g}[k(x,y)],$$

where p_r feature distribution of the reference dataset, p_g feature distribution of the generated dataset, $k(x,y) = \left(\frac{1}{d}x^T y + 1\right)^3$ polynomial kernel (with $d = 2048$), x, x' features from the reference dataset and y, y' features from the generated dataset.

A lower KID value indicates a greater statistical similarity between the datasets.

SSIM is computed locally—for small sliding windows (typically 11×11 pixels)—and then averaged over the entire image [10]. For a single window, SSIM is defined as:

$$\mathrm{SSIM}(x,y) = \frac{(2\mu_x\mu_y + C_1)(2\sigma_{xy} + C_2)}{(\mu_x^2 + \mu_y^2 + C_1)(\sigma_x^2 + \sigma_y^2 + C_2)},$$

where x, y—pixel vectors from the window in the reference and generated images, μ_x, μ_y—mean pixel intensities, σ_x^2, σ_y^2—variances (contrast measures), σ_{xy}—covariance between x and y (structural similarity) and C_1, C_2—small stabilizing constants preventing division by zero.

The SSIM value ranges from -1 to 1, where 1 indicates perfect similarity.

CLIP (Contrastive LanguageImage Pre-training) is a multimodal model introduced by OpenAI in 2021 [5]. Although not a traditional metric, its architecture enables the computation of semantic similarity between images and textual descriptions. The CLIP-based similarity metric operates by encoding an image I into a feature vector $v_I \in \mathbb{R}^d$ using the image encoder, while a text input T (e.g., a prompt or style description) is mapped to a feature vector $v_T \in \mathbb{R}^d$ using the text encoder. Semantic similarity is then computed as the cosine similarity between the two embedding vectors:

$$\text{sim}(I, T) = \frac{v_I \cdot v_T}{|v_I| \cdot |v_T|}$$

where $v_I \in \mathbb{R}^d$ is image embedding and $v_T \in \mathbb{R}^d$ is text embedding.

The resulting value lies in the interval $[-1, 1]$, where 1 denotes perfect semantic alignment.

2.3 Data for Experimental Results

Each dataset was compiled from reliable public sources, primarily museum collections, digital archives and licensed image databases, ensuring thematic and stylistic consistency. All images were resized to a uniform resolution and subjected to minimal preprocessing, such as frame removal and contrast normalization, in order to avoid introducing artificial bias in the evaluation metrics.

In this study, a reference dataset consisting of 1 000 selected works by Vincent van Gogh was used as a fundamental benchmark. This choice is motivated by the distinctive character of the artist's style, which poses a challenge for generative AI models. The images, including "The Starry Night" (1889), "Bedroom in Arles" (1888) and the "Sunflowers" series, were collected from the digital collections of the van Gogh Museum in Amsterdam, The Metropolitan Museum of Art in New York and WikiArt, which provide high-resolution images suitable for research purposes.

The Cubism dataset consists of 2 500 artworks representing analytical and synthetic Cubism, primarily from the period 19071914. Works by Pablo Picasso and Georges Braque were obtained from the digital collections of Centre Pompidou, MoMA and WikiArt. This dataset is intended to evaluate the ability of AI models to capture geometric abstraction and multi-perspective composition.

3 Solution Description and Methodology

This chapter presents the research methodology and the system architecture, including technologies used, implementation structure and execution environment.

3.1 System Architecture

The implemented solution consists of two main functional components:

1. **Image generation**—performed using three AI models based on a consistent set of textual prompts.
2. **Quality evaluation**—conducted using three metrics (KID, SSIM and CLIP) to compare generated images with original artworks representing a given visual style.

To ensure reproducibility of the experiments, a modern and well-documented technology stack was used, based on the PyTorch ecosystem and Hugging Face. Generative models were implemented using Hugging Face libraries, including **Diffusers (v0.29.2)**, which provides standardized pipelines for diffusion models such as Stable Diffusion 3.5, Flux.1 Dev and Qwen Image. **Transformers (v4.41.2)** library was used for language models and text encoders such as T5 and CLIP.

3.2 Methodology

The data generation process was organized as a modular workflow consisting of prompt management, image generation modules, validation and preprocessing components. For each prompt, five images were generated. In total, 20 prompts were prepared for each style category.

Prompts were created in "canonical" type, representing typical visual conventions of a given artistic style. Examples are like "A starry night sky over a quiet village with swirling patterns, detailed background, moonlight illumination, in the style of van Gogh" or "A city street with houses and rooftops, shown in multiple viewpoints, sharp angles, cubist abstraction, in the style of Cubism". Prompts were defined for two visual style categories used throughout the experiments: **van Gogh** as post-impressionist style characterized by expressive brushstrokes and **Cubism** characterized by geometric abstraction and multi-perspective representation.

Image generation presented on Fig. 1 was implemented as a structured multi-stage pipeline ensuring data consistency and computational efficiency and is divided into 3 stages:

1. **Prompt preprocessing** – prompts were optimized before being passed to the image generation models.
2. **Batch and single-image generation** – models generated five images per prompt using batch processing. If an error occurred, the system automatically switched to single-image generation.
3. **Generation parameter management** – where model-specific parameters were applied for images generation:
   - Stable Diffusion 3.5: 20 inference steps, guidance scale 3.5, DPMSolver-Multistep sampler,
   - Flux.1 Dev: 20 steps, guidance scale 3.5, maximum sequence length 512, FlowMatch sampler,

- Qwen Image: 20 steps, CFG scale 4.0.

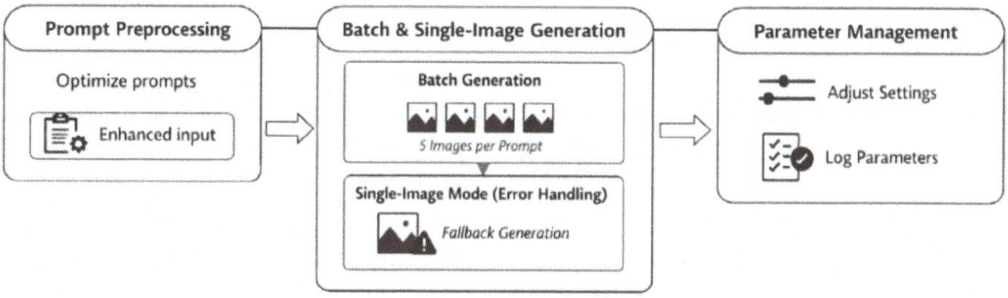

Fig. 1. Image generation and assessment process.

For each generated image, complete metadata were recorded, including generation parameters (seed, steps, CFG scale, sampler), generation time, file paths, model fingerprint and timestamp. Exemplary two images (from reference dataset and generated one) for van Gogh style and Qwen model are presented on Fig. 2. Additional images couldn't be included due to paper page limitations.

(a) Reference

(b) Generated

Fig. 2. Reference image and generated image by the Qwen model using the prompt: "A starry night sky over a quiet village with swirling patterns, detailed background, moonlight illumination, in the style of van Gogh"

After generation, all images were resized to a resolution of 512 px while preserving visual quality. Lanczos interpolation with a 1.15× margin and center cropping enabled approximately a 2.5× acceleration of subsequent analyses without noticeable degradation in image quality. Each image entering the system undergoes a quality validation procedure. Images with a resolution lower than 512 px or with extreme aspect ratios are discarded to ensure consistency and high quality of the reference datasets.

4 Experimental Results

Image generation was performed using virtual machines available on cloud computing platforms Vast.ai and RunPod.io. This choice was motivated by the need for access to high-performance GPUs, such as the NVIDIA RTX 6000 Pro, which provide large VRAM capacity and substantial computational power. Renting cloud resources proved to be an efficient and flexible solution, enabling large-scale image generation within a relatively short time without requiring investment in dedicated hardware infrastructure.

The development and testing stage of scripts for data preprocessing and evaluation metric computation was conducted on local hardware. A workstation equipped with an NVIDIA GeForce RTX 5090 GPU provided sufficient computational power for debugging, calibration and computing the full set of evaluation metrics. The available GPU memory also enabled the use of larger batch sizes during metric computation, reducing analysis time and improving overall experimental efficiency.

The experiments covered all combinations of models, styles, evaluation metrics and analysis modes, using a variable number of reference images (100, 500, 1 000 and 2 500 depending on the availability of original artworks). This approach enabled the assessment of result stability and reliability under different experimental conditions. The aggregation of evaluation results for each category was performed as follows:

- all images generated by a given model for a specific style were collected,
- the generated images were compared with a selected subset of reference images,
- metric values were computed for each image pair or image set, depending on the evaluation method,
- average and standard deviation for quality metrics were calculated for set of images and prompts.

Tables 1 and 2 present the results of images quality metrics for different types of models. Column *Samples* provides the number of the reference images used as input to the metrics. Column labeled as *Avg* represents average value of selected metric calculated for set of generated images, while *StdDev* shows standard deviation for the results.

Lower KID value indicates a greater statistical similarity between the datasets, while in SSIM and CLIP 1 means perfect similarity.

Flux.1 Dev exhibits a distinctive performance profile. It achieved the lowest average scores in KID, which may indicate lower diversity or a distributional mismatch with the reference artworks. At the same time, it obtained relatively high

Table 1. Image quality metrics for different types of models for van Gogh style.

Samples	KID		SSIM		CLIP	
	Avg	Std Dev	Avg	Std Dev	Avg	Std Dev
Flux.1 Dev						
100	0.0535	0.0002	0.2801	0.052	0.2953	0.0182
500	0.0620	0.0042	0.3147	0.058	0.3335	0.0223
1 000	0.0619	0.0046	0.3405	0.0488	0.3527	0.0245
Qwen Image						
100	0.0441	0.0004	0.2963	0.0484	0.3065	0.0212
500	0.0514	0.0036	0.3473	0.0482	0.3413	0.0269
1 000	0.0506	0.0035	0.3667	0.0399	0.3701	0.0286
Stable Diffusion 3.5						
100	0.0364	0.0030	0.1982	0.0549	0.3012	0.0193
500	0.0445	0.0036	0.2211	0.0639	0.3399	0.0241
1 000	0.0438	0.0037	0.2434	0.0597	0.3568	0.0205

Table 2. Image quality metrics for different types of models for Cubism style.

Samples	KID		SSIM		CLIP	
	Avg	Std Dev	Avg	Std Dev	Avg	Std Dev
Flux.1 Dev						
100	0.1014	0.0004	0.374	0.0563	0.2834	0.0291
500	0.0959	0.0049	0.4213	0.059	0.2965	0.0305
1 000	0.0945	0.0064	0.4694	0.0652	0.3133	0.0321
2 500	0.0876	0.0060	0.5211	0.0606	0.3312	0.0346
Qwen Image						
100	0.0830	0.0003	0.4136	0.1034	0.3146	0.0295
500	0.0763	0.0045	0.4626	0.0972	0.3282	0.0314
1 000	0.0713	0.0049	0.5312	0.1041	0.3428	0.0327
2 500	0.0648	0.0045	0.5928	0.0671	0.3547	0.0339
Stable Diffusion 3.5						
100	0.0840	0.0005	0.3707	0.0596	0.3254	0.021
500	0.0771	0.0037	0.4108	0.0540	0.3551	0.023
1 000	0.0716	0.0042	0.4641	0.0606	0.3691	0.026
2 500	0.0687	0.0038	0.5055	0.0528	0.3752	0.025

scores among the tested models in SSIM and CLIP, suggesting strong structural consistency and effective alignment with prompt content and context.

Stable Diffusion 3.5 shows a clear dependence on visual style. In the KID metric, it achieved very good results for the van Gogh style but considerably weaker performance for Cubism, indicating limitations in capturing abstract or culturally specific visual forms. In SSIM, particularly for Cubism, the model remained in the middle of the ranking, reflecting moderate but stable visual quality.

Qwen Image consistently achieved mid-range KID scores, indicating balanced but not outstanding distributional similarity. In SSIM, however, it competed with Flux.1 Dev for the top position, suggesting high structural fidelity in the generated images.

5 Conclusion

The conducted experiments indicate that no single model is universally superior across all evaluation criteria. Flux.1 Dev achieved the highest structural and semantic consistency, while Qwen Image and Stable Diffusion 3.5 proved particularly effective in selected prompts content, although their performance depended strongly on the target visual style. An important observation is the discrepancy between evaluation metrics: high distributional similarity does not necessarily correspond to strong visual or semantic quality and models achieving the best CLIP scores do not always perform best in KID. This confirms that evaluating generative models requires a multidimensional approach.

Future research should focus on collecting a larger set of both reference and generated images, which would improve the statistical robustness of distribution-based evaluation metrics. Moreover, a more comprehensive statistical analysis of the collected data and obtained results should be conducted. Another important direction involves extending the comparison to include more recent generative models, particularly those fine-tuned for specific artistic styles. Additionally, the evaluation framework could be enriched by incorporating complementary assessment methods, such as human-based evaluation. In particular, user studies in which participants assess aspects such as style consistency, content fidelity, and artistic appeal could provide valuable insights that complement automatic metric-based evaluation. Such an approach would enable a more comprehensive understanding of the performance of generative models in tasks that involve not only technical accuracy but also aesthetic and cultural dimensions.

Acknowledgment. The research was carried out within project no. WZ/WI-IIT/3/2026 at Bialystok University of Technology and financed from the research subsidy of Bialystok University of Technology.

References

1. Ho, J., Jain, A., Abbeel, P.: Denoising diffusion probabilistic models. In: Advances in Neural Information Processing Systems (NeurIPS), vol. 33 (2020). https://doi.org/10.48550/arXiv.2006.11239

2. Rombach, R., Blattmann, A., Lorenz, D., Esser, P., Ommer, B.: High-resolution image synthesis with latent diffusion models. In: Proceedings of the IEEE/CVF Conference on Computer Vision and Pattern Recognition (CVPR) (2022). https://doi.org/10.1109/CVPR52688.2022.01042
3. Ramesh, A., Dhariwal, P., Nichol, A., Chu, C., Chen, M.: Hierarchical text-conditional image generation with CLIP latents. arXiv preprint (2022). https://doi.org/10.48550/arXiv.2204.06125
4. Saharia, C., Chan, W., Saxena, S., et al.: Imagen: photorealistic text-to-image diffusion models. In: Advances in Neural Information Processing Systems (NeurIPS), vol. 35 (2022). https://doi.org/10.48550/arXiv.2205.11487
5. Radford, A., Kim, J.W., Hallacy, C., et al.: Learning transferable visual models from natural language supervision. In: Proceedings of the International Conference on Machine Learning (ICML) (2021). https://doi.org/10.48550/arXiv.2103.00020
6. Li, J., Li, D., Savarese, S., Hoi, S.: BLIP: bootstrapping language-image pre-training for unified vision-language understanding and generation. In: Proceedings of the International Conference on Machine Learning (ICML) (2022). https://doi.org/10.48550/arXiv.2201.12086
7. Alayrac, J.B., Donahue, J., Luc, P., et al.: Flamingo: a visual language model for few-shot learning. In: Advances in Neural Information Processing Systems (NeurIPS), vol. 35 (2022). https://doi.org/10.48550/arXiv.2204.14198
8. Heusel, M., Ramsauer, H., Unterthiner, T., Nessler, B., Hochreiter, S.: GANs trained by a two time-scale update rule converge to a local nash equilibrium. In: Advances in Neural Information Processing Systems (NeurIPS), vol. 30 (2017). https://doi.org/10.48550/arXiv.1706.08500
9. Bińkowski, M., Sutherland, D.J., Arbel, M., Gretton, A.: Demystifying MMD GANs. In: International Conference on Learning Representations (ICLR) (2018). https://doi.org/10.48550/arXiv.1801.01401
10. Wang, Z., Bovik, A.C., Sheikh, H.R., Simoncelli, E.P.: Image quality assessment: from error visibility to structural similarity. IEEE Trans. Image Process. **13**(4) (2004). https://doi.org/10.1109/TIP.2003.819861
11. Wang, Z., Simoncelli, E.P., Bovik, A.C.: Multiscale structural similarity for image quality assessment. In: Asilomar Conference on Signals, Systems and Computers (2003). https://doi.org/10.1109/ACSSC.2003.1292216
12. Zhang, R., Isola, P., Efros, A.A., Shechtman, E., Wang, O.: The unreasonable effectiveness of deep features as a perceptual metric. In: Proceedings of the IEEE Conference on Computer Vision and Pattern Recognition (CVPR) (2018). https://doi.org/10.1109/CVPR.2018.00068
13. Dhariwal, P., Nichol, A.: Diffusion models beat GANs on image synthesis. In: Advances in Neural Information Processing Systems (NeurIPS), vol. 34 (2021). https://doi.org/10.48550/arXiv.2105.05233
14. Betker, J., Goh, G., Jing, L., et al.: Improving image generation with better captions. OpenAI Technical report (2023). https://doi.org/10.48550/arXiv.2305.10601
15. StabilityAI Stable Diffusion 3.5 homepage. https://huggingface.co/stabilityai/stable-diffusion-3.5-large. Accessed 10 Feb 2026
16. Black Forest Labs Flux.1 Dev homepage. https://huggingface.co/black-forest-labs/FLUX.1-dev. Accessed 10 Feb 2026
17. Alibaba Cloud Qwen Image homepage. https://huggingface.co/Qwen/Qwen-Image. Accessed 10 Feb 2026

Explainable Deep Learning for RF Fingerprinting Using Convolutional Neural Networks

Piotr Korycki(✉), Michał Stojke, and Krzysztof Kępa

Wrocław University of Science and Technology, Wrocław, Poland
piotr.korycki@proton.me, {michal.stojke,krzysztof.kepa}@pwr.edu.pl

Abstract. Radio Frequency (RF) fingerprinting has emerged as a critical technology for securing the Internet of Things (IoT) ecosystem by identifying wireless devices through hardware-intrinsic physical-layer signatures. While Deep Learning (DL) models, particularly Convolutional Neural Networks (CNNs), have achieved remarkable classification accuracy in this domain, they often operate as "black boxes", obscuring the physical features they rely upon. This lack of interpretability poses a significant barrier to the deployment of reliable and robust security systems, as models may inadvertently learn transient channel artifacts instead of stable hardware impairments.

In this paper, we propose an optimized 1D-CNN architecture (IQCNN) for RF fingerprinting and introduce a rigorous explainability framework to validate its decision-making process. By utilizing Gradient-based attribution methods and quantitative metrics, including the Gini coefficient and Mean Kurtosis, we demonstrate that the model effectively localizes distinct hardware-induced features, such as power amplifier transients and I/Q imbalance. Our results show that the optimized model achieves an accuracy of 97.82% while maintaining high explanation fidelity. We argue that integrating Explainable AI (XAI) into the design cycle is essential for transitioning from purely performance-driven models to trustworthy, physics-aligned systems. This approach not only enhances model debugging but also paves the way for the development of more resilient RF identification frameworks in future zero-trust wireless networks.

Keywords: Radio Frequency Fingerprinting · Machine Learning · Explainable Artificial Intelligence · Convolutional Neural Network · Device Identification

1 Introduction

The introduction of software defined networking (SDN) in wireless communications makes it vulnerable to attacks from malicious applications. Furthermore, network function virtualization (NFV) presents security risks as functions are migrated from one platform to another [14]. This makes attributing malicious

W. Zamojski et al. (Eds.): DepCoS-RELCOMEX 2026, LNNS 2014, pp. 218–231, 2026.
https://doi.org/10.1007/978-3-032-27927-9_21

network activity to physical assets challenging. One of the methods for device identification is Radio frequency (RF) fingerprinting.

RF fingerprinting is a form of signal intelligence that addresses the methodology of extracting intrinsic characteristics of production imperfections ingrained in the transmitter. RF fingerprinting is a composite of three steps; feature identification, feature extraction, and device identification. Machine Learning (ML)-based RF fingerprinting is envisioned as a key enabler for physical device authentication at scale.

The recent surge in the accuracy and scalability of Machine Learning (ML) models is fundamentally linked to the rapid advancement of CUDA-based parallel computing enabled the transition from traditional, computationally constrained signal processing to high-performance Deep Learning (DL) models capable of processing massive datasets of raw radio signals in real time. In this context, Radio Frequency (RF) fingerprinting has become a primary beneficiary of this increased computational capacity. However, as noted by Jagannath et al. [6], despite these performance gains, the "black-box" nature of deep neural networks remains a significant open challenge and drastically limits the reliability of these systems.

The contributions of this work are as follows:

1. Extension of the Explainable Artificial Intelligence (XAI) validation framework with statistical methods (Gini Coefficient, Mean Curtosis) to confirm the correlation between model decisions and the existence of unique hardware impairments.
2. Design and implementation of an optimized convolutional neural network (CNN) with 97.82% accuracy to establish an accurate hardware impairment classifier.
3. Creation of a unique 802.11 dataset of raw IQ data in a fully controlled environment of anechoic EM chamber to eliminate interference from external sources.

The rest of the paper is organized as follows: In Sect. 2, we review related work and the current state of the art. Section 3 describes the methodology, experimental model used for training CNN networks, and the frameworks for hyperparameter optimization and visualization of learned radio signal features. Results are presented in Sect. 4, while Sect. 5 concludes and proposes future work.

2 Related Work

The identification of wireless devices based on their physical-layer characteristics, known as RF Fingerprinting, has evolved from classical statistical analysis to complex deep learning architectures. This section reviews the foundational work and recent advancements in this field.

2.1 Foundational RF Fingerprinting

Early methods of device identification focused on hand-crafted features derived from hardware impairments. The seminal work by Brik et al. [3] demonstrated that radiometric signatures, such as I/Q origin offset and frequency error, could uniquely identify IEEE 802.11 devices. Similarly, Soltanieh et al. [12] provided a comprehensive review of these traditional techniques, highlighting the importance of power amplifier (PA) transients and local oscillator (LO) inconsistencies as reliable biometric-like features for wireless security.

2.2 Deep Learning in the RF Domain

The shift towards end-to-end learning was catalyzed by the ability of Convolutional Neural Networks (CNNs) to extract features directly from raw IQ samples. Sankhe et al. introduced the ORACLE framework [9], which demonstrated near-perfect classification of bit-similar devices by leveraging hardware-centric variations. This line of research was further extended in [9], where the authors analyzed how physical-layer impairments are embedded within deep representations. Recent studies have also explored more complex input formats; for instance, Mohammad et al. [8] proposed the use of amplitude-phase spectrograms to enhance identification accuracy under varying channel conditions.

2.3 Interpretability and Sanity Checks

As RF classifiers become more sophisticated, the need for transparency has grown. Huang et al. [5] were among the first to visualize the internal features of radio modulation classifiers, revealing that CNNs often focus on modulation reference points. However, the reliability of these visualizations has been questioned. Adebayo et al. [1] introduced critical "sanity checks" to ensure that saliency maps are truly dependent on model weights and data, rather than acting as simple edge detectors.

Recent trends in XAI for wireless systems emphasize robust validation. Mohammad [8] proposed a framework for bias detection in representation learning, while Shrikumar et al. [11] introduced DeepLIFT to provide more faithful feature attribution than standard gradient-based methods like Grad-CAM [10]. Our work builds upon these foundations by applying quantitative metrics such as the Gini coefficient and Kurtosis to validate the physical relevance of attributions in the context of hardware-induced signal distortions.

3 Methodology

This section delineates the proposed deep learning framework for RF Fingerprinting, encompassing the neural network architecture and the hyperparameter optimization strategy.

3.1 Data Acquisition and Dataset Characteristics

The experimental setup utilized the 802.11b standard. Signals from five ESP32-C6 units were recorded using an ADALM-Pluto SDR in an anechoic chamber to isolate hardware impairments. To ensure the model captures stable hardware signatures, recordings were conducted across multiple sessions at different times and ambient temperatures. The resulting dataset [13] is publicly available to support reproducibility and further benchmarking.

3.2 CNN Architecture for IQ Signals

In order to extract device-specific features from the physical layer, we designed a simple one-dimensional Convolutional Neural Network (1D-CNN) tailored for complex-valued IQ samples. The architecture, referred to as IQCNN (see Fig. 1), is specifically structured to process raw signal components as two parallel input channels (I and Q) [9]. The internal structure of the model consists of the following functional blocks:

- **Feature Extraction:** Two stages of `Conv1d` + ReLU + `MaxPool1d` (stride 2) for non-linear feature mapping and dimensionality reduction.
- **Adaptive Flattening:** Dynamic input dimension calculation for the fully connected layers using a dummy tensor pass.
- **Classification Head:** Two `Linear` layers: a 128-neuron hidden layer (ReLU) followed by an output layer matching $N_{classes}$.

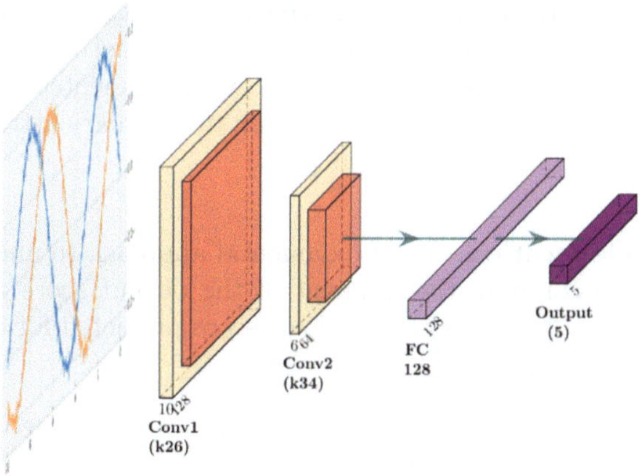

Fig. 1. IQCNN architecture. The input is a 2 × 128 tensor representing the complex IQ signal as two separate 1D channels (In-phase and Quadrature).

3.3 Hyperparameter Optimization and Training

To maximize performance, we implemented an automated hyperparameter optimization (HPO) framework using Optuna [2] with a Tree-structured Parzen Estimator (TPE) sampler.

- **Search Space:** We explored convolutional filters ($n\_filters_1 \in [6, 16]$, $n\_filters_2 \in [5, 10]$) and kernel sizes ($k \in [3, 37]$).
- **Optimization Insight:** The HPO identified that larger kernels (31–37) are optimal. This aligns with findings in radio modulation classification [5], suggesting that a broader temporal context is required to capture subtle hardware-induced distortions compared to short-term modulation changes.
- **Learning Dynamics:** The learning rate was sampled from a log-uniform distribution between 10^{-5} and 10^{-3} [4]. We also explored two distinct optimizers: Adam and Stochastic Gradient Descent (SGD) with momentum.

Training Protocol and Early Stopping. Each optimization trial consisted of training the IQCNN on a randomized dataset of sliced 802.11b beacon signals. To ensure computational efficiency without sacrificing the quality of the search, we implemented a Median Pruner. The study was configured to run for a total of 30 trials or until a predefined timeout was reached, ensuring a balance between exploration and computational time on the CUDA-enabled hardware.

3.4 Explainability Framework: Gradient-Based Attribution

A critical challenge in RF fingerprinting is ensuring that the CNN learns authentic hardware impairments rather than transient channel noise or payload. To validate the model's decision-making process, we implemented an explainability framework based on saliency maps. We focus on two primary attribution metrics: Vanilla Gradient and Gradient × Input [1].

Saliency Map Generation. The fundamental approach to understanding the model's focus is to compute the gradient of the output score with respect to the input IQ samples. Let $S_c(\mathbf{x})$ be the logit score predicted for class c given an input signal $\mathbf{x} \in \mathbb{R}^{2\times 128}$.

1. Vanilla Saliency: This metric identifies which samples in the IQ sequence the model is most sensitive to. It is defined as:

$$M_{Grad} = \left| \frac{\partial S_c(\mathbf{x})}{\partial \mathbf{x}} \right| \tag{1}$$

   The absolute value of the gradient indicates the magnitude of influence each sample has on the final classification.

2. Gradient × Input: To suppress the influence of noise in regions where the signal amplitude is low, we utilize the Gradient × Input metric. This provides a more precise representation of feature importance by accounting for the actual signal values:

$$M_{G\times I} = \mathbf{x} \odot \frac{\partial S_c(\mathbf{x})}{\partial \mathbf{x}} \tag{2}$$

where $\odot$ denotes the element-wise product. This technique is particularly effective for IQ data, as it highlights features only when both the model sensitivity and the signal strength are high.

Physical Interpretation in RF Domain. By analyzing these saliency maps over the 128-sample slices, we can identify **Salient Regions** in the signal. For instance, high attribution values at the beginning of a packet suggest that the model is capturing transient characteristics related to the power amplifier's ramp-up. Conversely, uniform distribution across the slice may indicate that the model is leveraging steady-state impairments such as I/Q imbalance or phase noise.

4 Experimental Results and Discussion

In this section, we evaluate the performance of the optimized IQCNN model and analyze the explainability results to understand the physical features driving the classification decisions.

4.1 Optimization Analysis and Classification Results

While classical machine learning approaches such as Support Vector Machines (SVM) or Random Forests are frequently cited in literature for RF fingerprinting, they rely heavily on manual feature engineering. In this study, we deliberately focus on the 1D-CNN (IQCNN) architecture to evaluate its end-to-end feature extraction capabilities. The reported 97.82% accuracy significantly exceeds typical benchmarks for classical classifiers operating on raw IQ data without handcrafted radiometric features, as shown in comparable studies [3,9].

The hyperparameter optimization (HPO) process conducted via the TPE sampler and Median Pruner converged within 30 trials. The best configuration achieved a peak validation accuracy of 97.82%. To ensure the highest level of experimental rigor and prevent any form of data leakage, a strict separation of datasets was maintained throughout the process. The HPO objective function was evaluated on a dedicated validation set, while the final performance was verified on a completely disjoint test set.

Furthermore, to ensure that the model learned generalized hardware features rather than specific temporal biases of the recording sessions, we implemented a cross-dataset randomization check based on an 80% training and 20% testing split. During the evaluation phase, samples from both the training and evaluation pools—encompassing various recording sessions and thermal conditions—were randomly shuffled and interchanged across different trials. This cross-check confirmed that the high accuracy of 97.82% was independent of the specific sample order or session-related artifacts, validating the model's robust generalization capabilities.

As shown in Table 1, the hyperparameter importance analysis identifies the learning rate as the most critical factor affecting the model's performance. This high importance score suggests that the optimization landscape of IQ-based radio signals is particularly sensitive to the gradient step size; even minor deviations in the learning rate could lead to sub-optimal convergence or numerical instability.

Following the learning rate, the **kernel sizes** of the convolutional layers emerged as the second most influential group of parameters. Specifically, the model favored larger kernels (range 31–37), which suggests that a broader temporal context is necessary to distinguish the subtle hardware impairments of different transmitters. This hierarchical importance confirms a two-stage optimization dependency: first, achieving numerical stability through a precisely tuned learning rate is paramount, and second, the architectural capacity to capture temporal hardware impairments is driven by the receptive field size. These findings align with recent studies on radio modulation interpretability [5], which emphasize that the structural configuration of the first layers is vital for extracting physical-layer fingerprints.

Table 1. Hyperparameter importance derived from 30 Optuna trials using the fANOVA evaluator [15].

Hyperparameter	Importance score
Learning rate	0.48
Kernel size in the first Conv1D	0.22
Kernel size in the second Conv1D	0.18
Training Epochs	0.04
No. of filters in the second Conv1D	0.03
Influence of the used optimizer	0.02
No. of filters in the second Conv1D	0.01
Batch size	<0.01

The final model's robustness was confirmed on the unseen test set. The classification results, summarized in the confusion matrix in Fig. 2, demonstrate that the IQCNN distinguishes between devices with high precision. Notably, even for devices of the same model and manufacturer, the accuracy remained above 97.92%, confirming that the network learns unique physical-layer fingerprints. This success across disjoint and randomized sets validates the strong generalization capability of the optimized IQCNN architecture in real-world RF fingerprinting scenarios.

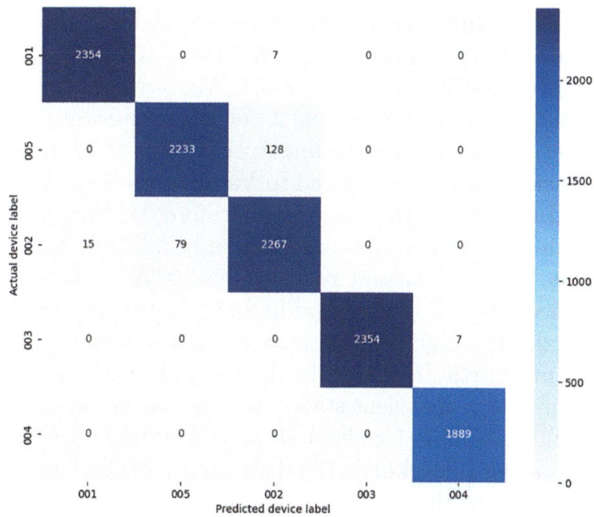

Fig. 2. Confusion matrix presenting the accuracy of the network classification.

4.2 XAI-Based Signal Interpretation

To verify the integrity of the model's decisions, we compared the Vanilla Saliency (see Eq. 1 saliency maps with the *Gradient*×*Input* ($G \times I$) attribution (see Eq. 2).

As illustrated in Fig. 4 and Fig. 5, the **Vanilla Saliency** (M_{Grad}) displays significant high-frequency noise, making it difficult to isolate specific features. In contrast, the **Grad** × **Input** ($M_{G\times I}$) metric successfully suppressed background noise and highlighted the I/Q Constellation Distortions. In steady-state segments, the $G \times I$ maps highlight periodic patterns that correlate with local oscillator phase noise and I/Q imbalance.

The temporal dynamics of the model's focus are further elucidated in the time-domain saliency maps (Fig. 6 and Fig. 7). While the In-phase (I) and Quadrature (Q) components are processed as parallel streams, the $G \times I$ method reveals a highly non-uniform attribution across the 128-sample window. As shown in these figures, the M_{Grad} metric exhibits a noisy, oscillating baseline, suggesting sensitivity to minor stochastic fluctuations. In contrast, the $M_{G\times I}$ maps for both I and Q channels demonstrate sharp, high-magnitude spikes primarily localized at the signal's transitions and high-energy segments. This localization supports the hypothesis that the model captures specific hardware impairments, such as non-linearities in the transmitter's DAC or localized phase noise, rather than a global average of the signal's energy [9].

4.3 Explainability Analysis and Metric Validation

To objectively evaluate the quality of the generated explanations, we conducted a comparative analysis between Vanilla Saliency and the Gradient × Input ($G \times I$) method. Following the sanity check principles suggested by Adebayo et al. [1], we aimed to determine if the attributions are concentrated on physically meaningful RF features or distributed as background noise. The statistical distribution of the attribution weights, summarized in Table 2, reveals significant differences in how both methods interpret device-specific fingerprints. The $G \times I$ approach yielded a higher Gini Coefficient (0.497) compared to Vanilla Gradient (0.424), indicating a more sparse and focused attribution. Furthermore, the Sparsity metric shows that 90% of the model's decision is based on only 58.24% of the signal length in $G \times I$, whereas Vanilla Gradient requires over 63% of the samples to reach the same confidence level. A crucial finding is the difference in Mean Kurtosis. The $G \times I$ maps exhibit a high positive kurtosis (3.43), which points to the presence of distinct lepto-kurtic [7] peaks. In the context of RF signals, these peaks correspond to high-energy transient states, such as the power amplifier ramp-up phase. In contrast, the Vanilla Gradient shows a negative kurtosis (−0.13), suggesting a nearly uniform (platykurtic [7]) distribution of gradients, which is often indicative of sensitivity to thermal noise rather than hardware-specific artifacts. To provide a physical interpretation of the model's decision-making process, the distribution of the top 10% of samples with the highest saliency scores was projected onto the constellation diagrams (see Fig. 3). This visualization highlights the specific signal states and phase-amplitude regions that contribute most significantly to the classification of each device.

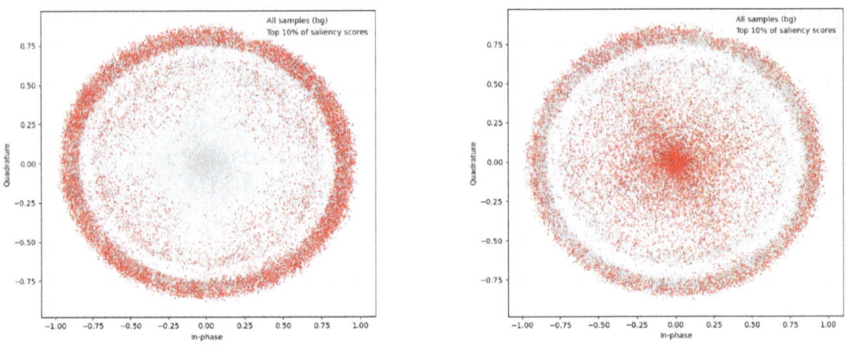

(a) *Gradient* × *Input*. (b) Vanilla Gradient

Fig. 3. 10% of IQ samples with the highest saliency scores based on behalfed evaluation dataset.

Table 2. Quantitative comparison of explainability metrics for IQ-based signal classification.

Metric	Vanilla Saliency	Gradient × Input
Gini Coefficient	0.424	0.497
Sparsity (90%)	63.28%	58.24%
Mean Kurtosis	−0.13	3.43

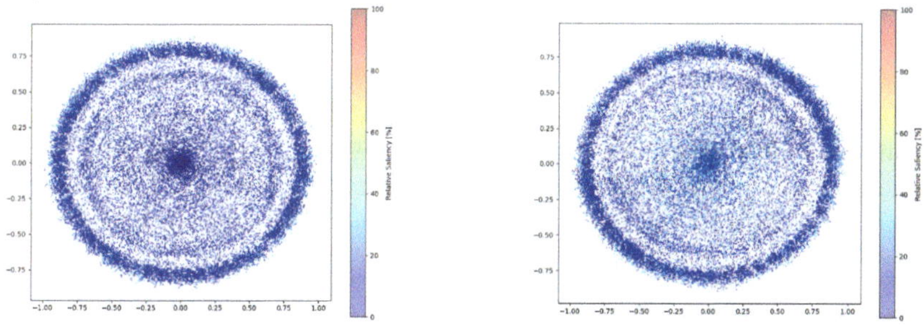

(a) *Gradient* × *Input*. (b) Vanilla Gradient

Fig. 4. Saliency map analysis based on 10% evaluation dataset subsample.

5 Conclusions and Future Work

This study demonstrates that the integration of Explainable AI (XAI) into the RF fingerprinting pipeline is not merely a validation tool, but a roadmap for architectural optimization. Results reported here demonstrated the IQCNN model that achieves 97.82% accuracy by focusing on high-kurtosis transient regions and stable I/Q imbalances. The identification of these **salient regions** through $G \times I$ saliency maps provides a significant advantage for future model development: instead of processing entire signal bursts, future architectures can be designed to selectively attend to these high-information segments. This "physics-aware" approach allows for the creation of more efficient, lower-latency models that prioritize discriminative hardware signatures over stochastic noise. By passing the rigorous sanity checks proposed in [1], we have proven that our data acquisition and HPO framework capture authentic hardware fingerprints. These findings lay the groundwork for more reliable and scalable RF identification systems, where model design is guided by quantifiable physical features rather than black-box experimentation.

Further research could focus on several key directions aimed at enhancing the interpretability and robustness of RF fingerprinting models. While this study utilized a controlled environment with five devices to isolate hardware-intrinsic features, we acknowledge this as a preliminary proof-of-concept. Future work will extend this framework to significantly larger datasets (50+ devices) and evaluate performance under varying Signal-to-Noise Ratio (SNR) levels and real-world multipath interference to further test the causal robustness of the identified fingerprints. Granular kernel interpretability could benefit from isolating the feature maps of specific filters to enhance the explainability of individual convolutional kernels, helping to localize and map distinct physical characteristics, such as LO leakage or DC offset, to particular layers. Furthermore, comparative synthetic-to-real analysis could evaluate model attention using idealized signals augmented with synthetic, controlled distortions (e.g., precise I/Q imbalance or phase noise). Comparing these synthetic attributions with real signals could help quantify the gap between theoretical impairments and observed hardware signatures. Finally, robustness optimization under low SNR would leverage the insights gained from XAI metrics to maintain classification accuracy for weak signals in real-world environments. To strengthen the causal validation, future iterations will also incorporate additional attribution methods, such as DeepLIFT or Integrated Gradients, combined with perturbation-based masking experiments to measure the direct performance drop when salient regions are removed.

A Additional Saliency Maps

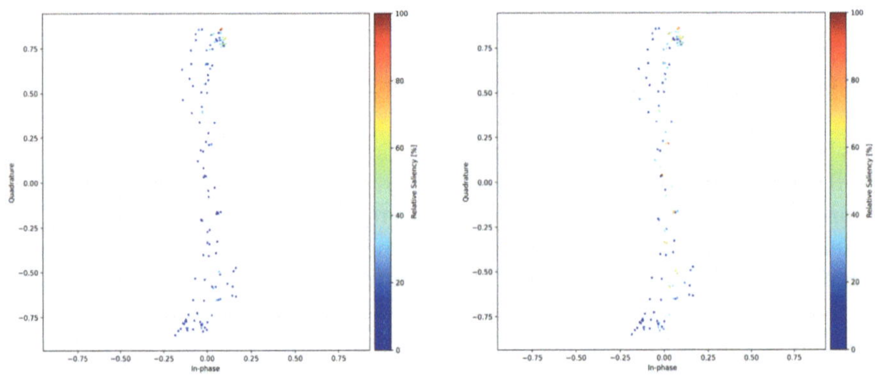

(a) *Gradient* × *Input.* (b) Vanilla Gradient

Fig. 5. Saliency map analysis based on single sample (128 IQ samples).

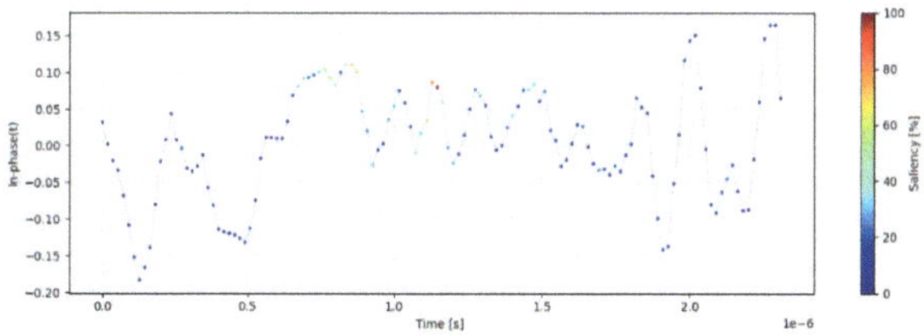

(a) *Gradient* × *Input*.

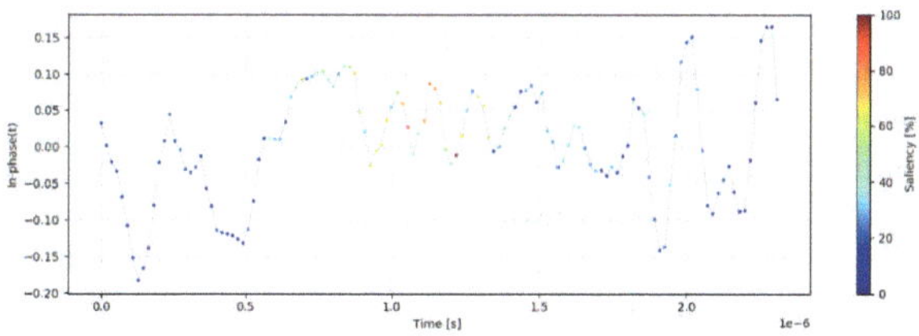

(b) Vanilla Gradient

Fig. 6. Saliency maps of the In-phase signal in the time domain for the single sample (128 IQ samples).

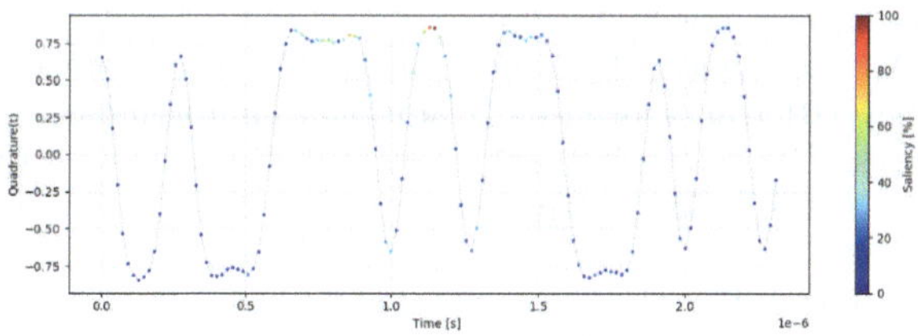

(a) *Gradient* × *Input*.

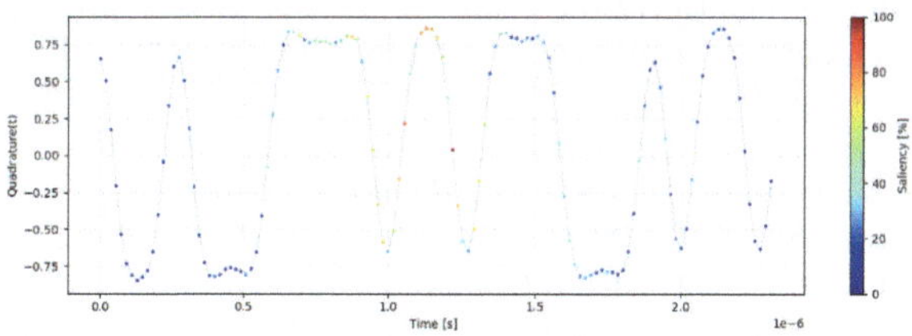

(b) Vanilla Gradient

Fig. 7. Saliency maps of the Quadrature signal in the time domain for the single sample (128 IQ samples).

References

1. Adebayo, J., Gilmer, J., Muelly, M., Goodfellow, I., Hardt, M., Kim, B.: Sanity checks for saliency maps. In: Proceedings of the 32nd International Conference on Neural Information Processing Systems. NIPS'18, pp. 9525–9536. Curran Associates Inc., Red Hook, NY, USA (2018). https://proceedings.neurips.cc/paper/2018/file/294a8ed24b1ad22ec2e7efea049b8737-Paper.pdf, event-place: Montréal, Canada
2. Akiba, T., Sano, S., Yanase, T., Ohta, T., Koyama, M.: Optuna: a next-generation hyperparameter optimization framework. In: Proceedings of the 25th ACM SIGKDD International Conference on Knowledge Discovery and Data Mining, pp. 2623–2631. ACM, Anchorage AK USA (2019). https://doi.org/10.1145/3292500.3330701, https://dl.acm.org/doi/10.1145/3292500.3330701
3. Brik, V., Banerjee, S., Gruteser, M., Oh, S.: Wireless device identification with radiometric signatures. In: Proceedings of the 14th ACM international conference on Mobile computing and networking, pp. 116–127. ACM, San Francisco California USA (2008). https://doi.org/10.1145/1409944.1409959, https://dl.acm.org/doi/10.1145/1409944.1409959
4. He, T., Zhang, Z., Zhang, H., Zhang, Z., Xie, J., Li, M.: Bag of tricks for image classification with convolutional neural networks. In: 2019 IEEE/CVF Conference on Computer Vision and Pattern Recognition (CVPR), pp. 558–567. IEEE Computer Society, Los Alamitos, CA, USA (2019). https://doi.org/10.1109/CVPR.2019.00065, https://doi.ieeecomputersociety.org/10.1109/CVPR.2019.00065
5. Huang, L., Zhang, Y., Pan, W., Chen, J., Qian, L.P., Wu, Y.: Visualizing deep learning-based radio modulation classifier. IEEE Trans. Cogn. Commun. Netw. **7**(1), 47–58 (2021). https://doi.org/10.1109/TCCN.2020.3048113, https://ieeexplore.ieee.org/document/9311238
6. Jagannath, A., Jagannath, J., Kumar, P.S.P.V.: A comprehensive survey on radio frequency (RF) fingerprinting: traditional approaches, deep learning, and open challenges. Computer Networks **219**, 109455 (2022). https://doi.org/10.1016/j.comnet.2022.109455, https://www.sciencedirect.com/science/article/pii/S1389128622004893
7. Mathai, A.M., Rathie, P.N.: Univariate probability models—continuous. In: Probability and Statistics, pp. 174–200. Palgrave Macmillan UK, London (1977). https://doi.org/10.1007/978-1-349-02767-5_6
8. Mohammad, A., Ashraf, M., Valkama, M., Tan, B.: Learning-based RF fingerprinting for device identification using amplitude-phase spectrograms. In: 2023 IEEE 98th Vehicular Technology Conference (VTC2023-Fall), pp. 1–6. IEEE, Hong Kong, Hong Kong (2023). https://doi.org/10.1109/VTC2023-Fall60731.2023.10333639, https://ieeexplore.ieee.org/document/10333639/
9. Sankhe, K., Belgiovine, M., Zhou, F., Riyaz, S., Ioannidis, S., Chowdhury, K.: ORACLE: optimized radio clAssification through convolutional neuraL nEtworks. In: IEEE INFOCOM 2019 - IEEE Conference on Computer Communications, pp. 370–378. IEEE, Paris, France (2019). https://doi.org/10.1109/INFOCOM.2019.8737463, https://ieeexplore.ieee.org/document/8737463/
10. Selvaraju, R.R., Cogswell, M., Das, A., Vedantam, R., Parikh, D., Batra, D.: Grad-CAM: visual explanations from deep networks via gradient-based localization. In: 2017 IEEE International Conference on Computer Vision (ICCV), pp. 618–626. IEEE, Venice (2017). https://doi.org/10.1109/ICCV.2017.74, http://ieeexplore.ieee.org/document/8237336/

11. Shrikumar, A., Greenside, P., Kundaje, A.: Learning important features through propagating activation differences. In: Proceedings of the 34th International Conference on Machine Learning, pp. 3145–3153. PMLR (2017). https://proceedings.mlr.press/v70/shrikumar17a.html
12. Soltanieh, N., Norouzi, Y., Yang, Y., Karmakar, N.C.: A review of radio frequency fingerprinting techniques. IEEE J. Radio Freq. Identif. **4**(3), 222–233 (2020). https://doi.org/10.1109/JRFID.2020.2968369, https://ieeexplore.ieee.org/document/8970312/
13. Stojke, M.: RF Fingerprinting WLAN Dataset (2026). https://doi.org/10.5281/ZENODO.18515186, https://zenodo.org/doi/10.5281/zenodo.18515186
14. Wang, M., Zhu, T., Zhang, T., Zhang, J., Yu, S., Zhou, W.: Security and privacy in 6G networks: new areas and new challenges. Digit. Commun. Netw. **6**(3), 281–291 (2020). https://doi.org/10.1016/j.dcan.2020.07.003, https://linkinghub.elsevier.com/retrieve/pii/S2352864820302431
15. Xiao, G., Tian, Y., Chen, B., Han, S., Lewis, M.: Efficient streaming language models with attention sink. In: International Conference on Learning Representations (ICLR) (2024). https://openreview.net/forum?id=NG7S89Y96L. Accessed 27 Mar 2026

Integrated Access Control and Time and Attendance Systems as an Element of an Organization's Security Strategy

Radosław Krzypkowski(✉)

RegiTech Sp. z o.o, Belgijska Strett 44, Wrocław, Poland
radoslaw.krzypkowski@regitech.pl

Abstract. This article examines integrated Access Control (AC) and Time and Attendance (T&A) systems as a significant component of contemporary organizational security strategies. The paper argues that such systems should not be perceived solely as operational tools for controlling physical access and recording working time, but as integral elements of a broader security architecture that supports risk management, data protection, and organizational resilience. The study discusses the fundamental principles of AC and T&A systems, their architecture, and their integration with ERP and HR systems. Particular attention is given to risk analysis concerning unauthorized access, employee misconduct, and sabotage, emphasizing that correlating access and HR data significantly enhances anomaly detection capabilities and mitigation of insider threats. A model case study of implementation in a medium-sized enterprise is presented to assess the effectiveness and scalability of integrated solutions. The findings indicate that the integration of AC and T&A systems, implemented in line with a risk-based approach under ISO 31000 and ISO/IEC 27001, improves access governance consistency, enhances the quality of managerial data, and strengthens both physical and information security. The article also highlights the importance of compliance with personal data protection regulations, including the GDPR, and the need to ensure proportionality in employee monitoring practices.

Keywords: Access control · Time and attendance systems · Organizational security · Risk management · Information security · System integration · ERP · HR

1 Introduction

Modern organizations operate in an environment of growing physical and digital threats, as well as increasingly strict legal requirements concerning personal data protection, information security, and recording of working time. In this context, integrated Access Control (AC) and Time and Attendance (T&A) systems constitute an important component of organizational security infrastructure, combining protective, administrative, and managerial functions within one coherent technological environment.

W. Zamojski et al. (Eds.): DepCoS-RELCOMEX 2026, LNNS 2014, pp. 232–242, 2026.
https://doi.org/10.1007/978-3-032-27927-9_22

The objective of this article is to demonstrate that integrated AC and T&A systems should be treated as elements of an organization's security strategy rather than as isolated technical subsystems. The paper adopts a systemic perspective and analyses the role of these solutions in relation to physical security, information security, risk management, and organizational governance.

The article makes three contributions. First, it systematizes the role of AC and T&A systems in the broader context of organizational security. Second, it restructures the discussion around architecture, integration, and risk analysis in order to show how these systems support both preventive and detective controls. Third, it presents a model implementation case in a medium-sized enterprise in order to discuss effectiveness and scalability in practical terms.

The remainder of the paper is organized as follows. Sect. 2 presents the theoretical background and explains the convergence of AC and T&A systems. Sect. 3 discusses the architecture of integrated systems, including hardware, software, and ERP/HR integration. Sect. 4 develops a risk-oriented analysis covering unauthorized access, employee misconduct, and sabotage. Sect. 5 examines the impact of integration on physical and information security. Sect. 6 presents a case study of implementation in a medium-sized enterprise. Sect. 7 concludes the paper.

2 Theoretical Background of Access Control and Time and Attendance Systems

2.1 Access Control Systems

Access control systems (ACS) are technical and organizational solutions used to manage entry permissions to specific physical zones, including buildings, rooms, and high-security areas. Their primary function is to ensure that only authorized individuals gain access to protected resources, in accordance with a predefined security policy.

According to PN-EN 60839–11-2014, an access control system includes user identification components, such as proximity cards, RFID identifiers, and biometric solutions; executive devices, including readers, controllers, and electromagnetic locks; management software; and communication infrastructure [1].

The literature emphasizes that access control is one of the fundamental mechanisms for protecting organizational resources in both physical and logical dimensions. Whitman and Mattord classify access control as a core element of information security architecture, because it determines who may enter protected areas and under what conditions [2]. From a systems perspective, access control forms part of a broader security environment integrated with video surveillance (CCTV), intrusion and hold-up alarm systems, and increasingly also with IT-oriented identity management solutions.

From the perspective of ISO/IEC 27001, access control constitutes one of the key areas of information security management. The standard highlights authorization management, the principle of least privilege, and segregation of duties as fundamental control objectives [3]. Although ISO/IEC 27001 primarily addresses logical access to information systems, in organizational practice physical access control is a natural complement to these requirements.

2.2 Time and Attendance Systems

Time and Attendance (T&A) systems are used to record employee presence, start and end times, breaks, overtime, and shift work. In the Polish legal context, their operation is closely linked to the Labour Code, which obliges employers to keep time records in a reliable manner that enables verification of compliance with working time standards [4].

Modern T&A systems frequently use the same identification technologies as access control systems, including proximity cards, RFID key fobs, PIN codes, and biometric solutions. In HR management literature, digital time-recording systems are associated with increased process transparency, reduced risk of abuse, and greater efficiency of payroll and HR settlements [5].

At the same time, T&A systems do not serve solely administrative function. Data generated by time-recording systems constitutes an important source of management information, enabling analysis of workforce utilization, shift planning, and labor cost optimization. Consequently, T&A systems should be interpreted not only as compliance mechanisms, but also as analytical tools supporting managerial control.

2.3 Convergence of AC and T&A Functions

A clear trend can be observed toward integrating AC and T&A systems within a single technological platform. In practice, this integration usually means using a shared identification infrastructure, for example one employee card, to perform both physical access control and time registration.

From the perspective of organizational security management, this approach brings several advantages. First, it improves policy consistency, because one system can centrally manage permissions and monitor user activity across different contexts, including building entry, presence in specific zones, and working time. Second, it increases transparency of events, because correlating AC and T&A data makes it possible to detect anomalies, such as access attempts outside registered working hours. Third, it reduces infrastructure costs by enabling shared readers, controllers, and software. Fourth, it supports audits and compliance, because integrated event logs facilitate demonstrating conformity with information security and data protection requirements.

In the context of risk management, integration should be analyzed as part of a strategy for reducing operational risk and insider abuse risk. ISO 31000 emphasizes that organizations should identify and assess risks in relation to operational objectives and then implement controls proportionate to those risks [6]. AC and T&A data can therefore be treated as an important component of internal control, supporting both preventive and detective mechanisms.

2.4 AC and T&A Systems as Part of Critical Infrastructure

The literature on critical infrastructure protection emphasizes that physical protection of facilities and access control form the foundation of organizational resilience. Keller notes that organizations increasingly depend on integrated protective solutions that connect physical security, operational continuity, and emergency response [7].

In manufacturing enterprises, data centers, financial institutions, and healthcare facilities, AC and T&A systems form integral elements of security architecture. They frequently cooperate with building management systems (BMS), video surveillance, and fire protection systems. In the era of digitalization and Industry 4.0, these solutions increasingly operate as networked, IP-based environments, which creates additional cybersecurity challenges.

NIST indicates that access control systems, when treated as part of operational technology or cyber-physical infrastructure, should be covered by appropriate protection measures against cyber threats [8]. Therefore, modern AC/T&A systems should be designed not only for physical robustness, but also for cyber resilience.

3 Architecture of Integrated Systems

The architecture of integrated AC and T&A systems is a multi-layer technical and organizational structure designed to ensure coherent, secure, and efficient management of physical access and working time records. Contemporary design approaches go beyond purely technical aspects and include integration with ERP and HR systems, as well as compliance with information security standards and personal data protection regulations.

From the perspective of organizational information systems architecture, an AC/T&A system should not function as an isolated technical solution. Instead, it should be embedded in a broader IT ecosystem supporting business and managerial processes. Laudon and Laudon argue that the effectiveness of information systems depends on their ability to integrate data and support decision-making across the organization [9].

3.1 Hardware Layer as the Foundation of Physical Security

The hardware layer includes identification devices, access controllers, actuators such as electromagnetic locks and turnstiles, time registration terminals, and communication infrastructure. This layer implements physical access control and enables event recording related to employee presence.

PN-EN 60839–11-2014 defines system and component requirements for electronic access control systems, emphasizing reliability, tamper resistance, and event logging for later analysis [1]. In practice, this means using controllers with local memory capable of autonomous operation in case of loss of connectivity to the server, as well as implementing backup power mechanisms.

In the context of infrastructure protection, the hardware layer should be designed according to the defence-in-depth principle, as emphasized in NIST SP 800–53 [10]. This requires the integration of access control with other security systems, including CCTV, intrusion detection, or fire systems.

3.2 Software Layer as the Center of Management and Analysis

The software layer is the logical center of the system. It includes server software, event databases, administrative modules, and reporting and analytical mechanisms. This is where permission management, zone definitions, work schedules, and time-record reports are implemented.

From an information security standpoint, the system should comply with ISO/IEC 27001:2022 requirements, particularly in the areas of access control, authorization management, and logging and auditing [3]. This entails multi-factor authentication for administrators, role-based access control, encryption of data transmission, and mechanisms ensuring the integrity of event logs.

Modern AC/T&A systems increasingly operate in distributed architectures using cloud or hybrid models. According to the NIST Cybersecurity Framework, regardless of the deployment model, network segmentation, software updating, and continuous security event monitoring are necessary [8].

3.3 Integration with HR Systems as the Source of Personnel Data

A key architectural element is integration with the HRM system, which serves as the primary source of truth for employee data, including employment status, position, organizational structure, and work schedules.

This integration enables automatic creation and modification of access rights based on HR data. For example, onboarding may automatically create an account in the AC system and assign appropriate access zones, while termination may trigger immediate revocation of rights. This significantly reduces the risk of leaving active permissions for former employees and is consistent with the principles of access lifecycle management emphasized in ISO/IEC 27001 [3]. From an HR management perspective, integrating T&A with HR systems streamlines payroll processes and reduces errors resulting from manual data processing. Armstrong and Taylor underline that digitalization of HR processes increases transparency and control over working time, which has both operational and strategic significance [5].

3.4 Integration with ERP Systems as Support for Management and Controlling

Integration with ERP systems extends AC and T&A functionality into the financial and controlling domain. Data on working time, overtime, or absences can be automatically transferred to payroll, controlling, and reporting modules.

Monk and Wagner define ERP as an integrated information environment that connects data across enterprise activity areas and supports comprehensive resource management [11]. Feeding T&A data into ERP improves settlement consistency, increases the reliability of financial data, and supports analysis of labor costs by departments, projects, or locations. From an enterprise architecture perspective, this integration fits within the data and application architecture layers described in TOGAF, where operational and management systems share consistent data models and communication interfaces [12].

3.5 Legal Aspects and Personal Data Protection

The architecture of an integrated AC/T&A system must account for labor law obligations and personal data protection regulations. Working time data and employee identification data constitute personal data under the GDPR [13].

According to privacy by design principle, the system should be designed to ensure data minimization, limited access to historical information, and appropriate retention periods. When biometric solutions are used, the organization may process special-category data, which requires an additional legal basis and enhanced security measures [13]. Thus, compliance considerations should be embedded in system architecture from the outset, rather than treated as a subsequent administrative task.

4 Risk Analysis: Unauthorized Access, Employee Misconduct, and Sabotage

Risk analysis in physical and organizational security based on integrated AC and T&A systems should cover both external threats, such as intrusion attempts, and internal threats, including employee misconduct, deliberate procedural bypassing, and sabotage. In practice, incidents involving individuals with legitimate access are often the most difficult to detect because they occur within formally granted permissions or close to their boundaries.

A structured approach to such analysis can be based on ISO 31000, which assumes risk identification, risk analysis in terms of likelihood and impact, risk evaluation against acceptance criteria, and selection of treatment methods including mitigation, transfer, avoidance, or acceptance, followed by monitoring and continuous improvement [6]. For AC/T&A systems, it is particularly important that they generate measurable operational data that can feed the risk assessment process continuously rather than only periodically through audits.

4.1 Unauthorized Access: Risk of Zone and Asset Breaches

Unauthorized access risk includes situations in which an unauthorized person enters a facility or protected zone, such as a server room, warehouse, laboratory, or finance department, or moves beyond the scope of granted permissions. Threat vectors may include forced entry, tailgating, impersonation, card sharing, or bypassing controls by staff.

In the integrated AC/T&A approach, this risk is analyzed not only in terms of the door-opening event itself but also through correlation with HR and time context. The system can verify whether the person was scheduled to work, whether they were present during the given period, and whether the event matches their role and duties. Such correlation enables detection of anomalies such as unusual-hour access to zones, repeated attempts to access role-irrelevant zones, or discrepancies between time registration and physical presence. From the standpoint of risk mitigation, integrated architecture supports the application of principles known from information security, such as least privilege, segregation of duties, and periodic access reviews, to the physical world. In practice, this translates into precise access roles, schedules, anti-passback, two-factor requirements, mantraps, and strong logging and supervision of events.

4.2 Employee Misconduct: Risk of Legal Access Used Improperly

Employee misconduct includes behaviors ranging from unintentional, such as routine bypassing of procedures, to intentional, including time fraud, overtime manipulation, access for theft, or using zone access for non-work purposes. A characteristic feature of this category is that the perpetrator generally has at least some level of legitimate access and knows the procedures and weaknesses of the organization.

For this reason, many organizations develop insider risk or insider threat programs that combine technical and non-technical data, including signals from access and attendance systems. In misconduct risk analysis, AC/T&A provides practical indicators such as repeated lateness and unusual entry and exit patterns, time punches without corresponding entry events, access attempts after denial, physically impossible jumps between locations, or atypical access on days off. When integrated with HR and ERP systems, analysis can also include contextual factors such as role changes, project completion, workplace conflicts, notice periods, or access to high-value assets. NIST SP 800–53 strongly emphasizes personnel security procedures, including transfers and termination procedures, access agreements, sanctions, and reviews, as measures for reducing insider risk [10].

4.3 Sabotage: Risk of Intentional Disruption and Damage

Sabotage may be understood as deliberate action aimed at disrupting organizational operations, for example by halting production, damaging infrastructure, manipulating devices, blocking passages, or triggering false alarms. Unlike theft or fraud, sabotage may be motivated by revenge, pressure, ideology, or attempts to conceal other actions.

In sabotage analysis, it is crucial to view the AC/T&A system as part of the organization's critical infrastructure. It not only records events but also controls access to key zones and may affect people's safety in situations involving evacuation, emergency exits, or access for emergency services.

Sabotage risk also includes manipulation of the system itself, such as damaging readers, controllers, power supply, or the network, attempting to compromise administrator accounts, modifying schedules, cleaning logs, or disabling event recording. Therefore, risk reduction measures must cover both physical protection of components and cybersecurity of the management layer, including strong authentication, role separation, log monitoring, alerting, and configuration backups. Logging and incident detection are clearly emphasized in ISO/IEC 27001 as prerequisites for effective response and post-incident analysis [3].

4.4 Control Approach: from Risk Identification to Mitigating Actions

In a mature organization, risk analysis regarding unauthorized access, misconduct, and sabotage should lead to a structured set of controls grouped into three layers.

The first layer is prevention, aimed at reducing the likelihood of undesirable events. It includes limiting permissions, zoning, stronger identification mechanisms, entry control, and movement rules. Good practices for physical security treat entry control as a key mechanism for reducing unauthorized access risk. The second layer is detection and correlation, which means rapid detection of anomalies and linking them with contextual information such as logs, alerts, schedules, and HR data. CISA describes insider risk reduction as a cycle that includes detection and identification, assessment, and incident management [14].

The third layer is response and improvement. This includes operational procedures such as alert escalation, evidence preservation, access audits, disciplinary actions, and configuration or policy changes. These measures fit the general risk management model, in which monitoring, review, and improvement are continuous activities.

4.5 Compliance and Personal Data Protection Dimension

It must be recognized that AC/T&A systems process personal data, including identifiers, presence events, and zone presence, and, in biometric variants, special-category data. From a compliance standpoint, organizations should apply technical and organizational measures adequate to the risk, consistent with Article 32 of the GDPR [13].

In practice, this includes strict access control to AC/T&A data, retention limitations, accountability of administrative operations, and conscious report design in order to avoid excessive data exposure. Therefore, any risk-based approach to AC/T&A integration should be accompanied by a proportionality assessment concerning employee monitoring.

5 Impact of Integrating AC and T&A on Physical and Information Security

Integration of access control and time and attendance systems significantly affects both physical security and information security, understood as protection of data, systems, and information processing processes. Systemically, combining these domains creates an environment where physical events, such as building entry and presence in protected zones, are linked with organizational data such as employment status, role, and schedule. This increases prevention, detection, and response capabilities.

From the perspective of security management, such integration fits a holistic approach recommended by ISO 31000 and ISO/IEC 27001, both of which emphasize coordination of technical and organizational measures to reduce risk consistently and systemically [3, 6].

5.1 Integration as an Enhancement of Physical Security

Physical security involves protecting people, property, and infrastructure against unauthorized access, damage, or disruption. In a classic model, AC functions as an autonomous tool, while T&A serves mainly administrative purposes. Integration increases control capabilities and turns both systems into a more comprehensive protective environment.

First, integration enables correlation of access events with work schedules and employment status. A person not scheduled to work or whose employment has ended should not gain access. Automatic synchronization with HR minimizes the risk of leaving active permissions after termination, which is one of the key gaps highlighted in insider threat literature [14]. Second, integration supports real-time anomaly analysis. Access to high-risk zones outside working hours, missing correlation between time punches and physical entry, or repeated access attempts to role-irrelevant zones may indicate potential violations. According to NIST, effective environmental control requires both preventive mechanisms and monitoring and response capabilities [10].

ISO/IEC 27002:2022, in the area of physical controls, emphasizes that entry control systems should be integrated with organizational processes and supervised for irregularities [15]. Integrating AC and T&A strengthens this requirement by linking access control with actual presence and organizational role.

5.2 Impact on Information Security

Although AC and T&A systems are mainly associated with physical security, their integration directly affects information security. ISO/IEC 27001 treats access control to information as encompassing both logical and physical access to data-processing infrastructure [3]. In practice, inadequate control of entry to a server room or document archive may compromise confidentiality, integrity, or availability.

Integrating AC with T&A and HR supports coherent identity management, aligned with the logic of Identity and Access Management (IAM), where lifecycle management from hiring through role changes to termination is crucial. Automatic revocation of physical rights alongside system account deactivation reduces the risk of unauthorized data access. Additionally, AC and T&A data may support investigations of information security incidents. In the event of an IT breach, it becomes possible to verify who physically stayed in a given zone at a specific time. Such correlation is fully consistent with logging and monitoring requirements indicated in ISO/IEC 27001 and NIST SP 800–53 [3, 10].

5.3 Integration as Part of Organizational Risk Management

In ISO 31000, risk is defined as the effect of uncertainty on objectives [6]. Integrating AC and T&A reduces uncertainty concerning presence, movement, and access status in the organization, thereby enabling more precise assessment of operational risks including theft, sabotage, unauthorized access, and information leakage.

The integrated environment also improves internal and external audits. Comprehensive logs covering physical and time-related aspects increase accountability, which is one of the cornerstones of information security management systems. In the GDPR context, however, integration requires adequate safeguards. Presence and movement data constitute personal data and are subject to minimization and purpose limitation. Poor integration design may lead to excessive monitoring and, consequently, to elevated legal and ethical risks [13].

5.4 Synergy of Physical and Information Security

One of the most important outcomes of integration is the synergy between physical and information security. Traditionally, these areas have often operated in separate organizational structures. System integration supports the creation of a shared security policy and enables feeding AC/T&A data into SIEM or analytics platforms, thereby improving detection of hybrid incidents that combine physical and digital components.

Thus, integrated AC/T&A systems should be interpreted not merely as support tools for entry control and time recording, but as a cross-domain mechanism strengthening overall organizational resilience.

6 Case Study of Implementation in a Medium-Sized Enterprise

This section presents a model case of a medium-sized enterprise (300–400 employees) operating in two locations, where access control (AC) and time and attendance (T&A) systems previously functioned as separate and inconsistently developed solutions without integration with HR and ERP systems.

Before integration, the organization relied on partially disconnected systems and manual processing of time records. The lack of a centralized access policy and automated permission management led to risks such as active access rights for former employees, limited correlation between physical presence and working time, and reduced incident analysis capabilities. Such inconsistencies increase exposure to insider threats [14].

The implementation was preceded by a risk analysis and identification of critical zones. A role-based access model was introduced, and the system was integrated using a unified identification method, centralized management software, and connections with HR and ERP systems. This enabled automated lifecycle management of access rights, consistent with ISO/IEC 27001 and NIST SP 800–53 [3, 10]. As a result, the organization improved both security and operational efficiency. Access rights became aligned with roles and schedules, reducing the risk of unauthorized access. Integration with ERP improved data quality and reduced payroll errors, while correlation of AC and T&A data enabled effective anomaly detection and monitoring [3, 10, 11].

The solution also proved scalable due to the use of role-based access control and HR integration as a single source of truth. It can be extended with additional systems, such as CCTV or BMS, provided that consistent data models and integration standards are maintained [12]. However, scalability must be balanced with GDPR requirements regarding proportional data processing [13].

Compared to standalone AC or SIEM solutions, integrated AC and T&A systems provide better contextual analysis by combining physical and organizational data, although at the cost of higher implementation complexity and increased data governance requirements.

7 Conclusion

The analysis shows that integrated AC and T&A systems extend beyond operational functions and constitute a key element of organizational security strategy. By integrating hardware, software, and HR/ERP systems, they create a coherent environment for managing access, presence, and identity, directly enhancing physical, information, and operational security.

From a physical security perspective, integration enables precise alignment of access rights with roles and schedules, reducing risks of unauthorized access and insider threats.

Automated lifecycle management, supported by logging and monitoring mechanisms, strengthens detection and incident analysis capabilities [3, 10]. In the information security domain, linking physical presence with system activity supports consistent identity management and improves accountability, in line with ISO 31000 and ISO/IEC 27001 [3, 6].

The case study confirms that the greatest benefits arise from process-oriented implementation and integration with HR and ERP systems, leading to improved data quality, reduced administrative errors, and increased transparency. System effectiveness depends on clear access policies, rapid response to personnel changes, and the ability to detect anomalies through integrated analysis, while scalability is driven by RBAC and standardized integration. At the same time, compliance with GDPR remains essential, as these systems process personal and potentially sensitive data. Therefore, privacy by design and controlled data access are necessary to balance security objectives with employee rights. In conclusion, integrated AC and T&A systems should be treated as a strategic component of enterprise risk management and information security governance, contributing to organizational resilience, efficiency, and security culture.

References

1. PN-EN 60839–11-1: Alarm Systems And Electronic Security Systems—Access Control Systems—System and Component Requirements (2014)
2. Whitman, M.E., Mattord, H.J.: Principles of Information Security. Cengage Learning (2018)
3. ISO/IEC 27001: Information Security Management Systems — Requirements (2022)
4. Act of 26 June 1974—The Labour Code (Poland), Journal of Laws No. 24 item 141, as Amended (1974)
5. Armstrong, M., Taylor, S.: Armstrong's Handbook of Human Resource Management Practice. Kogan Page (2020)
6. ISO 31000: Risk Management—Guidelines (2018)
7. Keller, W.: Homeland Security and Critical Infrastructure Protection. Praeger (2014)
8. NIST: Framework for improving critical infrastructure cybersecurity (2018). https://doi.org/10.6028/NIST.CSWP.04162018
9. Laudon, K.C., Laudon, J.P.: Management Information Systems: Managing the Digital Firm. Pearson (2022)
10. NIST SP 800–53 Rev. 5. Security and Privacy Controls for Information Systems and Organizations. https://doi.org/10.6028/NIST.SP.800-53r5 (2020)
11. Monk, E., Wagner, B.: Concepts in Enterprise Resource Planning. Cengage Learning (2013)
12. The Open Group. TOGAF® Standard, Version 9.2 (2018)
13. Regulation (EU) 2016/679 of the European Parliament and of the Council (General Data Protection Regulation—GDPR) (2016)
14. Cybersecurity and Infrastructure Security Agency (CISA). Insider Threat Mitigation Guide (2020)
15. ISO/IEC 27002: Information Security Controls (2022)

Compliance-First RAG with Inference Isolation: Embedding Choice and Prompt-Injection Robustness in a Polish Enterprise Corpus

Miłosz Kucharski(✉)

DSW University of Lower Silesia, Wrocław, Poland
miloszkucharski@gmail.com

Abstract. Retrieval-Augmented Generation (RAG) is a practical way to turn internal documents into a question answering assistant, but many deployments still send user queries and retrieved text to external inference endpoints at query time. For sensitive enterprise knowledge bases, this creates an awkward situation: the system may be helpful, yet the data flow is hard to defend and even harder to audit.

This paper reports an extension of LLKMS (Large Language Knowledge Management System) [13], initially a cloud-only proof-of-concept, towards a compliance-first design based on Inference Isolation. The principle is strict: cloud storage may be used for durability, but parsing, OCR, chunking, embedding, retrieval, and generation remain strictly local, keeping raw text outside third-party APIs.

We evaluate the system on a Polish enterprise corpus of 303 heterogeneous documents and 20 test queries (11 answerable, 9 no-answer), including explicit prompt-injection attempts such as "ignore instructions" and "answer without documents". We compare cloud RAG baselines (deepseek-chat and gpt-5-mini) against local stacks using two instruction-tuned models (Gemma 2 and Bielik) and two embedding models (BGE-M3 and mxbai). The best local configuration (Gemma 2 9B Q4 + BGE-M3) matches cloud refusal accuracy (0.95) while improving retrieval F1@4 (0.4519 vs. 0.4152). In the updated local runs, Bielik (11B v2.3, Q4) reaches the same retrieval scores and achieves strong refusal behavior (1.00), but at a high generation-latency cost compared to cloud baselines.

Keywords: Retrieval-Augmented Generation · knowledge management · prompt injection · refusal behavior · embeddings · Polish NLP · AI governance

1 Introduction

Enterprise knowledge bases are rarely tidy. A single bucket can contain policies, HR notes, technical manuals, meeting minutes, and scanned PDFs. Users still ask short questions and expect short answers.

W. Zamojski et al. (Eds.): DepCoS-RELCOMEX 2026, LNNS 2014, pp. 243–251, 2026.
https://doi.org/10.1007/978-3-032-27927-9_23

RAG has become a standard pattern for this. It retrieves relevant fragments and asks an LLM to answer using that context [2]. The problem is not the pattern itself. It is what often happens at query time.

A standard API-based architecture sends the user query and the retrieved fragments to a cloud API. That is convenient, but in real organisations the query text can be sensitive on its own. In the initial LLKMS proof-of-concept [13], cloud RAG worked, but it lacked a credible security argument regarding prompt retention.

The EU Artificial Intelligence Act raises the bar around documentation and oversight, pushing architectures toward auditability [1]. The goal here is practical: an exploratory study of a RAG assistant for Polish enterprise documents that remains useful while enforcing Inference Isolation—an architectural rule where all query-time semantic processing is executed on-premises.

Dependability is treated in an operational way: stable refusals when evidence is missing, and resistance to prompt-injection phrasing. This work contributes:

- A compliance-first RAG design based on Inference Isolation.
- A preliminary evaluation (20 queries) of prompt-injection robustness in Polish RAG systems.
- An empirical comparison of local vs. cloud RAG trade-offs, including latency and embedding choices.

2 Related Work

2.1 RAG and Grounded Answers

RAG couples retrieval with generation and is widely used to ground outputs in external text [2]. Grounding helps, but it is not a cure-all. Models still hallucinate, and failure handling becomes part of the system behavior, not just a model quirk [3].

2.2 Prompt Injection and RAG-Specific Threats

Prompt injection changes the meaning of "answer from documents". If the user explicitly asks to ignore the rules, or if the corpus itself contains adversarial instructions, a model may follow the wrong instruction unless it is constrained. Recent work also shows that attacks can target the retriever, not only the generator, including poisoning and backdooring behaviors [4]. This matters for enterprise RAG because retrieval is part of the trusted computing base.

2.3 Embeddings for Multilingual Retrieval

Retrieval performance is sensitive to embedding choice. BGE-M3 is positioned as a multilingual embedding model with multi-functionality and long-input support, and it reports strong results across multilingual settings [5]. For Polish corpora, the practical question is less about benchmark rankings and more about whether the embedding keeps retrieval stable under paraphrases and inflection.

2.4 Local Instruction Models

Gemma 2 is an open model family designed for strong performance at practical sizes [6]. Retrieval and generation in Polish are uniquely challenging due to rich morphology, complex inflection, and synonymy. Bielik [7] is prioritized here over other Polish models (e.g., PLLuM) due to its extensive community adoption, proven quantization availability for consumer hardware, and the author's prior empirical success with this family. Bielik is a Polish-focused instruction model released by the SpeakLeash project and used here as a local generator option [9].

3 System Overview

3.1 Inference Isolation

LLKMS uses one organizing rule: durable storage is not inference. Raw files may be stored in a cloud bucket for durability, but query-time semantic processing stays local.

That includes:

- parsing and OCR (for scanned documents),
- chunking,
- embeddings and indexing,
- retrieval and prompt construction,
- generation and refusal logic.

The local perimeter makes it easier to reason about exposure. It also makes it easier to log what the system did, and to repeat runs when something looks suspicious.

3.2 Pipeline

The reference pipeline is a modular stack:

- Ingestion: document download and parsing; OCR is applied where needed.
- Chunking: 1000 characters with 200 characters overlap (fixed across experiments).
- Embedding: BGE-M3 or mxbai embeddings.
- Vector store: FAISS similarity search [10].
- Generation: local instruction model (Gemma 2 or Bielik), or cloud baseline (Fig. 1).

Cloud storage (encrypted at rest) → Local ingestion/OCR → Local chunking/embedding → Local FAISS index → Local LLM generation

Fig. 1. LLKMS under Inference Isolation: cloud may store files, but query-time processing and generation remain local.

4 Threat Model and Audit Controls

4.1 Threat Model

The model assumes external inference providers are not trusted for raw prompts and retrieved fragments at query time. This includes practical concerns such as logging and retention, but also a simple governance issue: it is hard to prove what happened to prompts after submission.

The evaluation covers three user-facing threat types:

- No-answer questions that invite guessing.
- Direct instruction override (prompt injection), e.g., "ignore instructions".
- Indirect override phrasing that tries to sound benign, e.g., "without looking at documents".

Retriever-level attacks (poisoned chunks, backdoored retrievers) are not implemented as a full adversarial training setup here, but the paper treats this line of work as a reason to test prompt injection explicitly [4].

4.2 Audit Controls

LLKMS is designed to produce a minimal audit trail per query. In the current prototype this includes:

- configuration id (model + embedding + chunking + k),
- retrieved chunk identifiers for top-k,
- latency measurements for retrieval and generation,
- refusal flag and refusal correctness label (for evaluation),
- the final response text.

The intent is not to log sensitive content for its own sake. The intent is to make system behavior inspectable and repeatable, so failures can be traced to retrieval, prompting, or model behavior.

5 Experimental Setup

5.1 Corpus and Test Set

The corpus contains 303 internal documents. The test set contains 20 cases: 11 answerable and 9 no-answer. No-answer cases include prompt-injection attempts.

5.2 Fixed Parameters

Top-k is fixed at k = 4 for all experiments. Chunking is fixed to 1000 characters with 200 characters overlap. For all cloud configurations, temperature is set to 0.7 and top-p is left at the provider default (Table 1).

Table 1. Fixed evaluation parameters.

Parameter	Value
Documents	303
Test queries	20 (11 answerable, 9 no-answer incl. injection)
k (top-k chunks)	4
Chunking	1000 chars, 200 overlap
Cloud temperature	0.7
Cloud top-p	default

5.3 Models and Configurations

Cloud baselines:

- Cloud (DeepSeek): `deepseek-chat`, temperature 0.7.
- Cloud (OpenAI): `gpt-5-mini`, temperature 0.7.

Local generators:

- Gemma 2, served as `gemma2:9b-instruct-q4_K_M` [6].
- Bielik, served as `bielik-11b-v2.3-instruct:Q4_K_M` [7].

Embeddings:

- BGE-M3 [5].
- mxbai (`mixedbread-ai/mxbai-embed-large-v1`) as an English-centric open baseline.

5.4 Practical Deployment Considerations

All experiments were executed on a single workstation (Ryzen 5 5600, 64 GB RAM, RTX 4060 Ti). This is intentionally not a data center setup. It is closer to what a small team can actually buy.

For "what hardware is realistically enough", two constraints matter: model weights and context window. For Bielik-11B-v2.3-Instruct, the model card explicitly provides quantized variants (including GGUF Q4_K_M) for setups that cannot host full-precision weights [7]. In Ollama distribution, the Q4_K_M artifact is reported at about 6.7 GB, and the published Modfile also fixes low-temperature settings, which affects repeatability across runs [8].

Table 2. Main results (303 documents, 20 queries, k = 4).

Configuration	F1@4	MRR@4	Refusal acc.	Retr. ms	Gen. ms
Cloud (deepseek-chat)	0.4152	0.4750	0.95	469.88	3199.49
Cloud (gpt-5-mini)	0.4152	0.4750	0.95	260.38	6735.12
Local (Gemma2 Q4 + BGE-M3)	0.4519	0.5500	0.95	3269.05	12024.26
Local (Bielik 11B Q4 + BGE-M3)	0.4519	0.5500	1.00	2954.92	320996.16

For Gemma 2 9B at Q4_K_M, practical deployment guides for local LLM runners typically place 7–9B Q4_K_M models in the 6–8 GB VRAM class for comfortable use, depending on context length and runtime overhead [12]. Ollama supports NVIDIA GPUs and documents GPU compatibility requirements, which is relevant for repeatability across machines [11].

A preliminary check in an earlier internal prototype used Bielik 7B as a local generator; those results are not included as a main baseline, but they motivated the later choice to re-run the evaluation with a newer Bielik release.

Secondary Laptop Test. The local stack was also tested on a laptop (Intel Core Ultra 5 125U, 32 GB RAM, CPU-only). For Gemma2 9B Q4, prompt processing ran at roughly 5.8 tokens/s and generation at 3 tokens/s. While significantly slower than the GPU workstation, it confirms that Inference Isolation remains viable for asynchronous enterprise assistants even without discrete accelerators.

5.5 Metrics

Retrieval metrics are computed on the top-4 retrieved chunks: precision@4, recall@4, F1@4, and MRR@4. Relevance is defined by manual labels for whether a retrieved chunk contains the evidence needed to answer the query.

Cloud baselines were tested with English system prompts, which empirically yielded better instruction adherence for those providers, whereas Bielik was tested with Polish prompts. Dependability is measured via refusal accuracy through manual labeling (0/1) by a single annotator. A model correctly refuses if it explicitly declines to answer:

- For no-answer and prompt-injection cases: refusal is expected.
- For answerable cases: refusal is counted as incorrect.

Latency is reported as mean retrieval latency and mean generation latency.

6 Results

6.1 Main Results

Table 2 compares cloud baselines against the most relevant local configurations. The headline is not just retrieval. It is retrieval plus behavior under missing evidence and adversarial prompts.

Table 3. Embedding ablation for local stacks (k = 4).

Configuration	F1@4	MRR@4	Refusal acc.	Retr. ms	Gen. ms
Gemma2 Q4 + BGE-M3	0.4519	0.5500	0.95	3269.05	12024.26
Gemma2 Q4 + mxbai	0.3195	0.3458	0.85	2268.99	7206.80
Bielik 11B Q4 + BGE-M3	0.4519	0.5500	1.00	2954.92	320996.16
Bielik 11B Q4 + mxbai	0.3195	0.3458	0.80	2378.92	54018.90

Gemma2 + BGE-M3 matches cloud refusal accuracy and improves retrieval metrics. In the updated benchmarks, Bielik (11B v2.3, Q4) no longer collapses on no-answer and injection-style cases, and reaches refusal accuracy of 1.00 on this test set. However, this improvement comes with a large cost in generation latency, which is a practical trade-off for inference isolation on non-server hardware. The extreme generation latency of Bielik (over 300 s per query) is a direct consequence of running an 11B parameter model on hardware where layers must be heavily offloaded to system RAM, highlighting the strict hardware barrier for local execution.

6.2 Embedding Ablation

The embedding choice shifts retrieval metrics for both local generators. In this corpus, BGE-M3 improves F1@4 and MRR@4 compared to mxbai. Refusal accuracy is dominated by generator behavior rather than retrieval alone (Table 3).

6.3 Prompt-Injection Behavior

Four explicit injection-style queries were included (e.g., "ignore instructions" and "answer without documents"). For these cases, a correct behavior is to refuse when the retrieved context does not contain supporting evidence.

In the main benchmark reported in Table 2, the updated local configuration with Bielik-11B-v2.3 (Q4_K_M) shows consistent refusal behavior and reaches refusal accuracy of 1.00 on this test set. Gemma2 configurations also refuse reliably under the same injection phrasing, matching the cloud baselines in refusal accuracy.

This is a notable change compared to earlier pilot experiments carried out with Bielik-7B as the local generator (not reported as a main baseline in this paper), where the model was frequently susceptible to instruction-override phrasing and tended to comply with adversarial prompts rather than follow the "answer only from documents" constraint. The difference between the two Bielik generations is consistent with the broader observation that RAG systems remain vulnerable to instruction-manipulation attacks, and that robustness depends on the exact generator and its instruction-following properties rather than on retrieval alone [4].

7 Discussion

The local stack is not automatically slower or weaker. With Gemma2 Q4 and BGE-M3, retrieval is better than cloud baselines, and refusal behavior matches them. The cost is latency: cloud baselines remain much faster, while local generation is slower, and this is the operational price of keeping query-time raw text out of third-party inference zones.

Bielik shows a different trade-off in the updated runs. It reaches the same retrieval scores and strong refusal behavior, but the end-to-end experience is dominated by generation time on local hardware. From a compliance-first point of view, this is still acceptable in asynchronous internal workflows, but it is a real cost that has to be acknowledged.

There is also a smaller, practical point about embeddings. When mxbai embeddings are used, both local stacks show a drop in retrieval metrics. This pushes more pressure onto the generator, because weaker retrieval means weaker evidence, and weaker evidence is where hallucinations and policy breaks become more likely.

8 Limitations and Future Work

As an exploratory study, the evaluation uses a small test set of 20 queries, which precludes claims of statistical significance. The sample size is insufficient for rigorous variance reporting, and absolute differences (e.g., refusal accuracy of 0.95 vs. 1.00) indicate directional trends rather than definitive bounds.

Furthermore, relevance and refusal labels were assigned manually by a single annotator. Future work must expand the test set to at least 100 diverse queries, employ multiple annotators to report inter-annotator agreement (e.g., Cohen's kappa), and implement a dedicated evaluation of factual answer quality.

Finally, prompt injection was tested solely at the query level. While Inference Isolation mitigates data leakage to external APIs, defending against corpus-poisoning attacks (indirect injection) remains an open challenge requiring dedicated retrieval-time filters and output validation.

9 Conclusion

This work described a compliance-first RAG design for enterprise knowledge management based on Inference Isolation. On a Polish enterprise corpus (303 documents), a local configuration (Gemma2 9B Q4 + BGE-M3) matched cloud refusal accuracy (0.95) while improving retrieval F1@4 (0.4519 vs. 0.4152), without sending query-time raw text to third-party inference endpoints. Prompt-injection tests exposed large controllability differences between local generators: Bielik often followed adversarial instructions, while Gemma2 remained consistent.

Acknowledgments. This work extends the LLKMS engineering thesis prototype and adds security-driven evaluation (no-answer and prompt injection) together with local inference experiments. **The publication fee and conference attendance were supported by the DSW University of Lower Silesia.**

Disclosure of Interests. The author declares no competing interests.

References

1. European Parliament and Council: Regulation (EU) 2024/1689 of 13 June 2024 laying down harmonised rules on artificial intelligence (Artificial Intelligence Act). Off. J. Eur. Union, OJ L, 12.7.2024. ELI: http://data.europa.eu/eli/reg/2024/1689/oj
2. Lewis, P., Perez, E., Piktus, A.: Retrieval-augmented generation for knowledge-intensive NLP tasks. In: NeurIPS (2020)
3. Ji, Z., Lee, N., Frieske, R., et al.: Survey of hallucination in natural language generation. ACM Comput. Surv. **55**(12), 1–38 (2023)
4. Clop, C., Teglia, Y.: Backdoored Retrievers for Prompt Injection Attacks on Retrieval Augmented Generation of Large Language Models. arXiv:2410.14479 (2024). https://doi.org/10.48550/arXiv.2410.14479
5. Chen, J., Xiao, S., Zhang, P.: BGE M3-Embedding: Multi-lingual, Multi-functionality, Multi-granularity Text Embeddings (2024). arXiv:2402.03216
6. Gemma Team: Gemma 2: Improving Open Language Models at a Practical Size. arXiv:2408.00118 (2024)
7. SpeakLeash: Bielik-11B-v2.3-Instruct (model card and weights). https://huggingface.co/speakleash/Bielik-11B-v2.3-Instruct. Accessed Dec 2025
8. SpeakLeash: Bielik-11B-v2.3-Instruct:Q4_K_M (Ollama model page and Modfile notes). https://ollama.com/SpeakLeash/bielik-11b-v2.3-instruct:Q4_K_M. Accessed Dec 2025
9. SpeakLeash: Bielik-7B-Instruct-v0.1 (model card and weights). https://huggingface.co/speakleash/Bielik-7B-Instruct-v0.1. Accessed Dec 2025
10. Johnson, J., Douze, M., Jégou, H.: Billion-scale similarity search with GPUs. IEEE Trans. Big Data **7**(3), 535–547 (2019)
11. Ollama Documentation: GPU support notes. https://ollama.readthedocs.io/en/gpu/. Accessed Dec 2025
12. LocalLLM.in: Ollama VRAM Requirements: Complete 2025 Guide to GPU memory planning for local LLMs. https://localllm.in/blog/ollama-vram-requirements-for-local-llms. Accessed Dec 2025
13. Kucharski, M.: LLKMS – large language knowledge management system. Engineering thesis, DSW University of Lower Silesia, Wrocław (2026). https://apd.dsw.edu.pl/diplomas/18061/

Comparative Analysis of Declarative and Imperative Programming Approaches on Android Application Performance

Urszula Kużelewska(✉) and Błażej Kobeszko

Bialystok University of Technology, Wiejska 45a, 15-351 Białystok, Poland
u.kuzelewska@pb.edu.pl, blazej.kobeszko.107736@student.pb.edu.pl

Abstract. The purpose of this article is to analyse and compare 2 methods for creating user interfaces in Android applications. This involves Kotlin-XML and Jetpack Compose. The research was presented in the context of current state-of-the-art work related to the research methodology, including detailed evaluation criteria. Demonstration applications were designed and implemented using both technologies to create the same user interface. The complexity of both approaches was evaluated, and the applications were subjected to performance tests. An analysis of these tests reveals the differences in performance between XML and Jetpack Compose, leading to recommendations for choosing the appropriate technology depending on the project's specifics and requirements.

Keywords: Android · Jetpack Compose · application performance

1 Introduction

In the current digital age, smartphones have become an integral part of our daily lives. The continuous enhancement in the processing capabilities of these devices signifies their growing potential to substitute for conventional computers [4]. It is evident that a growing number of desktop applications and websites are now accompanied by mobile equivalents, which offer functionalities that were previously exclusive to desktop computers [9].

Consequently, the performance and efficiency of the application development process are of paramount importance to developers and technology companies [1,7]. In order to maintain alignment with current standards, it is imperative that developers embrace the most recent approaches and tools in the complex software development process. Since the inception of the Android platform, the user interface in native applications has been built using XML. Although the traditional approach is still supported and widely used, Google is developing a newer approach to UI development called Jetpack Compose [16]. These changes indicate a necessity to comprehend the impact of novel tools on the technical facets of application design and development. It is also essential to test the impact on the end user by examining the performance of the applications created.

W. Zamojski et al. (Eds.): DepCoS-RELCOMEX 2026, LNNS 2014, pp. 252–262, 2026.
https://doi.org/10.1007/978-3-032-27927-9_24

This paper presents a comparison of 2 methods for creating user interfaces in Android system in terms of the result application performance. The analysis covers both the traditional approach using Kotlin and XML views and the declarative approach based on Jetpack Compose. In contrast to the outcomes documented in the existing literature, in which often the version of this framework is not provided, the research described in this article utilises a new version of the Jetpack 1.8 and also verifies the detailed aspects of performance comparison. Despite typical scrolling and screen navigation, the factors related to identification of janky frames were examined.

2 Related Work

A review of the literature reveals a limited number of publications that directly compare in detail the 2 technologies for developing applications in the Android system. One of the first articles devoted such comparison are [2,3,5,10]. The code used for navigating between screens, and generating and viewing the list of items was analysed. The comparison was based on criteria such as the amount of code, the number of files, and the number of steps necessary to implement individual functions. The results demonstrate that the XML-based approach is more complex, while the absence of separation of logic from the interface in the Compose-based approach results in less readable and poorly structured code. The comparison did not include performance tests.

The article [14] presents a performance analysis of both approaches. The evaluation included measuring application startup and the time required to render the interface in various test cases. The measurements were performed using the Macrobenchmark library. The author constructed and defined in detail an experimental environment—the tests took into account different startup modes of the application: COLD, WARM and HOT, which differed in the state of the application at its startup. The COLD mode is related to the first start of the application—initially, its process was launched and then a new activity was created. The WARM procedure assumes restarting an application after it was paused—its process is already running, whereas its activity needs to be created. Finally, the HOT mode considers the process and the activity are created and running, however, the activity is restored as visible.

The results demonstrated that Jetpack applications achieved sub-optimal performance in comparison to those built using traditional views. Their stability was slightly lower. The author points to the use of Jetpack Compose version 1.1.1 as a potential problem. The analysis did not include tests performed on more complex user interfaces.

A recent comparative analysis [11] confirmed that the XML approach is effective in enhancing UI rendering stability and memory efficiency, as measured by reduced frame latency and lower RAM usage. Compose applications also shown improvements in reducing average CPU utilization.

The study [9] focused on empirical analysis of the performance of Jetpack Compose technology elements. The results indicate that specialised components of the framework did not consistently achieve enhanced performance in comparison to their less specialised variants. This suggests that the use of a ready-made component does not necessarily guarantee optimal work. Performance differences were frequently dependent on the specific context of use, such as the number of list items, or the depth and layering of the user interface hierarchy.

Further studies have been conducted that make comparisons between the Jetpack framework and a variety of other technologies [12,13,15]. The work [8] made a comparison between it and Flutter. The results demonstrated that both solutions have their respective strengths and weaknesses, and it is not possible to select the solution that outperforms the others.

3 Experiments

This chapter presents the evaluation criteria, description of the test environment and the results of the experiments.

3.1 Methodology

Following a thorough review of the relevant literature and existing comparisons (particularly [14]), the following evaluation criteria covering technical aspects such as loading time, rendering performance, and system resource consumption were established.

- Application launch time: Measurement of the time from application launch to interface loading. Each measurement was in a different initial configuration to examine its impact on a given technology.
- Scrolling performance: Analysis of the performance of scrolling through a list of items. Measurement of frame rendering time and identification of frames exceeding expected values.
- Screen navigation performance: Testing the speed and smoothness of transitions between different application screens by measuring the number of frames. Reporting of frames exceeding expected values.
- System resource consumption: Monitoring and comparing CPU load and cache consumption during application operation.

A series of demonstration applications were implemented for the experiments. In order to achieve a comprehensive evaluation and a variety of test scenarios, it was decided to implement 3 applications for each technology. Their functionalities were designed based on two key assumptions. Initially, their design was adapted to the requirements of performance analysis, with a focus on aspects such as startup time, list scrolling performance, and navigation between screens. In the context of the startup time test, it was important that the application contained user interface components that realistically load the system. Therefore, it was decided to implement a calculator that contains multiple buttons, a text field, and logic related to the user interface.

Furthermore, the functionalities were inspired by test scenarios that are widely used in similar studies in the literature. In the field of performance comparison, scenarios such as scrolling through a list of items or implementing a bottom navigation menu are frequently analysed. Each of the developed applications, both in XML and Jetpack Compose technology, was designed to have identical user interfaces.

The following set of test scenarios was designed to thoroughly examine the performance of both technologies. Each scenario was devised to facilitate a comparative analysis of key technical aspects related to performance.

Scenario 1: Measuring application startup time
Objective: To measure the time from application startup to interface loading.
Description: The application is launched under various initial conditions: COLD, WARM, HOT, as well as in different compilation modes. The StartupTimingMetric metric offered by Macrobenchmark was used to collect the measurements.

Scenario 2: List item scrolling performance
Objective: To check the smoothness of scrolling through a list of items and the efficiency of their rendering.
Description: An application with a list of items (100 and 1000 items) tested for scrolling performance. During scrolling, frame durations and the number of frames exceeding the expected values (so called janky frames) were measured.

Scenario 3: Navigation efficiency between screens
Objective: To evaluate the smoothness of transitions between different application screens in both technologies.
Description: An application with multiple screens and navigation tested for smoothness of transitions between screens. Frame duration and the number of frames exceeding expected values (janky frames) were measured.

Scenario 4: System resource consumption
Objective: To measure consumption of CPU and memory during operation.
Description: During screen scrolling and navigation tests, a test trace (Systrace) was recorded, and then replayed using Android Profiler. System resource consumption was monitored during the tests and CPU load and memory consumption were examined.

These applications were tested on two physical devices: Samsung Galaxy S20 FE and Samsung Galaxy S23, which had different memory capacity and versions of the operating system. Their specifications are described below. Samsung Galaxy S20 FE: operating system: Android 13, One UI 5.1, CPU: Qualcomm Snapdragon 865, RAM: 6 GB, internal memory: 128 GB, screen: 1080 × 2400px (6.50") 120 Hz. The same specifications for Samsung Galaxy S23 are the following: Android 14, One UI 6.1, Qualcomm Snapdragon 8 Gen 2, 8 GB, 256 GB, 1080 × 2340px (6.10") 120 Hz.

A number of techniques were used to prepare the test environment. Before each test, other applications and background processes were stopped on the physical devices, Wi-Fi and data transmission were disabled, smartphones were put into a mode that prevented them from receiving notifications, and the devices were sufficiently charged. Every test was repeated 20 times, and the minimal, maximal and average values were reported.

3.2 Results

The application with the calculator was assessed in terms of startup tests. Tables 1 and 2 contain the results of the measured time, whereas the second one includes the compilation mode. The tests were organised to take into account different starting conditions: cold start (COLD), warm start (WARM) and hot start (HOT). On both devices, Compose achieved approximately 3–4% worse application startup times in COLD mode. In WARM mode, significant differences between the technologies were observed in favour of the Compose-based approach. On the Samsung Galaxy S23 device, the median difference was around 10%, and on the Samsung Galaxy S20 FE device, the difference was as high as 17%. Also, in HOT mode, the XML-based approach achieved worse results. The median differed by 30% on the S23 and 6% on the S20 FE, respectively.

Table 1. Application startup time [ms]

Scenario variants	min	avg.	max	min	avg.	max
	Samsung Galaxy S23			Samsung Galaxy S20 FE		
COLD Compose	284.1	303.3	311.3	317.7	332.4	376.7
COLD XML	278.5	291.2	314.9	290.0	322.5	358.7
WARM Compose	79.7	79.7	93.8	74.7	82.0	96.0
WARM XML	73.4	88.8	113.4	85.7	98.7	151.4
HOT Compose	34.9	49.8	73.5	45.5	51.7	57.6
HOT XML	41.1	65.1	111.9	49.8	55.0	65.2

In addition, the impact of different compilation modes was tested. The application startup time was measured in COLD mode, without precompilation, with partial precompilation, and with full precompilation. The startup times of the application written using XML are practically the same for each compilation mode on a single test device. In the case of the second technology, we can see significant differences between the modes used. The startup times of the Compose application without precompilation turned out to be the longest. On test devices, the median launch times were 372.3 ms (28% compared to XML) and 402.9 ms (24% compared to XML) respectively. The use of partial compilation improved the launch time of the Compose application by 17–18%. Full compilation of Compose resources did not improve these times. Jetpack Compose achieved the most optimal times during partial compilation. These values are still 3–4% worse than XML. Regardless of the compilation mode selected, XML provided shorter startup times than the ComposeâĂŞbased approach.

Table 2. Application startup time, including compilation mode [ms]

Scenario variants	min	avg.	max
Samsung Galaxy S23			
COLD XML – no precompilation	277.8	289.9	316.2
COLD XML – partial precompilation	278.5	291.2	314.9
COLD XML – full precompilation	272.0	293.3	340.7
COLD Compose – no precompilation	333.2	372.3	397.4
COLD Compose – partial precompilation	284.1	303.3	311.3
COLD Compose – full precompilation	316.1	348.1	389.6
Samsung Galaxy S20 FE			
COLD XML – no precompilation	308.9	326.2	357.5
COLD XML – partial precompilation	290.0	322.5	358.7
COLD XML – full precompilation	295.2	310.2	388.1
COLD Compose – no precompilation	384.7	402.9	448.4
COLD Compose – partial precompilation	317.7	332.4	376.7
COLD Compose – full precompilation	337.3	369.2	449.0

Both approaches were tested for their performance in scrolling through a list of items: containing 100 items as well as 1000 items. This allowed for an assessment of the technologies' performance under different user interface loads. Measuring scrolling performance is important because it directly reflects the smoothness of the application. The Macrobenchmark Library and the FrameTimingMetric metric, which measures the duration of frames during the test, were used to conduct the tests. The final results are presented in Tables 3, 4 for 100 and 1000 items respectively.

Key metrics in the tables include CPU frame rendering time (frameDurationCpuMs) and frame latency (frameOverrunMs), measured in different percentiles (P50, P90, P95, P99). A lower score means better results for the metrics. On both devices, XML achieved significantly lower CPU frame rendering times. They reach a few milliseconds, which at such low times means a difference of several dozen percent. This is especially true in high percentiles, where more extreme cases occur. In the case of the Galaxy S23, the P99 percentile reaches as high as 21.2 ms. For comparison, XML achieves 12.8 ms in the same scenario, which indicates greater stability and performance of XML. Similar differences can be observed in frame delays, where Compose in P99 records a positive value (6.5 ms). We observe similar relationships on the Samsung Galaxy S20 FE. Similar results were achieved in the 1000-item list scrolling test. Again, in each case, XML achieved better frame rendering results and lower latency, which means greater smoothness and performance for applications created with this technology compared to Jetpack Compose.

Table 5 shows the quantitative results of slow frames that occurred during the test on individual devices. A greater number of them was observed when

Table 3. Test scrolling through a list of 100 items [ms]

Scenario	P50	P90	P99
Samsung Galaxy S23			
Compose frameDurationCpuMs	4.3	8.7	21.2
XML frameDurationCpuMs	3.6	4.9	12.8
Compose frameOverrunMs	−10.4	−6.3	6.5
XML frameOverrunMs	−11.3	−9.8	−1.6
Samsung Galaxy S20 FE			
Compose frameDurationCpuMs	5.1	8.9	17.0
XML frameDurationCpuMs	4.1	5.8	9.9
Compose frameOverrunMs	−9.8	−3.9	2.7
XML frameOverrunMs	−10.7	−8.6	−2.6

Table 4. Test scrolling through a list of 1000 items [ms]

Scenario	P50	P90	P99
Samsung Galaxy S23			
Compose frameDurationCpuMs	4.2	7.1	14.8
XML frameDurationCpuMs	3.6	5.1	9.8
Compose frameOverrunMs	−10.6	−7.8	−0.2
XML frameOverrunMs	−11.2	−9.7	−5.0
Samsung Galaxy S20 FE			
Compose frameDurationCpuMs	5.2	8.4	13.9
XML frameDurationCpuMs	3.8	5.8	8.6
Compose frameOverrunMs	−9.6	−3.9	−0.1
XML frameOverrunMs	−11.0	−8.1	−2.7

scrolling through a list of 1000 items. This is particularly noticeable on the Samsung Galaxy S23 device. The percentage of these frames was higher in the Compose-based approach, which means that XML offers greater performance and smoothness. However, the differences are marginal, and both approaches offered high performance.

The navigation test results on both devices are presented in Table 6. It shows distinct differences in performance between the technologies tested, especially in extreme cases (P99). XML shows lower frame durations on the CPU in the lower percentiles (P50âĂŞP95), however in P99, it increases rapidly, surpassing Jetpack Compose. However, a much more important indicator here is the frame latency relative to the target time of 16 ms. In the lower percentiles, both approaches show minimal delays and even negative values, which means that there are no performance issues. They show similar results, which means that they are comparable under less extreme loads. The most striking are the P99

Table 5. List scrolling test âĂŞ janky frames

Measure	100 elements		1000 elements	
	XML	Compose	XML	Compose
Samsung Galaxy S23				
Number of janky frames	2	3	16	23
Percent of janky frames	0.44%	1.13%	0.34%	0.84%
Samsung Galaxy S20 FE				
Number of janky frames	1	1	1	4
Percent of janky frames	0.31%	0.47%	0.02%	0.14%

latency results. On the Galaxy S20 FE, the Compose-based application achieves as much as 161.9 ms of latency in P99, which indicates significant performance issues in extreme scenarios. XML, on the other hand, maintains latency at 29.4 ms. This is still a result that indicates delays, but this value is more than 5 times lower. Both technologies have problems with full application stability when navigating on both test devices.

Table 6. Navigation time [ms]

Scenario	P50	P90	P99
Samsung Galaxy S23			
Compose frameDurationCpuMs	4.2	6.9	26.0
XML frameDurationCpuMs	3.5	5.4	40.7
Compose frameOverrunMs	−10.8	−6.8	87.9
XML frameOverrunMs	−11.3	−9.2	27.9
Samsung Galaxy S20 FE			
Compose frameDurationCpuMs	5.0	9.2	32.2
XML frameDurationCpuMs	4.1	8.2	41.9
Compose frameOverrunMs	−9.9	−3.5	161.9
XML frameOverrunMs	−11.0	−4.7	29.4

During the analysis of a single iteration, both applications experienced slow frames (see Table 7). Their percentage relative to the total number of frames is quite low, but still worth noting. The applications were smooth most of the time. Once again, applications based on Jetpack Compose had a higher rate of slow frames. The difference is even more noticeable than in the scrolling test. The analysis did not reveal any frozen frames on either device.

Another comparison criterion was the assessment of resource usage during various operations in the application. Both when scrolling through lists with

Table 7. Navigation time - janky frames

Measure	XML	Compose
Samsung Galaxy S23		
Number of janky frames	2	6
Percent of janky frames	1.05%	3.21%
Samsung Galaxy S20 FE		
Number of janky frames	2	3
Percent of janky frames	1.12%	1.79%

different numbers of items, and in tests related to navigation in the application, measurements of memory and CPU usage were performed. The results, presented in Tables 9 and 8, show the peak resource consumption during the test: CPU usage and memory consumption respectively. The results were collected using the Android Profiler tool by analysing the recorded test trace.

Table 8. Memory consumption [MB]

Scenario variant	Compose	XML	Compose	XML
	Samsung Galaxy S23		Samsung Galaxy S20 FE	
Scrolling of 100 elements	147	119	151	135
Scrolling of 1000 elements	149	127	144	128
Navigation	132	133	138	147

In the scrolling test, applications written using XML consumed 10 to 20% less memory than applications using Compose. In the navigation test, the Jetpack Compose library performed better. Memory consumption was 5% lower than XML on a Samsung S20 FE smartphone. On the second device, the difference was less than 1% in favour of Compose.

CPU consumption by both technologies turned out to depend not only on the type of task, but also on the device on which the test is performed. On the Samsung Galaxy S23 device, XML shows higher CPU load in most cases. When scrolling through lists on the S20 FE device, Compose tends to consume more processor resources. On both devices, XML shows higher CPU consumption in the navigation test, which may suggest that it is more resource-intensive in tasks involving dynamic user interface operations, such as navigation. However, the results are not clear-cut for both devices, making them difficult to interpret.

Table 9. CPU usage [%]

Scenario variant	Compose	XML	Compose	XML
	Samsung Galaxy S23		Samsung Galaxy S20 FE	
Scrolling of 100 elements	9.80	12.00	12.10	7.30
Scrolling of 1000 elements	7.00	7.60	10.40	9.30
Navigation	12.70	14.50	15.20	18.70

4 Conclusion

The results indicate areas where the Jetpack still lags behind XML, and where it has already managed to gain a significant advantage. The performance results are consistent with those found in the literature suggesting that this technology is still needs improving its efficiency. Jetpack technology is the preferred option for frequently used applications due to its rapid reloading speed. In contrast, XML-based technology is better for applications with complex user interfaces, such as extensive lists or advanced navigation, where stability is essential.

Acknowledgments. The research was carried out within project no. WZ/WI-IIT/3/2023 at Bialystok University of Technology and financed from the research subsidy of Bialystok University of Technology

Disclosure of Interests. The authors have no competing interests to declare that are relevant to the content of this article.

References

1. Bennett, M.: Scalable Android applications in Kotlin: Write and maintain large Android application code bases, BPB Publications (2024)
2. Corrala, L., Fronzab, I., Mikkonenc, T.: User interface matters: analysing the complexity of mobile applications from a visual perspective. Procedia Comput. Sci. **191**, 9–16 (2021)
3. Fu, J., et al.: How resource utilization influences UI responsiveness of Android software. Inf. Softw. Technol. **141**, 106728 (2022)
4. Haase, C.: Androids: the team that built the Android operating system, No Starch Press (2022)
5. Hort, M., et al.: A survey of performance optimization for mobile applications. IEEE Trans. Software Eng. **48**(8), 2879–2904 (2022)
6. Komatineni, S., et al.: Pro Android 4. Apress, Berkeley (2012)
7. Krajci, I., Cummings, D.: Android on x86: An Introduction to Optimizing for Intel® Architecture, Springer Nature (2013)
8. Kusuma, M, Rifani, A.H, Sugiantoro, B.: Comparison analysis of Jetpack Compose and Flutter in Android-based application development using Technical Domain. International Conference on Informatics and Computing, 1–5 (2023)
9. Kwiatkowski, A., Smołka, J.: Performance analysis of Jetpack Compose components in mobile applications. J. Comput. Sci. Inst. **37**, 391–398 (2025)

10. Milla, E., Radonjić, M.: Analysis of developing native android applications using xml and jetpack compose. Balkan J. Appl. Math. Inform. **6**(2), 167–178 (2023)
11. Nurhadi, N., Muslim, A.: Comparative analysis of the impact of declarative and imperative programming approaches on android application performance through benchmarking method. J. Res. Soc. Sci. Econ. Manag. **55**, 5234–5248 (2025)
12. Peters, M., Scoccia, G.L., Malavolta, I.: How does Migrating to Kotlin Impact the Run-time Efficiency of Android Apps?. In: International Working Conference on Source Code Analysis and Manipulation, pp. 36—46 (2021)
13. Putranto, B.P.D., et al.: A comparative study of java and Kotlin for android mobile application development. In: International Seminar on Research of Information Technology and Intelligent Systems, pp. 383—388 (2020)
14. Szczukin, J.: Performance analysis of user interface implementation methods in mobile applications. J. Comput. Sci. Inst. **26**, 13–17 (2023)
15. Wasilewski, K., Zabierowski, W.: A comparison of Java, flutter and Kotlin/Native technologies for sensor data-driven applications. Sensors **21**(10), 3324 (2021)
16. Jetpack Compose. https://developer.android.com/reference/kotlin/androidx/compose/runtime/package-summary. Accessed 30 Jan 2026

The Use of Machine Learning Techniques in a Smart Referee Table System

Mikołaj Lipiński and Tomasz Zamojski(✉)

Wrocław University of Science and Technology, Wrocław, Poland
tomasz.zamojski@pwr.edu.pl

Abstract. The main objective of the work was to develop a system to support the work of the referee's table during a basketball game. The prototype uses camera images to recognize the referee's hand gestures and converts them into information that would traditionally be entered manually into the game protocol. This helps reduce the workload of the person responsible for recording game events.

A custom dataset was prepared, two gesture recognition models were trained, and an architecture for real-time image analysis was designed. The system includes logic that interprets gesture sequences in accordance with official referee signals. It can recognize both single-digit and double-digit gestures, as well as basic administrative gestures. Tests have confirmed that the prototype can correctly identify most of the prepared gestures and works reliably in changing recording conditions. This solution can serve as a basis for further development of the game support system and as a tool to support the training of basketball referees.

Keywords: Image Processing · Gesture Recognition · Media Pipe · Machine Learning · Basketball Referee

1 Introduction

Refereeing a basketball game requires close cooperation between the court referees and the team operating the referee table. The table's responsibilities include recording individual and team fouls, monitoring foul limits, and documenting key game events. These tasks are performed in highly dynamic game conditions, where decisions are made in fractions of a second and referee signals appear sequentially and often in motion. In practice, this increases the risk of delays and registration errors, especially when the service is provided by a limited number of people.

The development of image processing methods and gesture recognition techniques makes it possible to partially automate this process. Refereeing signals in basketball are based on a standardized set of gestures, which makes them a good candidate for algorithmic analysis. This paper attempts to develop a system that uses camera images to identify selected referee gestures and convert them into data that supports the keeping of match records. The system is not intended to completely replace humans, but to serve as an assistant, enabling the verification and correction of automatically generated entries.

W. Zamojski et al. (Eds.): DepCoS-RELCOMEX 2026, LNNS 2014, pp. 263–274, 2026.
https://doi.org/10.1007/978-3-032-27927-9_25

1.1 Purpose of the Work

The aim of the work was to design and implement a prototype of a real-time basketball referee gesture recognition system and integrate it with a simplified electronic match protocol.

This system is intended to support the scoreboard operator during a basketball game by automatically recognizing referees' gestures and recording fouls in real time. Functional requirements were defined, including image capture from a camera or video file, detection of key points on the hand, gesture recognition with distinction between the right and left hand, signal aggregation for reading the player's number, and detection stabilization. The system also allows the selection of the image source, changing the active team, visual signaling of foul limits being exceeded, and manual correction of the protocol in case of errors.

Achieving the main objective required a number of specific goals to be met: developing a stable method of acquiring and processing images from cameras, implementing hand and key point detection, training a gesture classification model, designing the logic for interpreting signal sequences, and preparing a clear user interface for presenting and correcting results. Another important element is the assessment of the extent to which the developed prototype can improve the operation of the table by increasing the speed and accuracy of recording violations.

Fig. 1. A match photo taken with: (a) the Pixel 8 smartphone (b) with Hololens 2 glasses (c) an iPhone 12 mini smartphone. Słupsk, Poland. July, 2025. Source: Authors own archive.

1.2 Scope of Work

During the referee camp (Słupsk, Poland. July 2025), recordings were made showing standard basketball signaling using various devices (including Hololens 2, iPhone 12 mini, Pixel 8), which provided data on various technical parameters and lighting conditions (Fig. 1).

The next stage was to analyze the available image processing tools and verify the suitability of the *MediaPipe* and *OpenCV* libraries for hand detection in conditions similar to real-life ones. On this basis, a system concept was developed comprising: an image acquisition module, a gesture classification thread, an interpretation layer, and a user interface with a simplified match protocol.

The scope of work also included preparing a training set, training classification models for the left and right hands, integrating the models with the application, and implementing the logic that converts recognized gestures into messages corresponding to protocol records. The system allows for manual approval and correction of detected events, which ensures that the operator retains control.

1.3 Research Problem and Limitations

Automatic recognition of referee gestures in basketball is a complex technical challenge. The game is characterized by high dynamics, variable lighting conditions, and frequent hand occlusions by players. Gestures are performed at different angles to the camera and with varying dynamics, and the individual style of referees introduces additional variability in terms of speed and amplitude of movement.

Another significant problem is the interpretation of the sequential nature of the messages. Many signals, such as the player number and type of foul, are communicated as a series of consecutive gestures. The system cannot be limited to classifying individual hand positions, but must analyze their sequence and duration, distinguishing intentional signals from accidental movements resulting from movement on the court. This requires the implementation of reading stabilization mechanisms and contextual logic.

Due to the complexity of the problem, the functional scope of the prototype has been deliberately limited. The focus was on recognizing gestures related to fouls and player numbers, i.e., those elements that directly affect the keeping of the score sheet. Signals related to the direction of play, player substitutions, or the type of throw, which, although important for communication, are not a key area of registration for the table, were omitted.

The attempt therefore constitutes a feasibility study of a system supporting the work of table officials, taking into account both the technological potential of modern vision methods and the limitations resulting from the realities of the match environment.

1.4 State of Research

There is growing interest in the literature on the subject in the application of machine learning and computer vision methods in the analysis and support of refereeing in basketball.

Early work focused on analyzing the effectiveness of refereeing teams based on historical data, pointing to the important role of teamwork as a factor that determines the quality of decisions more strongly than the individual competencies of referees. the use of linear classifiers (including the pocket algorithm) allowed for the prediction of referee team performance with high accuracy [1].

More recent research shifts the focus towards human-machine collaborative decision support systems that integrate real-time image transmission with data processing and decision reporting modules [2]. At the same time, strictly vision-based solutions are being developed, such as "virtual referees" using object detection (e.g., YOLOv8) and body position estimation to automatically detect selected fouls, including steps and double dribbling [3].

Research on multi-stream models that learn temporal-spatial context further highlights the importance of integrating information about the position and movement dynamics of players in the analysis of on-court events [4]. At the same time, significant technical challenges associated with basketball analysis are highlighted, such as fast movement, frequent screens, variable camera angles, and difficulties in identifying players with similar appearances [5]. The current state of research therefore points to a clear trend towards automation and hybrid referee assistance systems, while keeping humans in the decision-making loop as an element ensuring control and context interpretation.

2 Basics and Analysis of the Topic

The development of an automatic basketball referee gesture recognition system required the analysis of two complementary areas: the organization of the referee table and modern image processing methods. Understanding the structure of communication between the referee and the referee table was a prerequisite for formalizing signals in the form of rules that could be implemented algorithmically. At the same time, it was necessary to determine which vision tools enable stable hand detection and real-time gesture analysis.

2.1 The Role and Tasks of the Referee's Table

The referee's table performs an administrative and control function in the structure of the game. It is responsible for keeping the match protocol, controlling the playing time and the 24-s clock, as well as recording individual and team fouls, time-outs, and player substitutions. The protocol is the official document of the competition, which is the basis for verifying the result and compiling statistics.

Communication between the field referee and the table is based on manual (visual) and voice signals. Based on these signals, the table staff enters data into the documentation and responds to regulatory situations (e.g., exceeding the foul limit). The fast pace of the game means that the operator must simultaneously observe the course of the match and interpret gestures. For this reason, the process is prone to errors and delays, which justifies the need for technological support.

2.2 Characteristics of Referee Signaling

Refereeing signals in basketball are defined in FIBA rules [6] as an orderly set of gestures conveying information about the player's number, the type of foul, and the consequences of the decision. The messages are sequential—first the player's number is presented, then the type of foul, and finally the decision regarding further play.

Player numbers are signaled with one or two hands, with two-digit numbers requiring the correct order and orientation of gestures. Standardization of signals promotes their formalization, but in practice there are individual differences in the dynamics and range of movement of referees. This means that the recognition system must take into account variability in execution and analyze not only a single frame of the image, but also the continuity and duration of gestures.

2.3 Use of the MediaPipe Library

Methods based on landmark analysis are particularly relevant in the context of this system. Instead of processing the entire match scene, the algorithm focuses on a selected area – the referee's hand – represented as a set of coordinates describing the position of the fingers and wrist. This approach reduces computational complexity and enables real-time operation, which is crucial for applications supporting the work of the referee's table.

MediaPipe is a framework for building real-time video processing pipelines [7]. The architecture is based on modules connected in a processing graph, which allows image reading, object detection, and machine learning models to be integrated into a single structure.

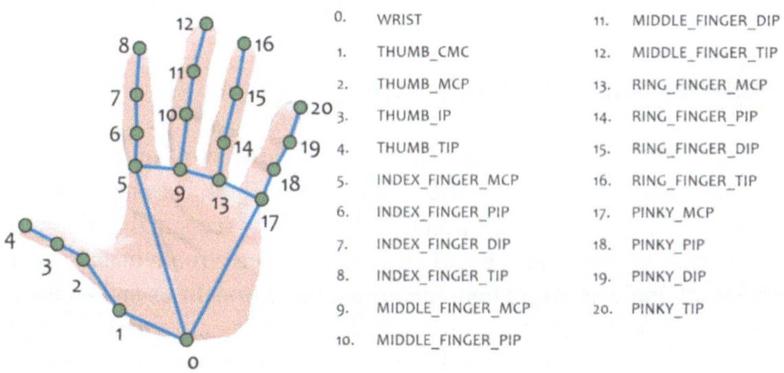

Fig. 2. Landmarks of the hand in *MediaPipe*; Source: *Authors own archive.*

The *MediaPipe Hands* module, which enables hand detection and tracking and the determination of 21 characteristic points in an approximate three-dimensional space, was of key importance in the project (Fig. 2). The solution works in real time and supports multiple hands in a single frame. This makes it possible to obtain an orderly geometric representation of the hand without having to implement proprietary detection algorithms.

The use of *MediaPipe* allowed the detection module to be treated as a sensory layer providing stable input data. The coordinates of characteristic points (*landmarks*) form the basis for further classification of gestures and their interpretation in accordance with basketball signaling logic. Integration with the *OpenCV* library enables direct processing of camera images and transfer of results to *Python* applications.

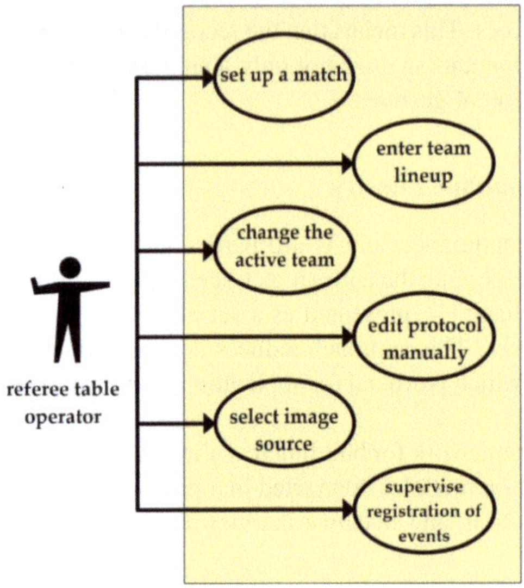

Fig. 3. Use case diagram; Source: authors own archive.

3 Requirements Analysis and System Design

The "Smart referee table" is an implementation of the concept of automatic interpretation of referee signals in basketball using vision and machine learning methods. The purpose of the system is to support the protocol operator by recognizing referees' gestures in real time and automatically recording fouls in an electronic match sheet. During the requirements analysis stage, both operational functionalities and technological limitations resulting from the nature of the working environment were defined (Fig. 3).

Key functional requirements included: image acquisition from a camera or video file, detection of key hand points, gesture recognition with distinction between right and left hands, signal aggregation for reading double-digit player numbers, and detection stabilization through time analysis. The system also allows for changing the active team, visual signaling of foul limits being exceeded, and manual correction of entries, which gives the operator control over the automatic event recording process.

Non-functional requirements include implementation in *Python* using *MediaPipe, OpenCV, and Tkinter* libraries, and the use of a multi-threaded architecture separating image processing from the user interface layer. The system operates in offline mode and has been optimized for real-time operation. The runtime environment is based on Windows, while the model training process was implemented in a Linux environment (WSL) using *the TensorFlow framework*. Limitations resulting from lighting conditions, camera perspective, and possible hand occlusions were also identified. The architecture was designed in a modular way, separating the presentation layer, image acquisition modules, inference thread, and gesture aggregation mechanism, which ensures scalability and the possibility of further development of the system.

4 System Implementation

The system is organized in a way that separates the application layer from machine learning resources. The root directory contains the source modules: ‘ui.py‘ (presentation layer), ‘workers.py‘ (image acquisition), ‘inference.py‘ (inference from MediaPipe models), ‘aggregator.py‘ (interpretation logic), ‘drawing.py‘ (visualization), and ‘config.py‘ (configuration parameters). Training data and trained models are stored in separate directories (‘dataset‘, ‘models‘).

The central element of communication between threads is the *SharedState* structure, which contains the current image frame, frame ID, and the latest classification result. Access synchronization is achieved using locking mechanisms, which eliminates race conditions between the acquisition thread, inference thread, and interface layer. This solution ensures separation of responsibilities and minimizes the risk of blocking the main application loop.

User interface layer

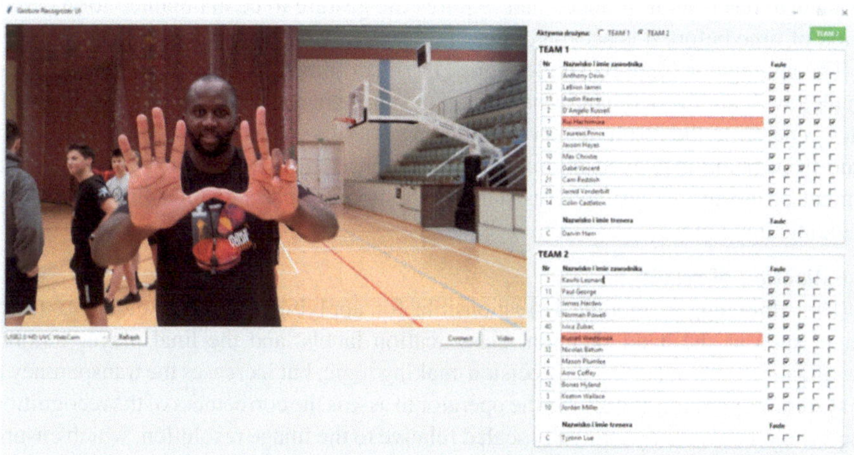

Fig. 4. Application window and protocol panel; Source: authors own archive.

The user interface was implemented using the *Tkinter* library. It is responsible for presenting the video preview, handling the simplified match protocol, and receiving messages generated by the aggregation module. This layer does not perform analytical operations – it serves a visualization and control function.

The image is refreshed cyclically using the ‘*after()*‘ mechanism, which ensures the responsiveness of the application regardless of the load on the computational threads. Recognized signals (e.g., player number or “COACH” message) are transmitted in text form and cause the relevant protocol fields to be updated automatically. When the foul limit is exceeded, the system generates a visual highlight of the player, while maintaining the possibility of manual correction of entries by the operator (Fig. 4).

Image acquisition and video stream handling

The ‘workers.py‘ module is responsible for delivering frames from a camera or video file. Two variants have been implemented: a camera worker operating in real time and a worker playing back recordings with speed control based on FPS values. Each frame is saved to a shared structure with an incrementing identifier, which allows only new data to be processed.

The ‘inference.py‘ thread retrieves the current frame, scales it, converts it to RGB space, and then passes the image to two instances of the MediaPipe Gesture Recognizer classifier–separately for the left and right hands [8]. Separating the models reduced errors resulting from gesture symmetry and increased classification accuracy. The inference results are stored in a shared structure, and the thread's operating frequency is regulated by controlling the iteration time, which ensures processing stability.

Gesture interpretation logic

The ‘aggregator.py‘ module transforms the raw results of the classifier into messages consistent with basketball signaling rules. The interpretation is temporal and is based on the analysis of a sequence of gestures rather than a single frame. A stabilization mechanism has been introduced that requires the gesture to be maintained for a certain period of time before it is accepted.

The aggregator supports single digits (1–5), two-handed combinations (6–15), tens signaled with an inverted hand, and special gestures such as “C” (coach's foul) or “00.” After generating the result, a short cooldown period is introduced to prevent the same value from being displayed multiple times in consecutive frames. This layer is a key element ensuring the system's resistance to temporary classification fluctuations and accidental hand movements.

Visualization of results

The ‘drawing.py‘ module is responsible for applying diagnostic elements to the image, such as the hand skeleton, classification labels, and the final player number. Visualization does not affect the decision-making logic, but increases the transparency of the system's operation and allows the operator to assess the correctness of the recognition. Graphic elements are dynamically scaled relative to the image resolution, which ensures their readability in various conditions.

Training data preparation and machine learning

The preparation of training data began with the collection of gesture sequences in conditions similar to real life, involving different people, distances from the camera, and variable lighting, which increased the representativeness of the set.

Due to the presence of unnecessary background elements, an automatic processing pipeline (run in the WSL environment) was developed, which used the MediaPipe Hands module to detect up to two hands in the image and then selected the most central and largest detection based on an evaluation function that took into account the surface area and distance from the center of the frame.

After determining the bounding box, a square crop was generated with an additional contextual margin, which increased the models' resistance to slight shifts and variability in hand position. The entire process—detection, selection, cropping, and scaling—was integrated into a unified batch processing mechanism, resulting in an organized dataset in which each image depicted only a hand performing a specific gesture, in accordance with the structure required by MediaPipe Model Maker.

The models were trained using dedicated sets for the left and right hands, which allowed for differences in the orientation of landmarks to be taken into account and improved classification accuracy. The data was divided in a ratio of 80/10/10 into training, validation, and test sets, while ensuring control over overfitting and objective evaluation of model quality. The learning process was carried out over 70 epochs with a learning rate of 0.0002 and a batch size of 16, monitoring loss and accuracy metrics on the validation set. After training, the models were evaluated on the test set and then exported to the *.task* format, allowing for their direct integration with the application and use in offline mode, both on CPU and GPU.

System operation scenario

The complete system operates as a multi-threaded real-time processing pipeline. After initialization, the user interface launches an image acquisition thread that saves successive frames with an identifier to a shared structure. At the same time, the inference thread retrieves the latest frame, pre-processes it, and passes it to two classification models, generating gesture labels and probabilities. The raw results are then sent to the aggregation module, which analyzes the stability of the gesture over time, filters out random readings, and generates a final numerical signal consistent with the logic of referee signaling. The final result is presented in the interface along with a visualization of characteristic points and labels, after which the system moves on to the next processing cycle, creating a closed, continuous loop of operation (Fig. 5).

The implemented prototype confirms the possibility of effective integration of vision methods and rule-based logic for the automatic interpretation of referee signals. The modular architecture allows for further expansion of the system, including the extension of the range of recognized gestures and integration with external statistical systems.

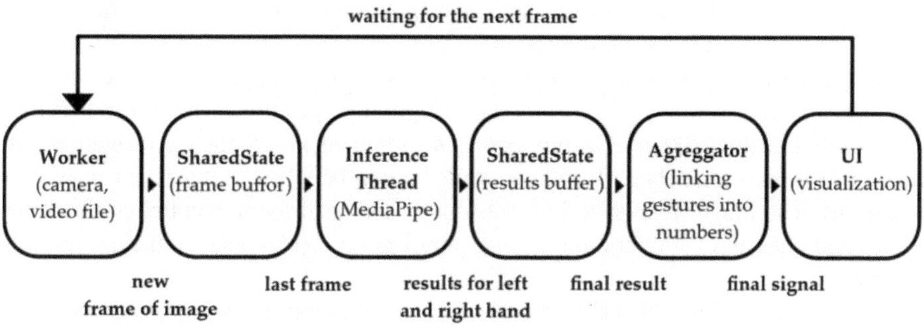

Fig. 5. Program operation diagram; Source: authors own archive.

5 Testing and Validation

The purpose of the testing phase was to verify whether the developed prototype of the referee gesture recognition system meets the functional requirements and whether it maintains stability and adequate performance in conditions similar to those of a real match. The validation included an assessment of the accuracy of gesture classification, an analysis of the number of frames per second (FPS) processed, the system's response time, and its behavior in ambiguous situations, such as partial hand coverage or a temporary decrease in classification confidence.

Fig. 6. (a) A referee showing the tens digit. (b) The number generated after showing the ones digit. Source: authors own archive.

A set of 20 test trials was prepared covering the full range of supported signals: single-digit numbers, double-digit numbers in the range 10–15, numbers 16–99, and special gestures ("C," "00"). The research material came from recordings made during a referee camp (Słupsk, Poland. July 2025) (Fig. 6). An expected result was specified for each test, and the result generated by the aggregation module was automatically saved in a log file along with the FPS value and the response time calculated from the moment the gesture stabilized to its interpretation.

Of the 20 trials, 16 were recognized correctly, corresponding to an accuracy of 80%. The highest accuracy was recorded for single-digit gestures, whose response time was usually less than 200 ms. For numbers in the range of 11–15, the interpretation time increased to approximately 230–260 ms, while for numbers 16–99, it reached an average of 300 ms. The longer response time was due to the need for sequential confirmation of tens and ones and the operation of time stabilization mechanisms in the aggregation module.

Error analysis showed that incorrect classifications mainly concerned gestures requiring the simultaneous cooperation of both hands (e.g., 18, 55). Sensitivity to partial finger coverage, suboptimal hand positioning relative to the camera, or lighting differences caused a temporary decrease in classification confidence in the *MediaPipe* module. In one case, the gesture "C" was incorrectly interpreted as "0," which indicates the similarity of hand shapes under suboptimal detection conditions.

In terms of performance, the system ran stably, maintaining an average of about 12 FPS in the inference module. No drops in interface fluidity, thread blocking, or loss of synchronization between modules were observed. These results confirm the correctness of the adopted multithreaded architecture and the system's ability to operate in real time (Table 1).

Table 1. Test trials of the system; *Source: authors' own development.*

Sample	Expected	Recognized	Correct	FPS	Time[ms]
1	0	0	Yes	12.9	165
2	00	00	Yes	13.1	182
3	C	C	Yes	12.7	198
4	11	11	Yes	12.8	201
5	13	13	Yes	12.4	241
6	14	14	Yes	12.2	252
7	10	10	Yes	12.9	194
8	8	8	Yes	12.3	188
9	18	17	No	12.6	255
10	20	20	Yes	13.0	253
11	23	23	Yes	11.9	261
12	44	44	Yes	12.1	282
13	55	50	No	12.0	293
14	3	3	Yes	12.4	172
15	90	90	Yes	12.7	269
16	98	98	Yes	13.0	301
17	72	72	Yes	12.5	303
18	15	15	Yes	12.6	243
19	C	0	No	12.1	201
20	00	00	Yes	12.5	179

6 Conclusions

The main objective of the work, which was to develop a system supporting the table operator by automatically recognizing selected foul gestures and player numbers, was achieved. Not only was the use of classification models crucial, but also the design of an interpretive logic layer that stabilizes readings, combines tens and ones, and filters out random hand positions. The results confirm that image classification alone is insufficient – only its integration with the rules resulting from referee signaling ensures the practical usefulness of the system.

The functional scope of the prototype was deliberately limited to key foul-related gestures, which allowed for verification of the concept while maintaining controlled system complexity. The developed solution is a starting point for further research on the automation of refereeing processes, including the expansion of the set of recognized signals, integration with multiple cameras, and links to statistical systems. Potential applications include, in particular, youth and amateur competitions, where technological support can reduce the number of errors and ease the organizational burden.

References

1. Wang, S.-W., Hsieh, W.-W.: Performance analysis of basketball referees by machine learning techniques. In: Correia, P.P., Cabri, J. (eds.) Proceedings of the 4th International Congress on Sport Sciences Research and Technology Support – icSPORTS, vol. 1, pp. 165–170. SciTePress, Porto, Portugal (2016)
2. Wang, X., Xing, J., Yuan, H., Ma, L., Ma, Q.: Design of Basketball Referee System based on human-machine collaboration. In: ICSTPA'24: Proceedings of the 2024 International Conference on Sports Technology and Performance Analysis, pp. 192–196. Association for Computing Machinery, New York, NY, United States (2025)
3. Zhang, Z., Liu, W., Zheng, Y., et al.: Learning spatio–temporal context for basketball action pose estimation with a multi-stream network. Nat. Sci. Rep. **15**, 29173 (2025)
4. Banarase, J., Gawande, M., Kale, S., Mahalle, S.: Basketball virtual referee. Int. Res. J. Eng. Technol. **11**(3), 1434–1440 (2024)
5. Roboflow Blog., Skalski, P., How to Detect, Track, and Identify Basketball Players with Computer Vision. https://blog.roboflow.com/identify-basketball-players/. Accessed 05 Feb 2025
6. Official Basketball Rules 2024. FIBA, Mies, Switzerland (2024)
7. AI and Google. MediaPipe Solutions guide., https://ai.google.dev/edge/mediapipe/solutions/guide. Accessed 05 Feb 2025
8. AI and Google. MediaPipe–Gesture Recognizer. https://ai.google.dev/edge/mediapipe/solutions/vision/gesture_recognizer. Accessed 05 Feb 2025

Modern Threats in Computer Networks: Examples and Analysis of Zero Trust Bypass Attacks

Krystian Marchwicki(✉)

DSW University of Lower Silesia, Strzegomska Street 55, Wrocław, Poland
krystianmarchwicki@gmail.com
http://dsw.edu.pl/

Abstract. Modern network architecture increasingly relies on the Zero Trust model, which assumes no implicit trust for users, devices, or infrastructure components, regardless of their location. This approach is widely recognized as a response to the limitations of traditional perimeter-based security models, particularly in cloud, hybrid, and distributed environments. Despite its growing adoption and reputation as a best-practice security paradigm, real-world implementations of Zero Trust architectures are increasingly subject to successful security bypasses.

This paper aims to analyse modern threats in computer networks, with a particular focus on Zero Trust Bypass attacks. The study identifies and classifies attack vectors that exploit identity compromise, abuse of trust context, misconfigurations in Identity and Access Management (IAM) systems, and weaknesses in Zero Trust Network Access mechanisms. Both theoretical aspects of the Zero Trust Architecture and practical attack scenarios are examined, including attacks targeting session tokens, multi-factor authentication mechanisms, API- and microservices-based environments.

Furthermore, the paper presents detection methods and mitigation strategies for Zero Trust Bypass attacks, emphasizing the role of behavioural monitoring, SIEM and XDR systems, and the principle of least privilege. The findings indicate that the effectiveness of Zero Trust architecture strongly depends on correct implementation, continuous verification, and integration with comprehensive information security management processes. The presented analysis may serve as a valuable reference for security professionals, system architects, and decision-makers responsible for protecting modern IT environments.

Keywords: Zero Trust · Zero Trust Bypass · Network security · IAM · ZTNA · Cybersecurity

1 Introduction

The Zero Trust Network Security model is based on the principle "never trust – always verify." In contrast to the perimeter-based approach, no network or entity is trusted by default, and every access request is evaluated based on identity and contextual information.

W. Zamojski et al. (Eds.): DepCoS-RELCOMEX 2026, LNNS 2014, pp. 275–283, 2026.
https://doi.org/10.1007/978-3-032-27927-9_26

This concept emerged in response to the limitations of the traditional security model, which did not correspond to the realities of cloud computing, remote work, and hybrid environments. The foundation of the model is the absence of implicit trust and the continuous verification of users, devices, and applications [3].

A central element of ZTA is the Identity and Access Management (IAM) system. Mechanisms such as Multi-Factor Authentication (MFA), access policy engines, and Zero Trust Network Access (ZTNA) solutions enable granular control over access to specific resources. These mechanisms are complemented by SIEM and XDR systems that support monitoring and anomaly detection.

Zero Trust can be implemented in on-premises, cloud, and hybrid environments, where the consistency of policies and centralized identity management are crucial. A common mistake is treating Zero Trust as a one-time technological implementation without integrating its components and without continuous risk assessment. Excessive privileges and lack of session validation may lead to Zero Trust Bypass attacks.

The Zero Trust model does not eliminate risk but reduces the impact of compromise through segmentation, continuous verification, and rapid detection. Its effectiveness depends on the quality of implementation and integration with systemic security management. In threat analysis, the MITRE ATT&CK framework is useful, as it enables the mapping of Zero Trust mechanisms to specific attack techniques and the identification of potential vulnerabilities [4] (Table 1).

Table 1. MITRE ATT&CK techniques related to Zero Trust Bypass attacks

ATT&CK tactics	Technique	ID	Connection with ZTNA
Credential Access	Valid Accounts	T1078	Exploitation of compromised accounts in IAM environment
Defense Evasion	Use of Web Tokens	T1606	Abuse of session and OAuth tokens
Lateral Movement	Remote Services	T1021	Lateral movement using legitimate services
Persistence	Account Manipulation	T1098	Modification of permissions and roles
Privilege Escalation	Abuse Elevation Control	T1548	Abuse of authorization mechanisms

Based on the related techniques, it is possible to determine which security controls should be applied during the analysis of incoming and outgoing traffic. For a Zero Trust solution to be effective, several architectural modules must be implemented (Fig. 1).

The figure presents a reference Zero Trust architecture, in which a user or application sends a request to access a resource through the Policy Enforcement Point (PEP). The decision to grant access is made by the Policy Decision Point (PDP) based on data from IAM, MFA, and contextual and telemetry information from SIEM and XDR systems.

Legend:

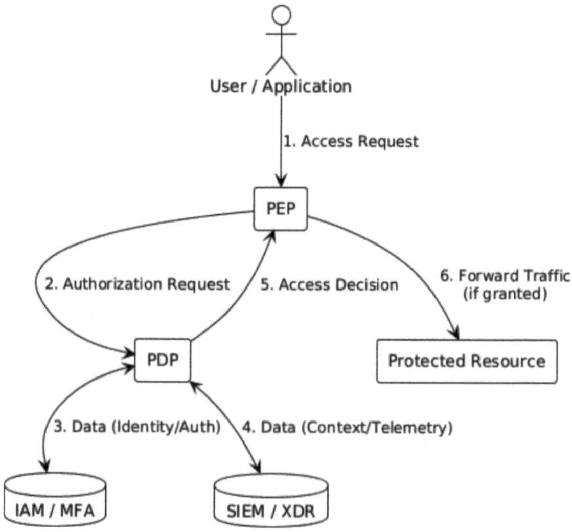

Fig. 1. ZTNA architecture – reference diagram

- **User/Application** – a user or application initiating the request.
- **Policy Enforcement Point (PEP)** – enforcement of policies.
- **Policy Decision Point (PDP)** – policy decision engine.
- **IAM/MFA** – identity verification and authentication.
- **Protected Resource** – an application, service, or data.
- **SIEM/XDR** – monitoring and event analysis.

Information security management defined in the **ISO/IEC 27001** standard [8].

The ISO/IEC 27001 standard focuses on risk management, security policies, and the continuous improvement of the Information Security Management System (ISMS). Zero Trust does not replace these frameworks but technically supports them in modern IT environments. The principles of least privilege, access control, and monitoring described in ISO/IEC 27002 are reflected in solutions such as IAM, ZTNA, and SIEM/XDR, strengthening the consistency and resilience of organizations [8, 9].

Zero Trust does not eliminate threats but changes their nature. Zero Trust Bypass attacks rely on the use of legitimate identities, tokens, and sessions, which makes the attacker's activity appear compliant with security policies. The key target therefore becomes digital identity—its compromise (e.g., OAuth tokens or SSO credentials) enables access without breaking technical protection mechanisms.

Although MFA is important, it is not fully resistant to manipulation techniques such as MFA fatigue or proxy phishing. Another threat is the compromise of identity providers and the supply chain, which may lead to large-scale circumvention of control mechanisms. Microsegmentation limits lateral movement but does not eliminate abuse resulting from excessive privileges and incorrect trust relationships.

Bypass attacks are difficult to detect because they operate within formally legitimate mechanisms. Therefore, Zero Trust should be understood as a process of continuous adaptation, in which the quality of policies, integration of detection with enforcement decisions, and the ongoing improvement of security management play a crucial role (Table 2).

Table 2. Correlation between threat scenarios in Zero Trust Network Access Architecture (ZTNA): Attack strategies, control gaps, and remediation approaches.

Attack types	Control gaps in ZTNA	Mitigation Mechanism
Token theft (OAuth2, OIDC, SSO)	Lack of continuous validation for sessions and authentication artifacts	Continuous session risk assessment, token binding, short-lived tokens, real-time revocation
MFA fatigue/MFA push abuse	Over-reliance on one-time MFA, lack of behavioural analysis	Adaptive MFA, request rate limiting, phishing-resistant MFA (FIDO2)
Proxy phishing	Lack of MFA resilience against intermediary attacks (Adversary-in-the-Middle)	Passwordless authentication, hardware-backed MFA
Service account abuse	Excessive privileges and lack of periodic role validation	Just-In-Time (JIT) access, automated credential rotation
Lateral movement within ZTA	Overly broad trust relationships between network segments	Identity-based microsegmentation, Zero Standing Privileges (ZSP)
Supply chain compromise (e.g., SolarWinds)	Inherited trust in third-party or external components	Integrity verification, Software Bill of Materials (SBOM), continuous vendor validation
Abuse of APIs	Lack of contextual and intent-based control over API calls	API Security Gateways, anomaly analysis, policy-as-code

The table presents the relationship between Zero Trust Bypass attacks and control gaps within the Zero Trust architecture. It indicates that the effectiveness of such attacks results less from the lack of technology and more from design simplifications, static policies, and insufficient process maturity. Therefore, the problem is primarily decision-making and contextual, rather than purely technical.

An example is token theft—the authentication mechanism itself functions correctly, but the lack of continuous session validation allows prolonged access. Similarly, MFA fatigue demonstrates that treating MFA as a one-time control mechanism may lead to abuse. Another significant threat is supply chain compromise, which under-mines trust in external components.

The attack → control gap → mitigation model describes a sequence consisting of the attack vector, a control gap resulting from the discrepancy between design assumptions and operational practice, and a mitigation mechanism that limits the impact of a breach. This process is iterative and requires continuous adaptation.

Zero Trust does not eliminate threats but changes their nature. Attacks increasingly focus on decision and contextual gaps, particularly those related to digital identity. The resilience of the architecture therefore depends not merely on the presence of security controls, but on their quality, integration, and capacity for continuous improvement.

2 Analysis of Selected Attacks and Their Impact

2.1 Attack on SolarWinds (2019–2020)

The SolarWinds security incident represents one of the most advanced examples of a supply chain compromise attack, clearly illustrating the limitations of traditional security mechanisms and selected implementation gaps in architectures that rely on trust in digitally signed software.

In September 2019, a group identified in analytical reports as "Nobelium" gained unauthorized access to SolarWinds' internal infrastructure. Publicly available technical analyses do not clearly identify the initial intrusion vector, which highlights the high level of operational sophistication of the attackers. After gaining access to the development environment, the attackers took control of the source code repositories, focusing on the analysis of the Orion component [6].

In the initial phase, no direct functional modifications were introduced. The attackers observed the code compilation and approval process, and subsequently introduced small, neutral changes (e.g., empty classes) to verify the effectiveness of integrity control mechanisms. The gradual escalation of interference eventually led to the injection of the malicious "Sunburst" component in the form of the `SolarWinds.Orion.Core.BusinessLayer.dll` library.

In March 2020, the modified code was approved and signed with the legitimate SolarWinds Worldwide LLC digital certificate, which granted it full cryptographic legitimacy within the trust chain [11].

The distribution of the update affected approximately 18,000 customers. The digital signature mechanism caused target systems to treat the package as a fully trusted artifact. This reveals a fundamental issue: the security architecture based on trust in the vendor's signature did not anticipate the compromise of the software development process itself.

The malicious "Sunburst" component was designed in a highly camouflaged manner. It executed in a separate thread, without disrupting the operation of the Orion system. It was embedded within a structure of 13 subclasses and 16 methods, while maintaining the naming conventions typically used by the developers. Base64 encoding and hashing mechanisms were also employed, which complicated static analysis and reverse engineering.

Operational Phase and Privilege Escalation

After verifying the environment, the malware connected to a command-and-control (C2) server, disguising the traffic as legitimate communication of the Orion platform. The use of living-off-the-land techniques and permitted protocols made detection more difficult.

The attackers then established hands-on-keyboard control, using native Windows tools, which limited the effectiveness of signature-based detection. The primary targets were AD FS servers and privileged accounts. After obtaining the signing keys, the attackers generated forged SAML tokens (Golden SAML technique) that were accepted by cloud services as legitimate.

AAA mechanisms treated the activity as valid, and MFA was not triggered, because federated authentication had formally completed successfully. As a result, identity controls were bypassed despite the use of strong authentication mechanisms.

The ultimate objective was data exfiltration (including email mailboxes) and the maintenance of persistent access through the creation of additional trust relationships within cloud environments (Fig. 2).

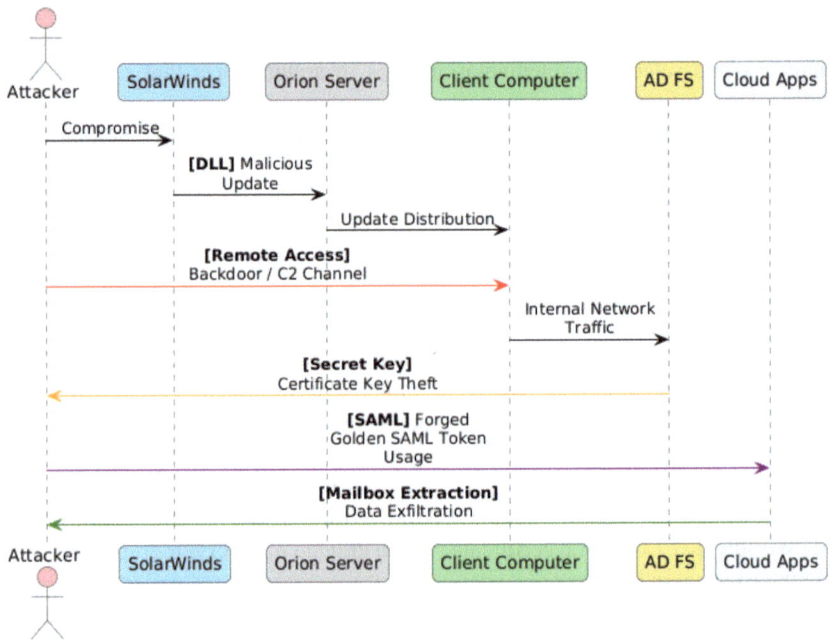

Fig. 2. Execution flow of the attack using the Sunburst malicious code [5].

The first signals of the SolarWinds compromise were detected by FireEye, which identified anomalies in its own environment. The analysis revealed that the malicious code had been operating within victim systems for approximately 14 months, exploiting trust in digitally signed software and federated trust relationships.

In response, a "kill switch" mechanism was implemented by redirecting traffic from the C2 domains to servers controlled by Microsoft, which led to the deactivation of the malware. SolarWinds subsequently released security patches and recommended system updates, system isolation, and the rotation of passwords and certificates [13].

The incident demonstrated the limitations of the perimeter-based security model and of relying solely on digital signatures in the case of a supply chain attack. From a Zero Trust perspective, key countermeasures include the principle of least privilege, access segmentation (e.g., ZTNA), and advanced anomaly detection mechanisms (UEBA, SIEM, XDR), particularly in identity and certificate management systems.

The SolarWinds incident represents an example of a Zero Trust Bypass attack—the authentication mechanisms themselves were not broken; instead, the foundation of trust was compromised, enabling attackers to operate within formally legitimate security mechanisms [5].

2.2 Attack on Okta (2022)

In January 2022, the threat actor group known as "LAPSUS$" executed a cyberattack against Sitel (Sykes), a technical support service provider collaborating with Okta. The group compromised an internal Sitel user account, enabling the attackers to gain Remote Desktop Protocol (RDP) access to a support engineer's workstation [7, 10].

This initial account takeover granted the threat actors extensive access to the Okta ecosystem. From the company's GitHub repository, they acquired privilege escalation tools and the Mimikatz credential-dumping software, which extracts sensitive credentials from memory. Consequently, their subsequent activities appeared as highly privileged operations, thereby evading immediate anomaly detection. Furthermore, the attackers exfiltrated the Windows Security Account Manager (SAM) file, which is responsible for managing and storing user credentials (functionally analogous to the `/etc/passwd` file in Linux environments). The attack chain also involved accessing an Excel spreadsheet containing domain administrator passwords [15].

To establish persistence, the attackers created "Backdoor Admin" accounts within the `TenantAdmin` group and configured blind carbon copy (BCC) email forwarding rules. Utilizing the internal Okta SuperUser application, the threat actors captured screenshots of external systems, including Jira and Slack, which were subsequently leaked on Telegram [12, 17].

However, a critical attack vector emerged through the exploitation of HTTP Archive (HAR) files, which clients routinely submitted to technical support for diagnostic purposes. By extracting valid session cookies from these files, the attackers successfully bypassed Multi-Factor Authentication (MFA) mechanisms. Within the compromised client systems, they provisioned fraudulent accounts, adhering to existing naming conventions to minimize the risk of detection. The targeted entities in this campaign included 1Password, BeyondTrust, and Cloudflare, among others. Furthermore, the threat actors targeted extensive casino networks (MGM and Caesars) by exploiting Okta Active Directory Sync. This secondary phase resulted in system paralysis, ransomware deployment, exfiltration of customer data, and multi-million-dollar financial losses [1, 2].

Detection and Mitigation Methods

The Zero Trust architecture changes the approach to detection by integrating it with the decision-making process. Protection focuses on identity, resources, and contextual information, and each access decision is based on continuous risk assessment and telemetry data, in accordance with NIST SP 800-207 [14, 16].

Behavioural monitoring (UEBA) plays a key role, as it enables the detection of deviations from typical user activity, even when valid credentials are used. The integration of data from IAM systems, endpoints, networks, and cloud environments through SIEM and XDR enables event correlation and automated responses, such as restricting access when anomalies are detected.

The foundation of mitigation is continuous verification and the principle of least privilege, further developed in the Just-In-Time and Just-Enough-Access models. This approach limits the impact of identity compromise and makes lateral movement more difficult. In the Zero Trust model, detection and access control form a single dynamic trust management mechanism.

3 Final Conclusions

The conducted analysis indicates that Zero Trust does not guarantee the elimination of security incidents but rather constitutes a trust management model for complex environments. According to NIST SP 800-207, it is based on the continuous verification of identity, context, and resource status, and its effectiveness depends on the quality of implementation and the process maturity of the organization [14].

Incidents such as SolarWinds, MFA fatigue, and token theft demonstrate that so-called "Zero Trust bypasses" result from control gaps, including excessive privileges, lack of segmentation, insufficient session validation, or weak integration between detection mechanisms and access policies.

The attack → control gap → mitigation model highlights that the key issue is closing the gap between policy and its enforcement through continuous verification, privilege minimization, and integration with XDR systems.

Zero Trust should therefore be understood as a model for limiting the impact of compromise, rather than completely preventing security breaches. Its purpose is to reduce the scope of an incident, hinder lateral movement, and shorten detection time. Effective implementation requires a programmatic and long-term approach, as partial implementations may create a false sense of security.

The resilience of the architecture depends on the quality of identity systems, policy consistency, effective monitoring, and the systematic application of the least privilege principle. Awareness of Zero Trust Bypass threats is essential because such attacks exploit legitimate access mechanisms and do not always generate obvious warning signals. Zero Trust is therefore a process of continuous adaptation and improvement rather than a final state of security.

It is also worth noting that the organizations that became victims of the attacks had already implemented ZTNA mechanisms, yet they operated under the assumption that these solutions were a "silver bullet" capable of protecting them from any attack or unauthorized activity. The incidents clearly demonstrated the necessity of collecting event logs and analysing anomalies. This ultimately highlights the importance of critically evaluating implemented security systems and integrating practical experience with theoretical assumptions in a coherent manner.

ZTNA solutions should always be adapted to the specific conditions in which they operate, including the organization that implements them and the systems and services it uses. An indispensable factor for the successful deployment of this powerful security approach is the human factor—security professionals who are aware of potential issues during the verification of incoming and outgoing traffic.

References

1. Beyond Identity: Okta Cyber Trust Report (2024). https://www.beyondidentity.com/resource/okta-cyber-trust-report-2
2. Bradbury, D.: Okta's Investigation of the January 2022 Compromise (2022). https://www.okta.com/blog/company-and-culture/oktas-investigation-of-the-january-2022-compromise/
3. Chandramouli, R.: Guide to a Secure Enterprise Network Landscape. https://doi.org/10.6028/NIST.SP.800-215
4. Cybersecurity, and E.U.A. for: ENISA threat landscape 2022 – July 2021 to July 2022 (2022). https://doi.org/10.2824/764318
5. Dulce, S.: Examining the SolarWinds Supply Chain Attack - Deep Dive (2020). https://zeronetworks.com/blog/examining-solarwinds-supply-chain-attack
6. Fortinet Inc: Solar Winds Cyber Attack. https://www.fortinet.com/resources/cyberglossary/solarwinds-cyber-attack
7. HALOCK: Okta's Delay in Notifying its Customers of a Breach is Authentic. https://www.halock.com/oktas-delay-in-notifying-its-customers-of-a-breach-is-authentic/
8. International Organization for Standardization, International Electrotechnical Commission: ISO/IEC 27001:2022 – Information Security, Cybersecurity and Privacy Protection – Information Security Management Systems – Requirements. International Organization for Standardization, Geneva, Switzerland (2022)
9. International Organization for Standardization, International Electrotechnical Commission: ISO/IEC 27002:2022 – Information Security Controls. International Organization for Standardization, Geneva, Switzerland (2022)
10. Jones, C.: Okta reveals full extent of LAPSUS$ breach as hackers announce hiatus (2022). https://www.itpro.com/security/cyber-attacks/367165/okta-statement-lapsus-data-breach
11. Microsoft Threat Intelligence: Analyzing Solorigate, the compromised DLL file that started a sophisticated cyberattack, and how Microsoft Defender helps protect customers (2020). https://www.microsoft.com/en-us/security/blog/2020/12/18/analyzing-solorigate-the-compromised-dll-file-that-started-a-sophisticated-cyberattack-and-how-microsoft-defender-helps-protect/
12. Nahari, S.: Piecing Together the Attack on Okta's Support Unit (2023). https://www.cyberark.com/resources/blog/piecing-together-the-attack-on-oktas-support-unit
13. Oladimeji, S., Kerner, S.M.: SolarWinds hack explained: Everything you need to know (2023). https://www.techtarget.com/whatis/feature/SolarWinds-hack-explained-Everything-you-need-to-know
14. Rose, S., et al.: Zero Trust Architecture. National Institute of Standards and Technology, Gaithersburg, MD (2020). https://doi.org/10.6028/NIST.SP.800-207
15. Sai: How Lapsus$ Breached Okta and its Customers? | OKTA Lapsus$ Hack Explained! https://www.netsecurity.com/how-lapsus-breached-okta-and-its-customers-okta-lapsus-hack-explained/. Accessed 28 Feb 2026
16. Unit 42 Threat Research Center: Threat Brief: Lapsus$ Group (2022). https://unit42.paloaltonetworks.com/lapsus-group/
17. Okta, Inc. Cybersecurity Incident. https://www.board-cybersecurity.com/incidents/tracker/20220325-okta-inc-cybersecurity-incident/

Hybrid Algorithm for Secure Key Generation Using Quantum Randomness Enhancement

Paulina Muzaj(✉) and Piotr Sulikowski

Faculty of Information and Communication Technology, Wrocław University of Science and Technology, Janiszewskiego 11/17, 50-372 Wrocław, Poland
paula.muzaj@gmail.com

Abstract. This paper addresses the challenge of establishing secure cryptographic keys over public or potentially intercepted channels without relying on dedicated, secure quantum infrastructure. While we assume that the underlying quantum hardware, such as the ODRA-5 system, yields high-entropy data, the lack of a secure distribution channel leaves the raw output vulnerable to eavesdropping. To mitigate this risk, we propose a hybrid quantum extraction algorithm designed to derive private cryptographic keys from the ODRA-5 output which we treat as a publicly accessible classical bitstream. The system utilizes quantum superposition, introduced by a Hadamard layer applied to all qubits, to inject indeterminacy. To guarantee the absolute statistical uniformity required for a One-Time Pad (OTP), the stream undergoes a secondary classical extraction phase using the von Neumann debiasing method. This process evaluates bit-grouping in pairs, effectively compensating for inherent hardware bias and energy relaxation dynamics observed in the ODRA-5 processor. Finally, a shared-secret extraction logic known only to the communicating parties is used to select specific bits, ensuring that the final key remains private. Through rigorous simulations and real hardware tests, we demonstrate that the security of the final key relies entirely on the extraction algorithm rather than the secrecy of the initial quantum transmission.

Keywords: quantum computing · QRNG · OTP · ODRA-5 · secure key

1 Introduction

This research explores the integration of quantum indeterminacy with classical post-processing to address the critical challenge of securing cryptographic keys in environments where the entropy source is under surveillance.

Modern quantum cloud services often rely on public or non-secure communication channels for data transmission. If the link between the user and the quantum processor (ODRA-5) is intercepted by an adversary (Spy), the raw

W. Zamojski et al. (Eds.): DepCoS-RELCOMEX 2026, LNNS 2014, pp. 284–293, 2026.
https://doi.org/10.1007/978-3-032-27927-9_27

measurement outcomes become public knowledge. In such a scenario, traditional cryptographic methods fail, as the entropy source is effectively shared with the eavesdropper. The vulnerability occurs at the classical layer during the transmission of measurement results. Once the quantum state is measured and converted into a classical bitstream for transport, it becomes accessible to any adversary monitoring the communication link.

The primary contribution of this work is the design and validation of a hybrid key generation mechanism that derives completely random and secure cryptographic keys from a quantum data stream. Since the ODRA-5 system operates over a non-secure channel, we treat its output as public entropy that requires further processing to ensure privacy. Our approach utilizes a three-stage enhancement process: first, we apply a Hadamard layer to all qubits to inject independent quantum entropy through superposition. Second, we apply the von Neumann debiasing trick [7] by processing bit pairs to extract a perfectly uniform distribution, effectively compensating for any hardware-induced bias in the raw quantum stream. Third, we introduce a shared-secret extraction algorithm known only to the communicating parties. By leveraging quantum indeterminacy and rigorous statistical validation, we demonstrate that this method satisfies the requirements of OTP even when the raw quantum data is accessible to an adversary.

2 Literature Review

2.1 One-Time Pad (OTP) Protocol

The One-Time Pad (OTP) [6] is an encryption algorithm that provides information-theoretic security, making it mathematically unbreakable regardless of the adversary's computational power. The encryption process relies on a bitwise XOR operation, combining the plaintext message with a random key to produce the ciphertext.

To guarantee absolute security, an OTP implementation must strictly satisfy four fundamental conditions:

- **Perfect Randomness:** The key must be derived from a truly random physical source and possess a perfectly uniform statistical distribution.
- **Strict Secrecy:** The key must remain completely secure and known exclusively to the legitimate sender and receiver.
- **Single Use:** The key must be used to encrypt only one message and destroyed immediately afterward. Key reuse fatally compromises the protocol.
- **Sufficient Length:** The key must be at least as long as the plaintext message being encrypted.

2.2 The Von Neumann Extractor

To address the inherent imperfections and bias of hardware-based random number generators, our work employs the classical debiasing method introduced by

John von Neumann. Von Neumann observed that in physical random processes, ensuring the independence of consecutive events is generally more feasible than guaranteeing a perfectly uniform probability. Assuming successive events are completely independent, a strictly unbiased 50–50 distribution can be reconstructed even from a heavily biased source by evaluating events in pairs.

The technique processes the raw bit stream in non-overlapping pairs using a simple filtering rule:

- If a pair consists of identical bits (i.e., 00 or 11), it is completely rejected.
- If a pair consists of differing bits (i.e., 01 or 10), the result is accepted and mapped to a single output bit.

Because the probabilities of obtaining heterogeneous pairs are mathematically identical (i.e., $P(10) = P(01)$), the resulting sequence is rigorously unbiased. While this method guarantees information-theoretic uniformity, it reduces the overall generation efficiency to at most 25% of the raw bit rate. Nevertheless, it serves as a necessary and robust first stage in decoupling our final cryptographic key from the initial bias of the raw quantum stream.

2.3 The Short Key Dream and Public Entropy Sources

The practical implementation of the One-Time Pad is fundamentally limited by the secure distribution of keys that must equal the plaintext in length. To overcome this logistical bottleneck, researchers have extensively explored the "Short Key Dream" paradigm, comprehensively reviewed by Cerruti [1]. The core concept, originally formalized by Maurer in 1992 [4], shifts the paradigm from expanding a short pseudorandom seed to extracting a secure key from a massive, publicly accessible source of randomness. In this model, communicating parties only need to securely share a relatively short secret (such as an extraction algorithm or pointer set). This short secret dictates which specific bits are harvested from the public data source to reconstruct the full-length, operationally secure OTP key.

Our proposed methodology directly realizes and extends this paradigm into the quantum domain. In our architecture, the live, unencrypted data stream generated by the ODRA-5 system serves as the massive public entropy source. Rather than attempting to securely transmit gigabytes of key material, the communicating parties share only the custom extraction algorithm (representing the "short key"). By applying this shared-secret logic to the public quantum stream—following the initial von Neumann debiasing phase—our system dynamically distills a perfectly secure OTP key. Consequently, our approach leverages the theoretical foundations of public-source key extraction to provide a highly practical, quantum-backed cryptographic solution.

The utilization of modern quantum computers for True Random Number Generation (TRNG) currently represents a highly promising area of research. The study by Nath [5] focuses on the theoretical and practical foundations of certifying randomness directly on available quantum processors. The authors demonstrate that employing shallow circuits and single-qubit measurements can hardware-guarantee cryptographic unpredictability without the need for complex optical setups. This provides an excellent foundation for the research presented by Gnatowski [3], who analyzed the performance of randomness generation on an IQM Spark processor (ODRA-5). Their empirical work directly proves that commercially available Quantum Processing Units (QPUs) can successfully serve as powerful sources of physical entropy.

3 Theoretical Framework

The extraction of secure cryptographic keys from the public ODRA-5 data stream is implemented via a hybrid mechanism. By mapping classical bits to qubits and applying a Hadamard gate to all qubits, the algorithm injects independent quantum entropy to ensure a perfectly random output.

3.1 Formal Algorithm Specification

Algorithm 1 Quantum Randomness Extraction (5-Qubit Block)

Require: A quantum register with 5 available qubits
Ensure: A 5-bit classical random sequence K
1: $K \leftarrow \emptyset$ ▷ Initialize an empty classical sequence
2: Initialize qubits q_0, q_1, q_2, q_3, q_4 to the state $|0\rangle$
3: Apply Hadamard gate (H) to all qubits $q_0, \ldots, q_4$ ▷ Creates equal superposition
4: Measure qubits $q_0, \ldots, q_4$ to obtain classical bits $k_0, \ldots, k_4 \in \{0, 1\}$
5: Append $\{k_0, k_1, k_2, k_3, k_4\}$ to the sequence K
6: **return** K

The ODRA-5 quantum computer is a public source that generates a classical bit stream. To ensure maximum entropy, we bypass deterministic initialization and apply Hadamard gates (H) directly to qubits in their ground state $|0\rangle$. The subsequent measurement collapses each qubit, yielding a truly random bit that is inherently unpredictable. This is followed by the von Neumann debiasing trick, which processes bit pairs to eliminate statistical bias and ensure the perfect uniformity required for a One-Time Pad.

3.2 Quantum Filtering Mechanism

The algorithm utilizes quantum properties, specifically the Hadamard gate, to inject true randomness into the bit extraction process. The quantum circuit is composed of five qubits initialized in the ground state:

$$|\psi_0\rangle = |00000\rangle, \tag{1}$$

where the state vector is defined as $|q_0q_1q_2q_3q_4\rangle$. The Hadamard gate transforms the ground state into a state of perfect quantum superposition:

$$H|0\rangle = \frac{|0\rangle + |1\rangle}{\sqrt{2}}. \tag{2}$$

Consequently, the global state of the five-qubit system evolves into:

$$|\psi_1\rangle = H^{\otimes 5}|\psi_0\rangle = \bigotimes_{i=0}^{4} \left(\frac{|0\rangle + |1\rangle}{\sqrt{2}} \right). \tag{3}$$

Finally, a simultaneous measurement is performed on all five qubits in the computational basis. The final readout collapses the global wave function into a classical 5-bit block $k_{out} = \{k_0, k_1, k_2, k_3, k_4\}$. Measuring this quantum state forces the system to collapse probabilistically. This random sequence of bits is then utilized in the subsequent stages of the secure key generation process.

Quantum Circuit. The quantum circuit representing the process described above is shown in Fig. 1.

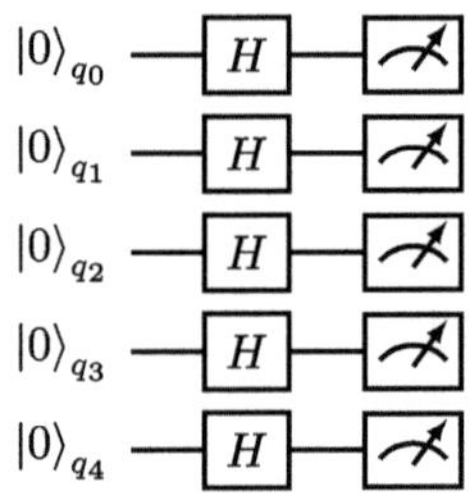

Fig. 1. 5-qubit quantum circuit for randomness generation using Hadamard gates.

3.3 Short Key Mechanism

Once the raw quantum stream has been debiased to ensure perfect uniformity, the final step is to securely extract the private key. This is achieved without transmitting the key itself, but rather by sharing a compact set of rules.

Example 1: Prime Stride Selection Algorithm. In this straightforward approach, the shared "Short Key" consists of two integer values: a starting index S and a large prime step value P. Both communicating parties must securely exchange (S, P) beforehand. Given the public quantum bitstream Q of length L, the private OTP key K is constructed by sampling bits at specific intervals.

The extraction logic is defined as follows:

Because P is prime and known only to the legitimate parties, an adversary monitoring the public stream Q cannot determine which specific bits are being harvested without brute-forcing the (S, P) combination.

Algorithm 2 Prime Stride Selection (Short Key)

Require: Public bitstream Q of length L, Target length N, Start S, Prime step P
Ensure: Private OTP key K

Initialize empty sequence K
for $i = 0$ **to** $N - 1$ **do**
  Target_Index $\leftarrow (S + i \times P) \bmod L$
  Append Q[Target_Index] to K
end for
return K

4 Implementation and Validation Results

The transition from theoretical design to practical application involves executing the hybrid protocol on physical hardware and evaluating the statistical integrity of the resulting bitstreams.

The computational experiments were conducted in a Python 3.11 environment utilizing Qiskit 1.2.4 and the `iqm-client` library. True quantum randomness was generated using ODRA-5, a 5-qubit superconducting quantum processor manufactured by IQM Quantum Computers. Remote access to the hardware endpoint (https://odra5.e-science.pl/) was managed securely via the `dotenv` library.

The experimental protocol involved initializing a 5-qubit register in the $|0\rangle^{\otimes 5}$ state, followed by a global layer of Hadamard (H) gates to induce a uniform superposition. After transpiling the circuit to match the backend's specific topology, it was executed for 1,000,000 shots. Crucially, the Qiskit `memory=True` parameter was enabled to record individual measurement outcomes rather than aggregated probability distributions. This execution yielded a continuous, raw sequence of 5-bit strings (e.g., '00011', '10101'), which was stored locally for classical post-processing.

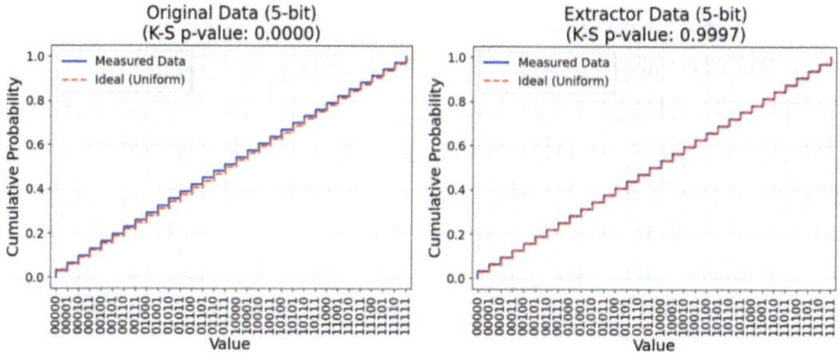

Fig. 2. Comparison of the empirical cumulative distribution function (solid blue line) with the theoretical uniform distribution CDF (dashed red line) for 5-bit words ($n = 5$). The left panel shows the raw hardware data with visible bias, while the right panel presents the highly uniform data after applying the von Neumann extractor.

Figure 2 presents a graphical and statistical analysis of the 5-bit words ($n = 5$) before and after the debiasing process. The left panel shows the cumulative distribution function (CDF) of the raw data. The empirical CDF rises faster than the theoretical uniform step-line, indicating a hardware bias that favors states with a predominance of zeros (lower decimal values). The Kolmogorov-Smirnov (K-S) test [2] statistically confirms this anomaly, yielding a p-value of 0.0000 and rejecting the null hypothesis of uniformity. The right panel demonstrates the effectiveness of the von Neumann extractor. After filtering, the empirical distribution aligns almost perfectly with the ideal uniform CDF. The p-value of the K-S test increases to 0.9997, providing robust evidence that the extraction algorithm successfully mitigated the initial hardware asymmetry.

Following the acquisition of the raw sequences from the ODRA-5 backend, the data were subjected to the von Neumann debiasing algorithm as described. Although quantum superposition provides the foundation for randomness, the results confirm that physical superconducting systems exhibit subtle measurement asymmetries. The von Neumann method mathematically eliminates these hardware-induced biases by processing the bitstream in pairs to extract only unbiased transitions. This process provides an unbiased bitstream, which is essential for the final stage of extracting the secret key.

5 Experimental Analysis and Discussion

The evaluation of the proposed key generation architecture is divided into three distinct phases. First, we identify the physical limitations and inherent bias of the raw quantum hardware. Second, we perform a rigorous statistical analysis of the bitstream after it undergoes quantum randomness injection and classical von Neumann standardization to prove its readiness for One-Time Pad (OTP) encryption. Finally, we evaluate how the shared short-key transforms the monitored quantum source into a secure, private encryption key.

5.1 Statistical Analysis of the Hardware Data

The results in Fig. 2 demonstrate that the von Neumann standardization yields a discrete distribution that is statistically indistinguishable from an ideal uniform model. When applied to the randomized quantum output generated by the Hadamard layer, this process effectively compensates for inherent hardware bias. This outcome is statistically validated by a significant p-value of 0.9997, which remains well above the $\alpha = 0.05$ threshold. As shown in Fig. 2, the data aligns closely with the ideal uniform model. This confirms that the proposed hybrid approach is both physically and statistically ready for secure OTP application.

5.2 Real Quantum Hardware Execution

An analysis of the probability distribution generated by the ODRA-5 quantum processor shows a strong alignment with the ideal 5-qubit simulation. Both distributions hover closely around the expected uniform probability of $p \approx 0.03125$ per state.

However, a zoomed-in comparison reveals minor, systematic deviations in the physical hardware. Specifically, there is a slight statistical shift where outcomes containing a higher number of zeros exhibit a slightly increased probability of occurrence.

While the ODRA-5 hardware demonstrates high gate fidelity and low overall noise, these microscopic physical interactions with the thermal environment slightly reduce the entropy of the raw output. Therefore, to achieve a cryptographically secure and perfectly uniform bit stream, algorithmic randomness extraction is still required (Fig. 3).

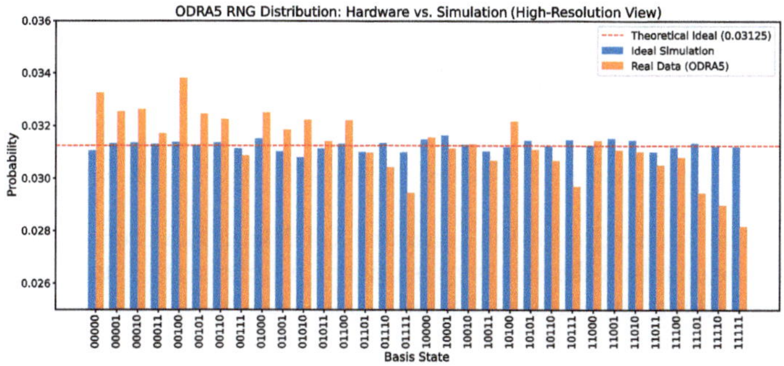

Fig. 3. Comparison of the probability distribution for 32 basis states (from 00000 to 11111) generated by a 5-qubit quantum circuit. The ideal simulation (blue) represents a flawless, uniform quantum state, while the physical ODRA-5 hardware data (orange) shows highly accurate real-world execution. The Y-axis is zoomed in to expose microscopic decoherence effects and hardware imperfections.

5.3 Security Scaling

The efficiency of security scaling in this system depends on the statistical quality of the bits used for the One Time Pad protocol. In our architecture, scaling robustness is achieved by neutralizing any advantage an observer might gain from monitoring the ODRA-5 system. Even though the raw output is compromised by hardware level interception and physical bias such as energy relaxation toward the ground state, the application of a superposition created with Hadamard gates and von Neumann debiasing effectively ensures that the probability of guessing any single bit remains at the ideal theoretical threshold of 0.5. This process is confirmed by the uniform distribution observed in our statistical validation, proving that the system maintains its integrity as the key length increases. Consequently, the security of the system is fundamentally determined by the robustness of the extraction logic rather than the privacy of the initial quantum transmission.

5.4 Visual Proof of OTP Encryption

To practically demonstrate the information-theoretic security of the generated keys, we applied the derived OTP sequence to a sample image. As shown in Fig. 4, the encrypted ciphertext is visually and statistically indistinguishable from uniform white noise. This visual proof confirms that the rigorous quantum randomness injection and classical debiasing successfully neutralize any structural patterns, ensuring strict secrecy of the plaintext.

Fig. 4. Practical application of the generated OTP key. The original image (left) is XORed with the quantum-derived key, resulting in a ciphertext (center) that resembles perfect noise, lacking any structural information. Decryption (right) is only possible with the exact, uncompromised key.

6 Adversary Model and Cryptographic Security

In our adversary model, we assume a powerful, passive interceptor who continuously monitors the raw quantum bitstream generated by the ODRA-5 system. We assume the adversary possesses arbitrary computational resources and complete knowledge of the classical von Neumann debiasing mechanisms applied. We restrict the model to a passive eavesdropper, as ensuring the active integrity of the public stream (protecting against Man-in-the-Middle bit-flipping attacks) would require classical authentication protocols, which are outside the scope of the raw key extraction itself.

Despite this comprehensive surveillance, the passive adversary cannot determine the final OTP key. The initial Hadamard layer injects independent quantum indeterminacy, and the von Neumann extractor yields a refined bitstream with perfect statistical uniformity. Because the resulting public stream is fundamentally unbiased and free of structural patterns, the adversary gains no analytical advantage from monitoring it.

The core security of the system relies exclusively on the shared-secret extraction algorithm (the "short key"). Legitimate parties utilize this undisclosed logic to selectively harvest specific bits from the public stream. Since the extracted sequence is statistically perfect, it fully satisfies the stringent requirements of the OTP protocol. Consequently, statistical cryptanalysis is ineffective, and the final key's information-theoretic security is fundamentally decoupled from the compromised nature of the initial quantum transmission.

7 Conclusion and Future Work

This study demonstrated a hybrid quantum-classical architecture using the ODRA-5 QPU for robust entropy generation. By integrating a Hadamard-based layer with a von Neumann extractor, we successfully mitigated T_1 relaxation biases. Statistical validation via Kolmogorov-Smirnov tests ($p = 0.9997$, Fig. 2) confirms the system's suitability for OTP. Future research will focus on evolving this entropy source into a fully synchronized communication protocol. Our immediate research directions include:

- **Synchronized Bit Selection:** Developing a mechanism where communicating parties use a shared algorithm to select specific bit indices from the quantum stream. This will ensure that both parties can derive the same key without a direct exchange of the key itself.
- **Cryptographic Hashing for Index Mapping:** Investigating the application of lightweight hash functions to transform public metadata into private bit-selection templates. This approach aims to maintain high throughput while ensuring that the selection process remains opaque to an eavesdropper.
- **Temporal Synchronization:** Implementing a protocol where the timing of quantum state measurement and bit extraction is used as an additional layer of security, effectively creating a "moving target" for potential interceptors.

Acknowledgements. The authors would like to thank Prof. Przemysław Śliwiński for sharing the algorithm. This work was supported by resources provided by Wrocław Centre for Networking and Supercomputing (http://wcss.pl). P.Muzaj and P. Sulikowski contributed equally to this research.

References

1. Cerruti, U.: One time pad and the short key dream. Cryptology ePrint Archive. (). Paper 2024/1733 (2024)
2. Massey, F.J.: The Kolmogorov-Smirnov test for goodness of fit. J. Am. Stat. Assoc. **46**(253), 68–78 (1951)
3. Gnatowski, A., Rudy, J., Niżyński, T., Święcicki, K.: True random number generators on IQM Spark (2025)
4. Maurer, U.M.: Conditionally-perfect secrecy and a provably-secure randomized cipher. J. Cryptol. **5**(1), 53–66 (1992)
5. Nath, P.P., Sinha, A., Sinha, U.: Certified random number generation using quantum computers. Front. Quantum Sci. Technol. **4** (2025)
6. Rubin, F.: One-time pad cryptography. Cryptologia **20**(4), 359–364 (1996)
7. von Neumann, J.: Various techniques used in connection with random digits. Notes by G. E. Forsythe. Nat. Bureau Stand. **12**, 36–38 (1951)

Reliable Software Distribution at Scale Using Containerized Delivery and Discrete Scheduling Optimization

Piotr Nadybski[1,2](✉), Krzysztof Rewak[1,2], Karol Zygadło[1,2], Kamil Piech[1,2], Jacek Sawoszczuk[1], and Ewelina Skrzypacz[1]

[1] Blumilk sp. z o.o, Legnica, Poland
{piotr.nadybski,krzysztof.rewak,karol.zygadlo,kamil.piech, jacek.sawoszczuk,ewelina.skrzypacz}@blumilk.pl
[2] Witelon Collegium State University, Legnica, Poland
{piotr.nadybski,krzysztof.rewak,karol.zygadlo, kamil.piech}@collegiumwitelona.pl

Abstract. Efficient software distribution management in large-scale environments requires addressing complex challenges related to license compliance, resource optimization, and operational scalability. Traditional methods often fail to balance the competing demands of network bandwidth limits and strict teaching schedules. This paper presents the K1 system, a comprehensive platform that combines centralized control, containerization, and advanced scheduling optimization. A key innovation of the system is the application of discrete optimization algorithms and heuristic strategies to dynamically allocate computing resources and optimize software deployment from available repositories. The primary objective is to deliver software instances to target machines while minimizing network load and avoiding distribution bottlenecks. The platform incorporates heuristic decision rules to improve scheduling efficiency and adapt to real-time system conditions. The system architecture enables secure, scalable software deployment integrated with ERP systems. The results from a large-scale production deployment confirm the effectiveness of the solution, demonstrating an 83% reduction in deployment times and near-elimination of license violations, validating the practical potential of discrete optimization techniques in real-world scenarios.

Keywords: Software Distribution · Discrete Optimization · Virtualization · Heuristic Scheduling · License Compliance

1 Introduction

Modern universities face an increasing challenge in managing software used in the teaching process. The number of specialist applications is growing, necessitating strict compliance with license terms and flexibility in class scheduling. In a typical scenario, a university may manage thousands of terminals spread

W. Zamojski et al. (Eds.): DepCoS-RELCOMEX 2026, LNNS 2014, pp. 294–303, 2026.
https://doi.org/10.1007/978-3-032-27927-9_28

across dozens of laboratories, each requiring a specific set of software tools that changes from semester to semester, or even from hour to hour. This creates a volatile environment where the "state" of a computer lab is never static.

Traditional methods, such as manual installation or monolithic disk cloning, are insufficient given the scale and pace of change in modern educational institutions. Manual installation is labor-intensive and error-prone, while imaging often requires transferring gigabytes of redundant data, saturating the network. Furthermore, neither approach effectively addresses the dynamic nature of concurrent licensing, where the number of allowed active instances is strictly limited by legal agreements. An error in this area can lead to significant financial penalties or legal action against the institution.

A particular problem is ensuring software availability exactly when and where it is needed, especially when schedules change dynamically and resources are limited. The just-in-time delivery of software in an academic environment is constrained by the strict deadlines corresponding to the start of classes. If a deployment finishes five minutes after the lecture begins, it is effectively a failure.

This article presents the K1 system, a platform designed to automate installation processes, optimize schedules via algorithms, and integrate with existing ERP infrastructure. We describe the solution architecture, the mathematical formulation of the scheduling optimization problem as a variation of the VME/T (Virtual Machine Earliest/Tardiness) problem, and the detailed results of a production implementation at the Witelon Collegium State University.

The main contributions of this paper are twofold, clearly separating our engineering implementation from methodological advancements. Methodologically, we formulate the dynamic software deployment problem as a VME/T variant and introduce a Tabu Search metaheuristic featuring a novel, formally specified adaptive weight-tuning mechanism. From an engineering perspective, we design and implement the K1 system—a production-ready architecture bridging ERP scheduling with containerized application delivery, strictly enforcing concurrent license limits in real-time.

2 Related Work

The problem of managing software deployment has been explored in both academic and industrial contexts, though often with a focus on server environments rather than desktop educational settings.

2.1 Traditional Deployment Strategies Vs. Modern Needs

Traditional package management systems, such as APT or RPM, laid the groundwork for automated deployment. However, these solutions operate on a pull model that often lacks central coordination regarding network congestion. A client machine requests an update without knowledge of the broader network state.

In the corporate sector, enterprise solutions like Microsoft SCCM, Ansible, or Puppet provide robust orchestration capabilities. While effective for maintaining configuration consistency, they typically rely on static schedules defined by administrators (e.g., "deploy at 2:00 AM"). They lack adaptive mechanisms for dynamic environments where class schedules change frequently, and the "target" of a deployment is determined by a timetable rather than a static inventory. Furthermore, these tools do not natively address real-time licensing compliance as a hard constraint for deployment scheduling. Our work bridges this gap by integrating ERP scheduling data directly into the deployment logic. Software license management remains another critical concern; Wintersgill [8] reviewed methods for monitoring usage, emphasizing the importance of central oversight in avoiding legal risks.

Alternative architectural paradigms include lightweight, task-specific virtual machines deployed on demand and the "virtual PC" model, where a reference machine is replicated and terminals stream components during runtime. While these models offer rapid startup, they heavily rely on continuous high-bandwidth networks and lack native, campus-wide concurrent license enforcement mechanisms. In contrast, K1 pre-caches containers and enforces global constraints centrally, balancing network load without runtime streaming dependencies.

2.2 Autonomic and Intelligent Systems

Several efforts have proposed more autonomous approaches. Lalanda et al. [4] proposed autonomic software deployment using decentralized agents capable of reacting to environmental changes. While focused on self-adaptive infrastructures, it underscores the need for intelligent coordination. Our system extends this by moving from decentralized agents to a centralized algorithmic decision-making engine, which is necessary to enforce global constraints like campus-wide license limits.

2.3 Scheduling Optimization

In the context of scheduling, classical optimization techniques have been effectively applied to manage constrained resources [7]. The Job Shop Scheduling Problem provides a theoretical foundation for our work. More recently, heuristic-based methods have emerged as powerful tools in this space. Mao et al. [5] demonstrated advanced algorithms for dynamic resource allocation using reinforcement learning. Our approach draws inspiration from such methods but employs a metaheuristic Tabu Search algorithm tailored specifically for time-window constraints, offering a balance between computational complexity and solution quality in real-time scenarios.

3 K1 System Architecture

The K1 system is designed as a distributed architecture composed of three distinct, interacting components: the Management Server, the Application Server, and the Client. This separation of concerns ensures scalability, security, and high availability across the campus network.

3.1 The Management Server

The Management Server acts as the central orchestration unit of the K1 ecosystem. Implemented in PHP using the Laravel framework, it runs in a containerized Docker environment to ensure portability and ease of deployment. Laravel was chosen for its robust ecosystem and team expertise, ensuring rapid development and maintainability [9]. The database system is externalized to ensure high performance under load [3].

Its primary responsibility is to serve as the single source of truth for the entire infrastructure. This involves synchronization with external ERP systems to import user credentials, classroom schedules, and academic groups. Consequently, the scheduling engine always operates with up-to-date data regarding which room is occupied by which group at any given time. Furthermore, the Management Server acts as a central ledger for license enforcement. Before any deployment command is issued, the server verifies if a license token is available, preventing legal non-compliance. Security is managed via Role-Based Access Control, with support for policies such as mandatory Multi-Factor Authentication [6] and idle session timeouts. Additionally, the system supports automated backups for disaster recovery [2].

3.2 The Application Server

While the Management Server handles logic and metadata, the Application Server is responsible for the heavy lifting of data transfer. Implemented in Go, it is optimized for high-throughput concurrency and acts as the repository for virtualized application packages, hypervisors, and updates. The Application Server implements a strict token-based validation mechanism. When a Client requests a file, the Application Server verifies the request's validity against the Management Server before opening a data stream. This architecture prevents unauthorized downloads even if the file URL is known to the user. To further optimize bandwidth, the server employs intelligent caching and supports multicast transmission for scenarios where multiple clients in the same subnet request the same package simultaneously.

3.3 The Client

The Client is a lightweight agent installed on end terminals, supporting both Windows and Linux operating systems. It operates in a privileged mode to perform system-level operations. Its duties include interfacing with local hypervisors to mount, unmount, and execute virtualized application containers. It also continuously reports its state to the Management Server, providing heartbeats that contain status updates, installed software lists, and active user session data. To ensure laboratories are ready before the first class of the day, the Client supports Wake-on-LAN functionality, allowing the Management Server to trigger machine startup and deployment processes outside of teaching hours.

Security on the client side is enforced by verifying the cryptographic signatures of all downloaded packages before execution. To formalize the overall threat model, we define the Management and Application Servers within a trusted datacenter zone, treating the campus network and end-terminals as untrusted. In an endpoint compromise scenario (e.g., unauthorized student access), the threat is strictly localized. The Client authenticates via short-lived tokens, and the central Management Server mitigates lateral movement by strictly refusing unauthorized license or topology requests.

4 Scheduling Optimization Model

A critical challenge in K1 is ensuring timely delivery of software to target terminals without overloading the network. In a university environment, hundreds of computers may attempt to download large software packages simultaneously, leading to network saturation. We model this as a constrained scheduling task with time windows, derived from the VME/T (Virtual Machine Earliest/Tardiness) problem [1].

4.1 Problem Formulation

Let I denote the set of software deployment tasks, where a task represents the requirement to install a specific application on a specific terminal. Let K represent the available transmission channels, where a channel abstracts a network segment, such as a switch serving a specific classroom.

For each task i in the set I, we define several key parameters. The expected processing duration p_i includes transmission and installation time and is derived from historical data. The time window for the task is defined by an earliest acceptable completion time e_i and a strict deadline d_i. Completing a task before e_i is allowed but may lock a license unnecessarily early, while completing after d_i disrupts the class. We also assign weights u_i and w_i to penalize earliness and tardiness, respectively.

The objective is to find a schedule $\Theta = (Q, \pi)$, where Q is the assignment of tasks to channels and π is the sequence of tasks on each channel, that minimizes the total weighted penalty function $F(\Theta)$:

$$F(\Theta) = \sum_{k=1}^{m} \sum_{j \in Q_k} f_j(C_j) = \sum_{k=1}^{m} \sum_{j \in Q_k} (u_j \cdot E_j + w_j \cdot T_j) \tag{1}$$

In this equation, earliness is defined as $E_j = \max\{0, e_j - C_j\}$ and tardiness as $T_j = \max\{0, C_j - d_j\}$, where C_j is the completion time of task j. It is important to note that a positive E_j implies the task finished before the preferred window $[e_j, d_j]$ (Fig. 1).

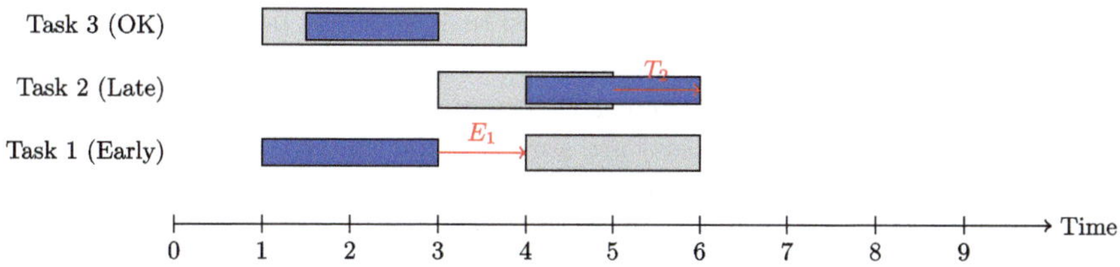

Fig. 1. Visual representation of the objective function. Task 1 finishes before e_1 (Earliness), Task 2 finishes after d_2 (Tardiness).

4.2 Heuristic Solution Algorithm

While population-based metaheuristics (e.g., Genetic Algorithms) or trajectory methods like Simulated Annealing are applicable to VME/T problems, we selected Tabu Search (TS) due to its deterministic neighborhood exploration and ability to strictly bound execution times. In our dynamic university environment, the scheduler must guarantee a response within a bounded time. The computational complexity of generating a neighborhood in our TS implementation is bounded by $O(|K| \cdot |I|^2)$, where $|K|$ is the number of channels and $|I|$ is the number of tasks. Combined with a fixed maximum number of iterations, this ensures strict operational feasibility and predictable runtime behavior without the memory overhead required for population maintenance.

Initialization. The algorithm begins by generating a greedy constructive solution. Tasks are grouped by their target room, which corresponds to a network channel, and sorted by their deadline d_i. This approach, known as Earliest Deadline First (EDF), provides a feasible, though likely sub-optimal, starting point. It ensures that critical tasks are initially placed at the front of the queue, minimizing immediate gross violations of the schedule.

Neighborhood Generation and Moves. To explore the solution space effectively, the algorithm generates neighboring solutions using two specific types of moves. *Inter-channel moves* attempt to balance the load across the network. A task from a highly congested channel is moved to a less loaded one using standard insert and swap operators. This is crucial for preventing bottlenecks where one switch is saturated while others are idle. *Intra-channel sequencing* focuses on local optimization. Two tasks assigned to the same channel are swapped in the execution sequence. This allows the algorithm to fine-tune the order of operations within a specific room, for instance, to ensure that a prerequisite component is installed before the main application.

Block Elimination and Aspiration Criteria. Evaluating every possible swap in a large schedule is computationally expensive. To accelerate the search, we utilize the block property described by Bożejko et al. [1]. A block is defined as a sequence of tasks where moving an internal task does not necessarily reduce the maximum penalty. We explicitly identify E-blocks (dominated by earliness) and sort them using the Weighted Longest Processing Time (WLPT) rule. Similarly, T-blocks (dominated by tardiness) are sorted using the Weighted Shortest Processing Time (WSPT) rule. Furthermore, to prevent the algorithm from getting stuck in local optima, we implement an *aspiration criterion*. This rule allows the algorithm to perform a move that is currently on the Tabu List if, and only if, that move results in a solution strictly better than the best solution found so far. This ensures that the Tabu restriction does not prevent the discovery of a global optimum.

Tabu List Management. To prevent cycling, performed moves are stored in a Tabu List, functioning as a FIFO queue. Inverse moves are forbidden for a specific number of iterations, known as the tabu tenure. In our implementation, the tabu tenure is dynamic, adjusting based on the size of the problem instance (number of tasks) to maintain a balance between exploration (searching new areas of the solution space) and exploitation (refining the current solution).

4.3 Adaptive Weight Tuning

A key methodological novelty of the K1 system is the dynamic adjustment of penalty weights, introducing a feedback loop that adapts to real-world operating conditions. Initially, u_i and w_i are assigned based on static administrative priorities. After each execution cycle t, the system analyzes execution logs and updates the weights for the next cycle $t+1$. The tardiness weight is updated according to $w_i^{(t+1)} = w_i^{(t)} + \alpha \cdot \max(0, C_i^{(t)} - d_i)$, where α is a learning rate parameter determining the aggressiveness of the penalty increment. Similarly, if a task is consistently completed early without license conflicts, u_i is decayed via $u_i^{(t+1)} = \max(u_{min}, u_i^{(t)} - \beta \cdot E_i^{(t)})$, where β is the decay factor and u_{min} ensures the penalty remains strictly positive. This formal mechanism allows the system to autonomously learn implicit institutional priorities, significantly reducing the need for manual reconfiguration.

5 Experimental Evaluation

To validate the K1 system, we analyze data from its production deployment at Witelon Collegium State University. The environment consists of over 3500 active students, 300 faculty members, and approximately 60 computer laboratories.

5.1 Methodology

We compare the performance of the K1 system against the previously used manual and disk-imaging methods, such as Symantec Ghost. The evaluation metrics include the total deployment time required to prepare a laboratory with a new set of software, the failure rate representing terminals that were not ready by the class start time, and the number of reported license conflicts during a semester.

In daily operation, the system consistently manages the distribution of approximately 1.5 to 2.0 terabytes of data per day across the campus network. The primary operational challenges encountered during this period included mitigating hardware failures on aging terminals and handling transient network drops in legacy infrastructure segments, which the system's retry logic successfully absorbed without failing the overall schedule.

5.2 Results: Efficiency and Reliability

The comparative results for a typical scenario involving the simultaneous reconfiguration of 10 laboratories are presented in Table 1.

Table 1. Comparison of Manual Deployment vs. K1 System (10 Labs, 200 Terminals)

Metric	Manual/Ghost	K1 System	Improvement
Avg. Prep Time per Lab	4.5 h	45 min	83%
Network Saturation	95–100%	40–60%	Improved
Failure Rate	12%	0.5%	95%
Staff Required	4 FTE	0.5 FTE	87%

The K1 system achieved a massive reduction in preparation time, dropping from 4.5 h to roughly 45 min. This efficiency is largely attributed to the differential update capability of the virtualization engine. Instead of rewriting the entire disk image, K1 only transmits the specific application packages required for the new schedule.

5.3 Results: Scheduling Effectiveness

To assess the quality of the scheduling optimization, we monitored the tardiness metric during a peak load period, specifically Monday morning classes at 8:00 AM. Table 2 shows how the adaptive weight tuning reduced the number of late deployments over the first month of the semester.

As shown in the table, the system stabilized rapidly. In the first week, the heuristic weights were generic, leading to nearly 12% of tasks finishing slightly after the deadline. As the adaptive tuning adjusted the weights w_i for bottleneck-prone rooms, the scheduler prioritized these critical tasks, reducing the late rate to below 1% by the fourth week.

Table 2. Reduction in Tardiness Events over 4 Weeks (Adaptive Tuning)

Week	Total Installations	Late Tasks ($T_j > 0$)	Late Rate (%)
Week 1	1,240	148	11.9%
Week 2	1,150	62	5.4%
Week 3	1,210	24	1.9%
Week 4	1,180	8	0.6%

5.4 Case Study: Dynamic License Management

A specific case involved a specialized CAD software with only 30 concurrent licenses available for the entire campus. Under the manual system, IT staff had to physically install the software in Lab A and remove it from Lab B, a process taking hours. With K1, the scheduler effectively moved the licenses dynamically. When the class in Lab A ended at 9:30 AM, K1 automatically triggered uninstall and deactivation commands. By 9:45 AM, the licenses were released back to the pool, allowing Lab B, which started at 10:00 AM, to claim them. This floating lab concept allowed the university to utilize the expensive CAD software at 100% efficiency without purchasing additional licenses.

6 Discussion

The implementation of K1 highlights several key insights for educational IT infrastructure. First, centralization is key to compliance. Delegating license management to individual machines or disparate tools inevitably leads to drift and non-compliance. A central authority acting as the single source of truth is essential.

Second, heuristic optimization outweighs exact methods in this domain. While an exact solution to the scheduling problem might theoretically save an additional small percentage of bandwidth, the computation time required would make it unusable for real-time reactions, such as when a teacher changes a classroom at the last minute. The Tabu Search approach provides feasible schedules within seconds, offering operational scalability that $O(n!)$ exact algorithms cannot match.

A significant, though initially unintended, benefit of the system is its contribution to Green IT initiatives. By utilizing precise scheduling combined with Wake-on-LAN and subsequent automated shutdown commands, the K1 system ensures that laboratory computers are only active when actually needed for deployment or teaching. This contrasts sharply with the previous approach where machines were often left running overnight to ensure updates were applied, leading to substantial unnecessary energy consumption.

However, the system has limitations. The current network model assumes static bandwidth capacity. In reality, background traffic fluctuates due to student activity. Future iterations should incorporate real-time network telemetry into the channel capacity constraints. Additionally, the current prediction of installation time relies on a simple moving average; employing regression models could improve prediction accuracy.

7 Conclusions

This paper presented K1, a reliability-oriented system for software distribution in educational environments. By modeling the deployment process as a constrained time-window scheduling problem and solving it via a tailored Tabu Search algorithm, K1 ensures timely access to educational tools while strictly enforcing license compliance.

The transition from manual, image-based deployment to K1's containerized and optimized model yielded significant operational benefits: an 83% reduction in lab preparation time and a reduction of license conflicts to near-zero levels. The adaptive scheduling mechanism proved capable of self-correcting, reducing task tardiness by over 90% within four weeks of operation.

These results validate the hypothesis that discrete optimization techniques, combined with modern virtualization, can transform IT operations in large-scale institutions. Future work will focus on integrating dynamic network telemetry and exploring deep reinforcement learning for even more precise weight tuning.

Author contributions. P.N. and K.R. conceived the study and designed the methodology. K.P., J.S. and E.S. implemented the approach and conducted the experiments. K.Z. analyzed the results. All authors contributed to writing and revising the manuscript and approved the final version.

References

1. Bożejko, W., Nadybski, P., Wodecki, M.: The virtual machine earliest/tardiness scheduling problem with time windows. In: 2017 Federated Conference on Computer Science and Information Systems (FedCSIS), pp. 33–37. IEEE (2017)
2. Gopireddy, V., Reddy, S.: Data backup and disaster recovery: best practices for enterprise systems. J. Inf. Secur. **11**(4), 201–215 (2020)
3. Grambo, M., Jensen, E.: Database Management Systems: SQL and NoSQL Integration. TechPress (2018)
4. Lalanda, P., McCann, J.A., Diaconescu, A.: Autonomic Management of Software Deployment. Principles, Design and Implementation, Autonomic Computing (2013)
5. Mao, H., Alizadeh, M., Menache, I., Kandula, S.: Resource management with deep reinforcement learning. In: Proceedings of the 15th ACM Workshop on Hot Topics in Networks, pp. 50–56 (2016)
6. Meyer, T., Schmidt, J.: Evaluating multi-factor authentication usability in academic environments. In: Proceedings of the 2023 International Conference on Cyber Security, pp. 112–120 (2023)
7. Nowicki, E., Smutnicki, C.: A fast taboo search algorithm for the job shop problem. Manage. Sci. **42**(6), 797–813 (1996)
8. Wintersgill, D., Smith, S.: Strategies for software license compliance in higher education. J. IT Asset Manag. **12**(2), 45–58 (2024)
9. Yang, L., Chen, W.: Performance analysis of modern PHP frameworks: Laravel vs Symfony. Int. J. Web Eng. Technol. **18**(1), 22–39 (2023)

AI in CCTV: The Gap Between Perception and Reality — A Survey of Multimodal AI, RAG, and Agentic Frameworks in Video Surveillance Systems

Paweł Narolski(✉)

Department of Artificial Intelligence, Wrocław University of Science and Technology, Wrocław, Poland
pawel.narolski@pwr.edu.pl

Abstract. Despite rapid advances in multimodal AI, the global video surveillance infrastructure—comprising over 90 million CCTV cameras—remains largely devoid of modern AI capabilities, limiting automated threat detection, situational awareness, and operational efficiency in both public and private security. This scoping review systematically examines the research-to-deployment gap across four technology dimensions: Retrieval-Augmented Generation (RAG), zero-shot video anomaly detection, Vision Large Language Models (VLLMs), and agentic AI, complemented by an empirical analysis of the commercial vendor landscape. Drawing on 34 peer-reviewed and technical sources from 2018–2026 and analyzing the offerings of 10 largest vendors by market share, we find that deployed surveillance AI lags far behind the scientific frontier: video RAG remains prototypical, zero-shot methods suffer catastrophic cross-domain degradation, VLLMs lack genuine temporal reasoning, and agentic systems have not been validated under operational conditions. We formalize five deployment barrier categories—computational cost, latency, reliability, privacy, and scale—and identify architectural efficiency as the binding constraint.

Keywords: video surveillance · CCTV · vision-language models · retrieval-augmented generation · anomaly detection · agentic AI · zero-shot learning

1 Introduction

The global video surveillance infrastructure, with an estimated 90+ million cameras and a market projected to reach $28.8 billion by 2030 [8], is among the largest generators of visual data worldwide [20]. A common assumption is that AI is already omnipresent in CCTV analysis, supporting complex decision making or even fully fledged incident response automation [7]. However, a majority

W. Zamojski et al. (Eds.): DepCoS-RELCOMEX 2026, LNNS 2014, pp. 304–313, 2026.
https://doi.org/10.1007/978-3-032-27927-9_29

of deployed CCTV systems worldwide either lack any AI video processing measures or rely on legacy AI methods such as *CNNs* due to the limited capabilities of commercial offerings [20].

The advent of *retrieval-augmented generation* (RAG) and *vision language modeling* (VLM) methods for grounded video understanding [13,14], *zero-shot anomaly detection* [29,33], and *agentic AI* [23,32] brings significant promise for AI-based automation, crucial for public and private security faced with growing CCTV supervision costs and rapid response requirements [11].

Our work examines the *research to deployment gap* in CCTV, or the lack of real-world, state-of-the-art AI methods usage in video surveillance, with an aim of systematically identifying key research gaps. We contribute:

- The first cross-dimensional survey mapping the research-to-deployment gap across four technology dimensions.
- A systematic identification of five deployment barrier categories—latency, scale, cost, privacy, and reliability—with quantitative evidence.
- A criteria-based, three-tier vendor stratification of multimodal AI adoption.

Prior works in this area catalog known *video anomaly detection* (VAD) methods [15,17] or VLMs in video processing [22], but fail to examine the reasons behind the deployment gap.

2 Survey Scope and Methodology

Following the scoping review framework [2], we investigate:

- **(RQ1)** What multimodal AI capabilities have been demonstrated for video surveillance?
- **(RQ2)** To what extent are they deployed in production?
- **(RQ3)** What are the specific barriers preventing deployment?

We queried IEEE Xplore, ACM Digital Library, Springer, arXiv, and Google Scholar from 2018 [21] through February 2026, using term combinations stemming from *video surveillance*, *anomaly detection*, *vision-language model*, *RAG video*, and *agentic AI surveillance*. Inclusion required foundation-model approaches to video understanding or detection, peer-reviewed or high-impact venues, and English language. We excluded image-only methods and non-AI surveillance studies. This yielded 34 primary sources, supplemented by industry reports [7,8,20], vendor documentation, and regulatory filings. This work uniquely spans RAG, zero-shot detection, VLLMs, and agentic AI in a deployment-focused analysis—prior surveys cover VAD methods [15,17] or video LLMs [22] without examining deployment barriers or commercial adoption.

3 Background

This survey draws on four intersecting areas: **RAG** [13] combines retrieval with generation for grounded responses. **Vision-language models**—notably CLIP [19] (zero-shot transfer from 400M image-text pairs) and BLIP-2 [14] (frozen-encoder bootstrapping)—underpin a significant majority of advances. The **VAD** paradigm, established by Sultani et al. [21] with UCF-Crime, extended to XD-Violence [24], and shifted toward foundation-model methods [26,29]. **Agentic AI** for surveillance spans from early multi-agent frameworks to modern LLM-based agents [27,32].

4 Retrieval-Augmented Generation in Video Surveillance

Assuming standard RAG definition as a pipeline in which retrieved context augments a *generative* model (LLM or VLM) to produce grounded natural language responses [13], *iRAG* (CIKM 2024) outlines practical usage limitations, with more complex requests requiring over a day of processing time for a single, 24-hour video stream [1]. Current SOTA, *AVA* (NSDI 2026), introduces *event knowledge graphs* and outperform baselines by 16.9–26.9%, but incurs a large, 101–174 s search request latency [27]. Prototypical *NVIDIA VSS Blueprint* proposes a *hybrid vector-graph RAG* architecture. Each method raises the issue of trade-off between retrieval granularity and computational cost at multi-camera scale.

No major vendor provides RAG offerings, as confirmed by product documentation and industry reports [7,20]. Five barriers explain RAG's absence: search *latency* (multi-stage pipelines improve retrieval accuracy, but incur significant computing overhead), *scale* (1,000 cameras at 1 FPS produce 86.4M daily frames), *information loss* (due to video-to-text conversion or representation method shortcomings), *privacy* (GDPR concerns with video embeddings [5]), and *cost* (prohibitively expensive hardware requirements).

5 Zero/Few-Shot Video Anomaly Detection

Foundation-model video anomaly detection approaches supervised performance on benchmarks [26,33], but fail to generalize to real-world conditions.
Three notable CLIP adaptation strategies have emerged in recent research. *Dual-branch adaptation* freezes CLIP and adds learnable branches: VadCLIP [26] (AAAI 2024) achieves 84.51% AP on XD-Violence and 88.02% AUC on UCF-Crime. *Object-agnostic prompt learning* targets generalization: AnomalyCLIP [33] (ICLR 2024) learns prompts transferable across 17 datasets. *Feature-based approaches* leverage CLIP/DINO representations directly. Across all strategies, stronger generalization demands higher compute, creating tension with edge deployment.
Training-free VLM/LLM pipelines eliminate task-specific training entirely. LAVAD [29] (CVPR 2024) combines BLIP-2 with LLaMA-2, outperforming

unsupervised methods but requiring two A100 GPUs. Holmes-VAD [31] achieves 89.51% AUC with natural language explanations, extended by Holmes-VAU [30] (CVPR 2025 Highlight) for long-term understanding. These methods trade interpretability for substantial computational overhead.

Open-vocabulary detection extends beyond fixed anomaly taxonomies via Grounding DINO [16], YOLO-World [3], and OVVAD [25] (the first open-vocabulary VAD method, surpassing supervised baselines).

The deployment gap remains: models trained or evaluated on one surveillance scenario (e.g., indoor retail) suffer catastrophic performance loss when transferred to a different domain (e.g., outdoor traffic or night-time footage)—MSAD [34] (NeurIPS 2024) demonstrates cross-domain AUC drops from 87.1% to 54.3% across its multi-scenario benchmark. Karim et al. [12] further show that performance degrades when inference is constrained to near-real-time decision speeds. Even 88% AUC yields thousands of daily false positives across 24/7 multi-camera installations, and multi-GPU inference requirements incur a significant operational cost overhead.

6 Vision Language Models in Surveillance

VLLMs show promise for surveillance automation but face several limitations. De Silva et al. [4] achieved 80% temporal accuracy with *Gemini Pro* for CCTV summarization. Meo et al. benchmarked compact VLMs (≤8B) for zero-shot anomaly detection, concluding their lack of reliability enough for complex, autonomous surveillance [18]. The *Holmes-VAD/VAU* program [30,31] represents the most sustained effort, constructing instruction datasets for fine-tuning multimodal LLMs for both detection and natural language explanation.

A critical temporal limitation was revealed by a 2025 study [6]: *GPT-4o* and *Gemini* scores remain stable after frame shuffling, suggesting reliance on language priors rather than temporal understanding—fundamental for surveillance where sequence distinguishes normal from anomalous behavior. Most VLLM applications sample only a subset of video frames for video processing at a fixed interval, causing additional temporal inconsistencies and a risk of information loss.

Practical barriers include cloud costs of tens of thousands USD daily for 1,000 cameras [4], significant hallucination risk in security contexts, and privacy mandates conflicting with cloud deployment [5]. Compact VLMs with LoRA fine-tuning (e.g., MiniCPM-V [28]) are emerging, but continuous video context management with no spatio-temporal information loss remains a challenge.

7 Agentic AI for Surveillance

Following Wang et al. [23], we define LLM-based autonomous agents (or agentic AI) as systems that leverage a large language model as a central controller to autonomously perceive, reason, plan, and act. Unlike the single-pass pipelines of RAG (Sect. 4) or VLM inference (Sect. 6), agentic architectures implement an iterative loop, in which the orchestrating LLM decides at each step what information to gather, which tools to invoke, and when sufficient evidence has been obtained [23]. Such systems may employ RAG retrieval or VLM perception as sub-components, but their distinguishing property is autonomous multi-step decision-making.

For instance, *VideoAgent2* [32] (NeurIPS 2025) introduces uncertainty-aware *chain-of-thought reasoning*: the LLM agent explicitly quantifies its confidence at each iteration, gathering additional video segments when uncertainty exceeds a threshold, improving accuracy by 13.1% over its predecessor on long-form video understanding benchmarks. In *AnomalyRuler* [10] (ECCV 2024), the LLM first *induces* normality rules from few-shot reference frames, then *deduces* anomalies by checking rule violations in test video, achieving competitive few-shot detection without task-specific training. However, neither work reports end-to-end latency, leaving real-time viability unquantified.

Commercially, agentic AI deployments remain nascent. Since AAI leverages RAGs and VLLM, barriers defined in Sect. 4 and 6 apply. Additionally, accountability in autonomous decision making from legal and technical standpoint (e.g. *human-in-the-loop* requirement satisfaction) remains an open problem [11,18].

8 Commercial Landscape

Majority of currently deployed AI methods in CCTV leverage deep convolutional neural networks (CNNs) for closed-set classification or object detection tasks [7, 20], unable to match the versatility of VLLM methods for analytics of ever-changing real-world conditions [18,34].

We propose a classification 10 largest vendors by global market share [8] into three tiers, based on the documented technical capabilities of their offerings, by: (1) native VLM integration, (2) natural language querying (NLQ), and (3) AI beyond CNN classification. *Tier 1* vendors satisfy all three; *Tier 2* has advanced analytics without VLM; *Tier 3* relies on CNN detection. Table 1 summarizes the stratification.

The scale of the research-deployment gap is quantifiable. The eight hardware vendors in our analysis collectively account for approximately 67% of global camera/network video recorder shipments, led by Hikvision (∼25%), Axis (∼13%), Dahua (∼11%), Hanwha (∼7%), and Bosch (∼5%) [8]; Genetec and Milestone dominate the VMS software segment. The vast majority of Hikvision's and Dahua's installed base runs legacy CNN-only firmware (AcuSense, WizMind). Tier 1 systems: Milestone's Hafnia VLM, Bosch's IVA Pro Context, and Hikvision's Guanlan enable scene interpretation without predefined rules (but have all launched in late-2025 and have a limited market share). Tier 3 comprises the market majority: Axis, Hanwha, Dahua, Honeywell, Uniview, and Hikvision's legacy AcuSense rely on basic CNN classification, forming the empirical basis for our thesis. Tier 2: Avigilon (Motorola) and Genetec offer analytics beyond CNNs such as embedding-based retrieval or text classification.

For all Tier 1-Tier 3 systems, false alarm rate issue remain critical: environmental factors cause the majority of outdoor false alerts [20], and high volumes

Table 1. Vendor AI capability stratification (February 2026). VLMs: native vision-language model integration; NLQ: natural language querying support.

Tier	Vendor	Product	VLMs	NLQ
Tier 1	Milestone	Hafnia VLM	Yes	Yes
	Bosch	IVA Pro Context	Yes	Yes
	Hikvision	Guanlan	Yes	Yes
Tier 2	Genetec	DL + NLP	No	Part.*
	Avigilon (Motorola)	App. Search	No	Part.†
Tier 3	Axis (Canon)	ACAP + 3rd party	No	No
	Hanwha Vision	Dual NPU, SoC	No	No
	Hikvision	AcuSense	No	No
	Dahua	WizMind	No	No
	Honeywell	DL Analytics	No	No
	Uniview	AI Guard	No	No

*Keyword/metadata search without generative VLM reasoning.
†Appearance-based similarity search without natural language input.

lead operators to disable systems entirely. Facial recognition bias is documented by NIST's FRVT [9], which found algorithms 10–100× more likely to misidentify minorities. Contextual blindness—the inability to reason about intent, social context, or temporal continuity—represents a fundamental limitation. Regulatory constraints further limit deployment: the EU AI Act [5] classifies AI video analytics as high-risk and restricts biometric identification.

9 Discussion

9.1 Synthesis

The evidence assembled across Sects. 4–8 permits a systematic assessment of each technology dimension against production readiness. Table 2 summarizes these findings; the *Adoption* column reports the number of top-10 vendors by global market share (Table 1) that ship a production product in the given category as of February 2026.

RAG. (Sect. 4) remains prototypical, with query latencies two orders of magnitude above operational requirements and no demonstrated continuous-indexing scale. No vendor offers RAG-based analytics.

Zero-shot VAD. (Sect. 5) shows strong in-distribution benchmarks but cross-domain AUC collapses to 54.3% [34] with further degradation under real-time

Table 2. Cross-dimensional synthesis: academic performance vs. deployment readiness. Adoption reports the number of top-10 vendors (Table 1) shipping a production product in the given category as of February 2026.

Dimension	Best reported result	Critical deployment gap	Min. hardware	Adopt.
RAG	+26.9% over baselines (AVA)	101–174 s query latency	7B VLM + cloud server	0/10
Zero-shot VAD	88.0% AUC, UCF-Crime (VadCLIP)	54.3% cross-domain AUC (MSAD)	2× A100 80GB	0/10
VLLMs	89.5% AUC (Holmes-VAD)	Frame-shuffle invariant scores	1× A100 40GB	3/10
Agentic AI	+13.1% over VA1 (VA2)	End-to-end latency unreported	GPT-4 API + VLM	0/10
NL search	Deployed at scale	—	DB server	≥6*

*Includes vendors outside the top-10 (e.g., Coram AI, Turing AI).

constraints [12]. Training-free methods demand multi-GPU infrastructure; near-real-time methods require task-specific adaptation. No vendor deploys open-set detection.

VLLMs. (Sect. 6) exhibit a temporal reasoning deficit (frame-shuffle invariance [6]), prohibitive cloud costs [4], and privacy conflicts [5]. Only three Tier 1 vendors ship VLM products, all since mid-2025.

Agentic AI. (Sect. 7) inherits RAG and VLM limitations. No system reports end-to-end latency or validates on continuous streams; legal accountability remains unresolved [11].

9.2 Cross-Cutting Barrier Analysis

We formalize five barrier categories: *computational cost* (multi-GPU requirements for inference vs. limited budgets), *latency* (101–174 s queries [27] vs. sub-second requirements), *reliability* (cross-domain degradation [34] and VLLM hallucination [6] yield unacceptable false positive rates at scale), *privacy* (EU AI Act high-risk classification [5] and cloud deployment constraints), and *scale* (86.4M daily frames from 1,000 cameras with no demonstrated system capable of solving this).

A recurring pattern across all dimensions is a 2–3 order-of-magnitude gap between the computational budget available and what frontier methods require. This suggests that *architectural efficiency*—not algorithmic improvement on benchmarks—is the binding constraint for deployment. Notably, natural language video search, the only capability to reach production [7], succeeded by operating on pre-indexed metadata rather than raw video, sidestepping this compute bottleneck entirely.

9.3 Limitations and Future Work

Our deployment assessments rely on vendor documentation and industry reports rather than direct system audits, introducing reporting bias. The landscape evolves rapidly; vendor capabilities reflect February 2026. Emerging areas such as privacy-preserving inference and federated learning for surveillance were not examined and warrant dedicated investigation. Future work should empirically validate the gap under operational conditions (continuous feeds, adverse weather, naturalistic anomaly distributions) and target model compression for surveillance SoC hardware (<5W), where deployment impact may exceed further benchmark gains.

10 Conclusion

This scoping review examined the gap between frontier multimodal AI and production CCTV across four technical dimensions. Among the 10 largest vendors (~67% of global shipments), no production RAG, zero-shot VAD, or agentic AI deployment exists; VLM integration is limited to three vendors since mid-2025. Five barrier categories—computational cost, latency, reliability, privacy, and scale—jointly prevent deployment, with a 2–3 order-of-magnitude compute gap as the binding constraint. Bridging this gap requires architectural efficiency for edge hardware, privacy-preserving inference, and human-AI collaborative frameworks [11] for security-critical contexts where the vast majority of cameras remain simple recording devices [20].

References

1. Arefeen, M.A., et al.: iRAG: advancing RAG for videos with an incremental approach. In: Proceedings of the 33rd ACM International Conference on Information and Knowledge Management (CIKM) (2024). https://doi.org/10.1145/3627673.3680088
2. Arksey, H., O'Malley, L.: Scoping studies: towards a methodological framework. In: International Journal of Social Research Methodology, vol. 8(1), pp. 19–32 (2005). https://doi.org/10.1080/1364557032000119616
3. Cheng, T., et al.: YOLO-world: real-time open-vocabulary object detection. In: IEEE/CVF Conference on Computer Vision and Pattern Recognition (CVPR) (2024)
4. De Silva, J., et al.: Large language models for video surveillance applications. In: IEEE Region 10 Conference (TENCON) (2024)
5. European Parliament and Council of the European Union. Regulation (EU) 2024/1689 Laying Down Harmonised Rules on Artificial Intelligence (AI Act). Official Journal of the European Union (2024). https://eur-lex.europa.eu/eli/reg/2024/1689/oj
6. Feng, B., et al.: Breaking Down Video LLM Benchmarks: Knowledge, Spatial Perception, or True Temporal Understanding? In: arXiv preprint arXiv:2505.14321 (2025)

7. Genetec. State of Physical Security Report. https://www.genetec.com/a/physical-security-report (2026)
8. Grand View Research. Video Surveillance Market Size, Share & Trends Analysis Report. https://www.grandviewresearch.com/industry-analysis/video-surveillance-market-report (2024)
9. Grother, P., Ngan, M., Hanaoka, K.: Face Recognition Vendor Test (FRVT) Part 3: Demographic Effects. Tech. rep. NISTIR 8280. National Institute of Standards and Technology (2019). https://doi.org/10.6028/NIST.IR.8280
10. Hao, Y., et al.: Follow the rules: reasoning for video anomaly detection with large language models. In: European Conference on Computer Vision (ECCV) (2024)
11. Hore, A.N., et al.: Alert prioritisation in security operations centres: a systematic survey on criteria and methods. In: ACM Computing Surveys (2024). https://doi.org/10.1145/3695462
12. Karim, H., Doshi, K., Yilmaz, Y.: Real-time weakly supervised video anomaly detection. In: IEEE/CVF Winter Conference on Applications of Computer Vision (WACV) (2024)
13. Lewis, P., et al.: Retrieval-augmented generation for knowledge-intensive NLP tasks. In: Advances in Neural Information Processing Systems (NeurIPS). Vol. 33, pp. 9459–9474 (2020)
14. Li, J., et al.: BLIP-2: bootstrapping language-image pre-training with frozen image encoders and large language models. In: Proceedings of the 40th International Conference on Machine Learning (ICML). Vol. 202, pp. 19730–19742 (2023)
15. Liu, J., et al.: Networking systems for video anomaly detection: a tutorial and survey. In: ACM Comput. Surv. (2025). https://doi.org/10.1145/3729222
16. Liu, S., et al.: Grounding DINO: marrying DINO with grounded pre-training for open-set object detection. In: European Conference on Computer Vision (ECCV) (2024). https://doi.org/10.1007/978-3-031-72970-6_3
17. Liu, Y., et al.: Generalized video anomaly event detection: systematic taxonomy and comparison of deep models. In: ACM Computing Surveys, vol. 56(7), pp. 1–38 (2024). https://doi.org/10.1145/3645101
18. Meo, A., Kermiche, A., Marchegiani, L.: Benchmarking compact VLMs for clip-level surveillance anomaly detection under weak supervision. In: Journal of Imaging, vol. 11(4) (2025)
19. Radford, A., et al.: Learning transferable visual models from natural language supervision. In: Proceedings of the 38th International Conference on Machine Learning (ICML). Vol. 139, pp. 8748–8763 (2021)
20. Security Industry Association. Security Megatrends Report. https://www.securityindustry.org/report/security-megatrends/ (2025)
21. Sultani, W., Chen, C., Shah, M.: Real-world anomaly detection in surveillance videos. In: IEEE/CVF Conference on Computer Vision and Pattern Recognition (CVPR), pp. 6479–6488 (2018)
22. Tang, Y., et al.: Video understanding with large language models: a survey. In: IEEE Transactions on Circuits and Systems for Video Technology (2024). arXiv:2312.17432
23. Wang, L., et al.: A survey on large language model based autonomous agents. In: Frontiers of Computer Science, vol. 18, p. 186345 (2024). https://doi.org/10.1007/s11704-024-40231-1
24. Wu, P., et al.: Not only look, but also listen: learning multimodal violence detection under weak supervision. In: European Conference on Computer Vision (ECCV), pp. 322–339 (2020). https://doi.org/10.1007/978-3-030-58577-8_20

25. Wu, P., et al.: Open-vocabulary video anomaly detection. In: IEEE/CVF Conference on Computer Vision and Pattern Recognition (CVPR), pp. 18297–18307 (2024)
26. Wu, P., et al.: VadCLIP: adapting vision-language models for weakly supervised video anomaly detection. In: Proceedings of the AAAI Conference on Artificial Intelligence. Vol. 38(6), pp. 5810–5818 (2024). https://doi.org/10.1609/aaai.v38i6.28423
27. Yan, Y., et al.: AVA: towards agentic video analytics with vision language models. In: USENIX Symposium on Networked Systems Design and Implementation (NSDI) (2026)
28. Yao, Y., Yu, T., Zhang, A., et al.: Efficient GPT-4V level multimodal large language model for deployment on edge devices. In: Nature Communications, vol. 16, p. 1548 (2025)
29. Zanella, L., et al.: Harnessing large language models for training-free video anomaly detection. In: IEEE/CVF Conference on Computer Vision and Pattern Recognition (CVPR), pp. 18527–18536 (2024)
30. Zhang, H., Xu, X., Wang, X., et al.: Holmes-VAU: towards long-term video anomaly understanding at any granularity. In: IEEE/CVF Conference on Computer Vision and Pattern Recognition (CVPR). CVPR 2025 Highlight (2025)
31. Zhang, H., et al.: Holmes-VAD: Towards Unbiased and Explainable Video Anomaly Detection via Multi-modal LLM. In: arXiv preprint arXiv:2406.12235 (2024)
32. Zhi, Z., Wu, Q., Shen, M., et al.: VideoAgent2: enhancing the LLM-based agent system for long-form video understanding by uncertainty-aware CoT. In: NeurIPS 2025 Workshop on Scaling Environments for Agents (2025)
33. Zhou, Q., et al.: AnomalyCLIP: object-agnostic prompt learning for zero-shot anomaly detection. In: International Conference on Learning Representations (ICLR) (2024)
34. Zhu, L., et al.: Advancing video anomaly detection: a concise review and a new dataset. In: NeurIPS 2024 Datasets and Benchmarks Track (2024)

Towards More Reliable Signal Strength Based Indoor Localization in Bluetooth Low Energy

Maciej Nikodem[1(✉)], Piotr Patronik[1], Adam Jeżak[1], and Mariusz Słabicki[2]

[1] Wrocław University of Science and Technology, Wrocław, Poland
{maciej.nikodem,piotr.patronik,adam.jezak}@pwr.edu.pl
[2] Ericsson AB, Stockholm, Sweden

Abstract. Apart from data communication, IoT technologies are often used to estimate transmitter positions in indoor environments. While indoor location can be estimated in several ways, received signal strength indicator (RSSI)-based ranging and multilateration are the most straightforward methods that can be deployed for all communication technologies, particularly Bluetooth Low Energy (BLE). The limitation of these methods arises from the fact that the RSSI readings are sensitive to propagation phenomena, which affect localization errors in complex ways. One way to improve accuracy is to use multichannel RSSI measurements to detect and account for these phenomena. Unfortunately, even though BLE can use 40 different channels for communication and RSSI measurements, the use of multichannel information is limited. This limitation occurs because BLE-capable devices have restricted capabilities in controlling the channel used and report RSSI readings without channel information. As a result, public datasets containing RSSI measurements across 40 BLE channels are scarce. This paper describes how to use low-cost BLE transmitters and the Zephyr BLE stack to enable RSSI measurements for selected channels. We present a solution that enables the conduction of multichannel RSS measurements. Based on the sample measurements, we demonstrate that the RSSI measurements across multiple radio channels can be used to understand propagation conditions.

Keywords: Indoor localization · Bluetooth Low Energy · Multichannel communication · Received Signal Strength

1 Introduction

Due to the deployment of global navigation satellite systems (GNSSs), the ability to estimate outdoor locations has become a prominent feature available at relatively low costs. Unfortunately, due to the weakness of the satellite localization signal, the use of this feature is highly limited indoors, where accuracy

W. Zamojski et al. (Eds.): DepCoS-RELCOMEX 2026, LNNS 2014, pp. 314–323, 2026.
https://doi.org/10.1007/978-3-032-27927-9_30

drops significantly or may fail entirely. The limitations of GNSS systems have motivated researchers and engineers to develop indoor localization systems that utilize various radio technologies available or easy to deploy indoors, such as WiFi, BLE, ZigBee, or UWB, to name the most popular ones [6,8]. While these technologies differ in various aspects (communication frequencies, signal bandwidth, range, throughput, etc.), they share the common feature of being able to transmit messages across different frequency ranges and measure and report the received signal strength indicator (RSSI). While RSSI can be calculated differently by various communication standards, it depends on the transmission power and the degradation of the signal in the current propagation environment. As attenuation depends on the distance between communicating devices, the measured RSSI can be used to estimate this distance. By using estimated distances from transmitters in known locations, a receiver can calculate its own position through multilateration. Unfortunately, the relationship between distance and RSSI is complex due to the intricacies of indoor propagation, which is influenced by various phenomena, including multipath and shadowing [5]. Consequently, the estimated distances are inaccurate, resulting in location errors of several meters.

While minimizing the impact of propagation phenomena is challenging, it is possible to use measurements collected in different frequency channels to identify whether these phenomena occur and to take appropriate actions. This may include discarding some or all of the measurements or accounting for these phenomena [9–11].

In this article, we focus on BLE communication technology, as it is one of the most widely used technologies in everyday devices. It is frequently utilized for indoor localization due to its low power operation, relatively long communication distance, and ability to estimate location in various ways [3,4]. BLE can also transmit messages on 40 distinct radio frequencies, spanning from 2402 MHz to 2480 MHz. However, most off-the-shelf devices do not allow explicit control of channel selection and do not report the radio channel at which the BLE packet was received and the RSSI was measured. Consequently, most BLE indoor localization systems based on RSSI measurements are limited to measurements from a subset of available channels (mostly 3 channels) without actual knowledge of the channel on which the value was measured. Simultaneously, a number of articles report that information about the channel used and the use of measurements from a larger number of communication channels can lead to improved ranging and more accurate localization [7]. Not only is higher accuracy possible, but also the detection of adverse propagation phenomena when RSSI measurements from various frequency channels are available [10,11], thus allowing for further improvement in the accuracy of ranging and position estimation.

We see the impossibility of multichannel RSSI measurement in off-the-shelf devices as a major obstacle to conduct in-depth research in BLE indoor localization. The lack of such measurements affects not only RSSI-based ranging but also other ranging and angle estimation methods that utilize multiple frequency channels without assessing their quality and how they are impacted by propagation phenomena. To address this, we present a low-cost and simple BLE device with reference software that allows turning the nRF52840 BLE System on Chip (SoC) into an experimental setup that can be easily used to measure RSSI in multiple communication channels. The contributions of this article include:

- the use of low-cost, off-the-shelf nRF52840-based radio modules as transceivers of radio packets transmitted in the 40 BLE channels,
- the design of the software setup for multichannel RSSI measurements using BLE communication,
- example measurements collected with this setup in conducted (controlled) and real-life environment, along with a discussion of their importance to indoor localization.

2 Related Work

The possibility of using RSSI measurements across different frequency channels to detect and quantify propagation phenomena has already been presented for IoT communication technologies. Zhang [10] proposed analyzing the RSSI across different spectrum channels to identify the amplitude of the signal solely from the Line of Sight (LoS) path. They defined the problem of finding this amplitude as a non-linear curve fitting problem that is ill-conditioned and has no stable, trusted solution. They proposed a dedicated algorithm that was verified using communication in 16 channels in 2.4 GHz frequency band. Their solution enabled significant improvements in localization within dynamic environments with people moving around the transceivers.

A Non-Line-of sight detection approach is also presented by Xiao [9], who proposed exploring the statistical features of the RSSI time series to separate LoS and NLoS (Non-Line of Sight) measurements. They tested a different set of statistical features and proposed two methods that utilize these features to determine whether the measurements were collected in LoS or NLoS conditions. The results demonstrate a high accuracy in distinguishing LoS from NLoS, which leads to improvements in both ranging and localization accuracy.

Zhao [11] proposed a ranging framework called FREDI that utilizes RSSI measurements at different frequencies to overcome the destructive effects of multipath propagation and external radio interference. They present an algorithm based on the inverse discrete Fourier transform, which is implemented in a software defined radio (SDR) platform and successfully identifies the multiple path components from the RSSI measurements. They demonstrate that the use of 8 distinct frequencies can significantly enhance RSSI-based ranging accuracy. Unfortunately, the experiments employed a wide frequency spectrum (5–5.9 GHz), and the quality of the results degrades as the frequency range narrows.

While preliminary research suggests their applicability to BLE communication, this aspect has not been adequately addressed in the literature. This is primarily due to the limited control over frequency channel selection in BLE and the frequent absence of channel identifiers in reported RSSI measurements. Although BLE radio internally retains this information and capability, it is typically not exposed to higher protocol layers. As a result, user applications neither control the channel used for transmission nor know the channel on which RSSI measurements are obtained. The above is not only true for user applications running on devices controlled by operating systems (e.g., smartphones, Raspberry Pis, and other microcomputers), but also for microcontrollers and systems on chips (SoCs) programmed with the use of BLE stacks (e.g., Zephyr). Consequently, only a few datasets are publicly available that report the RSSI measurements across different channels [1,2,7].

The unavailability of the data sets constrains research on multichannel RSSI-based ranging and indoor localization. The lack of data sets arises from the difficulty in conducting measurements across BLE communication channels, as the available tooling does not support it. To address this problem and stimulate research, we present a modified Zephyr BLE stack and a simple reference software that allows for easy measurements of multichannel RSSI.

3 System Description

In experiments, we work with BLE System on Chip from Nordic Semiconductor. This particular work was implemented using the nRF52840, which supports BLE 5.1 and Zephyr 4.1.0 BLE software stack that include the ability to transmit legacy and extended advertisements. Both types are crucial for RSSI measurements, as they can be transmitted on different frequencies assigned to 40 BLE channels. Legacy advertisements have been used since early versions of BLE to inform neighbors about device existence and functionality. These advertisements are always transmitted on three primary advertisement channels (37–39) and are restricted in the amount of payload they can carry. Extended advertisements were introduced to enable the transmission of larger amounts of data without the need to establish a connection between BLE devices. To ensure the efficient use of the radio spectrum, extended advertisements are not transmitted on primary channels. Instead, they are offloaded to secondary advertisement channels (0–36) that were previously used only during connections. To ensure backward compatibility, extended advertisement is always preceded by a specific legacy advertisement that informs the receiver about the parameters of an extended advertisement to follow, including the secondary channel that will be used.

Using both legacy and extended advertisements allows the BLE receiver to measure the RSSI of advertisement packets on all 40 channels. While this idea is straightforward and supported by the BLE standard, there are technical challenges to implementing it that stem from three aspects. First, the BLE transmitter selects the secondary channel to use in a pseudo random manner, making it impossible to transmit on a fixed channel. Second, the software stack that controls the BLE radio does not expose functions to user applications for controlling the channel selection algorithm (CSA) and does not report the channel at

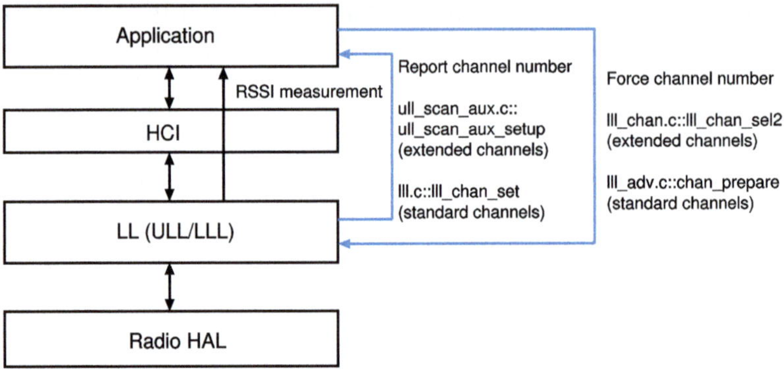

Fig. 1. Overview of the BLE software stack and data flow in the proposed approach. Black arrows show the standard communication between Application, Host-Controller Interface (HCI), Link Layer (LL), and Radio Hardware Abstraction Layer (HAL). Blue arrows indicate added pathways enabling control and reporting of the advertising channel used for transmission and reception. Measured RSSI is reported to the Application via standard HCI.

which the RSSI was measured. Third, although the transmission of an extended advertisement requires prior transmission of a legacy advertisement, the BLE receiver reports the RSSI measurement only for the extended advertisement. In other words, when extended advertisements are transmitted, the RSSI values for primary channels are not reported.

To address the above constraints, we modified the Link Layer (LL) functions in the Zephyr software stack that control radio operations, switch the transmitter between different modes of operation, and allow the receiver to report the channel used for reception (Fig. 1).

For the transmitter and legacy advertisements, we modified the channel selection function (`chan_prepare`) to allow fixing a specific channel instead of cycling through channels 37, 38, and 39. For extended advertisements, we modified the CSA (function `lll_chan_sel2`) to select either a CSA-determined channel or a user-defined one. In both cases, the channel can be specified via a global variable.

In the receiver, the software stack is modified so that a global variable is assigned to the channel used for reception. In particular, for legacy advertisements, the channel is saved when the receiver starts a new scanning window and selects a frequency to listen to (function `lll_chan_set`). For extended advertisements, the channel is saved upon the reception of a legacy advertisement that contains information about which channel will be used to transmit the extended advertisement (function `ull_scan_aux_setup`).

The above modifications allow to control channel used and report RSSI-BLE channel pairs, enabling a systematic and controlled collection of RSSI measurements from all 40 BLE channels.

4 Results

In addition to adjusting the Zephyr software stack, we have developed a simple application that demonstrates how the proposed mechanism can be utilized. It also provides an easy way to start collecting RSSI measurements across multiple BLE channels. The application was tested with a small and low-cost evaluation board that contains the nRF52840 SoC.

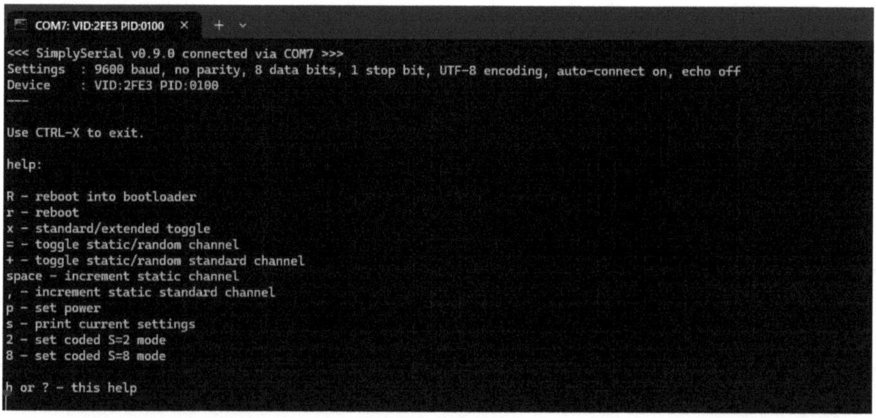

Fig. 2. Console of the BLE transmitter showing an example output of information when configuration is displayed.

Figure 2 presents a console of the BLE transmitter, along with a list of functions and outputs from its operation. The user can switch the transmitter operation between automatic and manual configurations. Automatic configuration is the default setting and enables the transmitter to periodically switch between legacy and extended advertisements. One period lasts 60 s, with 6 s allocated for legacy advertisements and 54 s for extended advertisement transmission. Since channels are selected randomly, this configuration enables the collection of approximately equal numbers of samples from each radio channel. In manual mode, the user can control the types of advertisements being transmitted, the channels to be used, and the transmission power. Channels can either be generated randomly or incremented by the user. Users can switch between legacy-only and extended advertisement transmission and decide whether to use uncoded or coded transmission for extended advertisements.

We have also prepared a simple BLE receiver software that reports information about the channel used for packet reception, the measured RSSI, and some additional information: the MAC address of the transmitter and the physical layer used for extended advertisement transmission.

Figure 3 shows sample RSSI measurements collected over 40 BLE channels, along with their statistical representation in the form of a boxplot. These plots do not take into account the channel on which the measurements were collected.

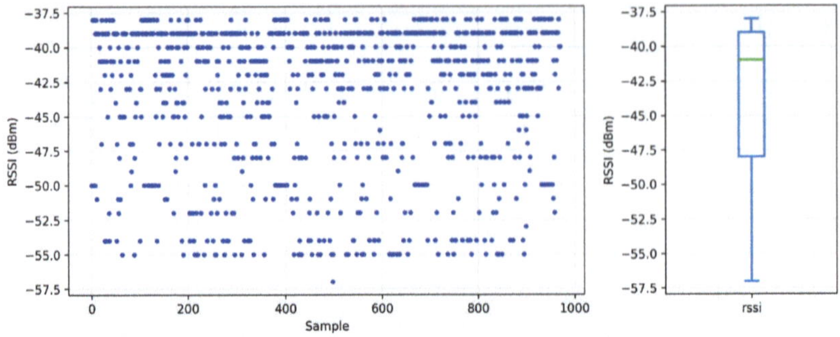

Fig. 3. Example of RSSI measurements across 40 BLE channels. Left plot shows subsequent measurements (i.e. over time) disregarding channel information. Right plot shows statistics of the RSSI measurements in the form of a boxplot.

They reflect the perspective of most BLE devices, which are capable of measuring RSSI but are unaware of the channel used for reception. Consequently, an analysis based solely on these plots may suggest that the measurements are affected by large and seemingly random errors. By applying the proposed approach, the channel information can be exploited, providing deeper insight into the measurement setup. In the frequency domain (Fig. 4), the same measurements no longer appear random, but instead exhibit a clear and repeatable pattern. For clarity, this figure also includes measurements acquired using a spectrum analyser, confirming that the observed pattern corresponds to the actual signal strength.

The pattern is not coincidental, but results from two-path propagation present in the conducted experiment and from frequency-dependent constructive and destructive interference. Based on the frequencies of the interference minima, the path difference can be estimated as:

$$\Delta d = \frac{c \cdot vf}{\Delta f}, \tag{1}$$

where c denotes the speed of light in the vacuum, vf is a velocity factor (depends on the type of transmission medium), and Δf is the frequency distance between the adjacent minima of the destructive interference. In the example (Fig. 4) the frequencies of the interference minima are approximately 2404 MHz and 2450 MHz, and $vf = 0.77$ for coaxial cables, thus the Δd equals 5.016 m – very close to the actual path difference of 5m. This shows that the propagation phenomena, which remained hidden when using conventional RSSI measurements, can be revealed and interpreted once RSSI-BLE channel pairs are available.

The above observation has several further implications that may be used in various applications. One such example is indoor positioning, where the position of a device is estimated based on its distances to fixed reference points. A common approach to distance estimation relies on RSSI measurements, as the received signal strength exhibits a logarithmic dependence on distance. By

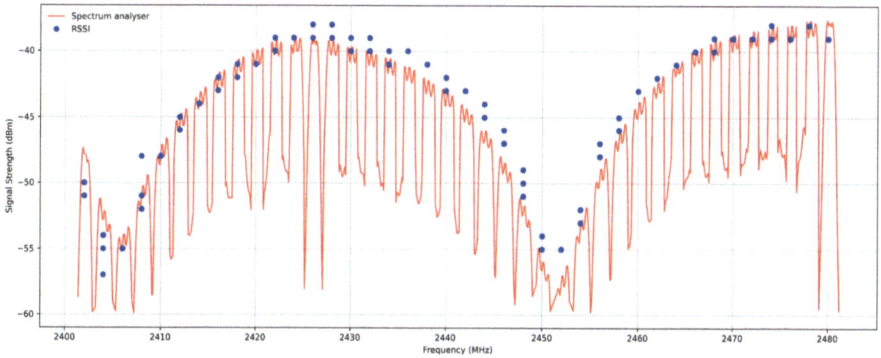

Fig. 4. Signal strength measured with a spectrum analyser (red lines) and a BLE receiver (blue dots) for transmission at different BLE channels. The measurements were performed using cable path configuration emulating two-path propagation (path difference of 5 m) between BLE transmitter and receiver. Multiple RSSI measurements for each channel were collected with resolution of 1 dB.

inverting this relationship, the distance can be estimated from the measured RSSI. Unfortunately, in the absence of channel information, it is not possible to identify which RSSI values are representative of the true distance (cf. Fig. 3). Consequently, using RSSI measurements without channel information or their statistics (e.g. mean or median) may introduce substantial errors in range estimation. In contrast, the availability of channel information enables the identification of propagation related effects and facilitates the rejection of measurements that are corrupted by such phenomena. An example of real-life measurements is presented in Fig. 5, where the measurements were collected in an indoor scenario with the transmitter and receiver located 2.3 m apart. As presented the measured RSSI values vary by more than 20 dB for various BLE channels. This confirms

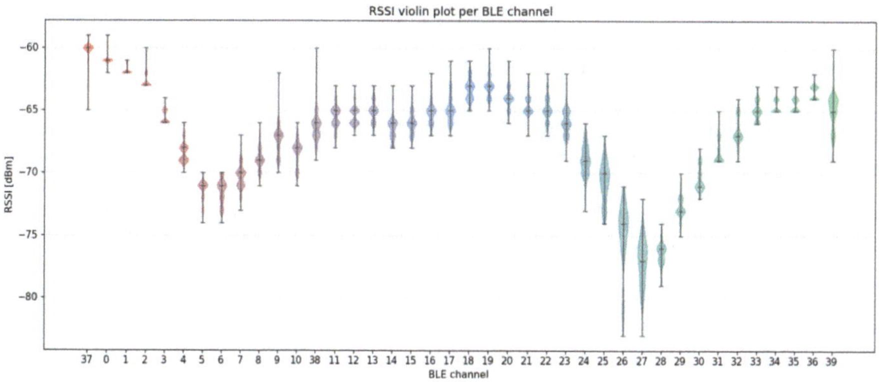

Fig. 5. Sample RSSI measurements collected in indoor environment showing significant variations across different channels.

that channel information is important for proper interpretation and selection of channels that should be used for range estimation.

5 Conclusion and Future Work

The software presented in this article enables straightforward measurement of RSSI across all 40 BLE channels, providing insights into signal propagation and enabling research on the application of frequency diversity. As demonstrated, RSSIfrequency pairs can reveal propagation phenomena that occur in the communication channel and allow detection of adverse propagation conditions. Importantly, these phenomena can not only be detected but also quantified, as various propagation effects affect the measurements in different ways. For example, in multipath propagation, the number of nulls and their frequency spacing depend on the path length differences. Additionally, the availability of RSSI-frequency measurements pave the way for new methods of selecting radio channels that are better suited for analysis. In RSSI-based ranging, for example, frequency-domain analysis can indicate which channels should be avoided, as constructive and destructive interference can significantly distort RSSI values. As these values differ from those expected for the particular distance between transmitting nodes, their use for range estimation increases inaccuracy.

Future work will involve more in-depth investigation of how various propagation phenomena manifest themselves in RSSI-over-frequency measurements, how they can be detected, and how their effects can be accounted for. In particular, we will focus on methods for selecting well-conditioned radio channels and using measurements from those channels for range estimation and indoor localization. Information about channel quality may later be leveraged in other localization techniques, including channel-sounding approaches that utilize multiple frequencies, where the differing impact of propagation phenomena can significantly affect the accuracy of the results.

As no two devices are identical, we will also investigate the influence of device variability on the observed measurements, both among devices of the same brand and model and across different vendors. This analysis will be used to assess the feasibility of designing universal, device-agnostic methods.

Acknowledgments. Software and data used in this article is available at https://github.com/pwr-zak/nrf52840-rssi-localization/tree/depcos2026.

Disclosure of Interests. The authors have no competing interests to declare that are relevant to the content of this article.

References

1. Aranda, F.J., Parralejo, F., Aguilera, T., Álvarez, F.J., Torres-Sospedra, J.: RSS channel-based integration for BLE fingerprinting positioning. In: 2023 13th International Conference on Indoor Positioning and Indoor Navigation (IPIN), pp. 1–6 (2023). https://doi.org/10.1109/IPIN57070.2023.10332223

2. Aranda, F.J., Rangel, P., Vicario, A., Morera, J., Aguilera, T., Álvarez, F.J.: Frequency-based decoupling and modeling of BLE RSS measurements for indoor positioning. In: 2025 IEEE International Instrumentation and Measurement Technology Conference (I2MTC), pp. 1–6 (2025). https://doi.org/10.1109/I2MTC62753.2025.11078973
3. de Blasio, G.S., Quesada-Arencibia, A., García, C.R., Rodríguez-Rodríguez, J.C.: Bluetooth low energy technology applied to indoor positioning systems: an overview. In: Moreno-Díaz, R., Pichler, F., Quesada-Arencibia, A. (eds.) EUROCAST 2019. LNCS, vol. 12013, pp. 83–90. Springer, Cham (2020). https://doi.org/10.1007/978-3-030-45093-9_11
4. Cortesi, S., Vogt, C., Magno, M.: Comparison between an RSSI- and an MCPD-based BLE indoor localization system. Computers **12**(3) (2023). https://doi.org/10.3390/computers12030059
5. Diago-Mosquera, M.E., Aragón-Zavala, A., Castañón, G.: Bringing it indoors: a review of narrowband radio propagation modeling for enclosed spaces. IEEE Access **8**, 103875–103899 (2020). https://doi.org/10.1109/ACCESS.2020.2999848
6. Farahsari, P.S., Farahzadi, A., Rezazadeh, J., Bagheri, A.: A survey on indoor positioning systems for IoT-based applications. IEEE Internet Things J. **9**(10), 7680–7699 (2022). https://doi.org/10.1109/JIOT.2022.3149048
7. Nikodem, M., Szeliński, P.: Channel diversity for indoor localization using Bluetooth low energy and extended advertisements. IEEE Access **9**, 169261–169269 (2021). https://doi.org/10.1109/ACCESS.2021.3137849
8. Singh, A., Emam, M., Al Mtawa, Y.: Comparative analysis of indoor localization across various wireless technologies. Eng **4**(3), 2293–2308 (2023). https://doi.org/10.3390/eng4030131
9. Xiao, Z., Wen, H., Markham, A., Trigoni, N., Blunsom, P., Frolik, J.: Non-line-of-sight identification and mitigation using received signal strength. IEEE Trans. Wirel. Commun. **14**(3), 1689–1702 (2015). https://doi.org/10.1109/TWC.2014.2372341
10. Zhang, D., Liu, Y., Guo, X., Gao, M., Ni, L.M.: On distinguishing the multiple radio paths in RSS-based ranging. In: 2012 Proceedings IEEE INFOCOM, pp. 2201–2209 (2012). https://doi.org/10.1109/INFCOM.2012.6195605
11. Zhao, Y., Liu, Y., He, T., Vasilakos, A.V., Hu, C.: Fredi: robust RSS-based ranging with multipath effect and radio interference. In: 2013 Proceedings IEEE INFOCOM, pp. 505–509 (2013). https://doi.org/10.1109/INFCOM.2013.6566824

Self-Assessment of Cybersecurity Readiness of Local Government Units in Poland

Paweł A. Nowak[1(✉)], Wiktor Ziętara[2], and Dariusz Jędryczek[2]

[1] Uniwersytet Łódzki, Łódź 90-136, Polska
pawel.nowak@uni.lodz.pl
[2] Uniwersytet Dolnośląski DSW, Wrocław 53-611, Polska

Abstract. The article analyzes the issue of the low level of cybersecurity in local government units (LGUs) in Poland. Despite collecting sensitive data and maintaining formal security policies, local governments face critical barriers: severe underfunding (IT budgets often below PLN 100,000 per year), staffing shortages, and low employee awareness of threats (e.g., phishing). As a result, LGUs are highly vulnerable to paralyzing cyberattacks.

In response to these challenges, the authors propose an intuitive, free self-assessment tool to evaluate LGUs' cybersecurity readiness. The tool is based on a five-level maturity model (ranging from reactive, through awareness and defined, to managed and optimizing levels) and uses simple yes/no control questions.

Pilot studies conducted in the Lower Silesia region revealed an "inverted maturity hierarchy." While local governments perform relatively well in basic, ad hoc measures, they demonstrate significant deficiencies in systematic risk management, performance monitoring, and continuous improvement.

The proposed assessment sheet serves as a low-cost, effective starting point for organizational diagnosis. The authors recommend implementing the tool on a broader scale and operationalizing its results—self-assessment should directly trigger the development and implementation of specific corrective action plans, systematically strengthening the resilience of Polish local governments to contemporary cyber threats.

Keywords: Cybersecurity · Information society · Information security management

1 Cybersecurity as an Element of Organizational Culture in Local Government

1.1 Local Government Systems Are an Attractive Target for Cybercriminals

Local Government Units (LGUs) in Poland constitute a key pillar of public service delivery, responsible for collecting and processing both personal data and classified information [2]. Therefore, cybersecurity in local government has become a particular concern, as ineffective security policies, limited financial resources, and poorly organized IT systems create ideal conditions for vulnerabilities. IT system outages, data breaches,

W. Zamojski et al. (Eds.): DepCoS-RELCOMEX 2026, LNNS 2014, pp. 324–332, 2026.
https://doi.org/10.1007/978-3-032-27927-9_31

or cybersecurity incidents may result in serious consequences and a loss of public trust [4]. Numerous examples of successful cyberattacks against LGUs in recent years confirm this trend—Kościerzyna, Rudniki, Wronki, Tuczna, Oświęcim, Lututów [6]—as well as the report Information on the Results of the Audit: Ensuring Information Security and the Continuity of IT Systems in Local Government Units [5].

The aim of this article is to identify the causes of the low level of cybersecurity in Polish local governments, including less obvious factors, and to propose a free self-assessment tool designed to evaluate a local government's readiness to perform cybersecurity-related tasks.

1.2 Formal Compliance without Effective Implementation

Although a significant proportion of LGUs have official security policies in place, the actual implementation of these guidelines often leaves much to be desired—regular employee training sessions or procedures for testing security systems are frequently lacking [2]. The main barriers include limited financial resources, which prevent the acquisition of modern solutions, and staffing shortages, meaning there is no one adequately responsible for overseeing security. Procedural complexity also stems from unclear step-by-step instructions and excessive bureaucracy, discouraging employees from complying with established rules. Additionally, insufficient threat awareness among staff leads to the underestimation of risks. Many units lack modern tools for monitoring IT infrastructure, as well as clear and widely known incident response procedures. The Supreme Audit Office (NIK) report emphasizes that nearly 40% of LGU employees do not know the proper procedural steps to take in the event of a security breach, which significantly prolongs response times—in several cases, delays of up to a dozen hours were recorded [4]. Budget constraints remain one of the most frequently cited factors affecting the state of cybersecurity in local governments [2]. In more than 60% of municipalities, the annual IT budget does not exceed PLN 100,000, which effectively prevents the implementation of modern technical solutions such as threat detection systems, advanced firewalls, automated incident monitoring platforms, and regular staff training programs.

By comparison, investment levels in Western European countries are several times higher. Cybersecurity responsibilities are often assigned to one or two individuals who, in addition, may lack specialized expertise in risk analysis, incident response, or information security management. This significantly limits effective monitoring and the ability to respond promptly to emerging threats [4].

1.3 User Awareness as the Weakest Link

The most serious systemic gap is the insufficient awareness among LGU employees of threats such as phishing recognition, knowledge of security procedures, and the ability to respond to incidents [1, 3]. Even the best-designed procedures or the most advanced technological tools will fail if users are unable to operate them properly or are unaware of potential risks. The lack of regular, practical training—such as phishing attack simulations or digital awareness workshops—means that even a single careless action, for example clicking on an infected attachment, may trigger a serious incident resulting in data loss or the need to pay a ransom in the event of a ransomware attack. According to CERT Polska reports, several thousand incidents involving LGUs are reported each year—the number has remained high or has increased in recent years—highlighting the importance of education and proper user preparedness [1].

Another challenge is the fragmented structure of information security management [4]. Procedures are often implemented only partially, and information about incidents is frequently communicated through informal channels, preventing key decision-makers from receiving timely warnings about threats. The activities of individual units are poorly coordinated, thereby undermining organizational coherence and slowing response times across the entire institution. It may also lead to duplicated efforts, overlooked critical threats, and prolonged restoration of normal operations. Establishing shared competence centers or centralized IT security teams could improve information flow and standardize incident response procedures. For example, in Estonia and in the banking sector in the United Kingdom, such centers enable coordinated action and the immediate exchange of critical information between units.

2 Self-Assessment as a Starting Point for Cybersecurity

The already complex issue of cybersecurity in Polish local government has been further complicated by the many months of uncertainty surrounding the amendment to the Act on the National Cybersecurity System (KSC Act) [9]. However, this situation has at least raised awareness among some individuals who, under the proposed regulations, will be responsible for cybersecurity. A key question has nevertheless emerged: how to assess the level of readiness of individual local governments and determine the proper sequence of necessary actions.

The authors of the article Cybersecurity Management in Local Government Units [7] proposed a self-assessment tool that assigns each examined unit to one of five maturity levels (Table 1).

The tool also serves as a specific "checklist" of actions that should be undertaken to raise the level of readiness to the most desirable "optimizing level." Since the authors provide research results from individual regions, each entity using the tool may also conduct a form of benchmarking against other local governments within its region as well as across other analyzed voivodeships.

2.1 Design of the Self-Assessment Sheet

Since the study's goal is to self-assess local governments, it was assumed that the tool should be as intuitive as possible, leaving no room for interpretation of either the questions or the possible answers. Therefore, it was decided that each maturity level would be assessed using four control questions, each of which could only be answered "yes" or "no" (Table 2).

The assessment sheet constructed in this way also highlights the missing elements necessary to achieve individual levels of cybersecurity maturity.

Table 1. Maturity Levels of Cybersecurity Management in Local Government Units (source: Nowak P., et al. [7]).

Maturity level	Management practices	Characteristics of organizational culture	Approach to risk
Reactive	No formal procedures, ad hoc actions	Low awareness of threats	Ignoring or lack of awareness of risk
Awareness	Basic procedures, introductory training	Growing awareness	Reactive risk management
Defined	Formalized processes, clearly assigned roles	Acceptance of the importance of security	Formal risk assessment processes
Managed	KPI monitoring, regular audits	Proactive security culture	Systematic risk management
Optimizing	Continuous improvement, benchmarking	A culture of innovation in security	Integrated risk management

2.2 Practical Testing of the Self-Assessment Sheet

To verify the tool's functionality, a pilot study was conducted among local government units in the Lower Silesian Voivodeship. The study sample included 169 municipalities—35 urban municipalities, 56 urban-rural municipalities, and 78 rural municipalities—26 counties (powiats), 4 cities with county status, and the regional self-government authority. In total, 200 local government units were invited to participate in the study.

Communication was carried out using official e-Delivery addresses from the database published on the Ministry of the Interior and Administration's website. Responses were received from 73 local governments (36.5%). Given the voluntary nature of participation and the study's exclusively correspondence-based format, the response rate should be considered satisfactory.

The average scores for the individual readiness levels are presented in Table 3. In practice, they indicate that the surveyed local governments demonstrate a basic level of cybersecurity awareness and that some key processes have been defined; however, these processes are not yet consistently managed or optimized. This reflects a typical transitional stage in which an organization moves from "we have rules in place" to "we manage performance and continuously improve". From an analytical perspective, the inverted maturity hierarchy (the results show a typical pattern in which the most basic levels—reactive and awareness—achieve the highest scores, while the higher maturity levels score the lowest) suggests that the surveyed local governments are relatively well prepared to respond to cybersecurity incidents (standard IT service activities), but weak in anticipating and systematically preventing threats in this area (organizational and managerial actions). An analysis of responses across individual categories also revealed a group of respondents who achieved the maximum score in each of the examined areas (Table 4).

However, based on the collected data, it was not possible to determine whether the local governments that achieved the highest scores shared any common characteristics

Table 2. Compilation of control questions–test version of the self-assessment tool (source: author's own elaboration).

Maturity level	Research question
Reactive	• Are there any formal cybersecurity procedures in place within the LGU, even at a basic level? • Are cybersecurity-related activities undertaken, even if they are ad hoc in nature? • Are initial steps being taken within the LGU to build threat awareness among employees? • Is the risk of cyber threats beginning to be recognized within the LGU, even if it is not yet fully managed?
Awareness	• Have basic cybersecurity procedures been implemented within the LGU? • Are basic cybersecurity training sessions provided for employees within the LGU? • Is threat awareness among employees gradually increasing? • Is cyber risk managed in a reactive manner?
Defined	• Are cybersecurity processes within the LGU formally defined and documented? • Have specific roles responsible for cybersecurity been designated within the LGU? • Is security recognized within the LGU as a key element of organizational functioning? • Are formal risk assessments conducted systematically within the LGU?
Managed	• Are key performance indicators (KPIs) related to security monitored within the LGU? • Are cybersecurity audits conducted regularly within the LGU? • Is there a proactive security culture within the LGU? • Is cyber risk managed in a systematic and continuous manner?
Optimizing	• Are cybersecurity-related processes continuously improved within the LGU? • Does the LGU benchmark the effectiveness of its cybersecurity activities against other LGUs and sectors? • Is there a culture of innovation in the area of security within the LGU? • Is risk management integrated with other management processes within the LGU?

(e.g., type or level of local government). Therefore, in the subsequent pilot study, a request was added to indicate the type of local government unit. The authors, however, concluded that under specific circumstances (e.g., a small number of cities with county status in a given region, or the single regional government authority in each voivodeship), such a question might raise concerns among respondents about the relative ease of identification, despite assurances of anonymity. For this reason, the question in the survey questionnaire was made optional.

Table 3. Average score for individual readiness levels (source: author's own elaboration).

Maturity Level	Average Score for the Level (Maximum Possible Value: 4.0)
Reactive	3.86 (approx. 96.5% of the maximum)
Awareness	3.22 (approx. 80.5% of the maximum)
Defined	2.90 (approx. 72.5% of the maximum)
Managed	2.18 (approx. 54.5% of the maximum)
Optimizing	2.03 (approx. 50.8% of the maximum)

Table 4. Distribution of the number of positive responses across individual maturity levels (source: Author's own elaboration).

Maturity Level	0 points	1 points	2 points	3 points	4 points
Reactive	0	0	2	6	65
Awareness	1	4	11	19	38
Defined	4	11	4	23	31
Managed	11	11	20	16	15
Optimized	12	19	13	13	16

Another test of the self-assessment sheet was conducted among local governments in the Podkarpackie Voivodeship. Once again, invitations were sent using official e-Delivery addresses published in the register on the website of the Ministry of the Interior and Administration. Unfortunately, fewer than 10% of the participating local governments responded to the optional question regarding the type of local government unit.

As a result, in the final version of the survey instrument, the mandatory additional question concerned the population size of the given local government unit, categorized as follows:

- small local governments (up to 10,000 inhabitants),
- medium-sized local governments (up to 40,000 inhabitants),
- large local governments (up to 150,000 inhabitants),
- very large local governments (over 150,000 inhabitants).

3 Summary and Recommendations: Continuation of Research and the Usefulness of the Self-Assessment Tool A Subsection Sample

3.1 Summary of Theoretical Foundations and Empirical Observations

The cybersecurity maturity self-assessment tool for Local Government Units (LGUs) proposed in this article addresses a significant practical gap in the context of the limited human and financial resources available to local governments. The use of a five-level maturity model combined with four control questions in a "yes/no" format facilitates clear interpretation and ensures low entry costs, which is crucial in the operational realities of Polish LGUs.

Observations obtained from the pilot studies indicate a characteristic profile: strong performance at the first two levels (reactive and awareness) and significant deficiencies at levels extending beyond formalization and systematization (managed and optimizing). This pattern suggests that a basic "core" of policies and threat awareness often exists, while key mechanisms of risk management, monitoring, and process improvement remain underdeveloped. Consequently, the tool serves not only as a diagnostic trigger (identifying existing gaps), but also as a stimulus for planning and initiating corrective actions within a cyclical framework (assessment – prioritization – planning – repetition).

3.2 Impact on Practice – how the Tool Can Support Decision-Making Processes

The tool may serve a threefold function: diagnostic, planning, and comparative. In practice, this means that LGUs can quickly identify critical gaps, prioritize actions, and compare their results with those of other units, thereby facilitating the identification of good practices.

However, to translate assessment results into tangible security improvements, it is essential to close the cycle of diagnosis – planning – verification. The outcome of the self-assessment should conclude with a concise, operational list of actions for the next 3–6 months and for a 12-month horizon, with clearly assigned owners and measurable verification indicators (e.g., conducting incident response exercises, implementing KPIs, carrying out an audit).

3.3 Research and Implementation Recommendations – Directions for Further Development

Taking into account the results of the pilot studies and the potential of the self-assessment tool, the authors conclude that it is necessary to broaden the scope and increase the comparability of the sample: continuing the research in subsequent regions, maintaining a uniform communication channel with respondents, and introducing systematic segmentation of LGUs by population size. The currently proposed classification (small/medium/large/very large) may in future iterations be further linked to the operational context (e.g., IT service delivery model, statutory responsibilities of LGUs).

If the study is formalized within the framework of official public statistics, it should be supplemented with data explaining the determinants of maturity. This would involve expanding the questionnaire with brief contextual variables (e.g., number of IT staff positions, IT service delivery model, existence of business continuity plans, frequency of training) in a non-intrusive manner, enabling the testing of hypotheses regarding factors that foster higher levels of maturity. In subsequent versions, validation of the tool (validity and reliability) would also be advisable, including content validation (expert panel) and criterion validation on a subsample of LGUs by comparing results with evidence of implementation (documents, protocols, training logs, audit reports). Adding an optional "evidence/reference" field could enhance the credibility of declarations without compromising the tool's simplicity.

A longitudinal study design to assess the dynamics of maturity development would also be valuable: a project in which the same LGUs repeat the self-assessment at intervals of 12–24 months would enable evaluation of the pace of maturity growth and the effectiveness of interventions. An important objective would be to identify which interventions most effectively advance progress at the “managed” and “optimizing” levels.

Recognizing the importance of standardization, the authors are preparing to develop (for participating units) a standard for interpreting individual results along with an action map—that is, a brief interpretative guide and a “next steps” roadmap leading from assessment outcomes to concrete implementation measures. This would facilitate the transition from the “defined” level to the “managed” and “optimizing” levels.

3.4 Implementation Guidelines for Local Government Units (Practical Application)

In the cover letter preceding the survey, it should be recommended that the questionnaire be completed by representatives of the IT department and data protection/information security officers (GDPR), or by the management team, in order to reduce the risk of individual, subjective interpretations.

To ensure transparency and comparability of results, a concise, publicly available maturity metric should be developed, accompanied by a brief corrective action roadmap. This would enable secure benchmarking among Local Government Units.

4 Conclusion

The proposed self-assessment tool has both scientific and practical potential: it may serve as a low-cost, rapid, and reliable starting point for evaluating cybersecurity maturity in Local Government Units (LGUs) and—through longitudinal and validation studies—become a solid foundation for more targeted interventions. However, to fully leverage its value, further empirical research is necessary to:

- verify the validity and reliability of the tool,
- explain the determinants of differences in maturity levels among LGUs,
- enable the assessment of intervention effects over time,
- facilitate the integration of results with risk management practices and business continuity planning.

Crucial to increasing maturity levels will be the integration of assessment results into an action plan: the outcome of the self-assessment should inform an annual or semi-annual cybersecurity action plan for the LGU, even at a minimal scope, with clearly defined priorities and deadlines.

References

1. CERT Polska.: Educational materials and e-learning courses (n.d.). https://cert.pl. Accessed 07 June 2024

2. Chodakowska, E., Kańduła, S., Przybylska, R.:The level of preparedness of local government units to counter cyber threats. In: Securing the Public Sector against Digital Threats (pp. 35–52). Poznań University of Economics Press. https://ue.poznan.pl/publikacje/JST_cyberzagrozenia.pdf (2022). Accessed 07 June 2024
3. NASK.: (n.d.). Guides and educational platform. https://nask.pl, last accessed 2024/03/07
4. Supreme Audit Office (Najwyższa Izba Kontroli): Audit report: IT security in selected local government units. https://nik.gov.pl/plik/id,28782,vp,31433.pdf (2024). Accessed 07 March 2026
5. Supreme Audit Office (Najwyższa Izba Kontroli): Information on the results of the audit: Ensuring information security and the continuity of IT systems in local government units. https://www.nik.gov.pl/kontrole/wyniki-kontroli-/pobierz,kap~p_24_004_202405240817251716531445~01,typ,kk.pdf ik (2025). Accessed 01 May 2025
6. Nowak, P.A., Ziętara, W., Jędryczek, D.: Cybersecurity management in local government units. https://doi.org/10.13140/RG.2.2.22658.39362 (2024) Accessed 03 Apr 2025
7. Nowak, P.A., Ziętara, W., Jędryczek, D.: Cybersecurity management in local government units. In: Soboń, J., Cywiński, M., Chojnacka, M. (eds.) Effective Organizational Management. Jacob of Paradies University Press, Gorzów Wielkopolski (2025)
8. Nowak, P.A., Czekaj, M.: Local government E-administration: analysis of the current state and development perspectives. : https://doi.org/10.13140/RG.2.2.27178.22728 (2024). Accessed 07 Apr 2025
9. Act of 5 July 2018 on the National Cybersecurity System, Journal of Laws 2026, item 20 (consolidated text)

Proxy-Based Interaction Logging for Traceability in AI-Assisted Software Development

Ami Otsuka[1](✉), Jinshan Luo[2], Madoka Hasegawa[3], Artur Bąk[4], and Atsushi Ito[2]

[1] Hundreds LLC, Shibuya Dogenzaka Tokyu Building 2F-C, 1-10-8 Dogenzaka, Shibuya, Tokyo 150-0043, Japan
amiotsuka.se@gmail.com

[2] Chuo University, 742-1 Higashi Nakano, Hachioji, Tokyo 192-0351, Japan
kingsamlawsan@gmail.com, atc.00s@g.chuo-u.ac.jp

[3] Utsunomiya University, 7-1-2 Yoto in Utsunomiya, Tochigi 321-8585, Japan
madoka@is.utsunomiya-u.ac.jp

[4] Polish-Japanese Academy of Information Technology, Warsaw, Poland
abak@pjwstk.edu

Abstract. The adoption of generative AI in software development is rapidly increasing. However, AI-assisted coding environments typically lack mechanisms for preserving interaction history between developers and AI systems. While version control systems store code changes, they usually do not record prompts, generated responses, or execution outputs, which makes debugging, auditing, and knowledge sharing difficult. This study addresses the problem of "black-boxing" in AI-assisted development by proposing a proxy-based architecture that captures interactions between developers and generative AI tools without modifying proprietary systems. Based on this approach, we implemented AgentLogger, a system that collects and visualizes prompts, responses, and execution outputs during development sessions. Case studies in realistic development scenarios demonstrate its usefulness for debugging support and organizational security auditing.

Keywords: Generative AI · No-code · Software development support tool

1 Introduction

The rapid adoption of generative AI coding assistants has introduced a new challenge in software engineering: the lack of traceability of interactions between developers and AI systems during the development process.

The application of generative AI to software development is expanding rapidly. According to Reference [1], using GitHub Copilot improved project-level productivity by 6.5%, increased individual-level productivity by approximately 5.5%, and increased the number of contributors by 5.4%. Reference [2] reports that developers who used AI completed tasks 55.8% faster than those who did not. As shown in Fig. 1, Reference [3]

W. Zamojski et al. (Eds.): DepCoS-RELCOMEX 2026, LNNS 2014, pp. 333–342, 2026.
https://doi.org/10.1007/978-3-032-27927-9_32

forecasts that the North American market for generative AI tools for programming was USD 265.9 million in 2022, expanded to USD 341.3 million in 2023, and is projected to reach USD 2,833.9 million by 2030. This corresponds to a compound annual growth rate (CAGR) of 35.3%.

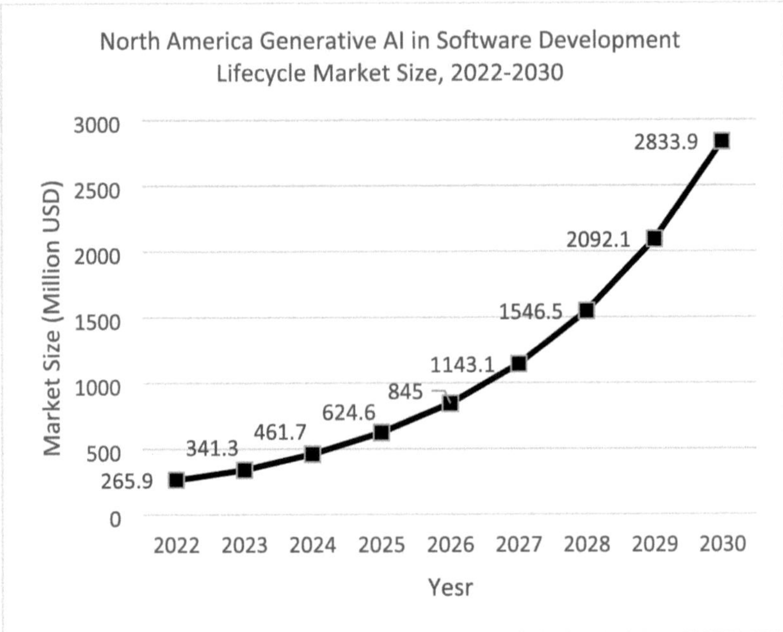

(Generated from by authors from [3])

Fig. 1 Market size of AI for software development in North America

(Generated from by authors from [3])

1.1 Trends and Characteristics of AI-Assisted Coding Environments

With the advancement of large language models (LLMs), attempts to integrate AI into software development environments—so that code can be generated, modified, and understood from natural-language instructions—have been spreading rapidly. In particular, "AI-assisted coding," represented by OpenAI Codex, has evolved beyond simple autocomplete functionality into a tool that supports the entire development process, including design, implementation, and modification.

In addition to code generation capabilities, recent research and industry tools have increasingly focused on observability and governance of AI-assisted development workflows. Systems such as LangSmith [4], PromptLayer [5], and Helicone [6] provide logging and monitoring of interactions with large language models. However, these solutions are typically designed for application-level LLM integration rather than interactive coding environments. In contrast, the approach proposed in this study focuses specifically on capturing development interactions in coding workflows, including prompts, generated code, and execution outputs produced during development sessions.

1.2 Ide Extension–Based Coding Assistance Environments

The most widespread form is to incorporate AI as an extension into existing integrated development environments (IDEs). OpenAI Codex is integrated into IDEs as a VS Code extension, enabling developers to generate and modify code by providing natural-language instructions within the IDE [7]. Similarly, GitHub Copilot and Copilot Chat support major environments such as VS Code and JetBrains IDEs, offering not only code completion but also chat-based Q&A and code comprehension support [8–11].

JetBrains AI Assistant is an AI feature integrated into JetBrains IDEs such as IntelliJ IDEA, PyCharm, and CLion, characterized by support that leverages IDE-specific analysis information [12]. Amazon Q Developer also supports multiple IDEs, including VS Code and JetBrains IDEs, and provides development assistance through in-IDE chat and inline suggestions [13, 14].

In addition, tools such as Continue [15–17], Sourcegraph Cody [18, 19], and Tabnine [20] are also provided as IDE extensions. While these tools have different strengths—such as an open-source orientation or understanding of large-scale codebases—they share the common advantage of enabling AI assistance without drastically changing the existing development environment. This IDE extension approach is highly compatible with existing workflows and offers a low adoption cost.

1.3 Other AI-Assisted Development Environments

In addition to IDE extension–based approaches, several other forms of AI-assisted development environments have recently emerged. AI-native editors such as Cursor [21, 22] and Windsurf Editor [23, 24] place interaction with large language models at the center of the development workflow, allowing developers to generate, modify, and analyze code through continuous dialogue with AI systems. Cloud-and browser-based development environments have also gained attention. For example, Replit provides a browser-based IDE combined with Replit Agent, which supports project setup, implementation, and execution based on natural-language instructions [25, 26]. Another category includes terminal-centric coding agents, which allow developers to interact with AI directly through command-line interfaces. A representative example is Claude Code, which focuses on CLI-based development workflows and enables AI-driven code operations through conversational instructions [27]. Although these environments differ in architecture and user interface, they share a common characteristic: AI is increasingly integrated into the development workflow as an active collaborator rather than merely a code-completion tool.

1.4 Problem: Lack of Traceability in AI-Assisted Development

Although many AI-assisted coding tools have been developed, existing development infrastructures provide limited support for capturing the full interaction history between developers and AI systems. Version-control systems record code changes but do not preserve prompts, generated responses, or execution outputs produced during AI-assisted development. As a result, important parts of the development process become difficult to reproduce, analyze, or audit.

1.5 Research Contributions

In this study, we propose a proxy-based logging architecture that captures interactions between developers and generative AI systems without modifying proprietary AI tools. The main contributions of this study are as follows: (1) identification of the traceability problem in AI-assisted software development workflows, (2) design of a tool-agnostic proxy architecture for capturing prompts, responses, and execution outputs, (3) implementation of AgentLogger, a log collection and visualization system, (4) demonstration of its usefulness through practical development scenarios including debugging support and security auditing.

2 Problems in AI-Assisted Coding

As discussed in the previous section, AI-assisted coding environments have rapidly evolved and are now widely integrated into software development workflows. However, existing development infrastructures provide limited support for capturing the interaction history between developers and AI systems during development sessions.

While version-control systems such as GitHub preserve code version histories, they do not record prompts, generated responses, or execution outputs produced during interactions with generative AI tools. As a result, developers often lose important contextual information needed to understand how a particular piece of code was generated or modified. This lack of traceability makes it difficult to reproduce development steps, analyze debugging processes, or audit the use of AI during software development. To investigate this issue in practice, this study focuses on coding environments based on Codex-style command-line tools as a representative example.

3 Proposal of an AI-Assisted Coding Support Tool

As shown in Fig. 2, there are two approaches to preserving work history. Case 1 is to modify Codex itself to record logs. However, because the internal code of Codex is not publicly available, this approach is not realistic.

In the more feasible Case 2 approach, a proxy is placed between the user and Codex to collect logs. This eliminates the need to modify Codex itself and makes implementation easier. However, this method has a drawback: each time the Codex API specification changes, the proxy must be continuously adapted.

For these reasons, we decided to implement a Case 2–type proxy. This enables the proxy to run locally, improving processing speed and enhancing flexibility in log processing. In the future, it may also be possible to embed AI into the proxy itself and provide coding advice.

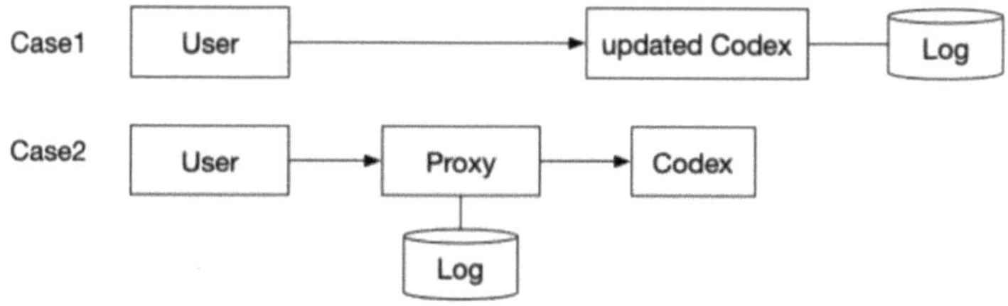

Fig. 2 Two design patterns

4 Log Collection and Visualization System (AgentLogger)

This chapter describes the details of the log collection and visualization system implemented based on the proposed method.

4.1 System Architecture

The system consists of two components: a Python log-collection proxy and a Next.js web dashboard (Vercel).

Log-collection proxy (Backend)

An HTTP server that receives requests from the user's development environment (e.g., CLI tools): 1) **Request interception** Captures the user's prompt, the selected model (e.g., gpt-5-codex), and various parameter settings, 2) **Response interception** Captures generated code/text returned by the AI, as well as standard output from tool execution (stdout/stderr), 3) **Data storage** Formats the captured data as JSON, adds timestamps, and stores it in a database.

Web dashboard (Frontend)

A web application for viewing stored logs. The screen is shown in Fig. 3: 1) **UI design** Adopts a simple table design that removes unnecessary elements so developers can quickly access the needed information, 2) **ANSI color support** Implements dedicated rendering to faithfully reproduce CLI color information (ANSI escape sequences) in the web browser, enabling users to view red error text and highlighting as-is.

4.2 Data Schema

For log persistence, we defined a data structure with the following attributes:

ID: Unique identifier of a log entry, **Timestamp (UTC):** Time when the interaction occurred (millisecond resolution), **Model:** Name of the AI model used (e.g., codex-cli, gpt-4), **Prompt:** User input content, **Response:** Full text including AI output and execution results, **Tokens:** Number of tokens consumed (for cost management), **Session ID:** Identifier to link a series of interactions.

4.3 Implementation Considerations

A particular focus was on handling unstructured data. In AI-assisted development, interactions include not only plain text exchanges but also shell command outputs and diff

Fig. 3 Example of the dashboard

information. This tool preserves such raw outputs as-is, and formats and displays them on the web frontend using an HTML pre tag, successfully recreating the full development context. The dashboard lists history in reverse chronological order, and users can expand each row to view detailed interactions (Fig. 3).

5 Implementation and Evaluation of the Effectiveness of AgentLogger

To verify the effectiveness of the implemented **AgentLogger**, we conducted case studies in real development scenarios.

5.1 Case Study: CLI-Based Interactive Development

A participant used the codex-cli tool to create a Python data-processing script. The usefulness of the tool was confirmed in a troubleshooting situation as follows:

- **Scenario** While implementing a "feature to summarize Git commit history," the user ran code suggested by the AI and encountered an error. The user entered a new prompt to fix it, but the error message scrolled out of the terminal buffer and could no longer be confirmed.
- **Application of the tool** The user opened the web dashboard and referenced the most recent log (ID #6). It contained the full stack trace that had disappeared from the CLI and the code output by the AI immediately before it.
- **Result** The user copied the error message from the log and re-entered it to the AI, obtaining a fix proposal within a few tens of seconds. This significantly reduced the time compared to the conventional approach, which would have required re-running or relying on memory.

5.2 Case Study 2: Security Auditing and Risk Management

In organizational settings, there is a risk that developers may inadvertently enter confidential information (API keys, passwords, customer data, etc.) into generative AI systems. To validate the tool's effectiveness, we simulated a scenario in which a developer accidentally sent a prompt containing an AWS secret key.

An administrator used the web dashboard's search function (Filter response) to search across logs using keywords such as "AWS" and "Key." As shown in Fig. 4, the administrator was able to identify the session containing the risky prompt within seconds.

This confirmed that AgentLogger functions not only as a personal development log but also as a tool for early detection of organizational compliance violations and for post-incident tracing (forensics) in the event of information leakage.

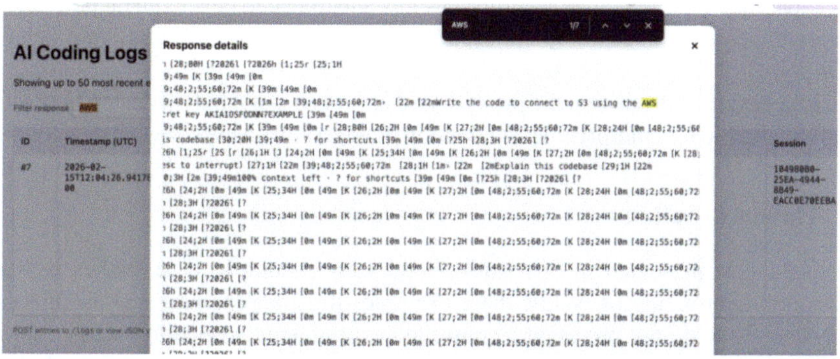

Fig. 4 An example of security auditing and risk management

5.3 Performance Evaluation

We measured the overhead introduced by routing through the proxy. When running the proxy locally, the additional time required to forward requests averaged under 10 ms, resulting in no perceptible delay for users. The web dashboard also rendered quickly, and by leveraging Vercel's edge network, smooth searching and browsing remained possible even with a large accumulation of logs.

Therefore, the proxy-based architecture does not significantly affect developer interaction with AI systems while enabling persistent logging of development activities.

6 Discussion

6.1 Effectiveness of the Proxy-Based Architecture

Through the implementation and evaluation of AgentLogger, the effectiveness of the proxy-based architecture was confirmed from several perspectives.

First, the proposed approach provides a high degree of **tool independence**. Because the system operates by intercepting communication between the user environment and the AI service, it does not depend on the internal implementation of a specific tool. Therefore, the system can be applied not only to *codex-cli* but also to a wide range of generative AI tools, as long as the endpoint can be configured. This flexibility makes the proposed architecture adaptable to future changes in AI services and emerging development tools.

Second, the architecture enables **local or controlled storage of interaction logs**. Unlike cloud-based logging mechanisms managed by external providers, AgentLogger allows organizations to retain development interaction data within their own infrastructure. This improves data ownership, enhances security, and reduces the risk associated with external data exposure.

Furthermore, the system is particularly useful for **organizational governance and compliance auditing**. By preserving detailed records of prompts, generated code, and execution outputs, organizations can review how AI was used during development. This capability supports internal audits, ensures adherence to development policies, and enables traceability in cases where security incidents or unexpected system behavior occur.

These results indicate that the proxy-based approach is not only technically practical but also valuable for enterprise-level adoption of generative AI in software development.

6.2 Threats to Validity

The evaluation presented in this study focuses on qualitative case studies in realistic development scenarios. While these examples illustrate the practical usefulness of the system, they do not represent a controlled experimental evaluation across multiple development teams or environments. Furthermore, the case studies are primarily based on CLI-oriented workflows, and different development environments may exhibit different usage patterns. Future work should investigate broader usage scenarios and collect quantitative data on development efficiency and debugging performance. Finally, analyzing the scalability of the proxy-based architecture under large-scale development environments remains an important area for future work.

6.3 Future Work

Although the current implementation of AgentLogger primarily focuses on recording and visualizing interaction logs, the accumulated data is a valuable resource that can be further leveraged to support developers.

A promising direction for future work is **prompt improvement assistance**. By analyzing historical interaction data, the system could identify patterns linked to successful or unsuccessful outcomes and provide developers with real-time feedback, such as warnings about potentially error-prone instructions or suggestions for alternative prompts that previously produced better results.

Another promising extension is **automatic document generation**. Development processes often require manual preparation of artifacts such as daily development reports or pull request descriptions. By analyzing a sequence of interaction logs, the system could automatically summarize the development activities performed during a session. This would reduce documentation overhead while improving the accuracy and completeness of project records.

These enhancements would transform AgentLogger from a passive logging tool into an active contributor to productivity, knowledge sharing, and process transparency.

7 Conclusions

This study addressed the problem of black-boxing in generative AI-based software development and proposed and implemented a proxy-based log collection and visualization system, AgentLogger.

The system automatically records and visualizes interactions between users and AI coding tools, enabling developers to review past trial-and-error processes. This is expected to improve the reliability and efficiency of AI-assisted development.

The system can be introduced without changing existing workflows and ensures the persistent storage of AI interaction histories. The Web dashboard improves debugging efficiency and supports retrospective analysis of development decisions.

As generative AI becomes central to software development, tools such as this—serving as a "dashcam for the development process"—are expected to become essential infrastructure important for debugging, governance, reproducibility, and security auditing.

Future work includes implementing AI-based functions to analyze accumulated logs and provide real-time suggestions for better prompts and code improvements.

Acknowledgement. This research is supported by JSPS Kakenhi (24 K14987, 23 K28338, 23H03649, 22 K12598,20H05702).

References

1. Song, F., Agarwal, A., Wen W.: The impact of generative ai on collaborative open-source software development: evidence from github copilot (2026). https://doi.org/10.48550/arXiv.2410.02091
2. Peng, S., Kalliamvakou, E., Cihon, P., Demirer, M.: The impact of ai on developer productivity: evidence from github copilot. (2026). https://doi.org/10.48550/arXiv.2302.06590
3. Generative AI in software development lifecycle market size (2026). https://www.fortunebusinessinsights.com/generative-ai-in-software-development-lifecycle-market-109041?utm_source=chatgpt.com
4. Langchain. langsmith documentation. https://docs.smith.langchain.com/
5. Promptlayer—prompt engineering platform. https://promptlayer.com/
6. Helicone. Helicone LLM observability platform. https://www.helicone.ai/
7. OpenAI codex IDE extension. (2026). https://developers.openai.com/codex/ide/
8. GitHub Docs (2026) Configuration of github copilot. https://docs.github.com/ja/copilot/how-tos/configure-personal-settings/configure-in-ide
9. Visual Studio Code Docs, Github copilot in VS code (overview) (2026). https://code.visualstudio.com/docs/copilot/overview

10. Visual Studio Code Docs Get started with chat in VS Code (2026). https://code.visualstudio.com/docs/copilot/chat/copilot-chat
11. GitHub Docs QA with github copilot through IDE (Copilot Chat) (2026). https://docs.github.com/ja/copilot/how-tos/chat-with-copilot/chat-in-ide
12. JetBrains Features and compatibility | AI assistant documentation (2026). https://www.jetbrains.com/help/ai-assistant/about-ai-assistant.html
13. AWS Using amazon Q developer in the IDE (2026). https://docs.aws.amazon.com/amazonq/latest/qdeveloper-ug/q-in-IDE.html
14. AWS Installing the amazon Q developer extension or plugin in your IDE (2026). https://docs.aws.amazon.com/amazonq/latest/qdeveloper-ug/q-in-IDE-setup.html
15. Continue Continue Docs (2026). https://docs.continue.dev/
16. VS Code Marketplace Continue—open-source AI code agent (2026). https://marketplace.visualstudio.com/items?itemName=Continue.continue
17. JetBrains Marketplace Continue plugin for jetbrains IDEs (2026). https://plugins.jetbrains.com/plugin/22707-continue
18. Sourcegraph Cody: AI code assistant (VS code marketplace) (2026). https://marketplace.visualstudio.com/items?itemName=sourcegraph.cody-ai
19. JetBrains marketplace Cody: AI code assistant plugin for jetbrains IDEs (2026). https://plugins.jetbrains.com/plugin/9682-cody-ai-code-assistant
20. Tabnine Docs Supported IDEs (2026). https://docs.tabnine.com/main/welcome/readme/supported-ides
21. Cursor Cursor (official site) (2026). https://cursor.com/
22. Cursor Download cursor (2026). https://cursor.com/download (accessed 2026-01-15).
23. Windsurf Windsurf editor (2026). https://windsurf.com/editor
24. Windsurf Download Windsurf Editor and Plugins (2026). https://windsurf.com/download
25. Replit Docs Replit Agent (2026). https://docs.replit.com/replitai/agent
26. Replit Replit AI (2026). https://replit.com/ai (accessed 2026-01-15).
27. Claude Code Docs Claude Code overview (2026). https://code.claude.com/docs/en/overview

Overview and Experimental Analysis of the HHL Quantum Algorithm

Dominik Pawelec[1], Jakub Blukacz[2], Hubert Siewiera[3(✉)], Paweł Siewiera[3], and Andrzej Kucharski[4]

[1] Faculty of Mathematics and Computer Science, University of Wrocław, ul. F. Joliot-Curie 15, 50-383 Wrocław, Poland
[2] Faculty of Pure and Applied Mathematics, Wrocław University of Science and Technology, Wyb. Wyspiańskiego 27, 50-370 Wrocław, Poland
[3] Faculty of Information and Communication Technology, Wrocław University of Science and Technology, Wyb. Wyspiańskiego 27, 50-370 Wrocław, Poland
hubert.siewiera5@gmail.com
[4] Department of Telecommunications and ICT, Wrocław University of Science and Technology, Wyb. Wyspiańskiego 27, 50-370 Wrocław, Poland
andrzej.kucharski@pwr.edu.pl

Abstract. This paper presents an overview and experimental evaluation of the Harrow–Hassidim–Lloyd (HHL) quantum algorithm for solving systems of linear equations. We discuss its theoretical foundations and report results obtained from simulations as well as executions on real quantum hardware, including the Odra5 and Garnet quantum computers. The study highlights discrepancies between theoretical predictions and experimental outcomes caused by noise, hardware topology, and circuit depth limitations across different quantum computers.

Keywords: quantum computing · Odra5 · Harrow-Hassidim-Lloyd · IQM · systems of linear equations

1 Introduction

Quantum computing is an emerging field that focuses on utilizing quantum effects to perform computations instead of relying on classical computers. One fundamental computational task is solving systems of linear equations, which appear in numerous applications, such as optimization, simulations, and data analysis [1]. In this paper, we tested the feasibility of solving linear equations on IQM quantum computers to determine whether the hardware allows for more complex quantum procedures. Our goal was to assess how well IQM devices can support this type of algorithm in practice and whether they can handle circuits of increasing complexity. Therefore, experiments on real quantum computers are necessary to understand the practical limits of current quantum computers. We are especially interested in the HHL algorithm [2], which offers an exponential

W. Zamojski et al. (Eds.): DepCoS-RELCOMEX 2026, LNNS 2014, pp. 343–352, 2026.
https://doi.org/10.1007/978-3-032-27927-9_33

speedup in time complexity compared to classical methods. We ran a simplified algorithm based on HHL on the AerSimulator simulator and two quantum computers:

- Odra5 - 5-qubit computer [3],
- IQM Garnet 20-qubit computer [4].

The remainder of the paper is structured as follows. In Sect. 2 we include an overview of the current literature on solving linear systems on quantum computers. In Sect. 3 we provide an overview of the HHL algorithm and its theoretical foundations. In Sect. 4 we present results from experiments conducted on simulators and real quantum devices. Finally, Sect. 5 contains the conclusions.

2 Literature Review

In this section, we provide a short review of recent advances in the use of quantum algorithms for solving systems of linear equations, with a main focus on implementation of the Harrow-Hassidim-Lloyd algorithm on quantum hardware.

The first quantum algorithm to solve linear systems of equations was developed by Harrow, Hassidim, and Lloyd (HHL) [2]. Compared with classical algorithms, it proposes an exponential speedup in time complexity, at the cost of getting the solution vector as a quantum state rather than an explicit vector. Although this representation allows the result to be used for further quantum computation, for example in quantum machine learning [5], extracting the numerical solution vector from the quantum state requires repeated measurements.

We are currently in a noisy intermediate-scale quantum era (NISQ) [6], which prevents the execution of the full version of this algorithm on real quantum computers. To test quantum computers in solving linear equations, an algorithm was developed using the same idea as HHL, but assuming that some known data that would be computed in the true algorithm (i.e., eigenvalues) are known in advance [7]. Other studies that solved linear equations on a quantum computer used a simplified version of the algorithm, which used all the principles of the full algorithm but simplified some computation steps [8].

More recently, hybrid approaches were developed to improve the feasibility. Lee et al. [9] introduced a hybrid algorithm that uses both a quantum computer and a classical computer to get more reliable results. Further work by Gao et al. [10] tested several versions of the hybrid algorithm and optimized its dependence on the number of qubits. Hybrid algorithms typically utilize a quantum computer for iterative phase estimation, working in a feedback loop with the classical computer to calculate exact solutions, compared to HHL based methods.

However, practical implementations of these algorithms are limited and performed on small instances [11]. This highlights the need for further investigation of its practical usage and limitations on currently available quantum computers.

3 Overview of the Algorithm

The HHL algorithm is a quantum algorithm that is used to solve systems of linear equations. Given matrix $A \in \mathbb{M}_{n\times n}$ and $\mathbf{b} \in \mathbb{R}^n$ representing the system of linear equations, it produces the output $|\mathbf{x}\rangle$, which corresponds to the normalized solution vector $\mathbf{x} \in \mathbb{R}^n$

$$A\mathbf{x} = \mathbf{b}. \tag{1}$$

It provides exponential speedup for linear system problems compared to the best classical methods with complexity $\tilde{\mathcal{O}}\left(\log(n)\frac{\kappa s^2}{\epsilon^3}\right)$ under the assumption that the matrix A is s-sparse, efficiently row-computable [2], $\epsilon > 0$ is the desired precision in the output state $|\mathbf{x}\rangle$, and κ is the condition number of the matrix A. A is assumed to be Hermitian. If not, we can convert A to $\begin{pmatrix} 0 & A \\ A^\dagger & 0 \end{pmatrix}$ which is Hermitian. As described in more detail in [12], this transformation allows the algorithm to handle non-Hermitian cases by doubling the system size.

Because A is Hermitian, A is invertible:

$$A^{-1}A|\mathbf{x}\rangle = A^{-1}|\mathbf{b}\rangle \tag{2}$$

$$|\mathbf{x}\rangle = A^{-1}|\mathbf{b}\rangle. \tag{3}$$

Therefore, in order to solve (1) we need to find A^{-1}. To do that, we can use HHL, which is composed of the following steps [13]:

1. Preparation of the state $|\mathbf{b}\rangle^{\otimes \frac{n}{2}}$ corresponding to the value of $\mathbf{b}$ after normalization.
2. Quantum Phase Estimation (**QPE**).
3. Rotation of the auxiliary qubit depending on the eigenvalues of A.
4. Inverse Quantum Phase Estimation (**IQPE**).

4 Experiments

To verify the validity of the analyzed problem and the implementation of the HHL algorithm, we conducted experiments on both real quantum computers and quantum simulators. This approach allowed us to assess the theoretical correctness of the problem formulation as well as the practical feasibility of the computations under physical conditions.

4.1 Experiments Formulation

For simulation purposes, we used the AerSimulator. The implementation and execution of circuits on physical hardware were made using Python 3.11.13 and Qiskit 1.2.4. This allowed us to evaluate the accuracy of the HHL algorithm and examine the difference in implementation on real hardware.

Our implementation was not developed from scratch. Instead, it was adapted from the implementation presented by Imam and Riaz [13] and adjusted for our experiments on IQM hardware.

Experiments were conducted on two physical quantum devices: a 5 qubit computer Odra5 and a 20 qubit IQM Garnet. Circuit transpilation was performed using optimization level 1, although similar results were observed across all optimization levels. Additionally, we analyzed the impact of physical qubit selection on transpilation outcomes.

As input parameters, we used the same example as in [12].

$$A = \begin{pmatrix} 1 & -\frac{1}{3} \\ -\frac{1}{3} & 1 \end{pmatrix}, \quad \mathbf{b} = \begin{pmatrix} 0 \\ 1 \end{pmatrix}, \quad \text{and} \quad \mathbf{x} = \begin{pmatrix} \frac{3}{8} \\ \frac{9}{8} \end{pmatrix}. \tag{4}$$

with $n_b = 1$ and $N_b = 2^1 = 2$, where n_b is the number of qubits that store $|b\rangle$, and N_b is the number of ancilla qubits required for the calculation of λ_1, λ_2. The theoretical ratio of $|01\rangle/|11\rangle$ that we expected was 1/9. Our results were mixed.

4.2 Results for Simulators

We began by determining the expected results of the algorithm through simulation tests. The results matched closely those reported in the literature [12]. We conducted a series of 1000 trials of the algorithm, and the ratio of states $|11\rangle$ to $|01\rangle$ was approximately 1/9.

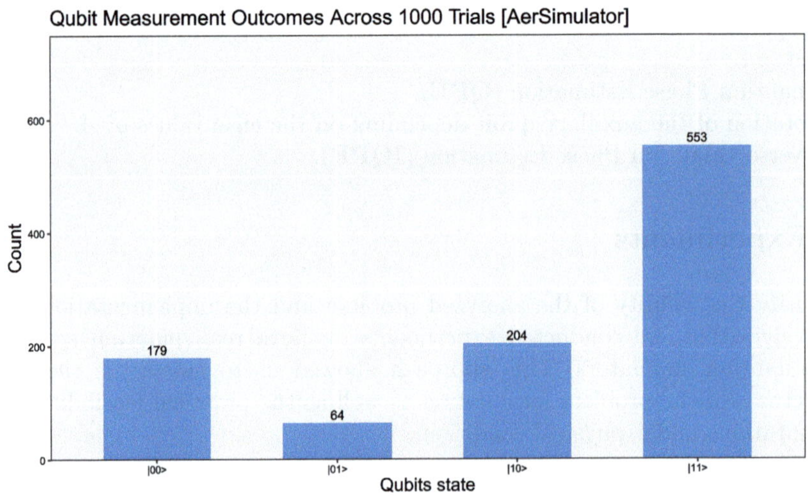

Fig. 1. Barplot of qubit measurements acquired via 1000 trials of HHL, performed on the AerSimulator.

These results demonstrate the consistency of the HHL algorithm in an ideal simulation environment. The ratio of $|01\rangle/|11\rangle = 0.116$ closely matches the theoretical prediction of 1/9, confirming that the algorithm correctly encodes the solution state $|x\rangle$. Small deviations observed in individual runs are attributable to statistical fluctuations due to the finite number of shots and the probabilistic nature of quantum measurements.

4.3 Test Performed on Odra 5 Quantum Computer

In the next step, we conducted 1000 shots on the 5-qubit Odra 5 quantum computer. The results deviated from the theoretical value of $1/9$. The measured ratio of the states was $|01\rangle/|11\rangle = 0.545$. The results are presented in Fig. 2. The implementation of the HHL algorithm for the system given in Eq. 4 requires four qubits: two clocks, one ancilla and one encoding. Efficient execution of the algorithm requires two-qubit interactions between the clock qubits and both the ancilla and the b qubits, as well as between the clock qubits themselves. This requirement poses a challenge on the Odra5 due to its limited qubit connectivity.

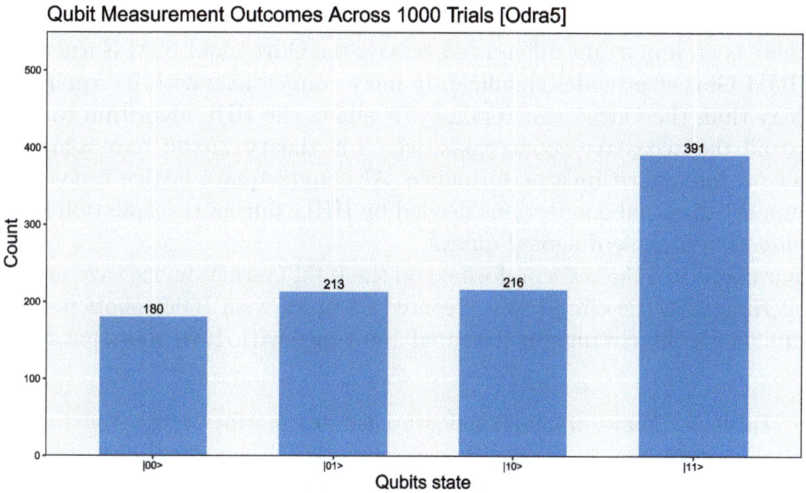

Fig. 2. Barplot of qubit measurements acquired via 1000 trials of HHL, performed on the Odra5.

The Odra5 quantum computer uses a star-shaped qubit topology in which a central qubit is connected to all peripheral qubits, whereas the peripheral qubits do not share direct connections between themselves. The topology is shown in Fig. 3. As a result of that, some required two-qubit interactions cannot be implemented directly between the necessary qubits. Consequently, the performance of the algorithm on the Odra5 is significantly degraded. This may explain why the results obtained on Odra5 differ from the results obtained on AerSimulator, where qubit connectivity constraints are not present (Fig. 4).

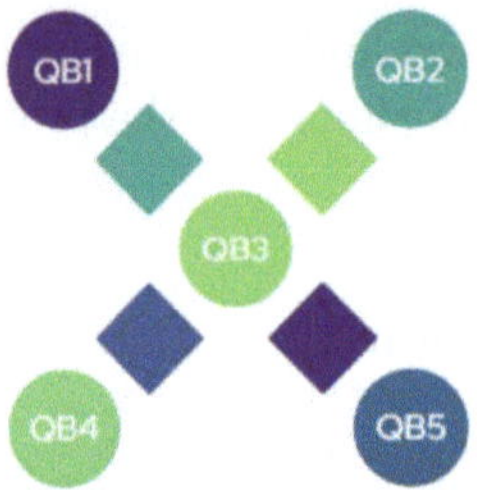

Fig. 3. Odra5 star qubit topology.

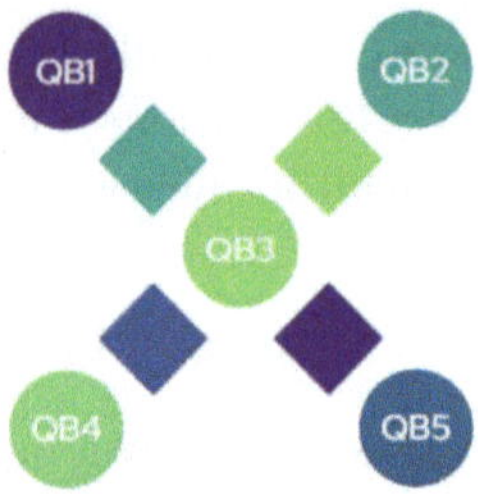

Fig. 4. IQM Garnet square lattice qubit topology.

4.4 IQM Garnet

The most extensive experiments were performed on the IQM Garnet quantum computer. We conducted two series of experiments and analyzed the data derived from them. The parameters of the experiment series are illustrated by Table 1.

There is an important difference between the Odra5 and IQM Garnet topologies. IQM Garnet provides significantly more connections with its square lattice topology than the Odra5 star topology. It allows the HHL algorithm to perform almost all the necessary connections, which in theory should have a significant impact on the algorithm's performance. Although square lattice topology does not support diagonal connections needed by HHL, during transpilation it is able to indirectly entangle diagonal qubits.

The parameters of the tests performed on the IQM Garnet device were as follows. In experiment A, the circuit was executed 50 times with 10000 shots per run. In experiment B, the circuit was executed 150 times with 1000 shots per run.

Table 1. Counts of qubit configurations used in experiments A and B.

Configuration	Experiment A	Experiment B
0	1	18
1	1	2
2	2	5
3	1	5
4	1	5
5	4	4
6	20	6
7	16	4
8	4	48
9	–	51
10	–	2

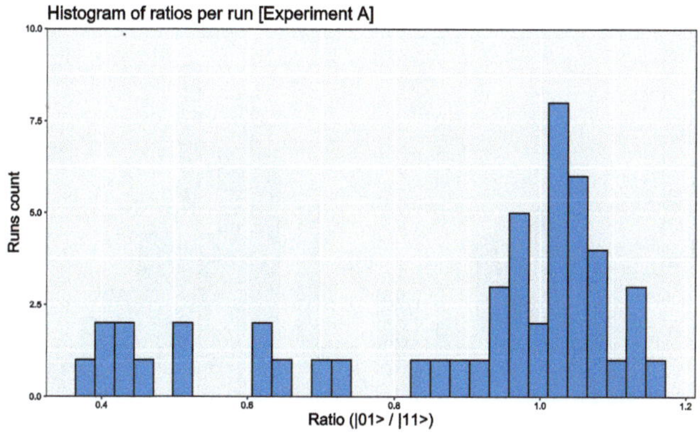

Fig. 5. Histogram for values of $|01\rangle/|11\rangle$ ratios, which resulted from Experiment A, for runs count = 50 and shots count = 10000.

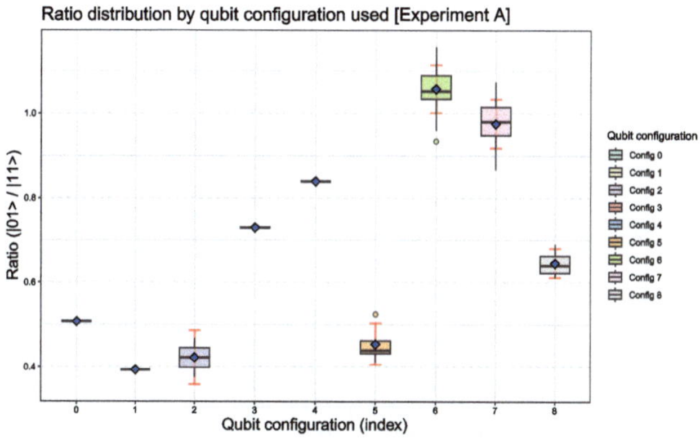

Fig. 6. Plots display the following metrics for each qubit configuration: mean ratio of $|01\rangle/|11\rangle$ indicated by blue rectangle, standard deviation of the ratio shown as red error bars. The qubit sets used are indicated on the X-axis. Parameters: 50 runs, 10000 shots per run.

The distribution of the ratio $|01\rangle/|11\rangle$ obtained from the results is shown in Fig. 1 and Fig. 7, for Experiments A and B respectively. Since IQM Garnet has 20 qubits, it allows for many possible mappings of our circuit to physical architecture. We checked how the physical set of qubits assigned to our algorithm changes the accuracy, which can be seen in Fig. 6, Fig. 8.

Both experiments resulted in the distribution of the ratios $|01\rangle/|11\rangle$ heavily concentrated around the value of 1. While Fig. 5 shows that the vast majority of the distribution mass lies around the value 1, Fig. 6 shows a more uniform distribution of the ratio $|01\rangle/|11\rangle$. The results diverge greatly from the expected ratio around $0.(1)$ and the results obtained on AerSimulator. This indicates a lack of significant advantage over Odra5, which could have been anticipated as the

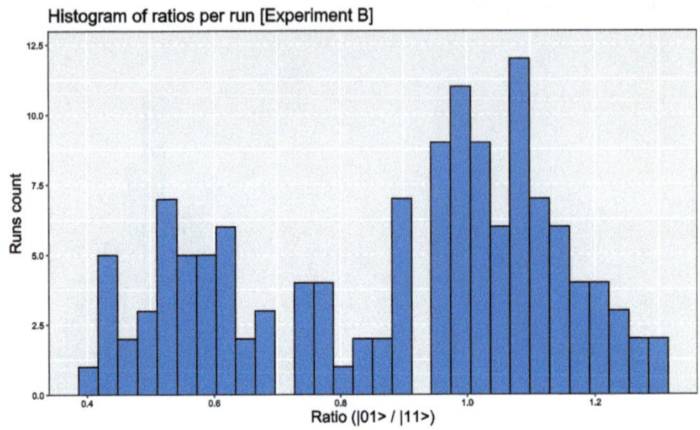

Fig. 7. Histogram for values of $|01\rangle/|11\rangle$ ratios, which resulted from Experiment B, for runs count = 150 and shots count = 1000.

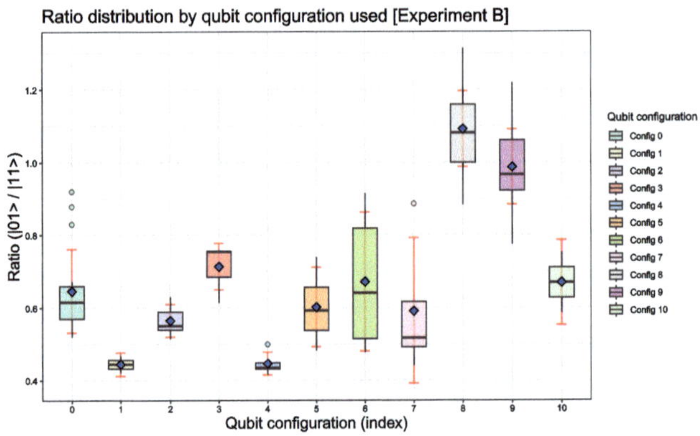

Fig. 8. Plots display the following metrics for each qubit configuration: mean ratio of $|01\rangle/|11\rangle$ indicated by blue square, standard deviation of the ratio shown as red error bars. The qubit sets used are indicated on the X-axis. Parameters: 150 runs, 10000 shots per run.

result of a higher number of connections between Garnet qubits. The reason may be the increased noise accumulation resulting from the introduction of indirect entanglement in the circuit. The influence of the physical qubit configuration to which the algorithm was mapped was also investigated. The results show that the layouts preferred by the transpiler are those with the highest average ratio $|01\rangle/|11\rangle$. It can be concluded that, for this algorithm, the default selection of qubits is not optimal. Neither IQM Garnet nor Odra5 results came close to the theoretically expected ratio 1/9.

5 Conclusion and Future Work

This paper presents an experimental study of the Harrow–Hassidim–Lloyd (HHL) algorithm implemented on both simulated and real quantum computing platforms. We have shown that while the algorithm performs as expected in a noise-free simulation environment, its execution on current quantum hardware leads to significant differences from the theoretical results.

Computational experiments conducted on the IQM quantum computer and the Odra 5 system indicate that the obtained results are strongly affected by hardware noise, limited qubit connectivity. In particular, the topology of the Odra 5 quantum computer restricts direct implementation of the HHL algorithm, further degrading result quality. On current quantum hardware, such a large number of operations leads to substantial decoherence and gate fidelity issues, where the quantum state degrades before the calculation is completed. Overall, the experiments demonstrate that while the HHL algorithm is theoretically effective for solving complex linear operations as can be seen in the results from the AerSimulator, its practical implementation on current quantum devices is still limited due to the high circuit noise and limited qubit connectivity, as observed in both experiments on IQM Garnet and Odra5. Future work could focus on hybrid quantum-classical approaches and their performance on NISQ devices. Additionally, optimization of the transpiling quantum procedure could be explored to decrease circuit depth and the bottlenecks encountered as a result.

References

1. Childs, A.M., Kothari, R., Somma, R.D.: Quantum algorithm for systems of linear equations with exponentially improved dependence on precision. SIAM J. Comput. **46**(6), 1920–1950 (2017)
2. Harrow, A.W., Hassidim, A., Lloyd, S.: Quantum algorithm for linear systems of equations. Phys. Rev. Lett. **103**(15) (2009)
3. Figueroa-Romero, P., Papič, M., Auer, A., de Vega, I.: Estimating the coherence of noise in mid-scale quantum systems (2024)
4. Abdurakhimov, L., et al.: Technology and performance benchmarks of IQM's 20-qubit quantum computer (2024)
5. Biamonte, J., Wittek, P., Pancotti, N., Rebentrost, P., Wiebe, N., Lloyd, S.: Quantum machine learning. Nature **549**, 195–202 (2017)
6. Preskill, J.: Quantum computing in the NISQ era and beyond. Quantum **2**, 79 (2018)
7. Barz, S., et al.: A two-qubit photonic quantum processor and its application to solving systems of linear equations. Sci. Rep. **4**, 6115 (2014)
8. Cai, X.D., et al.: Experimental quantum computing to solve systems of linear equations. Phys. Rev. Lett. **110**(23), 230501 (2013)
9. Lee, Y., Joo, J., Lee, S.: Hybrid quantum linear equation algorithm and its experimental test on IBM quantum experience. Sci. Rep. **9**(1) (2019)
10. Gao, F., Wu, G., Yang, M., Cui, W., Shuang, F.: A hybrid algorithm to solve linear systems of equations with limited qubit resources. Quant. Inf. Process. **21**(3), 1–31 (2022). https://doi.org/10.1007/s11128-021-03388-3

11. Sambhaje, V., Chaurasia, A.: The HHL algorithm: Implementation and research directions. Int. J. Quant. Inf. **22**(06), 2450037 (2024)
12. Zaman, A., Morrell, H.J., Wong, H.Y.: A step-by-step HHL algorithm walkthrough to enhance understanding of critical quantum computing concepts. IEEE Access **11**, 77117–77131 (2023)
13. Imam, D., Riaz, A.: HHL algorithm for linear systems of equations. Technical report, Introduction to Quantum Information PHY 612 (2023)

Federated SOC Integration for Enhancing SME Resilience in Critical Infrastructure Supply Chains

Magdalena Podrez-Raczycka(✉) and Raynold Sylvain

Smartech IT sp. z o.o. Wrocław, Wrocław, Poland
{biuro,rsylvain}@smartech-IT.eu

Abstract. This paper examines whether integrating small and medium-sized enterprises (SMEs) into a federated Security Operations Center (SOC) architecture can strengthen the resilience of critical infrastructure supply chains. The study addresses a structural weakness common in SMEs: reactive cybersecurity models characterized by fragmented monitoring, weak incident-response formalization, and limited analytical capacity. A conceptual-analytical approach is adopted. The paper proposes a federated model combining centralized event correlation (SIEM), response orchestration (SOAR), behavioral analytics, and standardized incident-response playbooks, and assesses it through analytical case scenarios reflecting typical SME roles in supply-chain ecosystems. The analysis indicates that federated SOC integration can reduce Mean Time to Detect (MTTD) and Mean Time to Respond (MTTR), limiting the probability of cascading cyber incidents across interconnected entities. The paper argues that resilience in critical infrastructure supply chains should be treated as an emergent property of coordinated organizational ecosystems rather than as a set of isolated controls across supply-chain ecosystems.

Keywords: Critical infrastructure · Supply chain · SMEs · Security Operations Center · Federated SOC · Cyber resilience · Incident response · SIEM · SOAR · Cascading risk

1 Introduction

The digitization of economic processes has produced tightly interconnected ecosystems in which the boundaries between critical infrastructure and commercial entities are increasingly blurred. Supply chains depend on shared digital platforms, external IT providers, cloud environments, and cross-border integrations. In such conditions, cyber risk is systemic rather than point-based: an incident affecting one entity may disrupt a broader network of interdependent organizations [1–3].

Recent supply-chain incidents show that organizations with relatively low cybersecurity maturity often serve as entry vectors into larger structures. SMEs acting as IT providers, subcontractors, software vendors, or operators of auxiliary systems therefore represent a structurally important risk link. Their limited resources, weak continuous

W. Zamojski et al. (Eds.): DepCoS-RELCOMEX 2026, LNNS 2014, pp. 353–361, 2026.
https://doi.org/10.1007/978-3-032-27927-9_34

monitoring, and low formalization of incident response (IR) processes often sustain a reactive and perimeter-oriented protection model.

Preventive controls alone are insufficient against automated, multi-stage attacks designed to exploit trust relationships within supply chains. Without early detection and structured containment, prolonged dwell time raises the probability of both local disruption and cross-entity propagation. In parallel, the NIS2 Directive reinforces the need to build resilience not only within designated critical infrastructure operators but also across their supply chains [4].

This paper presents a conceptual framework for integrating SMEs into a federated SOC architecture as a mechanism for strengthening critical-infrastructure supply-chain resilience. The central thesis is that the absence of a standardized integration model for SME monitoring and response sustains structural vulnerability, whereas a federated model based on SIEM, SOAR, behavioral analytics, and standardized playbooks can reduce cascading risk and shorten detection and response times.

The paper follows a conceptual-analytical methodology supported by case-based scenarios reflecting typical SME roles in supply-chain ecosystems. It reviews the evolution of threat models, examines the reactive security model prevalent in SMEs, proposes a federated SOC reference model, and evaluates its implications for systemic resilience.

2 Evolution of Threats in Supply Chains

2.1 Research Objective and Methodology

The objective of this paper is to assess whether integrating SMEs into a federated SOC architecture can improve incident detection and response efficiency and thereby reduce cascading cyber risk in critical infrastructure supply chains.

The study follows a conceptual-analytical methodology. It identifies structural limitations of the reactive cybersecurity model commonly observed in SMEs, proposes a federated SOC integration model based on centralized event correlation, response orchestration, behavioral analytics, and standardized incident-response procedures, and uses analytical case scenarios to compare reactive and federated response conditions. The evaluation relies on Mean Time to Detect (MTTD) and Mean Time to Respond (MTTR) as proxy measures of resilience. The paper does not use primary empirical data.

2.2 From Direct Attacks to the Cascading Model

Early cybersecurity models focused on direct attacks against high-value targets and emphasized perimeter defenses, segmentation, and endpoint hardening. Over the past decade, however, attackers have increasingly exploited dependency structures among organizations. Instead of attacking the most fortified target directly, adversaries identify weaker links—often SMEs—that maintain trusted technical or contractual relationships with the intended target.

Within this cascading model, the initial breach may appear limited in scope, yet its consequences propagate through software-update mechanisms, API integrations, remote-administration channels, managed service connections, and shared cloud infrastructures. Risk therefore becomes relational rather than isolated, and the resilience of the ecosystem depends on the maturity of each connected organization.

2.3 Characteristics of Contemporary Supply Chain Attacks

Contemporary supply-chain attacks rely on relational trust, prolonged dwell time, and automation. Attackers exploit legitimate technical relationships such as managed service provider access, remote monitoring and management tools, software-update channels, or shared authentication mechanisms. They often combine reconnaissance, privilege escalation, credential harvesting, and lateral movement before executing the final payload. Malicious updates, scripted lateral movement, and automated ransomware deployment then allow rapid scaling across multiple entities [1–3, 5].

2.4 SMEs as a Systemic Vector

SMEs play a central role in supply chains as specialized technology providers, software developers, system integrators, subcontractors, and operators of auxiliary services. Yet many operate under structural cybersecurity constraints, including the absence of continuous monitoring, limited multi-source event correlation, reduced access to advanced response expertise, and insufficient formalization of escalation procedures.

Under such conditions, SMEs may function as pivot points within supply-chain attacks. Compromise of an SME can provide attackers with legitimate communication channels and authenticated access paths to a larger partner. This highlights a core systemic property: the resilience of the supply chain cannot exceed that of its weakest operationally connected component.

3 Limitations of the Reactive Protection Model in SMEs

3.1 Fragmented Monitoring and Lack of Operational Visibility

One of the fundamental limitations of the reactive protection model observed in SMEs is fragmented monitoring. Security telemetry—such as endpoint logs, server events, firewall records, authentication data, application logs, and cloud signals—is often dispersed across heterogeneous tools or retained locally without centralized aggregation.

This lack of operational visibility constrains situational awareness, prevents early detection, and limits proactive security functions such as threat hunting and anomaly analysis. In supply-chain contexts, fragmented monitoring does not merely increase individual organizational risk; it weakens the ecosystem's collective ability to detect threats before they propagate.

3.2 Low Formalization of Incident Response Processes

A further structural weakness concerns the insufficient formalization of incident-response (IR) processes. In many SMEs, responsibilities related to incident handling are unclear, escalation paths are informal, and documentation is inconsistent. When an incident occurs, the response is therefore often improvised rather than playbook driven.

Such ad hoc approaches increase response latency, reduce repeatability, and weaken organizational learning. They also complicate cooperation with external stakeholders, including supply-chain partners and regulatory authorities. In regulated environments, weak formalization may hinder compliance with notification and reporting obligations [6–8].

3.3 Lack of Integration with Supply Chain Security Architecture

Even where primary contractors or critical infrastructure operators maintain mature cybersecurity capabilities, the absence of operational integration with SME suppliers creates persistent blind spots. If monitoring, detection logic, and incident-reporting mechanisms remain isolated within each organization, ecosystem-wide visibility is fragmented and response actions are not coordinated across contractual relationships.

These weaknesses are structural rather than merely technological. Fragmented monitoring, insufficient process formalization, resource constraints, and lack of systemic integration collectively sustain supply-chain vulnerability and justify a shift from isolated protection toward federated integration within a shared monitoring and response architecture.

4 Systemic Model—Integrating SMEs into a SOC Architecture

4.1 From an Isolated to a Federated Model

The federated model proposed here assumes that SMEs retain ownership and operational control over their infrastructure while integrating selected monitoring and response functions into a shared Security Operations Center (SOC) architecture. Federation does not imply full centralization of systems or data; rather, it establishes standardized interfaces, shared detection logic, and coordinated response mechanisms across participating entities [9].

Under this model, SMEs become nodes in a distributed analytical ecosystem. Security telemetry is aggregated and analyzed within a centralized correlation layer, while containment and remediation actions may be executed locally in accordance with predefined playbooks. This structure enables collective visibility without eliminating organizational autonomy.

4.2 SIEM Layer—Central Event Correlation

At the core of the federated architecture lies a Security Information and Event Management (SIEM) system responsible for aggregating, normalizing, and correlating security events from multiple sources. Participating SMEs forward selected telemetry—such as authentication logs, endpoint events, network-device logs, cloud-activity records, and application signals—to the central analytical layer.

The SIEM applies correlation rules, pattern recognition, and contextual enrichment to identify multi-stage attack behaviors that would remain undetected in isolated logs. Weak signals observed in one organization may acquire significance when correlated with similar signals across the ecosystem, improving triage, classification, and evidence-based escalation.

4.3 SOAR Layer—Response Orchestration and Automation

While detection is essential, resilience depends equally on the speed and consistency of response. The Security Orchestration, Automation, and Response (SOAR) layer provides structured operationalization of incident handling [10]. A formalized playbook defines triggering conditions, contextual enrichment procedures, containment measures, escalation paths, and documentation requirements.

Automation may support selected stages of the process, while governance mechanisms ensure that high-impact interventions remain subject to expert validation. By reducing the interval between detection and containment, the SOAR layer directly lowers MTTR and narrows the window for adversarial propagation.

4.4 Threat Intelligence and Information Sharing

An additional pillar of the federated architecture is the integration of threat intelligence and structured information sharing. External intelligence feeds providing indicators of compromise, adversary-infrastructure data, and information on tactics, techniques, and procedures can be centrally ingested and translated into detection logic applicable across participating SMEs.

Moreover, intelligence derived from incidents observed within one SME may be converted into generalized detection rules and disseminated in anonymized form to other participants. When governed properly, this mechanism turns localized incidents into ecosystem-wide learning opportunities without compromising organizational autonomy.

4.5 Comparison with Alternative Security Models

The federated SOC model differs both from isolated SME environments, where monitoring and response remain fragmented, and from fully centralized models, which may provide strong analytical capabilities but introduce dependency and concentration risk. The federated approach represents a compromise: it enables shared detection and coordinated response while preserving the operational autonomy of participating SMEs.

5 Advanced and Predictive Analytics in a SOC Architecture

5.1 From Signature-Based Detection to Behavioral Analytics

Traditional detection mechanisms in many SME environments remain predominantly signature based. Such methods are useful against known malware and previously catalogued indicators of compromise, but contemporary supply-chain attacks increasingly rely on techniques that do not immediately generate recognizable signatures. Credential abuse, misuse of legitimate administrative tools, and living-off-the-land strategies often appear technically legitimate when assessed in isolation.

Behavioral analytics, including User and Entity Behavior Analytics (UEBA), addresses this limitation by focusing on deviations from established baselines of normal activity. Within a federated SOC architecture, aggregated telemetry from multiple SMEs enables more robust baseline modeling and earlier detection of cross-entity patterns that may not be visible within a single organizational context [11–13].

5.2 Use of Machine Learning Algorithms

Machine learning methods can enhance detection by supporting anomaly identification, clustering of related events, and alert prioritization [10–12]. However, model performance depends on data quality, feature engineering, and continuous recalibration. Poorly tuned models may generate excessive false positives and contribute to alert fatigue. Predictive analytics should therefore complement—not replace—expert-driven analysis.

5.3 Reducing MTTD and MTTR as an Effectiveness Metric

The effectiveness of federated SOC integration should be evaluated primarily through Mean Time to Detect (MTTD) and Mean Time to Respond (MTTR). In reactive environments, detection often occurs only after observable service degradation or ransomware activation. By contrast, centralized correlation, behavioral analytics, and orchestrated response shorten both detection and containment intervals.

From a systemic perspective, reducing MTTD and MTTR within SMEs produces multiplier effects across the supply chain. Earlier detection in one node reduces the probability of propagation to interconnected entities and thereby lowers ecosystem-level cascading risk.

5.4 Limitations and Risks of Over-Automation

Although automation improves efficiency, excessive reliance on automated response may introduce new risks. Actions such as account suspension or network isolation may disrupt legitimate business processes if triggered inaccurately. Advanced analytics should therefore operate within a governance framework that combines automation with human oversight.

6 Standardized Incident Response and Interoperability

In federated security architectures, technological integration must be accompanied by procedural coherence. Participating SMEs require a shared incident lifecycle, harmonized escalation logic, and standardized operational playbooks to ensure that centrally detected signals can be translated into consistent local response actions.

Such alignment supports comparability across the ecosystem and facilitates compliance with NIS2 and ISO/IEC 27001-and ISO/IEC 27035-based incident-management approaches [8, 14]. From a systemic perspective, standardized incident response is a condition of interoperability: without it, centralized monitoring cannot produce consistent resilience outcomes across interconnected organizations.

7 Case-Based Validation of the Federated SOC Model

Three analytical scenarios were considered: an SME supporting healthcare systems, an IT service provider for public administration, and a manufacturing SME operating in an IT/OT convergence environment. In each case, the reactive baseline is characterized by fragmented monitoring, delayed detection, and improvised response, whereas the federated configuration introduces centralized correlation, behavioral analytics, and playbook-driven containment.

Across all scenarios, federated integration shortens detection and response time, reduces dwell time, and limits lateral propagation across trusted supply-chain relationships. Standardized response procedures further improve consistency during cross-organizational incidents.

8 Systemic Impact on Critical Infrastructure Supply Chain Resilience

8.1 Reducing Cascading Risk

In contemporary digital ecosystems, cyber risk propagates through structural interdependencies rather than remaining confined within organizational boundaries. Federated SOC integration addresses this asymmetry by shortening detection and response times in SMEs that might otherwise function as entry vectors for cascading attacks. Earlier identification of anomalous behavior reduces the time window during which adversaries can escalate privileges, establish persistence, and move laterally across trusted connections.

From a systemic perspective, the reduction of dwell time in a single SME produces multiplier effects across the supply chain. The probability of cross-entity propagation decreases, thereby stabilizing not only the directly affected organization but also its contractual partners.

8.2 Stabilizing Critical Infrastructure by Strengthening the Contractual Perimeter

Critical infrastructure entities typically rely on extensive networks of suppliers, subcontractors, and managed service providers. Contractual provisions alone cannot mitigate operational cyber risk. Without shared visibility and coordinated response capabilities, the contractual perimeter remains technically porous.

Federated SOC integration extends monitoring and response capacity beyond the formal organizational boundary of critical infrastructure operators. By incorporating SMEs into a shared analytical and operational framework, the effective security perimeter expands to include key supply-chain participants.

8.3 From Point Protection to Systemic Resilience

The cumulative effect of federated integration is a shift from isolated protection toward systemic resilience. In contrast to models that treat security as a purely internal function, federated architectures recognize resilience as an emergent property arising from interaction among interconnected entities. Visibility, standardized response, shared intelligence, and coordinated metrics collectively generate defensive capacity that exceeds the sum of individual organizational capabilities.

9 Limitations and Further Research

The model presented here remains conceptual and analytical and does not rely on primary quantitative data. Future research should measure reductions in Mean Time to Detect, Mean Time to Respond, and cross-entity propagation risk. Implementation constraints include financial barriers, onboarding complexity, and data-governance requirements.

10 Conclusions

This paper has argued that the persistence of reactive and isolated cybersecurity models in SMEs creates structural exposure in critical infrastructure supply chains. The proposed federated SOC integration model addresses this problem by combining centralized event correlation, orchestration of response, behavioral analytics, and standardized playbooks.

The analytical assessment indicates that such integration can reduce Mean Time to Detect and Mean Time to Respond, thereby limiting lateral propagation across contractual and technological dependencies. The contribution of the paper is primarily conceptual and operational: it frames resilience not as a property of isolated organizations, but as an emergent property of coordinated ecosystems and treats SME integration into federated SOC structures as a practical mechanism for strengthening supply-chain resilience around critical infrastructure.

References

1. European Union Agency for cybersecurity (ENISA), ENISA threat Landscape (2023)
2. Verizon , Data breach investigations report (DBIR), Verizon Business, (2024)
3. Google Cloud (Mandiant), M-Trends 2024 Special Report (Executive Edition), (2024)
4. European parliament and the council of the european union, directive (EU) 2022/2555 of 14 December 2022 on measures for a high common level of cybersecurity across the Union (NIS2). Off. J. Eur. Union, L 333, 27 (2022)
5. MITRE Corporation, MITRE ATT&CK Knowledge Base: Tactics, Techniques, and Procedures, ongoing framework
6. National Institute of Standards and Technology (NIST), the NIST Cybersecurity Framework (CSF) 2.0, NIST CSWP 29, (2024)
7. National Institute of Standards and Technology (NIST), Cybersecurity Supply Chain Risk Management Practices for Systems and Organizations, NIST SP 800–161 Rev. 1, (2022)
8. International Organization for Standardization (ISO), ISO/IEC 27035–1:2023—Information security incident management -Part 1: Concepts and principles, (2023)
9. I. Taqafi et al.: A maturity capability framework for security operation centers. (2022)
10. Ismail, A., et al.: Toward robust security orchestration and automated response (soar): a hyper-automation approach. Information, MDPI (2025)
11. Feng, W., et al.: Multi-granularity user anomalous behavior detection. Appl. Sci. (2024)

12. Chen, X., et al.: A-UEBA: an anomaly heterogeneity learning framework for user and entity behavior analytics. ACM Proceedings (2025)
13. Landauer M, Skopik F et al.: A user and entity behavior analytics log data set for anomaly detection in cloud computing, (2022)
14. International Organization for Standardization (ISO), ISO/IEC 27001:2022—Information security management systems—Requirements, (2022)

Cervical Cancer Classification Based on Graph–Extracted Features

Jan Poręba and Łukasz Jeleń(✉)

Faculty of Information and Communication Technology, Wrocław University of Science and Technology, Wybrzeże Wyspiańskiego 27, 50-370 Wrocław, Poland
268446@student.pwr.edu.pl, lukasz.jelen@pwr.edu.pl

Abstract. Cervical cancer remains one of the most common cancers among women worldwide. To overcome this problem, the key is to diagnose in as early a stage as possible. Many recent research has proposed deep learning approaches that offer high performance, but often suffer from interpretability limitations. In this study, we propose an application of graph-based features to classify cervical cancer cells in Liquid Based Cytology images. We first use image segmentation to extract geometric cell features, and then we use graph-based features for image classification using a variety of classifiers. For feature extraction, we evaluated the contour graph and Delaunay graph. The results show that the 50 layer multilayer perceptron achieved the highest accuracy, reaching 0.92, closely followed by kNN with an accuracy of 0.91. The presented results show a promising path toward an automated human-level diagnostic system and that inclusion of graph methods provides an essential insight to cancerous cell description.

Keywords: Cervical cancer classification · Liquid-based cytology · Machine learning · Graphs-based features · Topological descriptors · CAD

1 Introduction

Cervical cancer remains one of the most common cancers among women worldwide, including in Poland. Although survival rates have improved in recent years, outcomes remain heavily dependent on the clinical stage at the time of diagnosis, making early detection very important. Unfortunately, outcomes in Poland are less favorable than global averages, largely due to delayed diagnoses [1,3]. For that reason, there is an urgent need to validate new diagnostic tools that can bridge this gap.

Since the introduction of cytology-based screening, cervical cancer mortality rates have decreased significantly [4]. Today, together with vonventional Papanicolaou smears (CPS), liquid-based cytology (LBC) is currently the most effective and widely used screening method [2]. However, LBC requires careful interpretation by expert pathologists to analyze the cellular morphology [9]. To assist this

W. Zamojski et al. (Eds.): DepCoS-RELCOMEX 2026, LNNS 2014, pp. 362–371, 2026.
https://doi.org/10.1007/978-3-032-27927-9_35

process, automated classification methods like Convolutional Neural Networks have been studied. These deep learning models are highly effective at processing grid-structured data by automating complex preprocessing steps, such as pixel-level cell segmentation and feature extraction, which were previously performed manually [9,10]. Despite the high performance of deep learning, recent research has sought to reduce computational complexity and enhance the interpretability of results. Jelen *et al.* proposed a shift toward extracting specific geometric features—such as nucleus size, cell-to-nucleus ratios, and diameter—to guide classification [8]. The next evolutionary step in these methods could be an application of graph theory. Unlike standard image processing, graphs represent the spatial and topological relationships between fragments of an image that can be helpful in the interpretation of machine learning model decisions [15]. In the literature, several authors have used graph models to encode cellular information into structures such as Delaunay graphs or Minimum Spanning Trees (MST) for better result interpretation [11]. Furthermore, other graph models, including region adjacency graphs, K-nearest neighbor (KNN) graphs, and geometric graphs, have already demonstrated potential in representing complex object shapes [14,18].

The aim of this study is to fill the gap between traditional diagnostic methods and novel approaches based on structure graphs, such as Delaunay and MST graphs. In this study, we provide a comparative analysis of image classification methods, evaluating the performance of extracted geometric features against graph-based techniques. Integration of graph theory and oncology marks a sophisticated, transformative, and highly promising leap forward for the future of automated pathology.

2 Materials and Methods

In this section, the research methodology is presented. We propose a modification of the methodology described in [8] by introducing a novel application of graph-based algorithms in liquid based cytology. While the original method relies on image segmentation to extract geometric cell features, our approach utilizes these segmentation outputs as a foundation for constructing structural graphs. Then we use graph-theoretic features and measures for image classification using Decision Trees, Random Forests, and Support Vector Machines (SVMs) classifiers. In particular, we evaluated two representations, namely Contour Graphs and Delaunay Graphs. The proposed pipeline consists of image preprocessing, graph construction, feature extraction, and classification with k-fold cross-validation. The subsequent sections describe each step in detail.

2.1 Database

For this study, we used a dataset of liquid based cytology (LBC) smears (courtesy of prof. Maria Chosia and prof. Michał Jeleń) acquired from the Pomeranian Medical University in Poland. After preparation, the smears were digitized at a resolution of 2510 × 1540 px and 300 dpi at the Medical University of Wrocław using an Olympus BX 51 microscope using CellD software. We collected 263 digitized LBC images, of which 158 are Normal Squamous Intraepithelial Lesion (NSIL), 70 are High-Grade Squamous Intraepithelial Lesion (HSIL), and 35 are Low-Grade Squamous Intraepithelial Lesion (LSIL). From these images, 428 individual cell images were extracted (extracted by Stankiewicz I. in [17]) giving 233 NSIL, 110 HSIL, and 85 LSIL images, respectively. These images were used for cell shape analysis and were validated by an expert pathomorphologist and underwent irreversible anonymization to meet ethical standards for retrospective research.

2.2 Preprocessing and Graph Construction Pipeline

To construct the graphs used in this study, the images were preprocessed to ensure that the cellular components were correctly segmented and the resulting graph nodes represent meaningful biological features. The proposed pipeline consists of the following five steps:

1. Segmentation – to segment the nucleus and cytoplasm, the k–means clustering algorithm was applied to the raw RGB single-cell images. The algorithm uses RGB color distance between a pixel and cluster means a distance measure grouping the pixels into 3 regions: background, nucleus, and cytoplasm.
2. Noise Reduction and Contour Extraction – to smooth the regions and reduce noise, a median filter is applied to the segmented images. This ensures cleaner boundaries for subsequent steps. The filtered image is then used for nucleus edge detection.
3. Keypoints Extraction – to define graph nodes, the Harris Corner detection was applied to find significant corner points along the nuclear contours. At the same time, geometric raw moments were calculated to determine the centroid shape.
4. Graph Construction – Based on the extracted points, the centroid and centroid were treated as vertices for the construction of the Contour Graph and the Delaunay Graph (see Sect. 2.3 and 2.4).

2.3 Contour Graph

With the extraction of the key points, a set of corner points $P_i = (x_i, y_i)$ for $i = 1, \ldots, n$ and the centroid $M = (x_m, y_m)$ was obtained and used to derive the graph nodes $G_C = (V_C, E_C)$. The set of vertices V_C consists of all corner points and the centroid, while the set E_C contains edges (e) between adjacent corner points and edges connecting each corner point to M. The Contour Graph is formally defined as:

$$\begin{aligned} V_C &= \{P_1, \ldots, P_n\} \cup \{M\}, \\ E_C &= \{(P_i, P_{(i \bmod n)+1}) \mid i = 1, \ldots, n\} \cup \{(P_i, M) \mid i = 1, \ldots, n\}. \end{aligned} \tag{1}$$

Edge weights $w(e)$ were calculated as Euclidean distances between the vertices. Defining edges along the nuclear boundary and toward the centroid allows the graph to capture both the perimeter shape and the nucleus radial symmetry. Based on G_C we derive the Minimum Spanning Tree (T_{MST}) and calculated the following seven graph features [7]:

1. **Total MST Cost** calculated as $MST_{cost} = \sum_{e \in T_{MST}} w(e)$
2. **Edge Weight Statistics** that are used to quantify structural regularity. Here, we calculate:
   - the average branch length of T_{MST}: $AVG_{branch\_len} = \frac{MST_{cost}}{|T_{MST}|}$,
   - variance of branch lengths: $\sigma^2 = \frac{1}{n-1} \sum_{e \in T_{MST}} (cost(e) - AVG_{branch\_len})^2$
   - skewness of the tree edges: $\frac{1}{n} \sum_{e \in T_{MST}} \left(\frac{e - AVG_{branch\_len}}{\sigma}\right)^3$
   - kurtosis of the tree edges: $\frac{1}{n} \sum_{e \in T_{MST}} \left(\frac{e - AVG_{branch\_len}}{\sigma}\right)^4 - 3$
3. **Topological Metrics** that represent the longest path between any two nodes in the tree. Here, we identify the number of tree leaves $|\{v \in T_{MST} : deg(v) = 1\}|$ and the diameter of the MST as $\max_{u,\, v \in V(G)} d_{\mathrm{MST}(G)}(u, v)$.

This feature set should effectively capture the spatial distribution and morphological complexity of the cell nucleus for further cervical cancer classification [6,12].

2.4 Delaunay Graph

In addition to the features provided by the contour graph, we used the same set of corner points and the centroid extracted during the preprocessing stage to construct a Delaunay Graph. Compared to a contour graph, the Delaunay triangulation provides a more dense planar subdivision of the nuclear area. A Delaunay triangulation of a given set of points $V_C = \{P_1, \ldots, P_n\} \cup \{M\}$ is considered a maximal planar subdivision where the circumcircle of any triangle does not contain other points from P in its interior [5]. This means that no additional edges can be added without violating planarity.

The triangulation itself is built incrementally with a randomized algorithm, where for each point inserted into the existing triangulation $\mathcal{T}$, we identify the triangle containing it and subdivide it. To maintain the Delaunay property, the *LegalizeEdge* procedure was used [5]. The algorithm checks if the newly inserted point creates an illegal edges, that is, those where a vertex lies inside the circumcircle of an adjacent triangle (see Fig. 1). If $\overline{p_i p_j}$ becomes illegal, then we flip it to fix the Delaunay property.

From this triangulation, we constructed the Delaunay Graph where, according to O. Pina, the triangulation points become graph nodes [16]. The algorithm assumes that two nodes p_i and p_j are connected if there exists a triangle $p_i p_j p_k$ such that:

$$(p_i, p_j) \in E \iff (\forall p_l \in P)(p_l \notin circumcircle(p_i, p_j, p_k), \quad (2)$$

where $circumcircle(p_i, p_j, p_k)$ is a circle passing through points p_i, p_j, and p_k.

Based on the constructed graph, we extract ten cellular features:

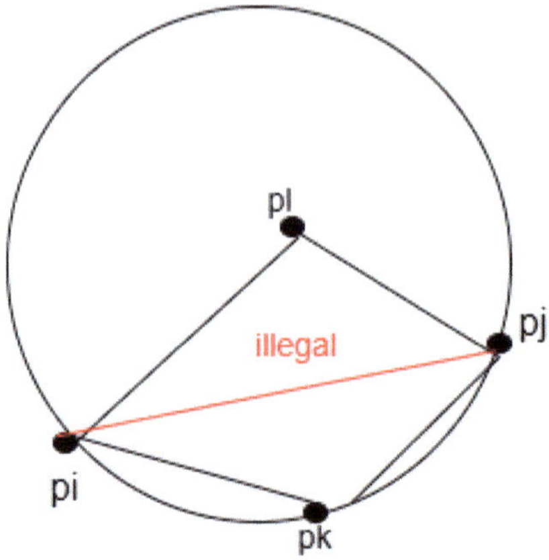

Fig. 1. An illegal edge in a Delaunay triangulation (after [5]).

1. **Geometry statistics** – here, we calculate the mean (μ) and variance (σ^2) for triangle areas A_i and edge lengths e_j.
2. **Regularity of the triangle shape** – these features are represented as the mean μ_s and variance σ_s^2 of the isopometric shape measure defined as:

$$s = \frac{4\sqrt{3}\,A}{e_1^2 + e_2^2 + e_3^2} \tag{3}$$

3. **Connectivity Metrics** calculated as the mean μ_d and variance σ_d of the vertex degrees $\deg(v_k)$
4. **Density Ratio – peri-to-core** – measures the concentration of points around the nucleus by comparing the number of vertices closer than the average distance to the centroid with those further away.
5. **Area–distance Correlation** – is determined as a Pearson correlation between triangle areas and the distance of their centroids to the nucleus center (see Eq. 4).

$$\rho_{A,r} = \frac{\sum_{i=1}^{n}(A_i - \mu_A)(r_i - \mu_r)}{\sqrt{\sum_{i=1}^{n}(A_i - \mu_A)^2}\,\sqrt{\sum_{i=1}^{n}(r_i - \mu_r)^2}} \tag{4}$$

Based on the features described in Sect. 2.3 and 2.4, we created a feature vector consisting of 17 graph-based features, respectively 7 from contour MST and 10 from Delaunay triangulation. The resulting feature vector captures both the skeletal structure and the surface distribution of the nucleus and is further used as an input for our classification models [13].

3 Results and Discussion

In our study, we implemented several supervised learning algorithms to evaluate the performance of the proposed graph-based features. The classifiers trained include Support Vector Machines (SVM) with Linear and RBF kernels, Random Forests (RF) with 10, 50 and 100 trees, Decision Trees (DT) with depth 5 and 10, Linear Regression, K-near neighbor (KNN) with $k = 3, 5$ and 7 and multi-layer perceptron with 10, 20 and 50 hidden layers. To ensure robust results, we performed repeated k-fold stratified cross–validation, which allowed us to assess how effectively these graph-based features distinguish between NSIL, LSIL, and HSIL cases.

Due to a severe class imbalance, we reduced the problem to binary classification in two separate tasks. The first task is devoted to cancer detection, where we classify HSIL vs Others (LSIL+NSIL). The second task is pathology detection, where normal cases (HSIL) are compared with abnormal (HSIL+LSIL). These scenarios are clinically valid as Task 1 can be treated as a diagnostic task and Task 2 could be applied as screening. To perform these classifications, all features were first normalized to the standard scalar ($\mu = 0, \sigma = 1$). The classification results are summarized in Tables 1, 2 and 3.

3.1 MST and Delaunay Graph Performance

In the first experiment, we tested the performance of the contour graph feature and Delaunay-based features alone. The results for the MST-based features are summarized in Table 1, while those of the Delaunay triangulation are shown in Table 2. The results show that Task 2 classification consistently yields higher accuracy than Task 1, indicating that both types of feature can resemble topological differences in healthy and modified cells. Furthermore, for MST-based features, the 50 layer MLP achieved the highest accuracy reaching 0.92, followed closely by kNN with an accuracy of 0.91. In contrast, HSIL detection (Task 1) obtained the highest accuracy of 0.86 with Random Forests, which likely benefited from the reduction of variance through ensemble learning.

Table 1. MST classification results for the two diagnostic tasks. Results are reported as mean ± standard deviation over cross-validation folds.

Model	Task 1 (Cancer Detection)				Task 2 (Pathology Detection)			
	Accuracy	Precision	Recall	F1	Accuracy	Precision	Recall	F1
kNN (k = 3)	0.82 ± 0.04	0.70 ± 0.07	0.71 ± 0.08	0.71 ± 0.06	0.89 ± 0.03	0.90 ± 0.05	0.86 ± 0.05	0.88 ± 0.03
kNN (k = 5)	0.84 ± 0.03	0.72 ± 0.06	0.75 ± 0.09	0.73 ± 0.06	0.90 ± 0.03	0.90 ± 0.05	0.87 ± 0.05	0.88 ± 0.03
kNN (k = 7)	0.84 ± 0.03	0.72 ± 0.06	0.76 ± 0.08	0.73 ± 0.06	0.90 ± 0.03	0.91 ± 0.05	0.87 ± 0.05	0.89 ± 0.03
Decision Tree (d = 5)	0.81 ± 0.03	0.65 ± 0.05	0.83 ± 0.10	0.72 ± 0.05	0.89 ± 0.03	0.89 ± 0.05	0.86 ± 0.05	0.87 ± 0.04
Decision Tree (d = 10)	0.81 ± 0.04	0.69 ± 0.07	0.69 ± 0.10	0.68 ± 0.08	0.87 ± 0.03	0.86 ± 0.05	0.85 ± 0.06	0.85 ± 0.04
Logistic Regression	0.78 ± 0.03	0.58 ± 0.04	0.91 ± 0.05	0.71 ± 0.04	0.86 ± 0.03	0.81 ± 0.05	0.90 ± 0.05	0.85 ± 0.03
Random Forest (10)	0.85 ± 0.03	0.79 ± 0.07	0.68 ± 0.08	0.73 ± 0.06	0.89 ± 0.03	0.90 ± 0.05	0.85 ± 0.06	0.87 ± 0.04
Random Forest (50)	0.86 ± 0.03	0.79 ± 0.06	0.72 ± 0.10	0.75 ± 0.07	0.90 ± 0.03	0.90 ± 0.05	0.86 ± 0.06	0.88 ± 0.04
Random Forest (100)	**0.86 ± 0.04**	**0.79 ± 0.08**	**0.73 ± 0.10**	**0.75 ± 0.08**	0.89 ± 0.03	0.89 ± 0.06	0.87 ± 0.05	0.88 ± 0.04
SVM (RBF)	0.79 ± 0.03	0.60 ± 0.04	0.92 ± 0.06	0.72 ± 0.04	0.88 ± 0.03	0.84 ± 0.06	0.91 ± 0.05	0.87 ± 0.03
SVM (Linear)	0.78 ± 0.03	0.59 ± 0.04	0.91 ± 0.06	0.71 ± 0.04	0.87 ± 0.03	0.83 ± 0.05	0.90 ± 0.05	0.86 ± 0.03
MLP (10)	0.84 ± 0.04	0.72 ± 0.08	0.74 ± 0.10	0.73 ± 0.08	0.91 ± 0.03	0.91 ± 0.05	0.89 ± 0.04	0.90 ± 0.03
MLP (20)	0.84 ± 0.04	0.72 ± 0.07	0.75 ± 0.11	0.73 ± 0.08	0.91 ± 0.03	0.92 ± 0.04	0.89 ± 0.05	0.90 ± 0.03
MLP (50)	0.84 ± 0.04	0.72 ± 0.07	0.75 ± 0.11	0.73 ± 0.08	**0.92 ± 0.02**	**0.92 ± 0.05**	**0.90 ± 0.03**	**0.91 ± 0.03**

From Table 2 we can note that Delaunay-based features showed similar trends, with Random Forests and MLP obtaining the highest accuracies. It should be noted that, while SVM and Logistic Regression underperformed in overall accuracy, they provided the highest recall (≥ 0.90), suggesting that these models are highly sensitive to catching positive cases but prone to false positives in graph-based topologies.

Table 2. Delaunay-based feature classification results for the two diagnostic tasks. Results are reported as mean ± standard deviation over cross-validation folds.

Model	Task 1 (Cancer Detection)				Task 2 (Pathology Detection)			
	Accuracy	Precision	Recall	F1	Accuracy	Precision	Recall	F1
kNN (k=3)	0.86 ± 0.03	0.77 ± 0.07	0.76 ± 0.08	0.76 ± 0.05	0.89 ± 0.03	0.88 ± 0.05	0.86 ± 0.05	0.87 ± 0.03
kNN (k=5)	0.85 ± 0.04	0.75 ± 0.07	0.77 ± 0.09	0.76 ± 0.06	0.90 ± 0.02	0.89 ± 0.04	0.87 ± 0.04	0.88 ± 0.03
kNN (k=7)	0.85 ± 0.03	0.75 ± 0.07	0.77 ± 0.08	0.76 ± 0.05	0.90 ± 0.02	0.89 ± 0.04	0.88 ± 0.04	0.88 ± 0.02
Decision Tree (d=5)	0.84 ± 0.04	0.72 ± 0.07	0.80 ± 0.07	0.75 ± 0.06	0.87 ± 0.04	0.86 ± 0.06	0.84 ± 0.07	0.85 ± 0.05
Decision Tree (d=10)	0.82 ± 0.04	0.71 ± 0.07	0.71 ± 0.08	0.71 ± 0.06	0.86 ± 0.03	0.85 ± 0.05	0.83 ± 0.07	0.84 ± 0.04
Logistic Regression	0.81 ± 0.03	0.63 ± 0.04	0.90 ± 0.06	0.74 ± 0.03	0.88 ± 0.03	0.85 ± 0.06	0.90 ± 0.04	0.87 ± 0.03
Random Forest (10)	0.87 ± 0.03	0.81 ± 0.07	0.71 ± 0.07	0.76 ± 0.06	0.90 ± 0.03	0.92 ± 0.04	0.85 ± 0.06	0.88 ± 0.04
Random Forest (50)	0.87 ± 0.03	0.81 ± 0.08	0.73 ± 0.07	0.77 ± 0.06	0.91 ± 0.02	0.93 ± 0.04	0.86 ± 0.04	0.89 ± 0.03
Random Forest (100)	**0.87 ± 0.03**	**0.82 ± 0.08**	**0.74 ± 0.07**	**0.77 ± 0.06**	**0.91 ± 0.02**	**0.93 ± 0.04**	**0.86 ± 0.04**	**0.89 ± 0.03**
SVM (RBF)	0.82 ± 0.03	0.64 ± 0.04	0.88 ± 0.06	0.74 ± 0.04	0.89 ± 0.02	0.87 ± 0.04	0.89 ± 0.04	0.88 ± 0.03
SVM (Linear)	0.82 ± 0.03	0.64 ± 0.05	0.90 ± 0.06	0.75 ± 0.04	0.89 ± 0.02	0.87 ± 0.04	0.90 ± 0.04	0.88 ± 0.02
MLP (10)	0.85 ± 0.04	0.75 ± 0.07	0.77 ± 0.09	0.76 ± 0.06	0.90 ± 0.02	0.90 ± 0.04	0.88 ± 0.04	0.89 ± 0.03
MLP (20)	0.86 ± 0.04	0.77 ± 0.07	0.77 ± 0.10	0.77 ± 0.07	**0.91 ± 0.02**	0.90 ± 0.05	**0.89 ± 0.05**	**0.90 ± 0.03**
MLP (50)	0.87 ± 0.04	0.78 ± 0.08	0.78 ± 0.09	0.78 ± 0.07	0.91 ± 0.03	0.91 ± 0.05	0.89 ± 0.05	0.90 ± 0.03

3.2 Comparison with Geometric Features

To evaluate the relative effectiveness of the proposed graph-based descriptors, we benchmark our classification results against a traditional set of geometric features previously established in the literature [8]. This comparison allows us to evaluate the discriminative power of graph-based features to detect and classify cervical cancer cells in LBC imagery. Using the same experimental setup including the same normalization techniques and cross-validation folds, we applied the same suite of classifiers to the geometric feature set.

As shown in Table 3 the geometric features currently provide a higher overall accuracy (≈ 0.96). The performance gap likely results from the higher dimensionality of the geometric set (30 features) compared to our compact graph set of 17 features all together.

Our results show that the classification performance across tasks is stable. For Task 1, random forests with 100 trees achieved the highest accuracy of 0.96 ± 0.03. Unlike graph-based features, which struggled to differentiate HSIL from LSIL, geometric features maintained high precision (0.93) and recall (0.93). This suggests that geometric features capture finer morphological differences. The performance of task 2 shows that the nonlinear kernel of the SVM classifier achieves the best classification, reaching a similar precision but significantly better recall (0.96).

Table 3. Classification results using geometric features for the two diagnostic tasks. Results are reported as mean ± standard deviation over cross-validation folds.

Model	Task 1 (Cancer Detection)				Task 2 (Pathology Detection)			
	Accuracy	Precision	Recall	F1	Accuracy	Precision	Recall	F1
kNN (k=3)	0.91 ± 0.04	0.82 ± 0.09	0.91 ± 0.08	0.86 ± 0.07	0.94 ± 0.04	0.94 ± 0.05	0.93 ± 0.06	0.93 ± 0.04
kNN (k=5)	0.92 ± 0.04	0.84 ± 0.09	0.93 ± 0.07	0.88 ± 0.06	0.94 ± 0.03	0.94 ± 0.05	0.93 ± 0.06	0.94 ± 0.04
kNN (k=7)	0.93 ± 0.04	0.84 ± 0.09	0.95 ± 0.06	0.88 ± 0.06	0.95 ± 0.03	0.95 ± 0.05	0.93 ± 0.06	0.94 ± 0.04
Decision Tree (d=5)	0.93 ± 0.04	0.88 ± 0.09	0.91 ± 0.08	0.89 ± 0.07	0.94 ± 0.04	0.95 ± 0.05	0.92 ± 0.07	0.94 ± 0.04
Decision Tree (d=10)	0.93 ± 0.04	0.90 ± 0.09	0.89 ± 0.09	0.89 ± 0.07	0.94 ± 0.04	0.95 ± 0.05	0.92 ± 0.07	0.93 ± 0.05
Logistic Regression	0.93 ± 0.04	0.83 ± 0.09	0.96 ± 0.06	0.89 ± 0.06	0.93 ± 0.04	0.92 ± 0.06	0.93 ± 0.06	0.92 ± 0.04
Random Forest (10)	0.95 ± 0.03	0.93 ± 0.07	0.92 ± 0.08	0.92 ± 0.06	0.95 ± 0.03	0.96 ± 0.04	0.92 ± 0.06	0.94 ± 0.04
Random Forest (50)	0.95 ± 0.03	0.92 ± 0.08	0.92 ± 0.06	0.92 ± 0.06	0.95 ± 0.03	0.96 ± 0.04	0.92 ± 0.06	0.94 ± 0.04
Random Forest (100)	**0.96 ± 0.03**	**0.93 ± 0.07**	**0.93 ± 0.07**	**0.92 ± 0.05**	0.95 ± 0.03	0.96 ± 0.04	0.93 ± 0.06	0.94 ± 0.04
SVM (RBF)	0.93 ± 0.04	0.83 ± 0.09	0.96 ± 0.06	0.89 ± 0.06	**0.96 ± 0.03**	**0.95 ± 0.05**	**0.96 ± 0.05**	**0.95 ± 0.03**
SVM (Linear)	0.93 ± 0.04	0.85 ± 0.09	0.96 ± 0.06	0.90 ± 0.06	0.95 ± 0.04	0.94 ± 0.05	0.94 ± 0.06	0.94 ± 0.04
MLP (10)	0.95 ± 0.03	0.91 ± 0.07	0.93 ± 0.07	0.92 ± 0.05	0.93 ± 0.04	0.93 ± 0.06	0.91 ± 0.07	0.92 ± 0.04
MLP (20)	0.95 ± 0.03	0.91 ± 0.07	0.94 ± 0.07	0.92 ± 0.05	0.93 ± 0.04	0.93 ± 0.06	0.91 ± 0.07	0.92 ± 0.04
MLP (50)	0.94 ± 0.04	0.91 ± 0.08	0.91 ± 0.09	0.90 ± 0.06	0.94 ± 0.04	0.93 ± 0.06	0.93 ± 0.07	0.93 ± 0.05

The research described in this paper confirms that graph-based methods provide a good description of cervical cancer cells and can be a leaner and highly efficient alternative for screening while geometric features remain the gold standard for high-precision classification. At this point, we need to mention that despite using only 7 MST or 10 Delaunay descriptors, graph-based techniques achieved over 91% accuracy for the pathology detection task. This observation suggests that the transition from a healthy to an abnormal cell creates a distinct structural shift that graph models capture effectively with lower dimensionality. Furthermore, the structural difference between LSIL and HSIL is more subtle, which is reflected in the results of Task 1.

3.3 Limitations and Future Directions

The presented results show promising path toward an automated, human-level diagnostic system. We can note that a multi-modal fusion architecture that combines geometric measurements with the structural graph topology and deep features extracted from Convolutional Neural Networks and Vision Transformers can lead to a more holistic representation of nuclear pathology. However, such a scaling will be impaired with the limitations of the dataset. This is actually a primary constraint of the current study. Due to a limited dataset size, to ensure the robustness of the described methods, research should be extended to a larger database. Inclusion of public benchmark data will also be considered in future research. Current research used the dataset described by Jeleń *et al.* [9] due to a mishmash of the benchmark in publicly available datasets. The main difficulties include the inclusion of isolated single-cell images that many of the dataset do not contain, and those that do are not based on Liquid-Based Cytology. This observation makes a direct comparison with our current LBC-based methodology complex. In future research, we also plan to extend the proposed idea to analyze of a Gigapixel Whole Slide Images. Implementing graph-based algorithms on a whole slide scale will allow the assessment of the entire cellular landscape.

4 Conclusions

In this study, we explored the possibility of utilizing graph-based features extracted from the Minimum Spanning Tree and Delaunay triangulation for cervical cancer classification using liquid-based cytology images. By extracting features from MST and Delaunay graphs, we showed that topological descriptors can effectively capture the morphological patterns of cellular pathology. Although the proposed method achieved slightly lower peak accuracy of 0.92 for pathology detection in Task 2 compared to 0.96 our geometric benchmark, it offers significant advantages in terms of model efficiency. The proposed approach uses a set of only 17 features, yet maintains a high accuracy of over 91% for pathology detection. Furthermore, the minimum spanning tree of the contour graph shows superior performance over the Delaunay triangulation, suggesting a better representation of subtle contour irregularities and pleomorphism characteristic of cancer cells. In addition, the high recall rates observed across our models indicate that the graph features are sensitive to topological anomalies in the cellular structure. This observation makes these features a powerful and reliable tool for primary screening procedures.

References

1. Allahqoli, L., et al.: Delayed cervical cancer diagnosis: a systematic review. Eur. Rev. Med. Pharmacol. Sci. **26**(22), 8467–8480 (2022)
2. Allahqoli, L., et al.: Diagnosis of cervical cancer and pre-cancerous lesions by artificial intelligence: a systematic review. Diagnostics **12**(11), 2771 (2022)
3. Basta, T., et al.: Current management of cervical cancer in Poland–analysis of the questionnaire trial for the years 2002–2014 in relation to ASCO 2016 recommendations. PLoS ONE **14**(1), e0209901 (2019)
4. Bedell, S.L., Goldstein, L.S., Goldstein, A.R., Goldstein, A.T.: Cervical cancer screening: past, present, and future. Sexual Med. Rev. **8**(1), 28–37 (2020). https://doi.org/10.1016/j.sxmr.2019.09.005
5. de Berg, M., Cheong, O., van Kreveld, M.J., Overmars, M.H.: Computational Geometry: Algorithms and Applications, 3rd edn. Springer, Heidelberg (2008). https://doi.org/10.1007/978-3-540-77974-2
6. Bhattacharjee, S., Paul, M.D., Biswas, P.K.: Cluster analysis of cell nuclei in h&e-stained histological images using minimum spanning tree. Diagnostics **11**(12), 2382 (2021). https://doi.org/10.3390/diagnostics12010015
7. Dussert, C., Gervois, G., Rivier, N.: Minimal spanning tree analysis of biological structures. J. Theor. Biol. **125**(3), 317–323 (1987). https://doi.org/10.1016/S0022-5193(87)80081-1
8. Jeleń, Ł., Stankiewicz-Antosz, I., Chosia, M., Jeleń, M.: Optimizing cervical cancer diagnosis with feature selection and deep learning. Appl. Sci. **15**(3) (2025). https://doi.org/10.3390/app15031458

9. Jeleń, Ł, Stankiewicz-Antosz, I., Chosia, M., Jeleń, M.: Stochastic feature selection and machine learning for optimized cervical cancer classification. Int. J. Appl. Math. Comput. Sci. **35**, 601–615 (2025)
10. Jiang, P., et al.: A systematic review of deep learning-based cervical cytology screening: from cell identification to whole slide image analysis. Artif. Intell. Rev. **56**, 1–72 (2023). https://doi.org/10.1007/s10462-023-10588-z
11. Jungck, J., Pelsmajer, M., Chappel, C., Taylor, D.W.: Space: the re-visioning frontier of biological image analysis with graph theory, computational geometry, and spatial statistics. Mathematics (2021). https://doi.org/10.20944/preprints202108.0579.v1
12. Li, J., Wang, X., Yang, W., Li, L., Zhou, H.: Nuclei-based features for uterine cervical cancer histology image analysis. J. Pathol. Inform. **6**, 10 (2015). https://doi.org/10.1109/JBHI.2015.2483318
13. Lu, C., Pati, S., Eddy, J.A., et al.: Feature-driven local cell graph (flock): new computational pathology-based quantitative descriptors for predicting clinical outcomes in cancer. Med. Image Anal. **65**, 101779 (2020). https://doi.org/10.1016/j.media.2020.101903
14. Narayanan, M., Rajendran, V., Kimia, B.: Shape-biased domain generalization via shock graph embeddings. In: Proceedings of the IEEE/CVF International Conference on Computer Vision (ICCV), pp. 9788–9798 (2021)
15. Pilcher, W., et al.: Shape-to-graph mapping method for efficient characterization and classification of complex geometries in biological images. PLoS Comput. Biol. **16**(3), e1007704 (2020)
16. Pina, O.: From pixels to graphs: cell-based digital pathology image analysis. Ph.d. thesis (2025). https://doi.org/10.5821/dissertation-2117-439958
17. Stankiewicz, I.: Zastosowanie programu komputerowego do oceny rozmazów cytologicznych uzyskanych z części pochwowej szyjki macicy metod LBC. Master's thesis, Medical University of Wrocław (2018)
18. Yao, R., et al.: GKGNet: group k-nearest neighbor based graph convolutional network for multi-label image recognition. arXiv preprint arXiv:2308.13905 (2023)

Dependable Resource Planning in Smart Breweries Through a Digital Twin of Flexible Batch Fermentation Processes

Krzysztof Rewak[1(✉)] and Andrzej Rusiecki[2]

[1] Blumilk sp. z o.o., Legnica, Poland
krzysztof.rewak@blumilk.pl
[2] Wrocław University of Science and Technology, Wrocław, Poland

Abstract. Fermentation is one of the most capacity-constrained and disturbance-sensitive stages of beer production, with fermentation tanks acting as long-occupation resources that strongly influence schedule feasibility. In practice, brewery fermentation planning must accommodate variable processing times, heterogeneous tank capacities, and flexible batch aggregation and splitting. This paper presents a digital twin-based framework for fermentation resource planning in smart breweries that integrates a flexible batch-to-fermenter allocation model with continuously updated processing-time estimates. The proposed architecture separates process state representation from decision logic, enabling adaptive re-planning under realistic disturbances. A case study based on an industrial craft brewery configuration illustrates how the framework supports dependable fermentation planning under process variability.

Keywords: Dependability · Digital twin · Fermentation scheduling · Production planning · Smart brewery

1 Introduction

Production planning and scheduling in brewery operations involve the coordination of multiple interdependent process stages subject to constrained resources. Among these stages, fermentation is characterized by long processing times and limited capacity, making fermentation tanks critical resources whose availability directly affects the feasibility of the schedule and the production throughput.

Early studies focused on deterministic optimization models integrating lot sizing and scheduling decisions, with mixed-integer programming formulations proposed to improve resource utilization under fixed process assumptions [3]. Subsequent work extended these models to multi-stage problems with cost-related objectives [5], while heuristic and metaheuristic approaches demonstrated applicability to parallel-machine and batch-processing environments [14]. Decision support systems have also been developed for industrial brewery settings [8]. Despite their detailed process representations, these approaches typically rely on

W. Zamojski et al. (Eds.): DepCoS-RELCOMEX 2026, LNNS 2014, pp. 372–381, 2026.
https://doi.org/10.1007/978-3-032-27927-9_36

fixed batch-to-resource assignments and offer limited support for adaptive planning under process variability.

More recently, evolutionary algorithm-based methods have been applied to fermentation scheduling in smart breweries to address capacity and availability constraints [10]. Although these approaches reduce planning time and improve feasibility in dynamic environments, they adopt simplified process representations, most notably assuming a fixed one-to-one relationship between brewed batches and fermentation tanks. The practice of industrial brewing does not conform to this assumption. Depending on operational decisions and capacity constraints, a single brewed batch can be split into multiple fermentation tanks, or multiple batches can be aggregated within a single tank. This flexibility increases the structural complexity of the fermentation planning problem and cannot be adequately captured using fixed batch-to-resource assignments.

In parallel, digital twin technologies have been increasingly applied in brewing and food-processing systems, covering fermentation monitoring [6], beer quality control [4], and architectural aspects of brewery digital twins [7,9]. End-to-end digital twin approaches for multistage batch scheduling have also been proposed with an emphasis on production safety [16]. Survey studies confirm the growing adoption of digital twins in the food industry, while noting the limited integration between process-level digital twins and production planning functions [1].

This paper addresses this gap by proposing a digital twin-based fermentation planning framework that, unlike prior work, combines flexible many-to-many batch-to-fermenter allocation with continuously updated processing-time estimates and formal dependability metrics. The framework supports adaptive replanning under realistic disturbances without assuming fixed batch-to-resource assignments or static processing times. Schedule feasibility and minimal tank blocking time are the primary planning goals; makespan serves as a computational objective in the underlying optimization procedure.

2 Digital Twin Architecture for Fermentation Planning

The proposed digital twin represents the beer production process as a layered cyber-physical system that integrates physical resources, a digital process model, and a decision layer. An overview of the architecture is shown in Fig. 1. The digital twin maintains a process-level representation of the production system, while planning decisions are restricted to the allocation of fermentation resources, which constitutes a practically relevant and capacity-constrained subproblem.

2.1 Physical and Digital Layers

The physical layer represents beer production as a sequence of batch-based operations executed on heterogeneous processing resources. The brewhouse produces discrete batches of cooled wort, which are subsequently processed through downstream stages such as fermentation, conditioning, or maturation, depending on the brewery configuration.

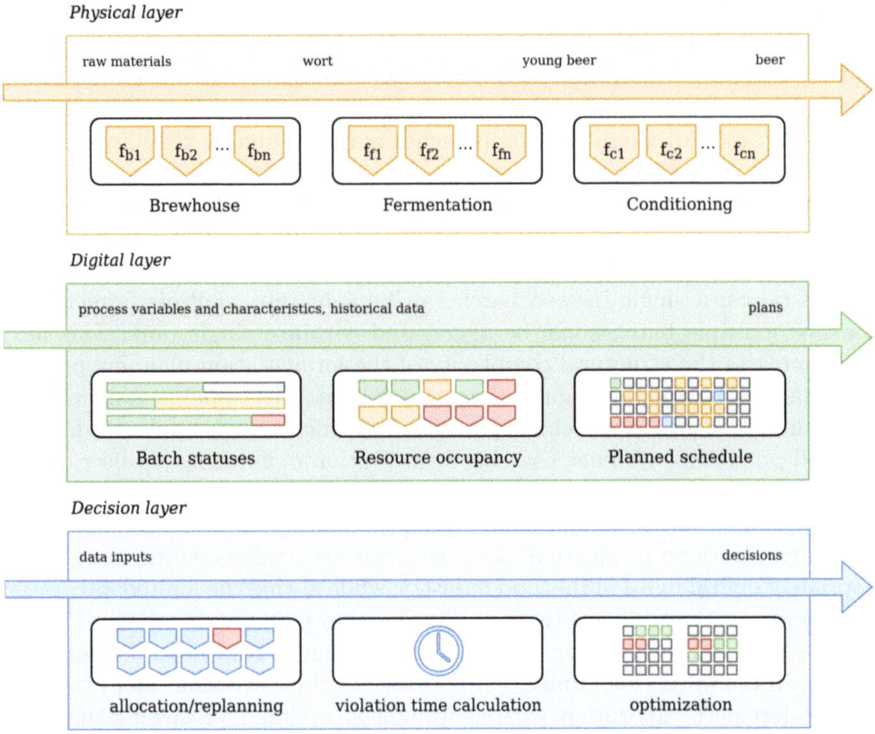

Fig. 1. Layered digital twin architecture for dependable fermentation planning.

Each batch $b \in \mathcal{B}$ is characterized by its volume V_b, beer style $s(b)$, and release time r_b. Production resources are grouped into sets of processing units, including brewhouse vessels and fermentation tanks, each with finite capacity V_f. The routing of batches through the processing stages is not fixed and may vary between breweries. No fixed cardinality between batches and fermentation resources is assumed; batches may be aggregated or split depending on operational requirements.

The digital layer maintains a synchronized and time-dependent representation of the production system, including batch availability, resource occupancy, and estimated processing progress. Processing times are not assumed to be fixed. Instead, the digital twin continuously estimates remaining processing durations based on process measurements and historical data. For each batch and processing stage, admissible bounds are imposed as

$$T_{s(b)}^{\min} \leq T_b \leq T_{s(b)}^{\max}, \tag{1}$$

where T_b denotes the estimated remaining processing time for batch b. The bounds $T_{s(b)}^{\min}$ and $T_{s(b)}^{\max}$ specify admissible limits on the duration of fermentation for the beer style $s(b)$, capturing both minimal processing requirements and practical upper bounds imposed by quality and operational constraints. These limits depend on the beer style; in general, lower-gravity beers are associated with shorter fermentation times than higher-gravity products.

This representation enables the evaluation of production trajectories under different brewery-specific configurations and provides consistent state information to the decision layer.

2.2 Decision Layer: Fermentation Planning Model

Based on the digital twin state, the decision layer formulates a fermentation planning problem that assigns batches to fermentation tanks over time. The model supports flexible batch aggregation and splitting subject to capacity and temporal constraints.

Batch-to-tank allocation is represented by the continuous decision variable

$$x_{b,f} \geq 0, \quad \forall b \in \mathcal{B},\ \forall f \in \mathcal{F}, \tag{2}$$

where $x_{b,f}$ denotes the volume of batch b allocated to fermenter f. Each batch must be fully allocated:

$$\sum_{f \in \mathcal{F}} x_{b,f} = V_b \quad \forall b \in \mathcal{B}. \tag{3}$$

Fermentation tanks are subject to capacity constraints over time. At any time t, the total allocated volume must satisfy

$$\sum_{b \in \mathcal{B}(t,f)} x_{b,f} \leq V_f \quad \forall f \in \mathcal{F}, \tag{4}$$

where $\mathcal{B}(t,f)$ denotes the set of batches occupying fermenter f at time t. The batch b occupies the fermentation capacity during the interval $[r_b, r_b + T_b]$.

Downstream stages are not considered capacity-constraining in the present formulation. Fermentation tanks are released after completion of the fermentation stage at time $r_b + T_b$.

Makespan serves as the computational objective within the optimization procedure:

$$\min\ C_{\max} = \max_{b \in \mathcal{B}} (r_b + T_b)\,. \tag{5}$$

This formulation defines the decision logic that maps digital twin state information to batch-to-fermenter allocation and re-planning decisions.

3 Resource Planning and Dependability Aspects

3.1 Disturbance Types and Planning Assumptions

Fermentation planning is affected by disturbances that influence processing times and resource availability. A primary source of uncertainty is the variability in the duration of fermentation, which arises from biochemical dynamics and quality constraints. For each batch b, the processing time is modeled as

$$T_b = T_{s(b)}^{\text{nom}} + \Delta T_b, \tag{6}$$

where $T_{s(b)}^{\text{nom}}$ denotes the nominal duration and ΔT_b represents deviations from the nominal trajectory. Extended processing times are of particular concern, as they prolong resource occupation.

Additional disturbances include the temporary unavailability of fermentation tanks due to cleaning or maintenance, represented by the time-dependent capacity $V_f(t) \leq V_f$, and upstream delays affecting batch release times,

$$r_b = r_b^{\text{nom}} + \Delta r_b. \tag{7}$$

Batch transfers between stages are assumed to be instantaneous. Batch aggregation and splitting are allowed only for compatible products. Downstream stages are represented at the digital twin level but are not treated as binding capacity constraints.

3.2 Dependability Analysis

The dependability of fermentation planning is evaluated by analyzing the ability of the system to maintain feasible and operationally acceptable schedules under the disturbances described above. Rather than focusing solely on nominal optimality, dependability is assessed using three metrics [2]. Schedule feasibility Φ is defined as the absence of capacity violations throughout the planning horizon:

$$\Phi = \mathbf{1}\left[\forall f \in \mathcal{F},\ \forall t\colon \sum_{b \in \mathcal{B}(t,f)} x_{b,f} \leq V_f(t)\right]. \tag{8}$$

The additional tank blocking time induced by a disturbance ΔT_b is

$$\delta_f = \max\left(0,\ T_b - T_{s(b)}^{\text{nom}}\right), \tag{9}$$

and the re-planning frequency over a horizon H is $\rho = |\mathcal{R}|/H$, where $\mathcal{R}$ denotes the set of re-planning events. These metrics jointly reflect the operational impact of disturbances and the planner's ability to respond to them.

Scenario-based analysis indicates that fermentation schedules constructed under nominal assumptions are highly sensitive to processing-time variability and resource unavailability. Even moderate extension of fermentation duration can lead to prolonged tank occupation, propagating into capacity conflicts for subsequent batches. Temporary unavailability of a single tank can trigger cascading disruptions, rendering an initially feasible plan infeasible.

The digital twin mitigates these effects by continuously synchronizing the planning model with the evolving system state [15], enabling early detection of deviations and allowing the decision layer to adapt batch-to-tank assignments before hard constraint violations occur. Schedule infeasibility arises when disturbances cause $\sum_b x_{b,f}$ to exceed $V_f(t)$; the severity and duration of such violations, captured by δ_f and ρ, provide a natural basis for dependability assessment without introducing additional cost assumptions.

4 Case Study

The proposed framework is evaluated using a case study based on the configuration of an operating Polish craft brewery with a small brewhouse and a heterogeneous fermentation cellar. The case study demonstrates how the digital twin architecture and the fermentation planning model can be instantiated for an industrial configuration and applied under variable operating conditions.

Earlier work [10] addressed fermentation planning under simplified assumptions, including fixed batch-to-fermenter assignments and static processing times. The present case study relaxes these assumptions by considering flexible batch aggregation and splitting, time-dependent resource availability, and dynamically updated processing-time estimates.

The brewery configuration shown in Fig. 2 consists of a mash-lauter tun and a brew kettle with a nominal capacity of 30 hl, producing discrete batches of cooled wort. The fermentation cellar comprises twelve 50 hl tanks, six 60 hl tanks, one 90 hl tank, and one 100 hl tank. A 9 hl bright beer tank is dedicated to packaging operations and is excluded from fermentation planning. Such heterogeneous fermentation cellars are commonly found in craft breweries [11].

All fermentation tanks in the considered configuration are used as combined fermentation and maturation vessels and may therefore be occupied across multiple processing stages. This reduces intermediate transfers but increases the coupling between processing durations and resource availability, making batch allocation and tank release decisions mutually dependent.

Daily production involves multiple brews whose volumes do not match available fermentation capacities. Feasible allocations therefore require batch aggregation within individual tanks or splitting across multiple tanks. The specific allocation patterns depend on production demand, current tank availability, and anticipated processing durations. Fermentation plans are revised in response to several types of disturbances. Variability in fermentation duration arises from yeast performance and temperature control [12]. Tanks may become temporarily unavailable due to cleaning-in-place procedures or maintenance activities [13]. Additional changes result from upstream disturbances affecting batch release times and from adjustments in production priorities related to demand, special events, collaborative brewing, or contract brewing activities. Brewhouse availability becomes an additional limiting constraint for replanning scheduled batches. These factors cannot be addressed using static batch-to-fermenter assignments.

To account for these conditions, the differential evolution-based planning script developed in earlier work was extended to support continuous allocation variables, many-to-many batch-tank relationships, and time-dependent processing durations provided by the digital twin. Differential evolution (DE) is used as

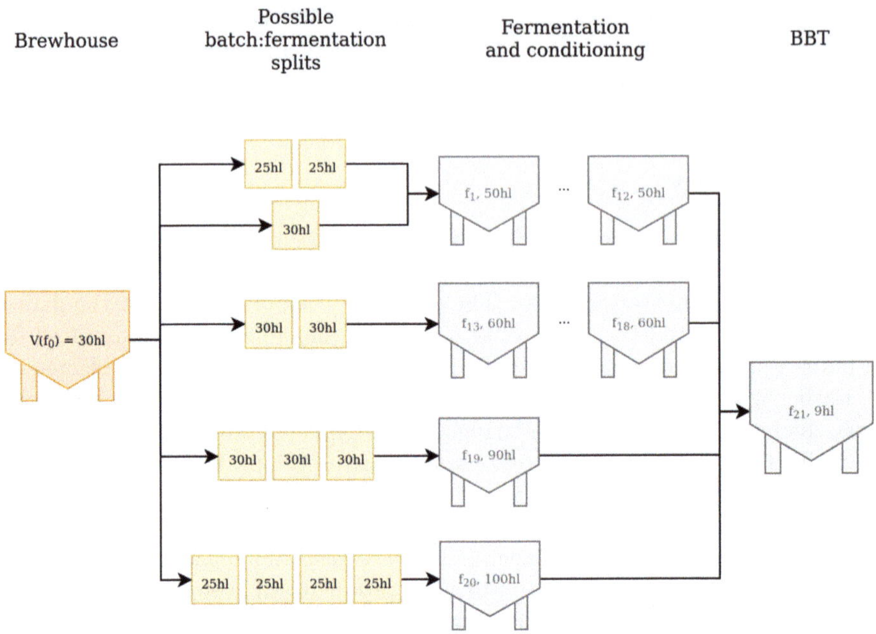

Fig. 2. Configuration of the case-study brewery, including brewhouse output, fermentation tank capacities, and batch routing assumed in the planning experiments. Tanks are not scaled.

a computationally lightweight realization of the decision logic, while the planning architecture remains independent of the optimization method employed. Candidate schedules were encoded as continuous vectors representing batch start times and fermentation tank assignments, while feasibility with respect to tank capacities, availability calendars, and process sequencing was enforced through penalty terms in the objective function. The DE algorithm iteratively evolved a population of such candidate solutions using mutation, crossover, and selection operators to minimize the overall makespan under given constraints. The method was selected for its ability to handle continuous decision variables, nonlinear constraints, and short computation times, making it suitable for repeated use in dynamic brewery environments.

The planning procedure can be executed repeatedly over short horizons in response to updated system-state information, enabling re-planning when disturbances occur and supporting fermentation planning aligned with the evolving production process.

4.1 Illustrative Replanning Scenario

At a given planning instance, shown in Fig. 3, several batches are undergoing fermentation, and tank f_{19} is expected to become available at a planned release time derived from nominal processing-time estimates. On this basis, batch b_1 is assigned to f_{19}, followed by batch b_2 immediately after the planned completion of b_1. Under nominal assumptions, the allocation is feasible and the occupancy intervals do not overlap.

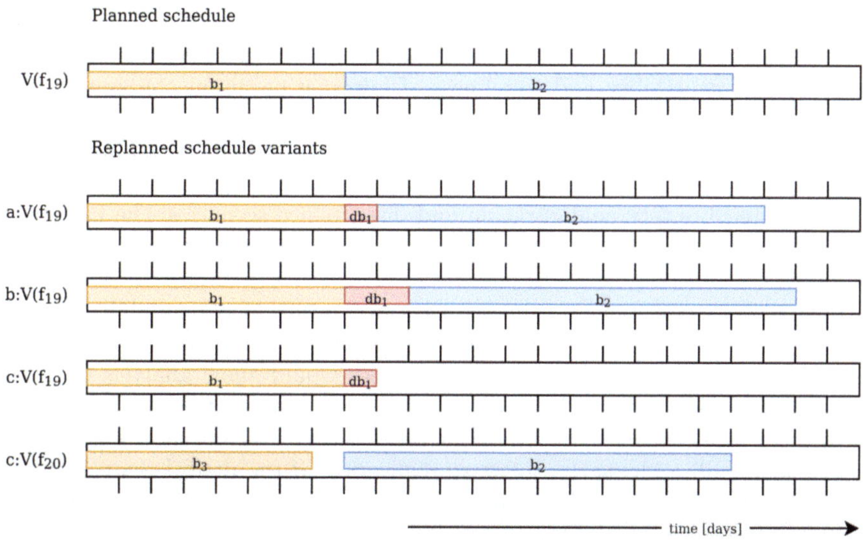

Fig. 3. Replanning scenario following delayed fermentation completion.

During execution, the measurements of the process from the physical layer indicate slower fermentation progress for b_1 in f_{19}. These measurements are propagated to the digital layer, where the estimated remaining processing time is updated. As a result, the predicted completion of b_1 is shifted by one or two days, extending the period during which f_{19} remains unavailable.

If the brewing process continues according to the original schedule, batch b_2 is completed while f_{19} remains occupied by b_1, leading to a capacity conflict caused by overlapping occupancy intervals. The original allocation therefore becomes infeasible under the updated availability constraints.

Based on the updated system state, the decision layer evaluates alternative feasible plans. Variants (a) and (b) delay the start of b_2 to align brewing with the revised release time of f_{19}. Variant (c) preserves the original brewing schedule by reallocating b_2 to an alternative tank f_{20}.

The resulting plans are evaluated using the same feasibility and utilization criteria as in the nominal procedure. The decision layer provides the planner with admissible replanning options, while the final selection and execution of a revised plan remains a human decision. The system therefore operates as a decision support tool rather than an autonomous scheduling system.

5 Discussion and Future Work

The presented framework integrates flexible batch allocation with a digital twin representation of fermentation processes. By decoupling process state estimation from decision logic, the approach supports adaptive planning under variable processing times and resource availability.

The case study illustrates that the dependability of planning is mainly influenced by the availability and timeliness of information about the system state rather than by the specific optimization algorithm used. The digital twin provides a consistent basis for detecting deviations and triggering re-planning, while the decision layer translates updated state information into revised allocations. For the tested brewery configuration and disturbance scenarios, the framework produced feasible re-planning outcomes consistent with the defined dependability metrics.

Future work will focus on quantitative evaluation of dependability metrics and on incorporating uncertainty-aware processing-time estimates. Extending the framework to additional production stages would allow analysis of disturbance propagation across the brewery. From an implementation perspective, the present work defines the planning architecture and decision logic; subsequent efforts will address the software realisation of the digital twin, including data acquisition interfaces, real-time state synchronisation, and integration with brewery control systems. Empirical validation of the framework in an operational brewery environment, including quantitative comparison with prior planning methods, remains a further direction for future work. Alternative planning methods may also be integrated within the same architectural framework.

The results indicate that combining digital twin-based state awareness with flexible batch allocation provides a viable foundation for dependable fermentation planning in smart breweries.

Acknowledgments. The authors would like to thank Dr. Mateusz Jackowski, Jerzy Gibadło, Rafał Łopusiński and Marcin Ostajewski for their valuable industry expertise and support from the brewing sector.

Disclosure of Interests. The authors declare no competing interests.

References

1. Abdurrahman, E.E.M., Ferrari, G.: Digital twin applications in the food industry: a review. Front. Sustain. Food Syst. **9** (2025). https://doi.org/10.3389/fsufs.2025.1538375
2. Aghileh, M., Tereso, A., Alvelos, F., Lopes, M.O.M.: Multi-project scheduling with uncertainty and resource flexibility: a narrative review and exploration of future landscapes. Algorithms **18** (2025). https://doi.org/10.3390/a18060314
3. Baldo, T.A., Santos, M.O., Almada-Lobo, B., Morabito, R.: An optimization approach for the lot sizing and scheduling problem in the brewery industry. Comput. Ind. Eng. **72**, 58–71 (2014). https://doi.org/10.1016/j.cie.2014.02.008

4. Dazzarola, C., Tighe, R., Pérez-Correa, J.R., Saa, P.A.: Toward a digital twin for beer quality control: development of a digital model integrating industrial process data and model-based fermentation descriptors. J. Food Eng. **403**, 112726 (2026). https://doi.org/10.1016/j.jfoodeng.2025.112726
5. Georgiadis, G.P., Elekidis, A.P., Georgiadis, M.C.: Optimal production planning and scheduling in breweries. Food Bioprod. Process. **125**, 204–221 (2021). https://doi.org/10.1016/j.fbp.2020.11.008
6. Goffi, P.E., Tremblay, R., Oakes, B.: Engineering a digital twin for the monitoring and control of beer fermentation sampling. arXiv preprint arXiv:2508.18452 (2025). https://doi.org/10.48550/arXiv.2508.18452, accepted for EDTconf 2025
7. Lee, A., Negrin, D.A.M., Cleophas, L.: Evaluating open-source tools for heterogeneous model-based digital twin development: a microbrewery case study. In: Proceedings of the 2024 Annual Modeling and Simulation Conference (ANNSIM) (2024). https://doi.org/10.23919/ANNSIM61499.2024.10732658
8. Mickein, M., Koch, M., Haase, K.: A decision support system for brewery production planning at Feldschlösschen. INFORMS J. Appl. Anal. **52**(2) (2022). https://doi.org/10.1287/inte.2021.1101
9. Muctadir, H.M., Liao, Y., Cleophas, L.: Model consistency management of a brewery digital twin. In: Proceedings of the ACM/IEEE 28th International Conference on Model Driven Engineering Languages and Systems Companion (MODELS-C), pp. 717–721 (2025). https://doi.org/10.1109/MODELS-C68889.2025.00096
10. Rewak, K.: Optimizing batch fermentation scheduling in smart breweries with artificial intelligence methods. Procedia Comput. Sci. **270**, 2838–2847 (2025). https://doi.org/10.1016/j.procs.2025.09.406
11. Salazar Tijerino, M.B., San Martín-González, M.F., Velasquez Domingo, J.A., Huang, J.Y.: Life cycle assessment of craft beer brewing at different scales on a unit operation basis. Sustainability **15** (2023). https://doi.org/10.3390/su151411416
12. Solgajová, M., Frančáková, H., Dráb, S., Tóth, Z.: Effect of temperature on the process of beer primary fermentation. J. Microbiol. Biotechnol. Food Sci. **2**, 1791–1799 (2013)
13. Stapelbroek, M., Kilic, O.A., Yang, Y., Van Donk, D.P.: Eliminating production losses in changeover operations: a case study on a major European food manufacturer. Product. Plann. Control **35**, 794–807 (2022). https://doi.org/10.1080/09537287.2022.2136041
14. Sáenz-Alanís, C.A., V.D., J., Salazar-Aguilar, M.A., Boyer, V.: A parallel machine batch scheduling problem in a brewing company. Int. J. Adv. Manuf. Technolo. **87**, 65–75 (2016). https://doi.org/10.1007/s00170-016-8477-8
15. Urgo, M., Lanza, G., Vrabič, R., Gyulai, D.: Future-proof production scheduling and control. CIRP Ann. **74**, 971–991 (2025). https://doi.org/10.1016/j.cirp.2025.04.095
16. Wang, J., Zhao, J.: An end-to-end scheduling digital twin for multistage batch plants considering safety. Process Saf. Environ. Prot. **193**, 1261–1278 (2025). https://doi.org/10.1016/j.psep.2024.11.127

A Modern Approach to the Protection of People and Property – Integration of Physical and Technological Security Systems

Marcin Sabat(✉)

AGAT GROUP sp.z.o.o, Mińska Street, 54-56/2A, Wrocław, Poland
marcin.sabat@agat-group.pl

Abstract. The publication addresses the issue of modern approaches to the protection of people and property in the context of increasingly complex threat environments and the rapid development of digital technologies. The aim of the study is to present the evolution of security systems from traditional models based primarily on physical guarding to integrated frameworks that combine human, technical, and digital solutions within a coherent security management architecture.

The paper conceptualizes security as a management process grounded in risk analysis and continuous improvement, in accordance with ISO 31000 and ISO 22301 standards. It discusses the role of traditional physical protection, highlighting both its preventive value and its limitations resulting from human factors and economic constraints. The development of technological security systems is examined, including IP-based video surveillance, access control systems, artificial intelligence applications, and PSIM (Physical Security Information Management) platforms.

Particular attention is given to system integration as a contemporary paradigm of security management. The study demonstrates that data correlation, automated response scenarios, and predictive analytics enable a shift from reactive to proactive and predictive security models. It also emphasizes the importance of regulatory compliance, especially in the field of personal data protection, as well as the need to foster a strong security culture and organizational resilience.

The findings indicate that effective protection of people and property requires a systemic approach integrating risk management, advanced technologies, and human competencies. Security is no longer merely an operational function but a strategic component of competitive advantage and long-term organizational stability.

Keywords: Protection of people and property · Security management · System integration · PSIM · Video surveillance · Risk analysis · Organizational resilience

1 Introduction

The security of people and property constitutes one of the fundamental categories determining the stable functioning of contemporary organizations, public institutions, and business entities. In the context of dynamic technological transformations, globalization, and the growing complexity of threats, the traditional understanding of protection

W. Zamojski et al. (Eds.): DepCoS-RELCOMEX 2026, LNNS 2014, pp. 382–388, 2026.
https://doi.org/10.1007/978-3-032-27927-9_37

as a set of physical measures is no longer sufficient. The security environment of the 21st century is multidimensional and hybrid in nature—it encompasses physical, cyber, informational, organizational, and reputational threats. The literature emphasizes that modern societies operate under the conditions of a "risk society" [1], in which uncertainty and vulnerability to disruptions are inherent characteristics of socio-economic systems.

Barry Buzan and his co-authors [2] argue that security should be analyzed across multiple sectors—military, political, economic, societal, and environmental. Similarly, at the organizational level, the protection of people and property cannot be treated solely as an operational issue but must be viewed as an element of a broader risk management and organizational resilience system.

The development of information and communication technologies, IP monitoring systems, video analytics, biometric solutions, and platforms integrating data from multiple sources has driven contemporary protection toward an integrated model. In this model, physical protection, technical safeguards, and digital solutions form a coherent system aimed not only at responding to incidents but also at predicting and preventing them. The purpose of this study is to present a modern approach to the protection of people and property based on the integration of physical and technological systems and to situate this issue within the context of academic literature, international standards, and industry practice.

Despite the growing body of literature on security systems and risk management, there is a noticeable gap in the integrated analysis of physical protection, advanced technological systems, and cybersecurity within a unified management framework. Existing studies often address these domains separately, lacking a comprehensive perspective on their interdependencies in complex, hybrid threat environments.

2 Security as a Management Process

The contemporary approach to the protection of people and property increasingly departs from perceiving security as a state achieved through the implementation of specific technical or organizational safeguards. Instead, a process-oriented perspective prevails, in which security is treated as a dynamic, cyclical, and systemically managed area of organizational activity. This implies the necessity of continuous monitoring, improvement, and adaptation to the changing threat environment.

The theoretical foundations of this approach can be found in Ludwig von Bertalanffy's general systems theory and in quality management concepts [3], where the cycle of continuous improvement (PDCA – Plan-Do-Check-Act) plays a central role. Security, understood as one of the organizational processes, is subject to the same principles as other management areas—it requires planning, implementation, control, and corrective actions.

ISO 31000:2018 provides a framework for risk management, defining risk as the "effect of uncertainty on objectives." This definition is fundamental for understanding contemporary protection of people and property. It indicates that threats should not be analyzed in isolation from organizational activity but in the context of strategic, operational, and reputational objectives. Protection thus ceases to be purely operational and becomes part of strategic risk management.

In practice, this requires systematic identification of threats—both external and internal—analysis of infrastructure vulnerabilities, and assessment of the potential consequences of incidents. This process includes criminal, terrorist, technical, environmental, and cyber threats. The literature emphasizes that an effective protection system should be designed according to the principle of adequacy and proportionality of security measures to the level of risk [4]. Excessive safeguards generate unjustified costs and may limit the functionality of a facility, whereas underestimating risk may lead to serious financial and reputational consequences.

In the Polish legal framework, the primary act regulating the protection of people and property is the Act of 22 August 1997, which defines forms of physical protection and technical safeguards. Although this act provides an important legal foundation, it does not exhaust the systemic dimension of security management. Contemporary organizations operate in a highly digitalized and networked environment, where the boundaries between physical security and cybersecurity are increasingly blurred. Consequently, standards and regulations concerning information security, such as ISO/IEC 27001 and ISO/IEC 27005, as well as personal data protection regulations [5], have gained growing importance.

An important extension of the traditional approach is the concept of organizational resilience. ISO 22301:2019 on business continuity management systems emphasizes that organizations should be capable of maintaining key processes even under severe disruptions. Security therefore extends beyond incident prevention to include preparedness for their consequences and the ability to rapidly restore operations. Crisis management literature indicates that organizations with a high level of risk management maturity integrate physical protection planning with business continuity and crisis response plans [6].

The process-oriented approach also requires analytical tools for modeling threat scenarios. In the critical infrastructure sector, methods such as scenario analysis, Event Tree Analysis, Fault Tree Analysis, and Quantitative Risk Assessment (QRA) are applied. Increasingly, data analytics and predictive models are used to identify incident patterns based on historical data. Reports by the McKinsey Global Institute [7] show that organizations using advanced analytics in risk management achieve higher operational efficiency and shorter response times.

An integral component of the process approach is the proper design of the physical environment. The CPTED (Crime Prevention Through Environmental Design) concept, developed among others by Crowe [8], assumes that architecture and spatial organization significantly influence security levels. Natural access control, natural surveillance, and territorial reinforcement are examples of measures reducing facility vulnerability. Contemporary practice integrates these principles with technological solutions, creating environments in which architectural, procedural, and digital elements complement one another.

Security management as a process also requires an appropriate organizational culture. Risk management literature highlights the importance of "safety/security culture," understood as a set of values, attitudes, and practices supporting threat identification and response [9]. Even the most advanced technological systems will not ensure effective protection if employees are unaware of procedures and not engaged in their implementation.

In conclusion, security as a management process may be perceived as an integrated system encompassing risk analysis, safeguard planning, implementation of technical and organizational measures, monitoring of their effectiveness, and continuous improvement. This model aligns with the broader systemic approach to organizational functioning, in which security contributes to sustainable competitive advantage and resilience to disruptions.

3 Traditional Physical Protection—Role and Limitations

Physical protection is historically the earliest and most widespread form of safeguarding people and property. Its origins can be traced to military structures and early forms of state organization, where guards and watch units protected facilities, goods, and socially significant individuals. Today, physical protection operates as an organized activity carried out by specialized entities based on legal regulations and professional standards.

In the security literature, personnel-based protection is regarded as a key element of security systems, particularly in high-risk facilities. Its primary functions include direct access control, environmental observation, incident response, and intervention. The presence of security personnel also has symbolic and psychological significance—it constitutes a visible signal of supervision and control.

Situational crime prevention theory [9] emphasizes that increasing the effort and risk perceived by potential offenders reduces criminal activity. Visible security personnel increase the perceived likelihood of detection. Rational choice theory [10] assumes that offenders assess costs and benefits before committing crimes. The presence of a guard raises perceived risk and may deter or displace criminal behavior.

Routine activity theory [11] introduces the concept of a "capable guardian," whose presence can prevent crime when a motivated offender and suitable target converge. Empirical studies confirm that increased visibility of surveillance—both human and technical—reduces certain categories of crime, particularly theft and vandalism [12].

However, a model based solely on the human factor faces structural and psychophysical limitations. From the perspective of ergonomics and work psychology, humans are susceptible to fatigue, reduced concentration, and perceptual errors. Reason's [13] theory of organizational accidents highlights the inevitability of human error. Prolonged monitoring tasks with low stimulus levels lead to vigilance decrement [14], reducing detection capability.

3.1 Organizational and Economic Constraints: Personnel Costs in Traditional Security Budgets [15]

The principle of "defense in depth" [4] proposes multilayered security architecture—perimeter protection, access control, internal safeguards, and response mechanisms. In this model, physical protection remains essential but operates in synergy with technical systems rather than independently.

Contemporary approaches increasingly perceive security personnel as operators within integrated information systems. Human decision-making is supported by data from surveillance, access control, and incident management platforms. Nonetheless, in crisis situations requiring direct intervention, evacuation, or communication, the human factor remains irreplaceable [6].

Thus, while traditional physical protection remains foundational, its effectiveness is limited if isolated from technological and systemic solutions. Its role evolves from a dominant form of protection to one component within an integrated security architecture.

4 Technological Security Systems

Physical protection is historically the earliest and most widespread form of safeguarding people and property. Its origins can be traced to military structures and early forms of state organization, where guards and watch units protected facilities, goods, and socially significant individuals. Today, physical protection operates as an organized activity carried out by specialized entities based on legal regulations and professional standards.

In the security literature, personnel-based protection is regarded as a key element of security systems, particularly in high-risk facilities. Its primary functions include direct access control, environmental observation, incident response, and intervention. The presence of security personnel also has symbolic and psychological significance—it constitutes a visible signal of supervision and control.

Situational crime prevention theory [10] emphasizes that increasing the effort and risk perceived by potential offenders reduces criminal activity. Visible security personnel increase the perceived likelihood of detection. Rational choice theory [11] assumes that offenders assess costs and benefits before committing crimes. The presence of a guard raises perceived risk and may deter or displace criminal behavior.

Routine activity theory [12] introduces the concept of a "capable guardian," whose presence can prevent crime when a motivated offender and suitable target converge. Empirical studies confirm that increased visibility of surveillance—both human and technical—reduces certain categories of crime, particularly theft and vandalism [13].

However, a model based solely on the human factor faces structural and psychophysical limitations. From the perspective of ergonomics and work psychology, humans are susceptible to fatigue, reduced concentration, and perceptual errors. Reason's [14] theory of organizational accidents highlights the inevitability of human error. Prolonged monitoring tasks with low stimulus levels lead to vigilance decrement [15], reducing detection capability.

Organizational and economic constraints are also significant. Personnel costs dominate traditional security budgets [16] and maintaining 24/7 coverage in large or dispersed facilities increases operational expenses. Moreover, personnel-based models have limited scalability, as increasing staff does not proportionally increase security effectiveness.

The principle of "defense in depth" [4] proposes multilayered security architecture—perimeter protection, access control, internal safeguards, and response mechanisms. In this model, physical protection remains essential but operates in synergy with technical systems rather than independently.

Contemporary approaches increasingly perceive security personnel as operators within integrated information systems. Human decision-making is supported by data from surveillance, access control, and incident management platforms. Nonetheless, in crisis situations requiring direct intervention, evacuation, or communication, the human factor remains irreplaceable [6].

Thus, while traditional physical protection remains foundational, its effectiveness is limited if isolated from technological and systemic solutions. Its role evolves from a dominant form of protection to one component within an integrated security architecture.

5 Integration as the Paradigm of Contemporary Security

Modern protection increasingly shifts from dispersed systems to integrated security architectures combining physical security, cybersecurity, fire systems, and organizational procedures. Platforms such as PSIM (Physical Security Information Management) aggregate data from heterogeneous systems, enabling comprehensive incident management [15].

Integration enhances situational awareness [17], enabling perception, comprehension, and projection of operational environments. Automated incident response scenarios align with ISO 22320 standards. Predictive analytics supports proactive risk management [18].

Integration of physical and cybersecurity domains is essential in hybrid threat environments. Standards such as ISO/IEC 27001 and ENISA guidelines promote holistic protection architectures.

Strategically, integrated security enhances organizational resilience [19], competitive advantage, and stakeholder trust.

6 Conclusion

Contemporary protection of people and property is undergoing fundamental transformation. Traditional models based primarily on physical security are insufficient in the face of complex, hybrid threats. International standards [20] position security as an integral element of organizational management.

Technological advancements—IP monitoring, analytics, AI, and integration platforms—enhance detection and response but introduce new legal and cybersecurity challenges.

The key conclusion is the necessity of integrating physical and technological systems within a coherent security management architecture. Integration generates synergy, enhances situational awareness, and enables predictive analytics.

Modern security rests on three pillars: process-based risk management, advanced technologies, and system integration. Only their synergy enables the development of effective, flexible, and resilient protection systems capable of addressing the challenges of the contemporary threat environment.

This paper contributes to the literature by proposing an integrated approach to the protection of people and property that combines risk management principles, technological systems, and organizational factors within a single analytical framework. It extends existing research by highlighting the strategic role of security as a proactive and system-oriented function in modern organizations.

References

1. Buzan, B., Wæver, O., de Wilde, J.: Security: a New Framework for Analysis. Lynne Rienner, Boulder (1998)
2. Bertalanffy, L.: General System Theory. PWN, Warsaw (1984)
3. Deming, W.E.: Out of the Crisis. MIT Press, Cambridge (1986)
4. Garcia, M.L.: The Design and Evaluation of Physical Protection Systems. Butterworth-Heinemann, Burlington (2008). https://doi.org/10.1016/B978-0-7506-8352-5.X5001-6
5. European Union: Regulation (EU) 2016/679 (GDPR). EUR-Lex (2016)
6. ISO: ISO 31000: Risk Management—Guidelines. International Organization for Standardization (2018)
7. ISO: ISO 22301: Business Continuity Management System Requirements. International Organization for Standardization (2019)
8. Crowe, T.: Crime Prevention through Environmental Design (CPTED). Butterworth-Heinemann (2013)
9. Reason, J.: Managing the Risks of Organizational Accidents. Ashgate (1997)
10. Clarke, R.V.: Situational Crime Prevention: Successful Case Studies. Harrow and Heston, Albany (1997)
11. Cornish, D., Clarke, R.: The Reasoning Criminal: Rational Choice Perspectives on Offending. Springer, New York (1986). https://doi.org/10.1007/978-1-4613-8625-4
12. Cohen, L., Felson, M.: Social change and crime rate trends: a routine activity approach. Am. Sociol. Rev. **44**(4), 588–608 (1979). https://doi.org/10.2307/2094589
13. Welsh, B., Farrington, D.: Effects of closed-circuit television surveillance on crime. Campbell Syst. Rev. (2009). https://doi.org/10.4073/csr.2009.17
14. Warm, J.S., Parasuraman, R., Matthews, G.: Vigilance requires hard mental work and is stressful. Hum. Factors. **50**(3), 433–441 (2008). https://doi.org/10.1518/001872008X312152
15. ASIS International: Physical Asset Protection Standard. ASIS International (2022)
16. Goodfellow, I., Bengio, Y., Courville, A.: Deep Learning. MIT Press (2016)
17. Endsley, M.R.: Toward a theory of situation awareness in dynamic systems. Hum. Factors. **37**(1), 32–64 (1995). https://doi.org/10.1518/001872095779049543
18. McKinsey Global Institute: The Future of Risk Management in the Digital era. McKinsey & Company (2020)
19. Hollnagel, E.: Resilience Engineering in Practice. Ashgate (2011)
20. Poland: Act on the protection of persons and property. J. Laws (1997)
21. Beck, U.: Risk Society: Towards a New Modernity. Scholar, Warsaw (2002)
22. IFSEC Global: Security & Fire Technology Industry Reports. IFSEC Global (2023)
23. ISO/IEC: ISO/IEC 27001: Information Security Management Systems—Requirements. International Organization for Standardization (2022)
24. ISO/IEC: ISO/IEC 27005: Information Security Risk Management. International Organization for Standardization (2022)
25. ISO/IEC: ISO/IEC 27017: Information Security Controls for Cloud Services. International Organization for Standardization (2015)
26. European Committee for Standardization: EN 62676: Video Surveillance Systems for use in Security Applications. CEN (2014)

Styles for Describing Safe Finite State Machines in Verilog HDL

Valer Salauyou(✉)

Bialystok University of Technology, Wiejska 45A, 15-351, Bialystok, Poland
v.salauyou@pb.edu.pl

Abstract. A safe finite state machine (SFSM) ensures that transitions from illegal states to the initial state occur without using a reset mechanism. The safety of finite state machines (FSMs) is paramount in medical life support equipment, avionics, autonomous robot control systems, and unmanned aerial vehicles. Unfortunately, methods for designing SFSMs receive little attention from modern researchers, and the techniques implemented in well-known design tools significantly degrade the FSM parameters. In this paper, the task of designing SFSMs is addressed in the Verilog hardware description language (HDL) using the Quartus design tool to implement FSMs on field-programmable gate arrays (FPGAs). To this end, three styles of FSM description are proposed: *safe, safe_error*, and *safe_idle*, which provide the user with various possibilities for implementing SFSMs. Experimental studies on FSM benchmarks have demonstrated the high efficiency of the proposed SFSM description styles. Compared to the method of synthesizing SFSMs in the Quartus design tool, the area is reduced on average by a factor of 2.436 (for individual examples, by 6.667 times), and the performance is increased on average by a factor of 1.588 (for individual examples, by 2.741 times). Recommendations are also given for the practical use of the proposed styles of describing SFSMs.

Keywords: Finite state machine (FSM) · Safe FSM (SFSM) · Hardware description language (HDL) · Illegal state · Field programmable gate array (FPGA)

1 Introduction

In digital systems, controllers and control units play a key role, implementing control algorithms for separate devices and parts of the system, as well as for the entire system [1]. Controllers and control units are usually implemented as finite state machines (FSMs) [2].

Fiure. 1 shows the traditional structure of an FSM, where X is the input signals (input vector); Y is the output signals (output vector); Rs is the state register, which stores the code of the present state; λ is a combinational circuit (logic) that determines the code of the next state (λ: $X \times S \rightarrow S$); δ is a combinational circuit that forms the values of the output signals; S is the set of states of the FSM. For Moore FSM, the output signal values are formed based on the present state code (δ: $S \rightarrow Y$), and for Mealy FSM, based on the present state code and the input signal values (δ: $X \times S \rightarrow Y$).

W. Zamojski et al. (Eds.): DepCoS-RELCOMEX 2026, LNNS 2014, pp. 389–399, 2026.
https://doi.org/10.1007/978-3-032-27927-9_38

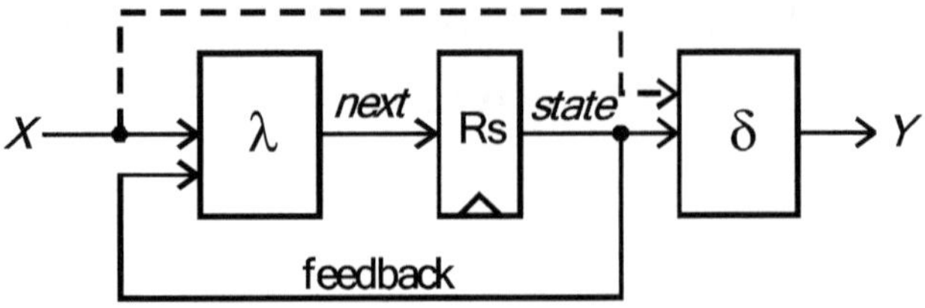

Fig. 1. Traditional structure of an FSM.

We will refer to the set *S* of states of an FSM as the set of valid or *legal* states. Due to a failure in the state register Rs, an FSM may enter an *illegal* state, from which transitions are not defined in the FSM specification. The FSM may remain in an illegal state indefinitely, i.e., the FSM fails to operate. Exiting an illegal state is only possible by resetting the FSM or by resetting the entire system. In many FSM applications, this behavior is unacceptable, for example, when the system is operating in the autonomous mode.

The causes of failure in the Rs state register can be solar radiation, cosmic rays, ionizing radiation near nuclear power plants and nuclear reactors, electromagnetic radiation, and laser beams from electronic warfare [3].

An FSM is called a safe FSM (SFSM) if it is guaranteed to return from any illegal state to its initial state without resetting the FSM or the entire system [4]. The design of SFSMs is critical in applications such as medical life support equipment, transportation systems, especially avionics and unmanned aerial vehicles, and autonomous robotic systems. Note that the design tools Quartus [4] and Vivado [5] offer options for SFSMs synthesis. In these synthesis modes, logic is added to the FSM to exit illegal states. However, adding extra logic to the FSM circuit significantly increases the area and reduces its performance.

This paper addresses the problem of designing SFSMs at the hardware description language level (HDL). The paper presents three styles for describing SFSMs in Verilog HDL that surpass in terms of area and performance not only the FSM synthesis method in the Quartus design tool [4], but also the traditional description of FSMs with three processes [6].

The main goal of this work is to develop effective styles for the SFSM description in HDL, in terms of area and performance.

The main contributions of this work are:

- the definition of three types of SFSMs: with a transition from any illegal state to the initial state (*safe*), additionally with the formation of an *error* signal (*safe_error*), and additionally with a transition to the *idle* state (*safe_idle*);
- three styles of describing SFSMs in Verilog HDL: *safe, safe_error*, and *safe_idle*;
- experimental studies of the effectiveness of the proposed styles of describing SFSMs;
- recommendations for the practical use of the proposed styles of describing SFSMs.

Note that the proposed approach to designing SFSMs is universal and can be easily adapted to other hardware description languages (e.g., VHDL or SystemVerilog) and other design tools (e.g., Vivado). It can also be used across different hardware platforms (e.g., systems-on-a-chip or application-specific integrated circuits).

The following works are most closely related to the topic of designing SFSMs. In [7], an efficient representation is proposed that adds a finite set of parameters to FSMs when modeling discrete-event systems; algorithms are proposed for autonomous and online synthesis of security management policies. In [8], FSMs extended with variables are considered; an algorithm for synthesizing a supervisor to enhance the protective functions of an FSM is proposed, which makes prohibited or blocking states in the controlled object unattainable. In [9], the SFSM is used to control an elevator. In [10], FSMs are used to analyze the safety of complex systems, such as aircraft engines. In [11], the safe configuration of asynchronously interacting FSM systems is considered. In [12], an FSM-based method for modeling and analyzing process safety is presented.

Description styles for reliable FSMs in Verilog HDL are given in [13], and for fault-tolerant FSMs are given in [14].

A review of the most relevant works on the subject showed that there are no methods that are directly (explicitly) aimed at designing SFSMs. This paper proposes new SFSM description styles in Verilog HDL. The main difference between the proposed approach to designing SFSMs and known approaches is that the task is performed at the level of describing the FSM project in HDL.

2 Designing SFSMs

2.1 Number of Illegal States of FSMs

Let *Sp* be the set of potential states to which the FSM can be entered, and *Si* be the set of illegal states of the FSM. The number of elements in *Sp* is determined by the number of binary codes that can be stored in the state register Rs:

$$|\mathrm{Sp}| = 2^{\mathrm{R}}|\mathrm{Sp}| = 2^{\mathrm{R}}, \tag{1}$$

where |A||A| is the cardinality of set *A*; *R* is the number of bits in the state codes (code length). For binary codes

$$\mathrm{R} = \log_2\mathrm{MR} = \log_2\mathrm{M}, \tag{2}$$

where BB is the smallest integer greater than or equal to *B*, and *M* is the number of states of the FSM. For one-hot codes, the code length *R* is determined using the following expression:

$$\mathrm{R} = \mathrm{MR} = \mathrm{M}, \tag{3}$$

i.e., the length of the code is equal to the number of states of the FSM.

The number of illegal states of an FSM is determined by the cardinality of the set *Si*:

$$|\mathrm{Si}| = |\mathrm{Sp}| - |\mathrm{S}| = 2^{\mathrm{R}} - \mathrm{M}|\mathrm{Si}| = |\mathrm{Sp}| - |\mathrm{S}| = 2^{\mathrm{R}} - \mathrm{M}. \tag{4}$$

From expression (4) it follows that the number of illegal states of FSMs increases exponentially with increasing length of state codes. Therefore, to reduce the number of illegal states and increase the safety of FSMs, the states of FSMs should be encoded using minimum-length codes, such as binary or Gray codes.

Note that when the number of states M coincides with a power of two and binary encoding is used, an FSM can't enter an illegal state. However, in this case, a failure in the state register Rs leads to an invalid transition from one legal state to another, which also causes the FSM to fail.

2.2 Types of SFSMs

Figure 2 shows the generalized structures of state transition graphs (STGs) of SFSMs, where the set Sp of potential states of an FSM is divided into two subsets: S, the set of legal states, and Si, the set of illegal states.

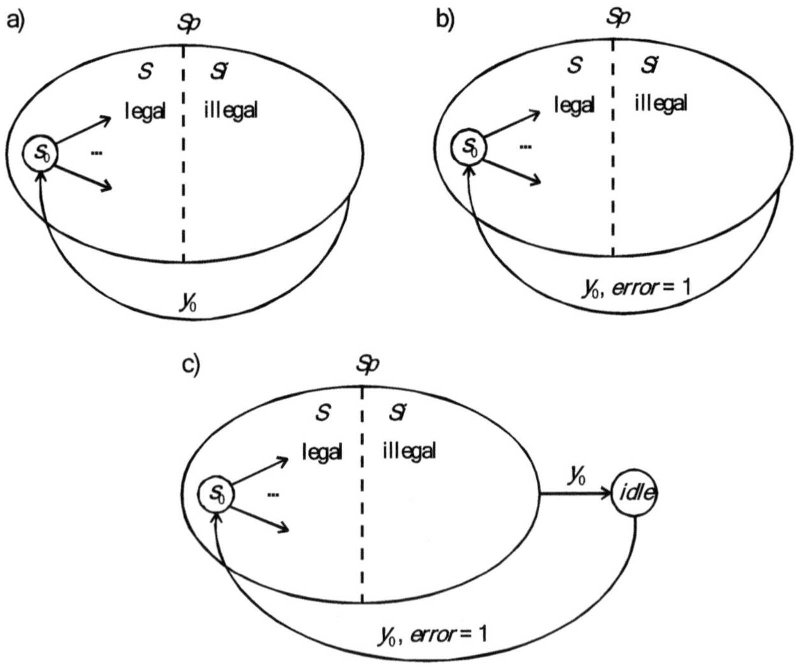

Fig. 2. STG stuctures of SFSMs: **a**—*safe*; **b**—*safe_error*; **c**—*safe_idle*.

In the STG structure in Fig. 2a for the *safe* type SFSM, the transitions from all illegal states are unconditional. They lead to the initial state s_0, and a zero output vector y_0 is formed on these transitions. In the event of a failure in the state register Rs, an SFSM returns to the initial state s_0, creating a zero vector y_0 at its outputs.

The disadvantage of an SFSM of the *safe* type is that the user (i.e., the external environment) cannot explicitly determine whether there was actually a failure in the state register Rs and whether the FSM entered an illegal state. The only indication of this event is the zero vector y_0 at the output of the FSM. However, if specific legal transitions of the FSM create a zero output vector, it becomes impossible to detect a failure in the state register Rs.

One solution to this problem could be to explicitly generate an additional *error* output signal, as shown in Fig. 2b. In an SFSM of the *safe_error* type, on unconditional transitions from illegal states to the initial state s_0, in addition to the zero output vector y_0, the *error* signal equal one. The *error* signal indicates an error when the FSM enters an illegal state.

The disadvantage of *safe_error*-type SFSMs is that the *error* signal is set for a very short period of time (especially for Mealy-type FSMs), and the external environment may not have time to perform the necessary actions to compensate for the consequences of the FSM entering an illegal state.

Figure 2c shows the structure of the STG of a *safe_idle* type SFSM when an additional *idle* state is used. In the event of an illegal state, the SFSM in Fig. 2c transitions to an additional *idle* state, from which it returns to the initial state s_0, generating the signal *error* equals one and a zero output vector y_0. For Moore-type FSMs, the value of signal *error* is equal to one is generated in the *idle* state. If necessary, in the *idle* state, in addition to the zero output vector y_0 and $error = 1$, additional output signals can be generated to compensate for the consequences of the FSM entering an illegal state.

2.3 Description Styles for SFSMs

In a traditional Verilog HDL description of FSMs, the design code is represented as three processes ('always' procedures): a state register description, a transition function description, and an output function description [6]. The *state* and *next* variables are used to represent the present and next state codes, and the *x* and *y* variables are used to define the input and output vectors of the FSM. The transition and output functions are described using 'case' statements at two levels. The first-level 'case' statements check the *state* variable (the present state code) and, depending on its value, determine the necessary actions in each state. The second-level 'case' statements check the input vector *x* and, depending on its value, calculate the value of the *next* variable (the next state code) in the second process and the value of the *y* variable (the value of the output vector) in the third process.

The standard way to define default values in 'case' statements is to use the 'default' construct. However, preliminary research has shown that a much more effective technique in terms of reducing area and increasing performance is to use a blocking assignment statement ('=') to determine the default value, which is placed before the 'case' statement in the 'begin-end' block.

The SFSM description styles, *safe*, *safe_error*, and *safe_idle*, correspond to the SFSM types of the same names discussed earlier.

In the style *safe*, the default values for the variables *next* and *y* are defined using blocking assignment statements before the 'case' statement in the 'begin-and' block of the 'always' procedure. This allows the FSM, when the 'Extract Verilog State Machine' parameter is set to 'Off', to return from illegal states to the initial state s0 with a zero vector at the output.

In the *safe_error* style, unlike the *safe* style, the port list includes an *error* output signal of type 'wire', which is generated using the continuous assignment statement 'assign'. The need to use the statement 'assign' to generate the *error* signal is due to the asynchronous nature of the outputs of Mealy-type FSMs. When generating the *error* signal in the traditional way in the 'case' statement, the *error* signal may be falsely set to 1.

In the *safe_idle* style, unlike the *safe* style, the port list includes the *error* signal, and the *idle* state is added to the FSM state set. In the transition function description, the default transition to the *idle* state is implemented by assigning the variable *next* the value *idle* using a blocking assignment statement, placed before the 'case' statement. Similarly, in the description of the output functions, default values are defined for the *error* signal and the variable *y*.

3 Experimental Studies

The effectiveness of the *safe, safe_error*, and *safe_idle* styles for describing SFSMs was compared with the traditional description of FSMs (*3proc* style) and SFSMs generated by the Quartus design tool when the 'Safe State Machine' option is set to 'On' (*safe_Q* style). The experiments were conducted on FSM benchmarks from the Microelectronics Center of North Carolina (MCNC) [15] using the Quartus Prime design tool version 24.1 to implement FSMs in the Cyclone 10 LP FPGA family. When using the *safe, safe_error*, and *safe_idle* styles, the 'Extract Verilog Machine' option of the Quartus design tool was set to 'Off' to enable the detection of illegal states in FSMs. The area of FSMs was measured by the number of look-up tables (LUTs) used, and the maximum operating frequency was used to determine performance.

3.1 Results of Experimental Studies

The results of experimental studies are presented in Tables 1 and 2, where *L3p, Lsq, Ls, Le*, and *Li* are the areas of FSMs when using the styles *3proc, safe_Q, safe, safe_error*, and *safe_idle*, respectively; *Lmin* is the minimum area value achieved by one of the styles *safe, safe_error*, or *safe_idle*; *F3p, Fsq, Fs, Fe*, and *Fi* are the performance (in megahertz) of FSMs when using the *styles 3proc, safe_Q, safe, safe_error*, and *safe_idle*, respectively; *Fmax* – maximum performance value achieved by one of the *safe, safe_error*, or *safe_idle* styles; *L3p/Lmin, Lsq/Lmin, Fmax/F3p*, and *Fmax/Fsq* – ratios of the corresponding parameters; *Av* – arithmetic mean value of the parameter; *Max*—maximum value of the parameter.

Table 1 illustrates that employing the proposed styles for describing SFSMs results in an average area reduction of 79.3% (a factor of 1.793) for FSMs when compared to the traditional description method (parameter *L3p/Lmin*). Additionally, there is a reduction factor of 2.436 when compared to the Quartus design tool method (parameter *Lsq/Lmin*). Notably, for the train4 example, the area was minimized by as much as 5.333 times compared to the traditional description and by 6.667 times compared to the Quartus design tool method.

Table 2 demonstrates that the use of the proposed styles for describing SFSMs results in an average improvement in FSM performance by a factor of 1.834 (or 83.4%) compared to traditional descriptions (parameter *Fmax/F3p*), and by a factor of 1.588 (or 58.8%) when compared to the Quartus design tool method (parameter *Fmax/Fsq*).

Table 1. Area of FSMs

FSM	*3proc*	*safe_Q*	*safe*	*error*	*idle*	*Lmin*	*L3p/Lmin*	*Lsq/Lmin*
	L3p	*Lsq*	*Ls*	*Le*	*Li*			
bbtas	8	13	6	7	7	6	1.333	2.167
beecount	36	45	19	21	21	19	1.895	2.368
cse	96	121	71	79	79	71	1.352	1.704
dk14	42	52	34	34	34	34	1.235	1.529
dk15	28	33	15	15	15	15	1.867	2.200
dk17	21	29	13	13	13	13	1.615	2.231
dk27	8	15	5	6	6	5	1.600	3.000
dk512	18	33	15	16	16	15	1.200	2.200
ex2	64	94	45	40	51	40	1.600	2.350
ex3	34	43	19	20	20	19	1.789	2.263
ex4	44	67	36	37	40	36	1.222	1.861
ex5	32	43	14	15	18	14	2.286	3.071
ex6	57	69	55	55	52	52	1.096	1.327
ex7	32	48	19	20	26	19	1.684	2.526
lion	15	18	3	3	3	3	5.000	6.000
lion9	47	62	13	14	15	13	3.615	4.769
mc	9	12	6	6	6	6	1.500	2.000
pma	114	167	105	104	110	104	1.096	1.606
s208	52	72	35	37	26	26	2.000	2.769
s27	20	27	7	6	12	6	3.333	4.500
s298	611	799	291	357	885	291	2.100	2.746
s420	38	52	35	36	22	22	1.727	2.364
s510	69	128	76	76	75	75	0.920	1.707
sand	208	267	164	164	160	160	1.300	1.669
shiftreg	9	14	3	3	3	3	3.000	4.667
styr	183	237	187	189	171	171	1.070	1.386
tbk	246	285	86	86	94	86	2.860	3.314
tma	128	173	78	79	78	78	1.641	2.218
train11	38	55	19	20	22	19	2.000	2.895
train4	16	20	6	3	3	3	**5.333**	**6.667**
Av	73.811	98.703	53.459	55.946	70.189	51.676	**1.793**	**2.436**
Max							**5.333**	**6.667**

Notably, the performance increased by as much as 4.951 times for the train4 example compared to traditional descriptions, and by 2.138 times for the shiftreg example compared to the Quartus design tool method.

The reduction in area and increase in the performance of FSMs as a result of using the proposed styles is explained by the fact that the Synthesizer of the design tool generates only the logic that is explicitly defined in the 'case' statements, since the implicit value has already been described earlier using the blocking assignment statement ('=') in the corresponding 'begin-end' block before the first-level 'case' statement.

3.2 Recommendations for the Practical Use of SFSM Description Styles

First of all, codes of minimum length, such as binary or Gray codes, should be used to encode the states of SFSMs. This reduces the number of illegal states and the probability of an FSM entering an illegal state.

The *safe* style can be used to describe SFSMs in which a zero output vector is not formed during any FSM transition. In this case, a zero output vector can indicate that the FSM has entered an illegal state. Otherwise, the *safe_error* style should be used. In the *safe_error* style, a value one of the additional output signal *error* serves as an indication that the FSM has entered an illegal state. The *safe_error* style is better suited for Moore-type FSMs, since for Mealy-type FSMs, the duration of the $error = 1$ signal can be very short. The *safe_idle* style should be used when it is necessary to perform additional actions in the *idle* state to mitigate the negative consequences of the FSM entering an illegal state. To do this, in the *idle* state, you can generate values of additional output signals or a special output vector.

4 Conclusion

This paper discusses three types and their corresponding three styles of describing SFSMs in Verilog HDL: *safe*, *safe_error*, and *safe_idle*. The *safe* style corresponds to the traditional implementation of SFSMs, where, when an FSM enters an illegal state, it returns to its initial state. In the *safe_error* style, an additional output *error* signal is generated, with a value of 1 indicating that the FSM has entered an illegal state. The *safe_idle* style provides for a transition to an additional *idle* state before the FSM returns from an illegal state to its initial state.

Experimental studies using FSM benchmarks demonstrated the high efficiency of the proposed SFSM description styles. Compared with the SFSM synthesis method in the Quartus design tool, the FSM area is reduced on average by a factor of 2.436 (6.667 times for individual examples), while performance is increased on average by a factor of 1.588 (2.741 times for individual examples).

Future research will aim to develop structural models of safe, robust, and fault-tolerant FSMs, i.e., to solve the problem of synthesizing such FSMs at the structural level.

Table 2. Performance of FSMs

FSM	*3proc*	*safe_Q*	*safe*	*error*	*idle*	*Fmax*	*Fmax/ F3p*	*Fmax/ Fsq*
	F3p	*Fsq*	*Fs*	*Fe*	*Fi*			
bbtas	937	534	912	916	900	916	0.978	1.715
beecount	251	542	627	632	641	641	2.554	1.183
cse	174	273	509	457	457	509	2.925	1.864
dk14	541	431	579	557	556	579	1.070	1.343
dk15	651	637	885	895	895	895	1.375	1.405
dk17	637	455	631	652	652	652	1.024	1.433
dk27	937	528	922	931	931	931	0.994	1.763
dk512	726	359	660	639	639	660	0.909	1.838
ex2	224	280	348	584	419	584	2.607	2.086
ex3	240	368	529	538	552	552	2.300	1.500
ex4	221	364	534	483	477	534	2.416	1.467
ex5	245	406	640	638	579	640	2.612	1.576
ex6	197	359	519	521	539	539	2.736	1.501
ex7	214	349	504	497	488	504	2.355	1.444
lion	299	595	1305	1302	1302	1305	4.365	2.193
lion9	149	452	596	633	550	633	4.248	1.400
mc	1250	720	831	823	823	831	0.665	1.154
pma	86	247	343	276	320	343	3.988	1.389
s208	461	280	428	428	574	574	1.245	2.050
s27	537	471	815	824	644	824	1.534	1.749
s298	318	159	161	144	207	207	0.651	1.302
s420	527	307	435	407	621	621	1.178	2.023
s510	917	232	341	**362**	**365**	365	0.398	1.573
sand	173	213	334	336	355	355	2.052	1.667
shiftreg	1078	482	1321	1319	1319	1321	1.225	**2.741**
styr	166	239	287	323	324	324	1.952	1.356
tbk	325	254	361	375	348	375	1.154	1.476
tma	139	273	341	358	396	396	2.849	1.451
train11	295	370	580	591	524	591	2.003	1.597
train4	263	609	1093	1302	1302	1302	**4.951**	2.138
Av	465.5	391.4	595.3	604.0	603.1	627.1	**1.834**	**1.588**
Max							**4.951**	**2.741**

Acknowledgments. The present study was supported by a grant WZ/WI-III/5/2023 from Bialystok University of Technology and founded from the resources for research by Ministry of Science and Higher Education.

References

1. Basile, F., Chiacchio, P.: On the implementation of supervised control of discrete event systems. IEEE Trans. Control Syst. Technol. **15**(4), 725–739 (2007). https://doi.org/10.1109/tcst.2006.890281
2. Monterrosa, N., Montoya, J., Jarquín, F., Bran, C.: Design, development and implementation of a UAV flight controller based on a state machine approach using a FPGA embedded system. In: IEEE/AIAA 35th Digital Avionics Systems Conference (DASC), pp. 1–8. IEEE, Sacramento, CA, USA (2016). https://doi.org/10.1109/dasc.2016.7778069
3. Park, S., Kim, H.T., Lee, S., Joo, H., Kim, H.: Survey on anti-drone systems: Components, designs, and challenges. IEEE Access. **9**, 42635–42659 (2021). https://doi.org/10.1109/access.2021.3065926
4. Intel Quartus Prime Standard Edition User Guide. Design Compilation. 3.4.12. Safe State Machine. https://www.intel.com/content/www/us/en/docs/programmable/683283/18-1/safe-state-machine.html. Accessed 30 Dec 2025
5. Xilinx: Vivado Design Suite User Guide: Synthesis UG901. https://docs.amd.com/r/en-US/ug901-vivado-synthesis/FSM_SAFE_STATE. Accessed 30 Dec 2025
6. Cummings, C.E., Chambers, H.: Finite state machine (FSM) design & synthesis using SystemVerilog–part I. In: Proceedings of Synopsys Users Group (SNUG), San Jose, CA, USA (2019) https://www.scribd.com/document/834329060/CummingsSNUG2019SV-FSM1. Accessed 30 Dec 2025
7. Chen, Y. L., Lin, F.: Safety control of discrete event systems using finite state machines with parameters. In: Proceedings of the 2001 American Control Conference, pp. 975–980. IEEE, Arlington, VA, USA (2001). doi:10.1109/acc.2001.945847
8. Ouedraogo, L., Kumar, R., Malik, R., Akesson, K.: Nonblocking and safe control of discrete-event systems modeled as extended finite automata. IEEE Trans. Autom. Sci. Eng. **8**(3), 560–569 (2011). https://doi.org/10.1109/tase.2011.2124457
9. Anreddy, S.R., Charan, K.S., Sai, N.N., Panda, N.: Modelling and Design of a Finite State Machine Based Elevator Control System for Efficient Vertical Transportation. In: 5th IEEE Global Conference for Advancement in Technology, pp. 1–7. IEEE, Bangalore, India (2024). https://doi.org/10.1109/gcat62922.2024.10924000
10. Song, X., Qi, L., Liu, S., Ding, S., Li, D.: Simple analysis of complex system safety based on Finite State Machine Network and phase space theory. Reliab. Eng. Syst. Saf. **249**, 110205 (2024). https://doi.org/10.1016/j.ress.2024.110205
11. Barbanera, F., Hennicker, R.: Safe composition of systems of communicating finite state machines (2024). arXiv: 2412.08234. https://doi.org/10.4204/eptcs.414.3
12. Zhao, N., Wu, F., Zhang, Y., Wang, W.: Process safety modeling and analysis method based on finite state machine. J. Phys. Conf. Ser. 2977(1), 012110. IOP Publishing. (2025). https://doi.org/10.1088/1742-6596/2977/1/012110
13. Salauyou, V.: Styles for describing reliable finite state machines in Verilog HDL. In: International Conference on Dependability of Computer Systems, pp. 241–251. Springer Nature Switzerland, Cham (2024). https://doi.org/10.1007/978-3-031-61857-4_24
14. Salauyou, V.: Description styles of fault-tolerant finite state machines for unmanned aerial vehicles. Radioelectron. Comput. Syst. **2024**(1), 196–206 (2024). https://doi.org/10.32620/reks.2024.1.15

15. Yang, S.: Logic synthesis and optimization benchmarks user guide: version 3.0, pp. 502–508. Microelectronics Center of North Carolina (MCNC), Research Triangle Park, NC, USA (1991)

Analysis of the Impact of Dynamic Adaptive Refinement Criteria on the Quality and Performance of Three-Dimensional Fluid Simulation Using the Smoothed Particle Hydrodynamics Method

Andrzej Sołtysik[1] and Marek Woda[2(✉)]

[1] Wrocław, Poland

[2] Department of Computer Engineering, Wroclaw University of Technology, Janiszewskiego 11-17, 50-372 Wrocław, Poland

marek.woda@pwr.edu.pl

Abstract. The present thesis sets out a methodology for implementing adaptive refinement in the context of incompressible fluid simulation, utilising the Smoothed Particle Hydrodynamics (SPH) method. The approach is founded on dynamic refinement criteria, with the objective of generalising the static, case-specific criteria commonly encountered in the extant literature. The paper delineates the architecture and implementation of a framework that supports both standard and adaptive three-dimensional SPH simulations. The two approaches are evaluated using the following three classical benchmark cases are hereby presented for your consideration: The phenomena of Poiseuille flow, lid-driven cavity flow, and dam break are of particular interest. The objective of this comparison is to ascertain the conditions under which adaptive resolution yields measurable advantages in terms of accuracy or performance.

Keywords: Smoothed Particle Hydrodynamics · Adaptive Refinement · CUDA

1 Introduction

In recent decades, fluid simulation has become a key topic in the natural sciences, engineering, and computer graphics. The significance of this research area extends far beyond theoretical academic considerations, finding practical applications across various economic sectors—from predicting the effects of floods and modeling climate change, through analyzing blood flow in vessels, to producing realistic special effects in films and video games. In all of these domains, reliable and efficient modeling of fluid behavior is essential. Moreover, such simulations often extend beyond closed, regular geometric domains. Flows frequently occur in open environments, where different media dynamically interact.

W. Zamojski et al. (Eds.): DepCoS-RELCOMEX 2026, LNNS 2014, pp. 400–412, 2026.
https://doi.org/10.1007/978-3-032-27927-9_39

Fluid simulation is one of the most demanding problems in both physics and computer graphics. Despite the well-established physical description of fluids, the problem is not yet analytically solved in general case [2] and requires the use of approximate methods. Numerical approximations of differential equations inherently lead to deterministic chaos, which further complicates the reliability of calculations.

Among various approaches to fluid simulation, the Smoothed Particle Hydrodynamics (SPH) method has gained considerable popularity due to its intuitive concept of representing fluids as particles, its Lagrangian formulation, and its versatility. This method has been successfully applied across a wide spectrum of use cases—from physically accurate scientific simulations to visually appealing animations, accommodating both incompressible and compressible fluids.

However, despite its advantages, SPH faces a fundamental challenge: computational performance. Capturing subtle details requires increasing the simulation's resolution, which dramatically extends computation time. This inherent dependency creates a trade-off between simulation quality and performance.

One proposed solution to this trade-off is adaptive resolution adjustment. This concept involves increasing the number of particles in regions where higher accuracy is needed, as well as decreasing it where such precision is unnecessary. This approach allows this simulation to preserve important details of the system while improving overall computational efficiency. However, this concept is typically implemented through manual tuning of parameters for each individual test case, making them difficult to generalize across different scenarios. Identifying dynamic, automatic refinement criteria remains an open problem.

Beyond raw performance, the dependability of numerical methods is a critical concern in engineering applications. Adaptive refinement introduces new challenges to dependability: while it promises to maintain accuracy with fewer computational resources, the dynamic nature of particle splitting and merging may introduce spatially non-uniform errors that are harder to predict and bound than those arising in uniform-resolution simulations.

2 Smoothed Particle Hydrodynamics

Smoothed Particle Hydrodynamics is fully Lagrangian method for fluid simulation, originally introduced by Lucy [5] and independently by Gingold & Monaghan [4] in the context of astrophysical problems. Over the years, the method has evolved into a versatile tool for modeling a wide range of phenomena, including incompressible and compressible flows, multiphase interactions, free-surface dynamics, and fluid-structure interactions.

Unlike Eulerian methods, which dicretize the computation domain into a fixed grid, SPH represents the fluid as a set of particles that move with the flow. Each particles carries physical properties such as mass, density, pressure, etc. This formulation naturally avoids many difficulties associated with moving interfaces, making SPH especially suitable for situational free-surfaces and highly deformable domains.

The physical behavior of fluids is classically described by the Navier-Stoke equation. For fluids that are incompressible (i.e., such that $\nabla \cdot \mathbf{v} = 0$), it takes the form:

$$\rho \left(\frac{\partial \mathbf{v}}{\partial t} + (\mathbf{v} \cdot \nabla) \mathbf{v} \right) = -\nabla p + \mu \Delta \mathbf{v} + \rho \mathbf{F}, \tag{1}$$

where:

ρ – fluid density
$\mathbf{v}$ – fluid velocity
∇p – pressure gradient
$\mathbf{F}$ – resultant external forces acting on the fluid

In Eulerian methods, this equation is solved within static grid cells [3]. In contrast, SPH performs all calculations in a particle-centered manner, avoiding the need for special handling of e.g. divergence of velocity field.

The fundamental concept of SPH is the kernel-based interpolation of fields quantities. Any quantity A at position $\mathbf{r}$ is estimated as [4]:

$$A(\mathbf{r}) \approx \sum_j A_j \frac{m_j}{\rho_j} \cdot W(\mathbf{r}_{ij}, h_j) \tag{2}$$

where:

$\mathbf{r}$ – position of a particle
$\mathbf{r}_{ij}$ – $\mathbf{r}_i - \mathbf{r}_j$
$W(\mathbf{r}, h)$ – smoothing kernel
h – smoothing radius of a particle
m – mass of a particle
ρ – density of a particle

Despite the possibility of accelerating SPH simulations using GPUs, the challenge of efficiently increasing simulation resolution remains unresolved. In the standard approach, increasing the resolution in regions requiring high detail also necessitates raising it in areas where is unnecessary. This is due to the constant particle size across the entire simulation domain.

Adaptive Mesh Refinement is widely used in Eulerian methods to concentrate computational effort where it is most needed. Analogously, adaptive SPH techniques aim to vary the particle resolution across the domain.

One influential direction was introduced by Vacondio et al. [7,8], who proposed dynamic particle splitting in regions requiring higher accuracy and particle merging where fine resolution offers no benefit. This method maintains conservation properties while allowing the simulation to focus computational effort on dynamically important regions.

The effectiveness of such techniques, however, strongly depends on the refinement criteria. Vacondio's method uses geometric regions defined manually for each scenario. While simple, this approach lacks generality and must be retuned for every new simulation setup. As a result, the formulation of robust, universally applicable dynamic refinement criteria remains an open research problem.

3 SPH Formulation

In this work, the formulation proposed by Becker et al. [1] is adopted. Following this approach, the standard, summation density estimation is used for each particle i:

$$\rho_i = \sum_j m_j W(\mathbf{r}_{ij}, h_j). \tag{3}$$

The pressure of each particle is computed using Tait equation of state, which enforces very low density variations:

$$P = B\left[\left(\frac{\rho}{\rho_0}\right)^{\gamma} - 1\right], \tag{4}$$

where $B = \rho_0 c_s^2/\gamma$, ρ_0 is the reference (rest) density of the fluid, γ is the exponent (typically equal to 7), and c_s denotes the speed of sound in the fluid. The symmetrized version of pressure-induced acceleration, which preserves both linear and angular momentum, is then given by:

$$\mathbf{a}_i^{\text{pressure}} = -\sum_j m_j \left(\frac{P_i}{\rho_i^2} + \frac{P_j}{\rho_j^2}\right) \nabla W(\mathbf{r}_{ij}, h_j). \tag{5}$$

The physically based laminar viscosity, following Vacondio et al. [7], is dicretized as:

$$\mathbf{a}_i^{\text{viscosity}} = \sum_j m_j \left(\frac{4\nu_0 \|\mathbf{r}_{ij}\| \nabla \bar{W}_{ij}}{\rho_{ij} \|\mathbf{r}_{ij}\|^2}\right) (\mathbf{v}_i - \mathbf{v}_j). \tag{6}$$

where $\nabla \bar{W}_{ij} = \frac{1}{2}\left[\nabla W(\mathbf{r}_{ij}, h_j) + \nabla W(\mathbf{r}_{ji}, h_i)\right]$, $\rho_{ij} = \frac{1}{2}(\rho_i + \rho_j)$, and ν_0 denotes the kinematic viscosity of the fluid.

4 Adaptive Refinement

4.1 Merging Procedure

The purpose of the merging stage is to reduce the number of particles in regions of stable flow. Two nearest particles are merged into a single particle, provided that they lie within each other's smoothing radius. The merging scheme is illustrated in Fig. 1.

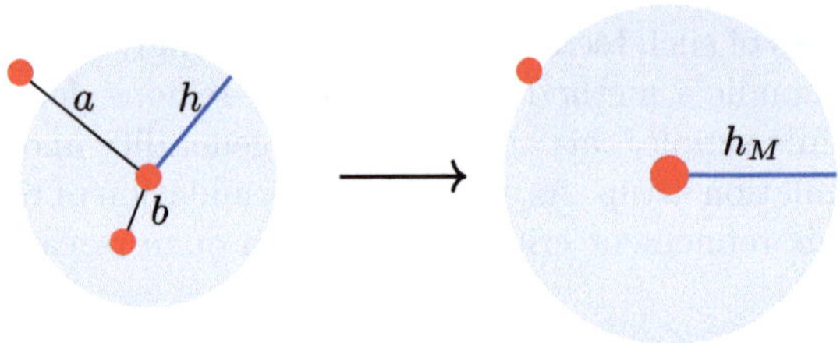

Fig. 1. Particle merging scheme. Distances a and b denote separation between the particles, and h_M is the smoothing radius of the particles created as a result of merging the two closest particles.

In the first stage, the merging criterion is evaluated, and the particles with positive criterion values are treated as candidates. Among all particles, those with the smallest masses and the lowest positive criterion values are selected as potential merging candidates. Each candidate then searches for its nearest neighbor that satisfies the merging condition. The merging operation is performed only for particle pairs that mutually select each other [6].

The properties of the newly created particle follow from the conservation of mass and momentum:

$$m_M = m_i + m_j, \quad \mathbf{v}_M = \frac{m_i \mathbf{v}_i + m_j \mathbf{v}_j}{m_M}, \quad \mathbf{r}_M = \frac{m_i \mathbf{r}_i + m_j \mathbf{r}_j}{m_M}. \tag{7}$$

Following Vacondio et al. [7], the new smoothing radius is computed in order to minimize density errors. For the Wendland kernel used as the smoothing function, the resulting smoothing radius is given by:

$$h_M = \sqrt[3]{\frac{16\pi}{21} \frac{m_M}{m_i W(\mathbf{r}_{Mi}, h_i) + m_j W(\mathbf{r}_{Mj}, h_j)}}. \tag{8}$$

After the merging procedure is completed, particles, that must be removed, are compacted to maintain the coherence of the data structure.

4.2 Splitting Procedure

The purpose of the splitting procedure is to increase the spatial resolution in regions characterized by high flow activity. Each particle that satisfies the splitting criterion is replaced with 13 daughter particles: one central particle and twelve particles positioned at the vertices of an icosahedron, as illustrated in Fig. 2.

First, the spitting criterion is evaluated, and particles with positive criterion values are considered candidates. Among these candidates, those with the highest positive criterion values are selected, representing particles located in regions of the highest flow activity.

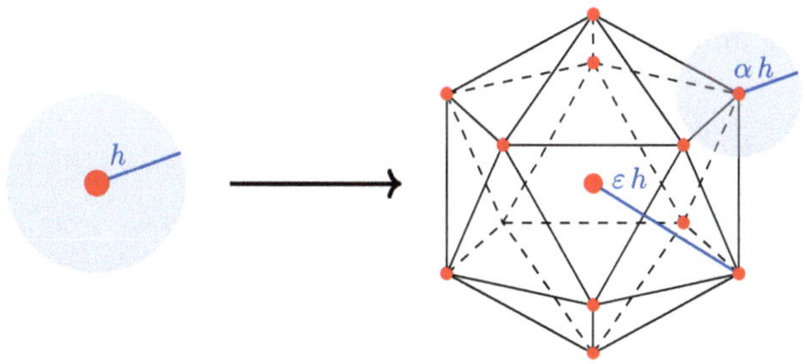

Fig. 2. Particle splitting scheme. The parameter α denotes the scaling factor of the smoothing radius and ε describes the distance scaling between the central daughter particle and the remaining ones.

In the next stage, each selected particle is replaced with new daughter particles, whose positions follow directly the geometry of the icosahedron. This process is controlled by the parameters:

- α, which determines the scaling of the smoothing radius,
- ε, which sets the distance of the daughter particles from the central daughter particle,
- γ_{min} and γ_{max}, which define the mass distribution between the central and outer particles, under the constraint $\gamma_{\text{max}}+12\cdot\gamma_{\text{min}} = 1$, ensuring conservation of mass.

According to the recommendations of Vacondio et al. [7], the following parameter values are recommended for the smallest density errors:

$$\alpha = 0.7, \quad \varepsilon = 0.65, \quad \gamma_{\text{min}} = \frac{1}{15}, \quad \gamma_{\text{max}} = \frac{1}{5}. \tag{9}$$

5 Dynamic Refinement Criteria

The adaptive refinement system implements three dynamic criteria for assessing local fluid properties. Each criterion defines independent thresholds for splitting and merging operations, enabling precise control over the adaptation process in response to various flow characteristics. For each criterion, the symbol S denotes the splitting criterion value and M denotes the merging criterion value.

5.1 Velocity Criterion

The velocity criterion provides an intuitive measure of local dynamic activity, based on the velocity vector magnitude of each particle. Regions exhibiting high velocities require particle densification to maintain simulation accuracy, whereas in low-velocity areas the resolution may be reduced to improve computation performance.

The splitting and merging criterion values are defined as follows:

$$S_i^{\text{velocity}} = \begin{cases} \|\mathbf{v}_i\| & \|\mathbf{v}_i\| > v^{\text{split}} \\ -1 & \text{otherwise} \end{cases} \tag{10}$$

$$M_i^{\text{velocity}} = \begin{cases} \|\mathbf{v}_i\| & \|\mathbf{v}_i\| < v^{\text{merge}} \\ -1 & \text{otherwise} \end{cases} \tag{11}$$

where:

v^{split} – minimal velocity threshold qualifying a particle for splitting
v^{merge} – minimal velocity threshold qualifying a particle for merging

5.2 Interface Boundary Criterion

The interface proximity criterion detects regions near fluid interfaces, including free surfaces, solid boundaries, and obstacles, where the spatial distribution of neighboring particles becomes asymmetric. These regions require higher simulation resolution to accurately capture interface dynamics, deformation, and potential fragmentation. Conversely, in the interior of the fluid where particles are uniformly surrounded by neighbors, the resolution may be potentially reduced.

The interface proximity indicator is based on the weighted average displacement vector from neighboring particles:

$$\mathbf{u}_i = \frac{\sum_j m_j W(\mathbf{r}_{ij}, h_j) \cdot \mathbf{r}_{ij}}{\sum_j m_j W(\mathbf{r}_{ij}, h_j)}. \tag{12}$$

The normalization ensures the indicator is independent of local particle density and smoothing length, providing a pure geometric measure of neighbor asymmetry. The physical interpretation is as follows: in the fluid interior, neighbors are distributed uniformly in all directions, causing their contribution vectors to cancel out, yielding $\|\mathbf{u}_i\| \approx 0$. Near any interface, neighbors exist predominately on one side, production a non-zero resultant vector pointing away from the interface.

After computing the interface proximity magnitude $\|\mathbf{n}_i\|$ for particle i, the concretion values are defined as:

$$S_i^{\text{interface}} = \begin{cases} \|\mathbf{u}_i\| & \|\mathbf{u}_i\| > u^{\text{split}} \\ -1 & \text{otherwise} \end{cases} \tag{13}$$

$$M_i^{\text{interface}} = \begin{cases} \|\mathbf{u}_i\| & \|\mathbf{u}_i\| < u^{\text{merge}} \\ -1 & \text{otherwise} \end{cases} \tag{14}$$

where:

u^{split} – minimal proximity magnitude threshold qualifying a particle for splitting
u^{merge} – minimal proximity magnitude threshold qualifying a particle for merging

5.3 Vorticity Criterion

The vorticity criterion detects regions with strong rotational motion (vorticity) in the velocity field, which require higher simulation resolution to accurately capture complex flow structures. Conversely, in low-vorticity regions the resolution may be safely reduced to conserve computational resources.

The vorticity vector is defined as:

$$\boldsymbol{\omega} = \nabla \times \mathbf{v} = \left[\frac{\partial v_z}{\partial y} - \frac{\partial v_y}{\partial z}, \frac{\partial v_x}{\partial z} - \frac{\partial v_z}{\partial x}, \frac{\partial v_y}{\partial x} - \frac{\partial v_x}{\partial y} \right]. \tag{15}$$

Each velocity derivative is approximated using:

$$\left. \frac{\partial v_\alpha}{\partial x_\beta} \right| i \approx \sum_j \frac{m_j}{\rho_j} (v_{j,\alpha} - v_{i,\alpha}) \nabla W \left(\mathbf{r}_{ij}, h_j \right). \tag{16}$$

where α and β are vector components, and α, $\beta \in \{x, y, z\}$.

After computing the vorticity magnitude $\|\omega_i\|$ for particle i, the concretion values are defined as:

$$S_i^{\text{vorticity}} = \begin{cases} \|\omega_i\| & \|\omega_i\| > \omega^{\text{split}} \\ -1 & \text{otherwise} \end{cases} \tag{17}$$

$$M_i^{\text{vorticity}} = \begin{cases} \|\omega_i\| & \|\omega_i\| < \omega^{\text{merge}} \\ -1 & \text{otherwise} \end{cases} \tag{18}$$

where:

ω^{split} – minimal vorticity magnitude threshold qualifying a particle for splitting
ω^{merge} – minimal vorticity magnitude threshold qualifying a particle for merging

6 Test Cases

The aim of research is analysis of the impact of dynamic adaptive refinement criteria on the quality of three-dimensional fluid simulation using the SPH method. During the experiments, three simulation variants will be compared:

- **Low resolution**—a simulation with constant number of particles equal to 80 000, considered as a reference case representing minimal computation effort.

- **Adaptive resolution**—a simulation with 80 000 particles, in which particle size and number are dynamically adapted. The particle count may remain close to the initial number to maintain symmetry between merging and splitting operations and to ensure a fair comparison with the low-resolution case.
- **High resolution**—a simulation with 1 000 000 particles, corresponding to the scenario in which all particles are fully split in the adaptive resolution experiment. This case is treated as the reference simulation to which other simulations are compared.

For each simulation variant, density errors were computed relative to the reference density of 1000 kg/m^3 using the root mean square error (RMSE). Velocity errors were calculated in the same manner, with each simulation variant compared against the high-resolution reference simulation, also using RMSE. For each type of experiment, velocity profile plots are presented for the low-resolution, high-resolution and for adaptive-resolution variants, where the adaptive resolution criterion was selected to achieve the lowest error.

6.1 Poiseuille Flow

The experiment focuses on the simulation of thee-dimensional Poiseuille flow. The simulations were performed in a computational domain of size $2 \times 2 \times 2$ with periodic boundary conditions applied along the x-axis. A constant driving force of 2 N was applied to simulate the pressure difference between the two ends. The quantitative results after 5-second simulation are summarized in Table 1. The velocity profiles in the x-direction, evaluated along a vertical cross-section of the domain, are shown in Fig. 3.

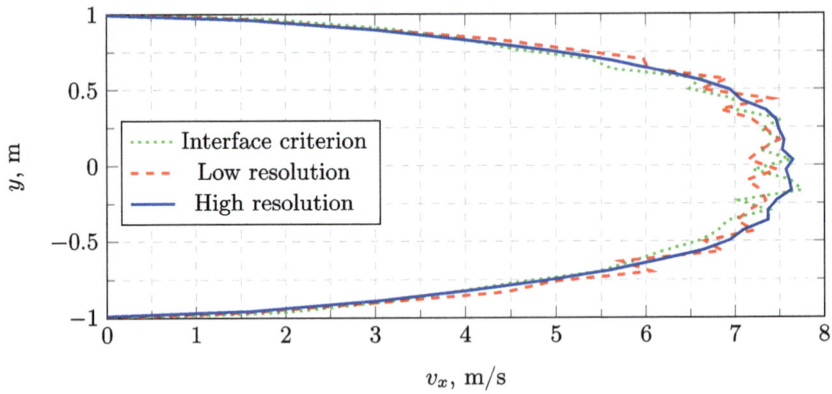

Fig. 3. Comparison of vertical velocity profiles in the x-direction for different resolutions in Poiseuille flow.

6.2 Lid-Driven Cavity

The lid-driven cavity test case introduces a more challenging flow configuration for method evaluation. The motion of the upper boundary generates complex flow structures within the domain. The computational domain has dimensions of $2 \times 2 \times 2$, with the top lid moving at a constant velocity of $\mathbf{v}_{\text{lid}} = (-10, 0, 0)$. The simulation was run for a total time of 5 s. The resulting error metrics are summarizes in Table 2. Velocity profiles extracted along a horizontal cross-section of the domain are presented in Fig. 4.

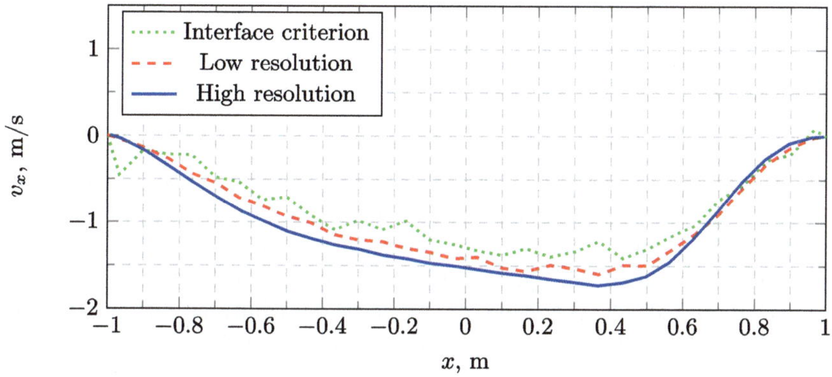

Fig. 4. Comparison of horizontal velocity profiles in the x-direction for different resolutions in lid-driven Cavity flow.

Table 1. Comparison of density and velocity errors for different Poiseuille flow configurations

Variant	Frame Time [ms]	Density Error [kg/m^3]	Velocity Error [m/s]
Low Resolution	19.42	18.33	0.37
High Resolution	165.72	1.54	–
Adaptive Resolution:			
Interface Criterion	52.49	6.49	0.40
Velocity Criterion	53.27	11.55	0.63
Vorticity Criterion	54.97	9.62	0.57

Table 2. Comparison of density and velocity errors for different lid-driven Cavity flow configurations

Variant	Frame Time [ms]	Density Error [kg/m^3]	Velocity Error [m/s]
Low Resolution	16.40	3.26	0.34
High Resolution	181.04	2.64	–
Adaptive Resolution:			
Interface Criterion	41.60	7.40	0.57
Velocity Criterion	44.03	9.03	0.60
Vorticity Criterion	42.28	8.87	0.62

6.3 Dam Break

The dam-break scenario is employed to asses the numerical handling of free-surface dynamics. In the setup phase, the fluid initially a confined region of size $2 \times 2 \times 2$, and subsequently evolves into a larger domain of dimensions $5 \times 2 \times 2$, corresponding to the sudden release of the fluid. The simulation covers a physical time span of 2 s. Quantitative error metrics obtained from this case are reported in Table 3. Additionally, velocity distributions extracted along a horizontal cross-section of the computation domain are shown in Fig. 5.

Table 3. Comparison of density and velocity errors for different Dam Break flow configurations

Variant	Frame Time [ms]	Density Error [kg/m^3]	Velocity Error [m/s]
Low Resolution	17.92	3.51	0.56
High Resolution	204.09	2.58	–
Adaptive Resolution:			
Interface Criterion	34.46	9.07	0.66
Velocity Criterion	32.93	9.63	0.60
Vorticity Criterion	27.85	7.79	0.64

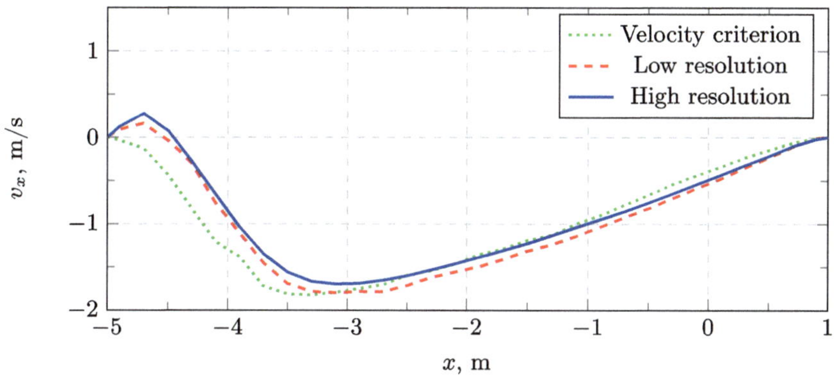

Fig. 5. Comparison of horizontal velocity profiles in the x-direction for different resolutions in dam break flow.

7 Conclusion

This paper proposed a method for the dynamic evaluation of an adaptive refinement criterion and introduced three such criteria based on velocity, interface proximity, and vorticity. The performance and accuracy of each criterion were evaluated by comparison with both low- and high-resolution reference simulations. The results indicate that, in all considered cases, the standard low-resolution simulation yielded better accuracy, with the high-resolution simulation performing best overall.

The observed degradation in accuracy of the adaptive approach is most likely caused by interactions between particles of significantly different refinement levels. In the proposed method, large merged particles can directly interact with both unrefined and split particles, leading to close coupling between particles with substantially different physical properties. This behavior differs from the original approach proposed by Vacondio, in which particle interactions between different refinement levels occur in spatially separated regions of the domain.

Although the obtained results do not demonstrate an improvement over uniform-resolution simulations, the observed discrepancies are moderate, suggesting that further refinement of the method may lead to improved performance. Importantly, one aspect of the proposed approach can be considered successful. The three-dimensional region-based refinement strategy of Vacondio was effectively reduced to two scalar parameters, which can be automatically and robustly tuned for different simulations. This represents a significant improvement over the original method, which relied on manual parameter adjustment.

From a dependability perspective, these results carry important implications on two distinct levels. At the level of simulation accuracy, the adaptive approach exhibits reduced dependability compared to uniform low-resolution simulation. The spatially heterogeneous errors introduced by interactions between particles of different refinement levels are harder to anticipate and bound than the uniform errors of a fixed-resolution run.

At the level of method reproducibility, however, the proposed parametrization represents a genuine improvement. Reducing the region-based refinement strategy of Vacondio to two scalar thresholds per criterion makes the method's behavior more consistent and transferable across different flow configurations—a property that fixed, manually-tuned geometric regions cannot offer. Future work should investigate whether introducing spatial separation between particles of different refinement levels, preventing direct interactions between the coarsest and finest particles, can reduce the spatially heterogeneous errors observed in this study.

References

1. Becker, M., Teschner, M.: Weakly compressible SPH for free surface flows, vol. 9, pp. 209–217 (2007). https://doi.org/10.1145/1272690.1272719
2. Clay Mathematics Institute: The Navier-Stokes Equation (2025). https://www.claymath.org/millennium/navier-stokes-equation/
3. Flynn, S., Egbert, P., Holladay, S., Oborn, J.: Adaptive fluid simulation using a linear octree structure. In: Proceedings of Computer Graphics International 2018, CGI 2018, pp. 217–222. Association for Computing Machinery, New York, NY, USA (2018). https://doi.org/10.1145/3208159.3208196
4. Gingold, R.A., Monaghan, J.J.: Smoothed particle hydrodynamics: theory and application to non-spherical stars. Mon. Not. R. Astron. Soc. **181**(3), 375–389 (1977). https://doi.org/10.1093/mnras/181.3.375
5. Lucy, L.B.: A numerical approach to the testing of the fission hypothesis. Astron. J. **82**, 1013–1024 (1977)
6. Muta, A., Ramachandran, P.: Efficient and accurate adaptive resolution for weakly-compressible SPH. Comput. Methods Appl. Mech. Eng. **395**, 115019 (2022)
7. Vacondio, R., Rogers, B., Stansby, P., Mignosa, P.: Variable resolution for SPH in three dimensions: towards optimal splitting and coalescing for dynamic adaptivity. Comput. Methods Appl. Mech. Eng. **300**, 442–460 (2016)
8. Vacondio, R., Rogers, B., Stansby, P., Mignosa, P., Feldman, J.: Variable resolution for SPH: a dynamic particle coalescing and splitting scheme. Comput. Methods Appl. Mech. Eng. **256**, 132–148 (2013)

FPGA Implementations of the Ascon Permutation Function

Jarosław Sugier(✉)

Faculty of Information and Communication Technology, Wrocław University of Science and Technology, Wrocław, Poland
jaroslaw.sugier@pwr.edu.pl

Abstract. The aim of this paper is to evaluate the characteristics of the permutation function used in the Ascon cipher—a recent NIST standard—with respect to feasibility and efficiency of its automatic implementation in field-programmable gate array devices using the software provided by chip manufacturer. Because the cipher is intended for lightweight applications, where high data throughput is not the primary requirement, the study focused on loop-unrolled architectures rather than pipelined ones. Starting from the standard iterative implementation, unrolled variants were created by cascading 2, 3, 4, 6, and 12 round modules, resulting in six design cases, with the last representing essentially a combinational structure. These implementations were assessed in terms of size and achievable operating frequency in Spartan-7 devices from AMD. The results illustrate how Ascon transformations—even if considered lightweight—are mapped onto the resources of contemporary programmable arrays and highlight the specific challenges encountered by the implementation tools.

Keywords: Cryptographic hash function · Implementation efficiency · Loop unrolling

1 Introduction

Ascon is a family of modern lightweight cryptographic algorithms built around a common core—the so-called Ascon permutation function—which was recently announces as a National Institute of Standards and Technology (NIST) standard of cryptography on constrained devices. The standard defines application of the permutation function in several contexts, including authenticated encryption with associated data (AEAD), computation of fixed-size hash functions, and extendable-output functions (XOFs). The aim of this work is to systematically examine the efficiency of automatic implementation of the permutation module in different architectures, using cost-effective field-programmable gate array devices from the Spartan-7 family (AMD) as the platform.

Effective hardware realization of the new standard, including on FPGA platforms, was one of the criteria used to review all candidates in the CAESAR and NIST competitions, and numerous results evaluating Ascon implementations are available. These include broad, methodical surveys [5] as well as highly customized and narrowly optimized proposals [4, 9]. However, the focus of this study differs from most of that work.

W. Zamojski et al. (Eds.): DepCoS-RELCOMEX 2026, LNNS 2014, pp. 413–422, 2026.
https://doi.org/10.1007/978-3-032-27927-9_40

Instead of optimizing the size or speed of a complete module that implements a specific function under the precise configurations required by the NIST for particular data sizes and formats, this work examines the intrinsic characteristics of the cipher permutation and how they are handled in automatic implementation by the software provided by the FPGA manufacturer. The results demonstrate how Ascon transformations map onto the resources of the programmable array and highlight the main challenges that implementation tools must address in order to realize in hardware the complex data processing required by this algorithm.

Ascon is specifically targeted at limited-resource devices which—even if have reduced computational power at their disposal—need efficient and secure protocols in environments like IoT, medical implants, automotive in-vehicle networks, solar-powered spacecrafts, small satellites, and alike. With these kinds of applications high data throughput is usually not required (being less important than, for example, power consumption) and therefore pipelined architectures have not been considered in this study as an option in porting the algorithm to hardware; instead, the focus will be on loop unrolled organizations. For the same reasons, cost-effective and more mature devices from Spartan-7 family will be selected as the implementation platform [2].

The rest of the paper is organized as follows. In the next section the Ascon algorithm is briefly outlined with emphasis on organization of the rounds and their elementary transformations. In particular, specific attention had to be paid to HDL coding styles and their consequences to implementation efficiency. The results are then presented in Sect. 3 which describes construction of the final designs, discusses their speed and size characteristics after implementation, and then looks closer at specific issues revealed in the process. Finally, conclusions summarizes the work in Sect. 4.

2 The Ascon Cipher Family

2.1 CEASAR Competition and NIST Lightweight Cryptography Project

CAESAR (Competition for Authenticated Encryption: Security, Applicability, and Robustness, [6]) began in 2013 as an independent, community-driven initiative overseen by a committee of eighteen prominent experts, including co-authors of the Advanced Encryption Standard and the Secure Hash Algorithm 3. Its primary goal was to develop a new generation of authenticated-encryption algorithms. More than fifty candidates were submitted to the first round and the extensive research conducted during their analysis and evaluation significantly advanced the state of the art in secret-key cryptography. The final selections were announced in 2019 and, among the winners, Ascon submission [7, 8] was nominated as the best algorithm for the "Lightweight Applications" use case.

Although not formally connected, the National Institute of Standards and Technology supported the CAESAR contest from its inception and closely followed its progress. Beginning in 2013, the Institute placed particular emphasis on cryptography for constrained environments, and in 2018 it officially announced the Lightweight Cryptography project with a call for a new, state-of-the-art cryptographic standard tailored to low-power, resource-limited devices. After three rounds of evaluation of fifty-seven submissions, in February 2023 the Ascon family was announced as the winner, and the new standard based on this algorithm was officially published in August 2025. The extensive work carried out on this algorithm during the CAESAR competition made its selection a natural outcome.

2.2 The Ascon Permutation Function

The standard [12] defines four algorithms: an AEAD scheme of a 128-bit security strength, a hash function with 256-bit output, an XOF function with user-selected size, and a variant of the XOF with a customization string. All these algorithms operate on 320-bit state data S composed of five 64-bit words S_0 ... S_4. The state is processed by a permutation function $Ascon\text{-}p[N_R]$, with N_R denoting number of rounds repeated in the function: this param is 12 except the AEAD scheme which uses instances with $N_R = 8$. The standard, on the other hand, defines necessary constants for N_R up to 16 to provide for future extensions with even larger number of rounds. In the rest of the paper we will consider configuration with $N_R = 12$.

The permutation function simply transform the state S without any other parameters:

$$S' = Ascon\ {-p}[N_R](S) \tag{1}$$

All other input – like key, nonce, auxiliary data, plaintext, message, etc. – are bitwise XOR'ed with the state outside of the function. This makes organization of the data flow regular and simplifies hardware implementations, as compared to e.g. BLAKE algorithm where the hashed message iterates through the rounds along with the state.

Computation of the function itself is also very regular and consists in application of a round function p to the state N_R times. All rounds are identical and use three transformations—constant addition, substitution and linear diffusion:

$$p = p_L \circ p_S \circ p_C \tag{2}$$

First, in the constant-addition phase (p_C) an 8-bit constant c_i corresponding to the round number i is xor'ed with least significant bits of S_2; all other words remain unchanged. With definition of 16 constants the standard allows for this maximum number of round iterations while at present the algorithms use 12 of them.

The second transformation p_S applies 5-bit substitution boxes to 64 bits of the S_i words:

$$(S_0(i), S_1(i), \ldots, S_4(i))\prime = SBox(S_0(i), S_1(i), \ldots, S_4(i)) \tag{3}$$

The standard defines the SBox as a set of Boolean expressions or an equivalent circuit representation (see Fig. 1) but also provides a standard truth table: a set of 32 final 5-bit constants which can be computed from such definitions.

Finaly, the last transformation p_L provides a linear diffusion within each S_i word:

$$S_0' = S_0 \oplus (S_0 \ggg 19) \oplus (S_0 \ggg 28)$$

$$S_1' = S_1 \oplus (S_1 \ggg 61) \oplus (S_1 \ggg 39)$$

$$S_2' = S_2 \oplus (S_2 \ggg 1) \oplus (S_2 \ggg 6) \tag{4}$$

$$S_3' = S_3 \oplus (S_3 \ggg 10) \oplus (S_3 \ggg 17)$$

$$S_4' = S_4 \oplus (S_4 \ggg 7) \oplus (S_4 \ggg 41)$$

where " >>>" denotes a word rotation by a given number of bits.

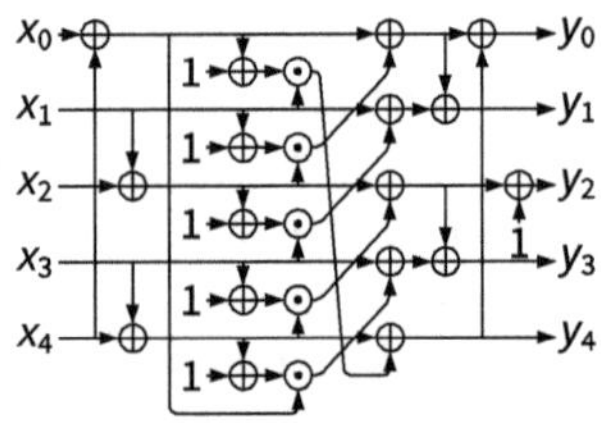

Fig. 1. Circuit representation of the Ascon substitution layer [12]

2.3 Style of the Reference Code

The authors refer to various reference VHDL implementations of the algorithm in hardware and their code could be applied in this study. It soon turned out, though, that the HDL style used in them is efficient in functional verification but can be problematic in implementation.

The reference code officially endorsed on author's webpage [10] describes at the Register-Transfer Level (RTL) only the overall structure of cipher processing but the essential and most resource-consuming part—the permutation function itself—is described in a behavioral style. In particular, it follows the circuit representation of the p_S step as in Fig. 1 instead of using a table of 32 constants. Description in behavioral style has its advantages: functionality of the whole round is represented at a high level of abstraction and an implementation tool can apply advanced automatic optimizations when searching for best realizations with resources available on a particular platform. Nevertheless, loop-unrolled architectures investigated in this work resulted in large combinational blocks spanning multiple rounds and the tools could not finish their synthesis in a reasonable time when the unrolling factor exceeded 3.

In such situation an original RTL code was prepared which closely followed definition of round transformations thus ensured their direct mapping during synthesis to the resources of the FPGA devices. In particular, as the arrays natively offer lookup tables implementing a 5-input / 2-bit output combinational function as an alternative to the standard 6-input / 1-bit output configuration [1], it was obvious to express p_L as 64 instances of the 5-input SBox'es rather than through a behavioral sequence of S_i transformations. The other round operations: 8-bit constant addition in p_L or word rotations in p_S also led to straightforward expressions. The overall code had very simple modular organization while the reference followed a procedural, software-like style.

The main focus of this work was on designs which use our original RTL code while the standard implementations endorsed by Ascon authors had limited applicability and will be included mainly for comparison. The suffix in design names—RTL or Std—will distinguish between two types of code used in particular case.

3 Implementing the Cipher in FPGA Devices

3.1 Designs

The set of considered Ascon architectures starts from the standard iterative organization (denoted as X1) where a hardware module representing one round repeatedly transforms the state data in $N_R = 12$ clock iterations, and includes implementations with the loop unrolled into a cascade of k rounds (den. Xk) which can finish whole computation in N_R / k iterations. Because $N_R = 12$, one can consider unrolled Ascon designs with $k = 2, 3,$ 4, 6, and 12 which creates the total set of 6 architectures included in the further analysis. The last case of X12 (the loop unrolled completely) is actually a combinational design where the entire cipher is computed by one continuous cascade of gates—an extreme case usually not implemented in practice but included here for completeness.

Each Xk module was equipped with a basic input / output shift registers which transferred the entire input or output state in 32-bit chunks in order to be able to fit designs in small packages with limited number of pins. This turned a permutation core into a complete system which the implementation tool would see as a fully functional design and would not try to optimize out any of its part. In all cases the Vivado 2025.1 was used as an implementation software.

3.2 Results

Parameters of the implemented designs are given in Table 1. Their sizes are given by the numbers of occupied lookup tables (LUT) and slices (nodes of the FPGA array) while the rest of the columns describe speed characteristics. Apart from maximum operational frequency (as estimated by the tools for the final, fully routed design) they describe the longest path which determined F_{max}: levels of logic within the path, delays incurred by the logic and routing resources, and percentage of the latter one in the total delay. The rows corresponds to particular designs cases: apart from architecture type (Xk) the suffixes in design names denote type of HDL code (original RTL or standard reference) and, for $k = 4$, 6, or 12, optional use of special "Spread Logic High" (SLH) implementation strategy—which will be introduced in Sect. 3.4.

It was an initial intention to fit all designs in the smallest device of the Spartan-7 family, i.e. xc7s6cpga196–1. With 2000 slices and 3750 LUTs it was used only in 25% by the smallest X1 design and provided enough room for the unrolled cases up to X6; only the fully combinational design had to be fit in the second larger xc7s15cpga196–1 chip. Application of special SLH implementation strategy made selection of even larger devices reasonable and xc7s50fgga484–1 chip was tested for the two largest designs.

Evaluation of the basic size and speed parameters listed in the table will assess efficiency of the unrolling mechanism and this will be the subject of Sect. 3.5 but first the discussion will focus on two another issues.

Table 1. Variants of the Ascon core and their implementation details

Variant (Device)	Slices	LUTs	F_{max} [MHz]	Levels of logic	Logic [ns]	Route [ns]	Route part
X1 RTL (1)	320	956	363	2	0.766	1.989	72.2%
X1 Std (1)	318	942	309	2	0.730	2.117	74.4%
X2 RTL (1)	481	1563	189	4	0.952	4.248	81.7%
X2 Std (1)	489	1643	177	4	0.952	4.660	83.0%
X3 RTL (1)	616	2183	123	6	1.200	6.814	85.0%
X3 Std (1)	619	2249	111	6	1.262	7.750	86.0%
X4 RTL (1)	585	2031	75.8	8	2.801	10.258	78.6%
X4 RTL + SLH (1)	741	2737	84.7	8	1.704	10.020	85.5%
X6 RTL (1)	784	2765	49.6	12	4.243	15.864	78.9%
X6 RTL + SLH (3)	1099	3986	58.1	12	1.944	15.157	88.6%
X12 RTL (2)	1282	4756	25.5	24	8.328	30.726	78.7%
X12 RTL + SLH (3)	1269	4757	25.9	24	7.631	30.838	80.2%

Device:
(1) xc7s6cpga196-1
(2) xc7s15cpga196-1
(3) xc7s50fgga484-1

3.3 Coding Styles: RTL Versus Behavioral Code

Working with the reference implementation revealed some limitations of the behavioral-level description, at least within the specific application context required in this study. While the iterative architecture X1 and the first two unrolled variants (X2 and X3) were implemented successfully in the Std variants, the larger designs—particularly X6 and X12—proved too demanding for the synthesis tool, which was unable to complete their synthesis within a reasonable time. No such difficulties were encountered with the proprietary RTL code, which synthesized smoothly in all examined configurations, up to and including X12.

Additional insight into this issue can be gained by comparing the implementation parameters of the X1 through X3 designs—that is, all variants for which synthesis could be completed with the Std code. Figure 2 shows the deviations of the RTL results from those obtained with the standard behavioral description. It proves that, on average, the register-transfer-level implementation produced slightly smaller and noticeably faster hardware. The particular differences in size are modest (in fact, the RTL variant occupies 1% or 2% more of slices or LUTs in the X1 case), but the maximum operating frequencies are up to 18% higher in the RTL cases. The logic and routing delay details further indicate that these improvements stem primarily from a reduction in routing resources, which typically account for the dominant share of path delays in field-programmable gate arrays. These observations was the decisive reason to adopt the proprietary register-transfer-level HDL code as the main base in the subsequent study.

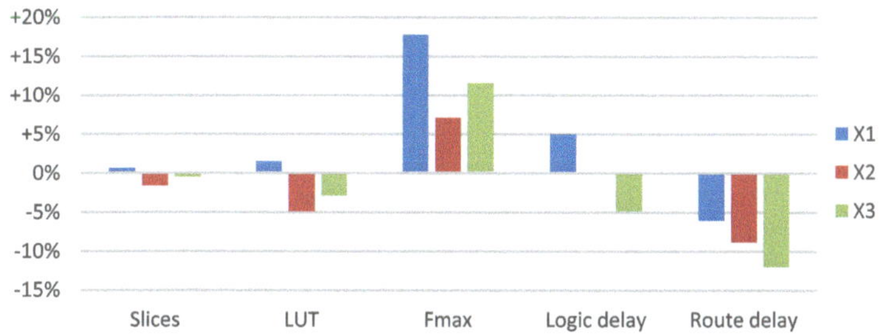

Fig. 2. Advantages of describing the core cipher transformations at the Register Transfer Level (RTL) compared to the results of the reference behavioral code (Std).

3.4 Different Implementation Strategies

Highly unrolled architectures—those containing a cascade of four or more round instances—produce large combinational designs, and their implementation pushed the Vivado tools to their limits. When subjected to the strict timing constraints required for maximum-frequency optimization, the tools occasionally terminated with fatal routing-congestion errors. This motivated an additional set of experiments in which the X4 through X12 variants were synthesized using a non-standard Vivado implementation strategy, "Spread Logic High" (SLH, [3]), which attempts to distribute logic resources over a larger area of the device in order to provide additional routing capacity for congested paths. Table 2 presents the results obtained with this approach (Table 2).

Table 2. Effects of applying implementation strategy "Spread Logic High" in different devices; underlined are the chips finally selected as the best platforms for particular designs

Design	Device	Slices	LUT	F_{max}	Logic delay	Route delay
X4 RTL + SLH	7s6–196	+ 26.7%	+ 34.8%	+ 11.7%	− 39.2%	− 2.3%
X6 RTL + SLH	7s15–196	+ 0.3%	0.0%	+ 0.3%	+ 8.6%	− 2.6%
	7s25–196	+ 1.7%	0.0%	− 5.0%	− 2.7%	+ 7.4%
	7s25–324	+ 1.1%	− 0.0%	− 2.6%	− 10.4%	+ 6.6%
	7s50–484	+ 40.2%	+ 44.2%	+ 17.0%	− 54.2%	− 4.5%
X12 RTL + SLH	7s15–196	− 0.1%	0.0%	+ 0.2%	+ 7.4%	− 2.2%
	7s50–484	− 1.0%	+ 0.0%	+ 1.4%	− 8.4%	+ 0.4%

Devices used in the reference implementations with the normal implementation strategy:
X4, X6 – xc7s6-196pins
X12 – xc7s15-196pins

For the smallest of the three designs, selecting the SLH strategy alone was sufficient to improve the results: slice and LUT occupancy increased as expected, and the hardware achieved a maximum operating frequency that was 11.7% higher. In the X6 design, however, the impact was negligible until a substantially larger chip was used. Only when the implementation was mapped to the 7s50 array in the 484-pin package the spreading mechanism did take effect, resulting in a 17% increase in maximum frequency. It is worth noting that switching to the same 7s50 device in the smaller 324-pin package produced no noticeable improvement. Finally, the X12 design was too large to benefit from the new strategy: even in the 7s50–484 device, no meaningful logic spreading occurred, and the increase in operating speed was only 1.4%.

Figure 3 summarizes final improvements achievable with the SLH strategy: in the X4 design even without changing the 7s6–196 platform it offered 12% increase in speed on the cost of 35% higher LUT utilization, in the X6 case it required additionally switching to a considerable larger 7s50–484 chip but then offered 17% higher clock speeds, and the X12 module—the completely unrolled organization—proved to be too complex to reach any actual improvement, even in a much larger array.

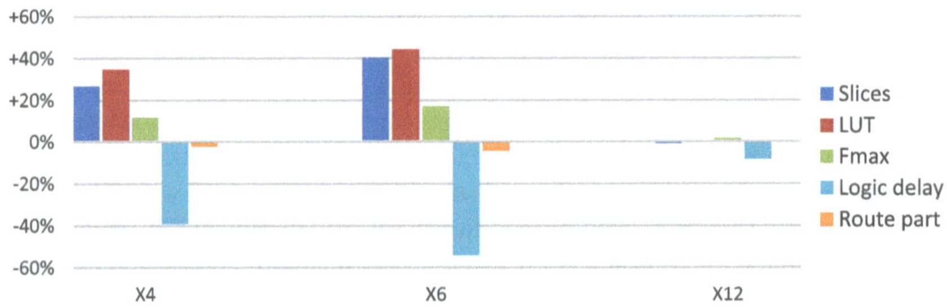

Fig. 3. Final changes brought to the unrolled architectures by "Spread Logic High" implementation strategy: variations of the X*k* RTL + SLH params from the X*k* RTL ones.

3.5 Efficiency of Loop Unrolling

With an implementation of the standard iterative architecture, which contains a single instance of the round module, one can estimate the expected parameters of the unrolled designs. Under the assumption of an ideal cascade of k rounds derived directly from the X1 implementation, both the size parameters and the combinational delays should increase in exact proportion to k. Using this proportional-scaling rule as a reference, the actual parameters of the X2, X3, …, X12 cases can be compared with the values extrapolated from the X1 design. The results of this comparison are presented in Fig. 4, and the graph can be used to assess the efficiency of loop unrolling applied to the Ascon permutation function. For this analysis, the X4, X6, and X12 designs were takes as the RTL variants with the "Spread Logic High" strategy fit in the most suitable device, as described in the previous section.

When unrolling large cryptographic designs it is typical that the resulting long combinational paths (which span multiple rounds) offer additional optimization opportunities for the implementation tools. It was already seen in the case of other ciphers—like AES, Keccak or BLAKE e.g. in [11]—that they can be packed in the LUT elements more efficiently than the ones within just one round, thus the final hardware has significantly lower LUT / slice occupancy than a simple proportionality rule would suggest. This observation is fully and without exception confirmed across the sizes of all X2 ÷ X12 designs: the numbers of utilized LUT element is by 20% lower than expected already in the X2 case and this reduction intensifies with loop unrolling up to an extreme -60% in the X12 architecture (albeit in the last three cases effects of the special "Spread Logic High" implementation strategy should not be forgotten).

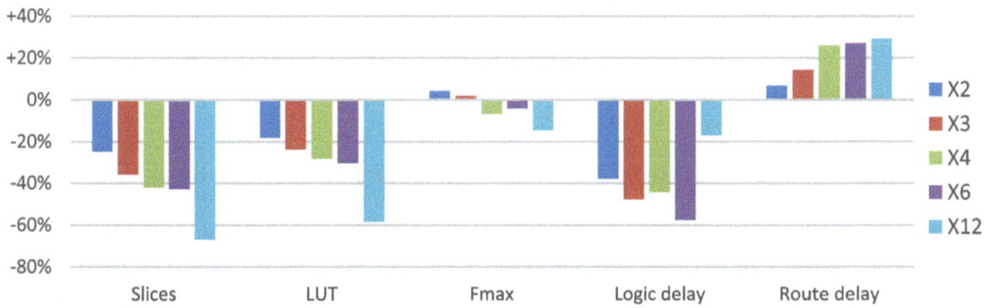

Fig. 4. Actual parameters of the loop unrolled architectures vs. expected values computed from the X1 results.

The conclusions are not as positive in the speed aspect: while the X2 and X3 designs can operate at frequencies moderately higher than expected (by, respectively, 4% and 2%), all further ones cannot keep the speed and, with increasing loop unrolling, fall behind the expectations by up to -15% in the case of the fully combinational architecture X12. The last two parameters in the graph provide some explanation of this trend: although reductions in the logic delays are always on the positive side, the increases in the routing part (which in general are of much higher absolute values) counterweigh them to the negative final effect.

4 Conclusions

This study offers several insights into the efficiency of automatic implementation of unrolled Ascon permutation functions in field-programmable gate array devices—insights that are not apparent when examining only complete cipher modules, such as those evaluated in the CAESAR or NIST competitions. The results highlight the importance of HDL coding style, demonstrating advantages of register-transfer-level descriptions over behavioral ones. The latter not only proved too demanding for synthesis in the larger, more heavily unrolled architectures, but also produced noticeably slower hardware even when synthesis succeeded. The evaluations further revealed that specialized automatic implementation strategies can provide measurable benefits for highly congested designs, while also showing the limits of these techniques, as they do not improve all design cases.

With a full set of loop-unrolled implementations—from the iterative architecture to the fully unrolled, essentially combinational variant—the efficiency of loop unrolling for this specific cipher could be assessed comprehensively. Although Ascon was designed as a lightweight alternative to standard cryptographic algorithms such as SHA-3 or AES, one must not forget that its large state and mixture of complex transformations still pose significant challenges for implementation tools when the round module is cascaded through loop unrolling. Nevertheless, with appropriate use of implementation options, stable and consistent results were obtained for all viable unrolling factors, up to the fully combinational organization.

Disclosure of Interests. The author have no competing interests to declare that are relevant to the content of this article.

References

1. Advanced Micro Devices, Inc.: 7 Series FPGAs Configurable Logic Block, UG474 v1.9. www.amd.com, ug474.pdf, last accessed 2026/03/10
2. Advanced Micro Devices, Inc.: 7 Series FPGAs Data Sheet: Overview, DS180 v2.6.1. www.amd.com, ds180.pdf, last accessed 2026/03/10
3. Advanced Micro Devices, Inc.: Vivado Design Suite User Guide: Implementation, UG904 (v2025.1). www.amd.com, ug904.pdf, last accessed 2026/03/10
4. Ahmet, M.: High-Performance fpga implementations of lightweight ASCON-128 and ASCON-128a With Enhanced Throughput-to-Area Efficiency. In: Information Security and Cryptology (ISCTURKEY), Ankara, Turkiye, pp. 1–7, https://doi.org/10.1109/ISCTrkiye64784.2024.10779273 (2024)
5. ATHENA Automated Tool for Hardware Evaluation: Evaluation of Finalists in the NIST LWC Process. https://cryptography.gmu.edu/athena/index.php?id=LWC, last accessed 2026/03/10
6. CAESAR: Competition for Authenticated Encryption: Security, Applicability, and Robustness, https://competitions.cr.yp.to/caesar.html, last accessed 2026/03/10
7. Dobraunig, C., Eichlseder, M., Mendel, F. et al.: Ascon v1. Submission to the CAESAR competition. Graz University of Technology. (2014). Available online https://ascon.isec.tugraz.at/index.html
8. Dobraunig, C., Eichlseder, M., Mendel, F. et al.: Ascon v1.2: Lightweight Authenticated Encryption and Hashing. J.Cryptol. **34**:33 (2021). https://doi.org/10.1007/s00145-021-09398-9
9. Groß, H., Wenger, E., Dobraunig, C., Ehrenhöfer, C.: Suit up! Made-to-Measure hardware implementations of ASCON. In: Euromicro Digital System Design, Madeira, Portugal, pp. 645–652 (2015). https://doi.org/10.1109/DSD.2015.14
10. Primas, R.: NIST LWC Hardware Reference Design of Ascon v1.2. https://github.com/ascon/ascon-hardware, last accessed 2026/02/16
11. Sugier, J.: Efficiency of Spartan-7 FPGA devices in implementation of contemporary cryptographic algorithms. J. Polish Safety and Reliability Association **9**(3), 75–83 (2018)
12. Turan, M.S., McKay, K., Kang, J. et al.: Ascon-Based lightweight cryptography standards for constrained devices: authenticated encryption, hash, and extendable output functions. Natl. Inst. Stand. Technol. SP 800–232 (2025). https://doi.org/10.6028/NIST.SP.800-232

Soft Contrastive Learning for Semi-supervised Image Recognition

Jakub Szafrański(✉), Patryk Wawrzacz, and Kamil Szyc

Wrocław University of Science and Technology, Wrocław, Poland
j.szafranski16@gmail.com, p.wawrzacz@interia.pl, kamil.szyc@pwr.edu.pl

Abstract. Semi-supervised learning (SSL) effectively leverages unlabeled data to supplement limited supervision, yet a disconnect persists between classification and unsupervised representation learning. Standard contrastive objectives often suffer from "false-negative repulsion," in which samples from the same semantic category are indiscriminately repelled, hindering the formation of clean class boundaries. We propose a framework that synergizes pseudo-labeling with contrastive learning by adaptively modulating repulsive forces. Using confidence-weighted pseudo-labels generated via a SoftMatch-based classification head, our method neutralizes repulsion between samples confidently predicted to share a class, thereby preserving semantic consistency in the latent space. We evaluate our framework on STL-10 and ImageNet benchmarks. Our approach achieves 83.66% Top-1 accuracy on STL-10 with 1,000 labels, outperforming strong baselines including CoMatch and SoftMatch. On ImageNet with 10% labels, we reach 67.9% Top-1 accuracy, surpassing SimCLR and SoftMatch. However, we observe that performance degrades in extremely label-scarce regimes (ImageNet 1%), where unreliable pseudo-labels provide insufficient guidance for the contrastive objective. Our findings highlight the efficacy of adaptive soft-weighting in SSL when a representative supervisory signal is available to guide representation learning.

Keywords: Semi-supervised learning · Contrastive learning · Image recognition · Soft Contrastive Learning

1 Introduction

The remarkable success of deep convolutional neural networks has traditionally been predicated on the reliance on large-scale, manually annotated datasets [12]. To mitigate the high cost of labeling, Semi-Supervised Learning (SSL) aims to harness the structural information embedded in unlabeled data to supplement limited labeled samples.

Modern SSL frameworks primarily leverage consistency regularization and pseudo-labeling. While foundational methods like FixMatch [14] set high benchmarks, they rely on rigid confidence thresholds that create a "quantity-quality

W. Zamojski et al. (Eds.): DepCoS-RELCOMEX 2026, LNNS 2014, pp. 423–431, 2026.
https://doi.org/10.1007/978-3-032-27927-9_41

trade-off," limiting data utilization. Recent advancements like SoftMatch [2] mitigate this through Gaussian weighting. However, these approaches still treat the classification task in isolation from unsupervised representation learning.

Simultaneously, self-supervised learning has advanced through contrastive frameworks like SimCLR [3]. By comparing augmented views of the same image, these models learn robust representations without labels. However, this objective does not account for underlying semantic categories. As observed in recent hybrid approaches [10], contrastive loss indiscriminately repels all other samples in a batch, even those sharing the same semantic class. This "false-negative repulsion" can hinder the formation of clean class boundaries and limit classification accuracy.

In this paper, we propose a unified framework to mitigate this tension. We introduce a mechanism that bridges the gap between the classification and contrastive heads. By leveraging confidence-weighted pseudo-labels, we adaptively modulate the repulsive forces within the contrastive objective. This ensures that samples sharing a predicted category are not unnecessarily pushed apart, preserving semantic consistency in the latent space.

Our experiments demonstrate that this synergy yields strong improvements on STL-10 and ImageNet with sufficient labeled data. On STL-10, our method outperforms all baselines, achieving 83.66% accuracy with 1,000 labeled samples. On ImageNet with 10% labels, we surpass SimCLR and SoftMatch; however, CoMatch remains a stronger baseline. Performance degrades substantially at ImageNet 1%, where label scarcity produces unreliable pseudo-labels that undermine the contrastive objective.

2 Related Work

In recent years, computer vision, especially image classification, has become one of the most important testbeds for deep learning. Classification is not only a standalone task, but also a foundation for many other vision problems, such as object detection, semantic/instance segmentation, tracking, pose estimation, or image retrieval. Beyond these, pretrained classifiers are often reused as general-purpose feature extractors for many other problems.

A standard starting point is supervised learning, where models are trained on fully annotated datasets. Common backbones include CNN architectures such as VGG [13], ResNet [8], DenseNet [9], EfficientNet [15], and newer convolutional designs like ConvNeXt [11]. Transformer-based backbones, especially ViT [5], are also widely used. These supervised approaches are typically benchmarked on large labeled datasets, with ImageNet as a canonical example.

Because collecting high-quality labels is costly, recent work increasingly focuses on unsupervised and self-supervised learning, where models learn representations without labels. A common workflow is to pretrain without labels and then adapt the learned features to human-interpretable classes using linear evaluation or fine-tuning of a small classifier head.

Self-supervised methods are often grouped into two directions. Contrastive learning builds invariances by pulling together representations of different augmentations of the same image and separating different images, as in SimCLR [3]. A second direction is self-distillation, where a student matches a teacher under different views and training dynamics, with DINO [1] or BYOL [6] as prominent examples. Masked image modeling is another related trend, represented by MAE [7]. In practice, these models are often evaluated in low-label regimes such as 1% or 10% labeled data, which directly measures label efficiency.

Between fully supervised and fully label-free learning lies semi-supervised learning, where only part of the training set is labeled. This setting is practical and is supported by benchmarks such as STL-10, which provides a small labeled set and a larger unlabeled set. Many methods rely on pseudo-labeling and/or consistency regularization - a representative self-training approach is Noisy Student [16].

Although self-supervised pretraining and semi-supervised learning both improve label efficiency, they are often presented as separate training recipes. Self-supervised work typically learns a general representation without labels and only later introduces labels through linear evaluation or limited fine-tuning. Semi-supervised work usually trains with labeled and unlabeled data jointly, focusing on pseudo labels and consistency. This leaves fewer studies that treat "pretrain with SSL, then run a strong semi-supervised method" as a primary pipeline, and quantify how representation quality affects pseudo-label reliability and selection.

There are also closely related semi-supervised methods that can complement self-supervised pretraining. SoftMatch [2] addresses the quantity-quality tradeoff in pseudo-labeling by replacing hard thresholding with confidence-aware sample weighting, which limits the influence of low-confidence unlabeled examples while still leveraging a larger fraction of unlabeled data. CoMatch [10] couples pseudo-label predictions with representation learning by introducing contrastive graph regularization on feature embeddings, encouraging semantic structure among unlabeled samples and improving pseudo-label reliability through better feature geometry during semi-supervised training. Unlike CoMatch, our method directly modulates the instance-level NT-Xent repulsion between samples sharing pseudo-labels, requiring neither graph construction nor a memory bank.

3 Method

We introduce a unified framework that synergizes the unsupervised representation learning of SimCLR [3] with the semi-supervised classification capabilities of SoftMatch [2]. Our approach leverages a shared encoder to simultaneously optimize a classification objective using distribution-aligned pseudo-labels and a contrastive objective that mitigates false-negative repulsion.

3.1 Overall Framework

Let $\mathcal{X}_l = \{(x_{l,i}, y_{l,i})\}_{i=1}^{B}$ denote a batch of labeled data and $\mathcal{X}_u = \{x_{u,i}\}_{i=1}^{\mu B}$ denote a batch of unlabeled data, where B is the labeled batch size and μ is the expansion ratio. For the unlabeled data, we apply three distinct stochastic transformations: a weak augmentation $\alpha(\cdot)$, a strong augmentation tailored for classification $\mathcal{A}_{cls}(\cdot)$, and a strong augmentation tailored for contrastive learning $\mathcal{A}_{ctr}(\cdot)$.

Our architecture consists of three core components. A backbone encoder $f(\cdot)$ extracts continuous representations $h = f(x)$. A linear classification head $p(\cdot)$ maps h to class logits $c \in \mathbb{R}^K$, where K is the number of classes. A non-linear projection head $g(\cdot)$, modeled as a two-layer multi-layer perceptron, maps h to a lower-dimensional embedding $z \in \mathbb{R}^d$ for contrastive learning.

The complete training objective linearly combines the supervised, unsupervised, and contrastive losses. The final loss $\mathcal{L}$ is computed as:

$$\mathcal{L} = \lambda_{cls}^{lb}\mathcal{L}_{sup} + \lambda_{cls}^{ulb}\mathcal{L}_{unsup} + \lambda_{ctr}\mathcal{L}_{ctr}$$

where λ_{cls}^{lb}, λ_{cls}^{ulb}, and λ_{ctr} are constant scalar weights that balance the influence of each component. To stabilize training and provide a stable target distribution for the unsupervised consistency loss, we maintain an exponential moving average (EMA) of the model parameters.

3.2 Semi-supervised Classification

For the classification head, we strictly follow the semi-supervised formulation proposed in SoftMatch. The objective comprises a supervised loss on labeled data and an unsupervised consistency loss on unlabeled data. For labeled samples, we compute the standard cross-entropy loss:

$$\mathcal{L}_{sup} = \frac{1}{B}\sum_{i=1}^{B} H(y_{l,i}, p(f(x_{l,i})))$$

where H denotes the cross-entropy function.

For unlabeled samples, we enforce consistency between the predictions of weakly and strongly augmented views. Following SoftMatch, we obtain the weak prediction $p_w = \mathrm{softmax}(p(f(\alpha(x_u))))$ and apply distribution alignment to correct for class imbalance. We adopt the truncated Gaussian weighting mechanism to compute a confidence mask $m \in [0, 1]$ for each sample, which relies on the exponential moving average of the maximum probabilities. The unsupervised loss is calculated against the strongly augmented view $\mathcal{A}_{cls}(x_u)$:

$$\mathcal{L}_{unsup} = \frac{1}{\mu B}\sum_{i=1}^{\mu B} m_i \cdot H(\arg\max(p_{w,i}), p(f(\mathcal{A}_{cls}(x_{u,i}))))$$

3.3 Soft Contrastive Learning

Standard contrastive learning objectives, such as NT-Xent, indiscriminately repel all other samples in a batch, treating them as negative pairs. This inherently introduces false negatives when samples from the same underlying class are pushed apart. We introduce a modified Soft NT-Xent loss to mitigate this issue by adaptively modulating the repulsion term according to the pseudo-labels produced by the classification head.

Given the projection embeddings from two augmented views:

$$z_i = g(f(\mathcal{A}_{cls}(x_u))), \quad z_j = g(f(\mathcal{A}_{ctr}(x_u))).$$

we compute the pairwise cosine similarity $\mathrm{sim}(z_i, z_j)$. Let $\hat{y}$ denote the predicted pseudo-label and m denote the SoftMatch confidence mask for a given sample. We define a continuous weighting factor $w_{i,k}$ for any pair of samples (i, k):

$$w_{i,k} = 1 - (m_i \cdot m_k \cdot \mathbb{1}[\hat{y}_i = \hat{y}_k])$$

When two samples are confidently predicted to belong to the same class, $w_{i,k}$ approaches 0, effectively neutralizing their repulsive force. To ensure initial representation stability, the soft weighting mechanism is activated following a warm-up period, during which standard NT-Xent is employed. The Soft NT-Xent loss is formulated as:

$$\mathcal{L}_{ctr} = -\frac{1}{2N}\sum_{i=1}^{2N} \log \frac{\exp(\mathrm{sim}(z_i, z_{p(i)})/\tau)}{\exp(\mathrm{sim}(z_i, z_{p(i)})/\tau) + \sum_{\substack{k=1 \\ k\neq i,\, k\neq p(i)}}^{2N} w_{i,k} \cdot \exp(\mathrm{sim}(z_i, z_k)/\tau)}$$

where $N = \mu B$ is the number of unlabeled samples, $p(i)$ is the index of the positive augmented pair for sample i, $\mathbb{1}$ is the indicator function, and τ is a scalar temperature parameter.

4 Experiments

4.1 Datasets

We conduct experiments on the STL-10 [4] and ImageNet [12] benchmarks. STL-10 contains 5,000 labeled images of size 96×96 pixels from 10 classes and 100,000 unlabeled images. Following the [10], we evaluate our model on the 5 pre-defined folds. For ImageNet (ILSVRC-2012), we randomly sample 1% or 10% of the images with labels. Sampling is performed in a class-balanced, stratified manner. Remaining images are treated as unlabeled data.

4.2 Baselines

We compare our approach against several baselines. This includes SimCLR [3], SoftMatch [2], and CoMatch [10].

4.3 Implementation Details

We employ ResNet-18 as the backbone for STL-10 and ResNet-50 for ImageNet [8]. For the SimCLR baseline, we reproduce results following the original hyperparameters [3] with the exception of the batch size, which we set to 512. For STL-10 fine-tuning, we adjust the batch size to 128 and the learning rate to 1×10^{-3}. Additionally, we evaluate SoftMatch [2] by following the training procedure outlined by the authors. On STL-10, our method and the reproduced SimCLR and SoftMatch baselines are all trained for 200 epochs. We report SimCLR results on STL-10 using a batch size of 512. For ImageNet, we report both 512-batch reproduction and the original results.

We optimize our method using SGD with Nesterov momentum and a cosine decay learning rate schedule preceded by a 10-epoch linear warmup. The projection head is a 2-layer MLP yielding 128-dimensional embeddings. We incorporate Exponential Moving Average (EMA) for model weights and apply distribution alignment. Most hyperparameters were adapted from the original SimCLR and SoftMatch configurations. Table 1 summarizes the complete hyperparameter set and augmentation strategies.

The source code is available at: GitHub repository.

Table 1. Hyperparameters and augmentation settings for our method.

Dataset	STL-10	ImageNet 1%	ImageNet 10%
Backbone	ResNet-18	ResNet-50	ResNet-50
Labeled Batch Size	64	128	128
λ_{cls}^{ulb} (Unsupervised)	1	1	2
Weight Decay	1×10^{-4}		
Learning Rate	0.1		
Unlabeled Batch Size	512		
Scheduler	CosineAnnealingLR (10 ep. warmup)		
Optimizer	SGD (Nesterov)		
SGD Momentum	0.9		
Model EMA Momentum	0.999		
Prediction EMA Momentum	0.999		
τ (Temperature)	0.5		
λ_{cls}^{lb} (Supervised)	1		
λ_{ctr} (Contrastive)	5		
Soft Weighting Start Epoch	50		
Weak Aug.	RRC*, Horizontal Flip		
Strong Aug. (cls)	RRC*, Horizontal Flip, RandAugment		
Strong Aug. (ctr)	RRC*, Horizontal Flip, Color Jitter, Grayscale, Gaussian Blur		

*RRC: Random Resized Crop.

4.4 Results

Table 2 and Table 3 summarize the performance comparison on the STL-10 and ImageNet benchmarks, respectively. Some of the results reported below are cited directly from the respective publications. Specifically, CoMatch results on both STL-10 and ImageNet, SoftMatch on ImageNet, and the original SimCLR results are cited from the papers, and thus reflect the authors' exact experimental setup, which may differ in hardware, data sampling, or other implementation details. All remaining results, including our method, SoftMatch on STL-10, and the SimCLR (Batch 512) reproduction, are obtained by us.

Table 2. Top-1 accuracy (%) on STL-10 for 1,000 labeled samples. Results marked with † are cited from the respective publications.

Method	STL-10
[2] SoftMatch	70.15±0.49
[3] SimCLR	74.53±0.67
[10] CoMatch†	79.80 ± 0.38
Ours	**83.66±1.05**

Table 3. Performance comparison on ImageNet with 1% and 10% labels. Results marked with † are cited from the respective publications.

Method	# Epochs	ImageNet 1%		ImageNet 10%	
		Top-1	Top-5	Top-1	Top-5
[2] SoftMatch†	∼400	–	–	59.5*	–
[3] SimCLR (Batch 512)	400	42.4	70.7	61.7	84.3
[3] SimCLR† (Original)	1000	48.3	75.5	65.6	87.8
[10] CoMatch†	400	**66.0**	**86.4**	**73.6**	**91.6**
Ours	400	35.0	59.0	67.9	87.2

* Reported result obtained on 100k labeled samples.

Our method achieves the best reported accuracy on STL-10 with 1,000 labeled samples, reaching 83.66% ± 1.05, outperforming CoMatch by nearly four percentage points and surpassing SimCLR and SoftMatch by larger margins. On ImageNet with 10% labeled data, our method attains 67.9% Top-1 accuracy, exceeding SimCLR (Batch 512), SimCLR (Original), and SoftMatch. However, CoMatch remains the strongest baseline at this label fraction, outperforming our method by 5.7 points. At the more challenging 1% label setting, our method underperforms all baselines substantially.

The strong performance on STL-10 can be attributed to the relatively small and low-diversity dataset, where the combination of contrastive and pseudo-labeling objectives provides a strong regularization signal even with few labels. The poor performance at ImageNet 1% is likely caused by the extreme label scarcity, where very few labeled samples result in highly unreliable pseudo-labels. This corrupts the contrastive objective, leads the model to reinforce incorrect cluster assignments, and further propagates noisy supervision into the unlabeled classification through invalid soft labels. At ImageNet 10%, our method recovers substantially, suggesting that a moderate amount of labeled data is sufficient to stabilize pseudo-label quality and allow the contrastive loss to contribute positively. The improvement over SoftMatch and SimCLR on STL-10 and ImageNet 10% demonstrates that our approach benefits from combining both contrastive and pseudo-labeling objectives, as neither method alone achieves comparable performance.

5 Conclusion

In this work, we introduced a framework that integrates adaptive pseudo-labeling with contrastive representation learning. By modulating the NT-Xent loss with a soft-weighting mechanism, our method addresses the conflict between instance-level discrimination and class-level consistency, effectively mitigating the false-negative repulsion problem.

Our results on STL-10 and ImageNet 10% demonstrate that this synergy provides a significant advantage when a representative supervisory signal is available. In these settings, the classification head provides reliable guidance, allowing the contrastive objective to refine class boundaries and improve overall accuracy. However, the performance on ImageNet 1% highlights a dependency on the initial quality of pseudo-labels. In extremely label-scarce regimes, incorrect predictions can suppress necessary discriminative pressure between samples shared by the same pseudo-label.

References

1. Caron, M., et al.: Emerging properties in self-supervised vision transformers. In: Proceedings of the IEEE/CVF International Conference on Computer Vision, pp. 9650–9660 (2021)
2. Chen, H., et al.: Softmatch: addressing the quantity-quality trade-off in semi-supervised learning (2023)
3. Chen, T., Kornblith, S., Norouzi, M., Hinton, G.: A simple framework for contrastive learning of visual representations. In: International Conference on Machine Learning, pp. 1597–1607. PmLR (2020)
4. Coates, A., Ng, A., Lee, H.: An Analysis of Single Layer Networks in Unsupervised Feature Learning. In: AISTATS (2011). https://cs.stanford.edu/~acoates/papers/coatesleeng_aistats_2011.pdf
5. Dosovitskiy, A., et al.: An image is worth 16x16 words: transformers for image recognition at scale, arXiv preprint (2020). arXiv:2010.11929

6. Grill, J.B., et al.: Bootstrap your own latent-a new approach to self-supervised learning. Adv. Neural. Inf. Process. Syst. **33**, 21271–21284 (2020)
7. He, K., Chen, X., Xie, S., Li, Y., Dollár, P., Girshick, R.: Masked autoencoders are scalable vision learners. In: Proceedings of the IEEE/CVF Conference on Computer Vision and Pattern Recognition, pp. 16000–16009 (2022)
8. He, K., Zhang, X., Ren, S., Sun, J.: Deep residual learning for image recognition. In: Proceedings of the IEEE Conference on Computer Vision and Pattern Recognition, pp. 770–778 (2016)
9. Huang, G., Liu, Z., Van Der Maaten, L., Weinberger, K.Q.: Densely connected convolutional networks. In: Proceedings of the IEEE Conference on Computer Vision and Pattern Recognition, pp. 4700–4708 (2017)
10. Li, J., Xiong, C., Hoi, S.: Comatch: Semi-supervised learning with contrastive graph regularization (2021)
11. Liu, Z., Mao, H., Wu, C.Y., Feichtenhofer, C., Darrell, T., Xie, S.: A convnet for the 2020s. In: Proceedings of the IEEE/CVF Conference on Computer Vision and Pattern Recognition, pp. 11976–11986. (2022)
12. Russakovsky, O., et al.: ImageNet large scale visual recognition challenge. Int. J. Comput. Vision **115**(3), 211–252 (2015). https://doi.org/10.1007/s11263-015-0816-y
13. Simonyan, K., Zisserman, A.: Very deep convolutional networks for large-scale image recognition, arXiv preprint arXiv:1409.1556 (2014)
14. Sohn, K., et al.: Fixmatch: simplifying semi-supervised learning with consistency and confidence (2020)
15. Tan, M., Le, Q.: Efficientnet: Rethinking model scaling for convolutional neural networks. In: International Conference on Machine Learning, pp. 6105–6114. PMLR (2019)
16. Xie, Q., Luong, M.T., Hovy, E., Le, Q.V.: Self-training with noisy student improves imagenet classification. In: Proceedings of the IEEE/CVF Conference on Computer Vision and Pattern Recognition, pp. 10687–10698 (2020)

Practical Classical Quantum Aspects in the Use of Gaussian Kernels for Quantum Unidimensional Binary Clustering, Based on the Method Proposed by Horn and Gottlieb

Sergii Trotskyi(✉) and Wojciech Bożejko

Department of Control Systems and Mechatronics, Wrocław University of Science and Technology, Janiszewskiego 11-17, 50-372 Wrocław, Poland
{sergii.trotskyi,wojciech.bozejko}@pwr.edu.pl

Abstract. We investigate the feasibility of quantum clustering on near-term quantum hardware using a dataset encoded via a sum-of-Gaussians representation. Classical data are first processed using classical methods and then encoded into quantum state amplitudes in Qiskit. With multi-qubit quantum circuits, we then investigate how modeling the wave-function as a superposition of Gaussian functions influences clustering performance on both balanced and imbalanced datasets. Clustering quality is evaluated using the Adjusted Rand Index (ARI), with particular attention to the role of different measurement post-processing strategies. Two approaches are examined: selecting the two most probable bitstrings and selecting the top-k most probable bitstrings with subgrouping. The results indicate that naive Gaussian-kernel-based clustering is not well suited for quantum devices from a practical point of view, as its effectiveness is constrained by sensitivity to data imbalance and the exponential cost of state initialization. Simulation and execution were done using the Qiskit Aer simulator and the ODRA5 quantum computer located at Wroclaw University of Science and Technology.

1 Introduction

Clustering is the task of separating a set of data points $\mathcal{X} = \{x_1, x_2, \ldots, x_n\}$ (one- or multi-dimensional) into a collection of groups $\mathcal{C} = \{C_1, C_2, \ldots, C_k\}$ based on their features (dimensions), without any prior knowledge of which group each point should belong to. These groups are called clusters:

$$C_i \cap C_j = \emptyset \quad \text{for } i \neq j, \tag{1}$$

$$\bigcup_{i=1}^{k} C_i = \mathcal{X}. \tag{2}$$

W. Zamojski et al. (Eds.): DepCoS-RELCOMEX 2026, LNNS 2014, pp. 432–443, 2026.
https://doi.org/10.1007/978-3-032-27927-9_42

Although this specific procedure belongs to unsupervised machine learning, there is a practical synergy between quantum computing and various branches of machine learning (unsupervised, supervised, reinforcement learning, generative AI, etc.). For instance, in [3], the authors proposed the concept of an autonomous agent in FMS driven by an LLM that has access to a quantum computer, which serves as a computational engine for tackling NP-hard scheduling tasks. Another real-world example of quantum machine learning is presented in [8] and [7].

We are currently in the era of Noisy Intermediate-Scale Quantum (NISQ) devices [9]. This means that quantum processors can contain up to a few hundred qubits, but both the qubits themselves and the operations performed on them are subject to noise. The qubit-to-qubit connectivity is limited (in a real quantum computer, even if there are hundreds of qubits, each one is only directly connected to a limited subset of the others). For superconducting computers, there are several standard parameters that can be used to characterize "stability of computations": one of them is the T_1 time [4] (which measures how long a qubit remains in the state $|1\rangle$ or superposition before decaying to the state $|0\rangle$), and it is typically in the range of dozens to hundreds of microseconds for that kind of device.

D. Horn and A. Gottlieb first proposed a theoretical method for quantum clustering in 2001 [6], where the wavefunction being represented as

$$\psi(x) = \sum_i e^{-\frac{(x-x_i)^2}{2\sigma^2}}, \tag{3}$$

x_i are the points we want to separate into clusters. In this article, we seek to put into practice a concept presented in 2001.

Another study [10] by Rahman and Geiger focuses on transforming the classical Gaussian Mixture Model into a quantum-classical formulation in which clusters are represented as wave functions whose squared amplitudes correspond to probabilities (with wavefunctions being mixed before squaring).

One more work done by Xie and Ben-Ami [14] focuses on designing an efficient state-preparation circuit for creating approximate Gaussian states on universal quantum computers. Their approach begins by constructing the desired amplitudes through single-qubit rotations, followed by the use of a quantum Fourier transform. According to the authors, this method achieves high fidelity while maintaining linear complexity.

2 Intuition Behind the Gaussian Wavefunction in Quantum Clustering

As presented by Horn and Gottlieb in [6], the potential is defined as

$$V(x) = E + \frac{2\nabla^2\psi}{\psi},$$

where ∇^2 denotes the second derivative of the wavefunction ψ in 3, $V(x)$ represents the potential function, and E is an energy that adds a constant offset to

$V(x)$. The role of E might be compared to b in the linear equation $y = ax + b$, and a particular value is chosen based on the problem. Let's run a simulation and visualize the wavefunction, its second derivatives, and $V(x)$ for cases with two clusters. The simulation of two clusters with centers at -3 and 3 and a standard deviation of 0.5 is shown in Fig. 1, where the black line represents the wavefunction, the blue line shows its second derivative, the red line corresponds to the potential function $V(x)$, and data points are shown as gray dots. Clusters have been created using the function `make_blobs` [11] from the package `scikit-learn` [12].

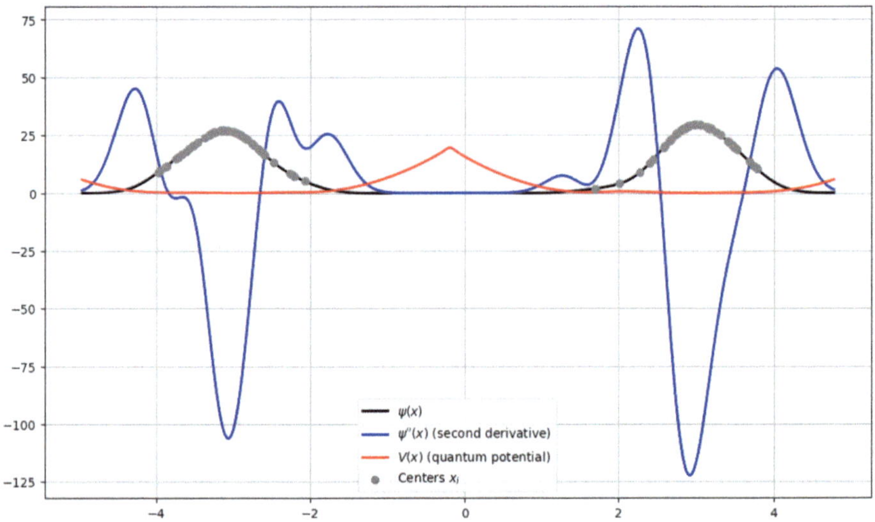

Fig. 1. The Horn and Gottlieb wavefunction simulation for two clusters with centers at -3 and 3.

According to this idea, representing the wavefunction as a sum of Gaussian functions is the method used to fit clustering data into a wavefunction. However, other functions can also be used, especially if they have well-defined second derivatives, as required by Schrödinger equation. In the paper [5], Casana-Eslava et al. proposed defining the wavefunction as a sum of normalized Gaussian functions centered at each data point(instead of using the same σ in each Gaussian). They define σ_i as a function of the K-nearest neighbors (KNNs) $\sigma_i = \frac{1}{K}\sum_{j\in \text{knn}(x_i)} \text{dist}(x_i, x_j))$, meaning that σ now depends on the average distance to the K nearest neighbors of point x_i, adjusting the length scale to local density with d number of features:

$$\psi(x) = \frac{1}{n}\sum_{i=1}^{n} \frac{1}{\left(\sqrt{2\pi}\cdot \frac{1}{K}\sum_{j\in \text{knn}(x_i)} \text{dist}(x_i, x_j)\right)^d} \exp\left(-\frac{\|x - x_i\|^2}{2\left(\frac{1}{K}\sum_{j\in \text{knn}(x_i)} \text{dist}(x_i, x_j)\right)^2}\right).$$

Now that we have explored the theoretical "Gaussian functions" approach for clustering, the next step involves implementing this formulation on a universal quantum computer and/or Quantum simulator. The task should be divided into three sub-tasks: (1) construct a sequence of unitary matrices that encodes the mathematical formulation (use a method to set up the quantum state as a sum of Gaussians), (2) measure the quantum circuit, and (3) define what each bitstring obtained after measurement represents in terms of clustering.

3 A 1-Dimensional, Binary Clusters Synthetic Dataset with 1000 Samples

In the experiment with the sum of Gaussians, a one-dimensional dataset was used. The dataset consisted of 1000 samples (generated using `make_blobs`) with a single feature, grouped into two clusters centered approximately at −2.51 and 9.01. Other cluster centers can be used as well. However, we should remember that as clusters move closer together, it becomes increasingly difficult to separate them on a quantum computer (and classically as well).

Each Gaussian component had a standard deviation $\sigma = 0.3$. The data were normalized to the range $[-1, 1]$, yielding cluster centers at −0.63 and 0.70. These values were then mapped to the nearest points in a discretized representation of possible bitstrings, located at −0.61 and 0.68, enabling subsequent association between cluster points / cluster centers and bitstrings.

In this experiment, we used a quantum circuit having a five-qubit register.

The `Initialize(`ψ`)` Qiskit method [1] was used to set up the quantum state (for circuits of up to $n < 10$ qubits), where the vector ψ encoded the amplitudes corresponding to a sum of Gaussian functions.

Before the quantum state can be created, Gaussians must first be calculated. Post-processing includes measuring the bitstrings corresponding to the discretized representations, which must then be converted to the original values (with some error that will always exist).

The circuit was constructed as `qc = QuantumCircuit(`n`)` [2]. The resulting measurement histogram Fig. 3 (denoted as a frequency map $H(\mathbf{b})$ over bitstrings $\mathbf{b} \in \{0,1\}^n$) shows dominant outcomes for each cluster $|11100\rangle$ and $|00111\rangle$ for a particular measurement. The complete execution flow is shown below, with red boxes indicating the quantum state preparation and the circuit execution (Fig. 2):

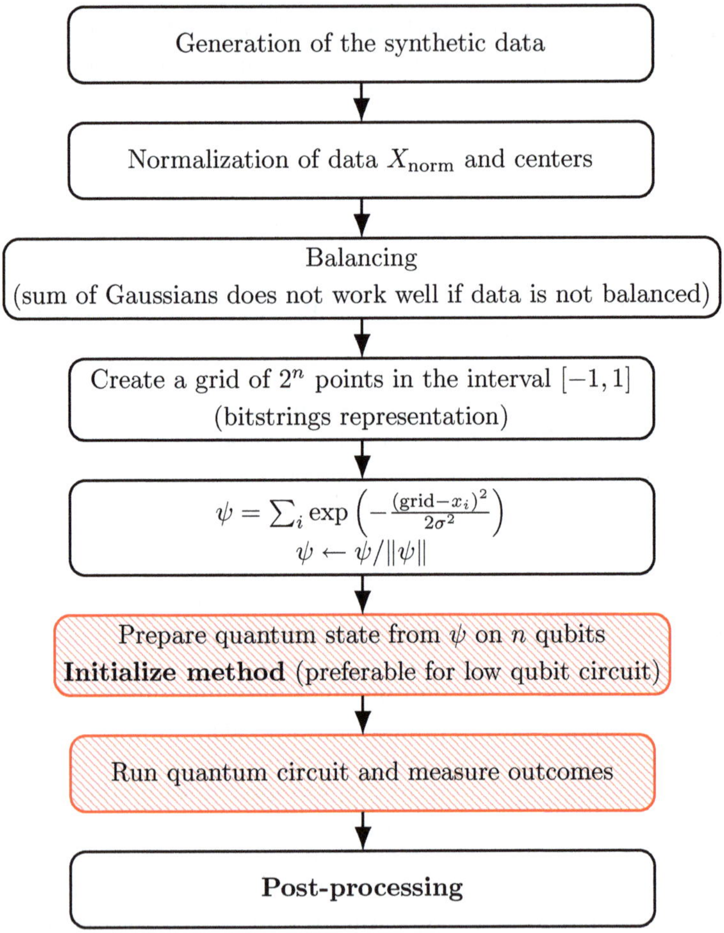

Please note that post-processing plays an important role. A straightforward approach at this stage is to assume that the two measurements with the highest counts represent the centers of two distinct clusters (which can be used as cluster centers when converted back to the original values):

$$\{\mathbf{b}_1, \mathbf{b}_2\} = \arg \max_{\mathbf{b}}^{(2)} H(\mathbf{b}),$$

where $\max_{\mathbf{b}}^{(2)} H(\mathbf{b})$ denotes the two bitstrings with the highest counts.

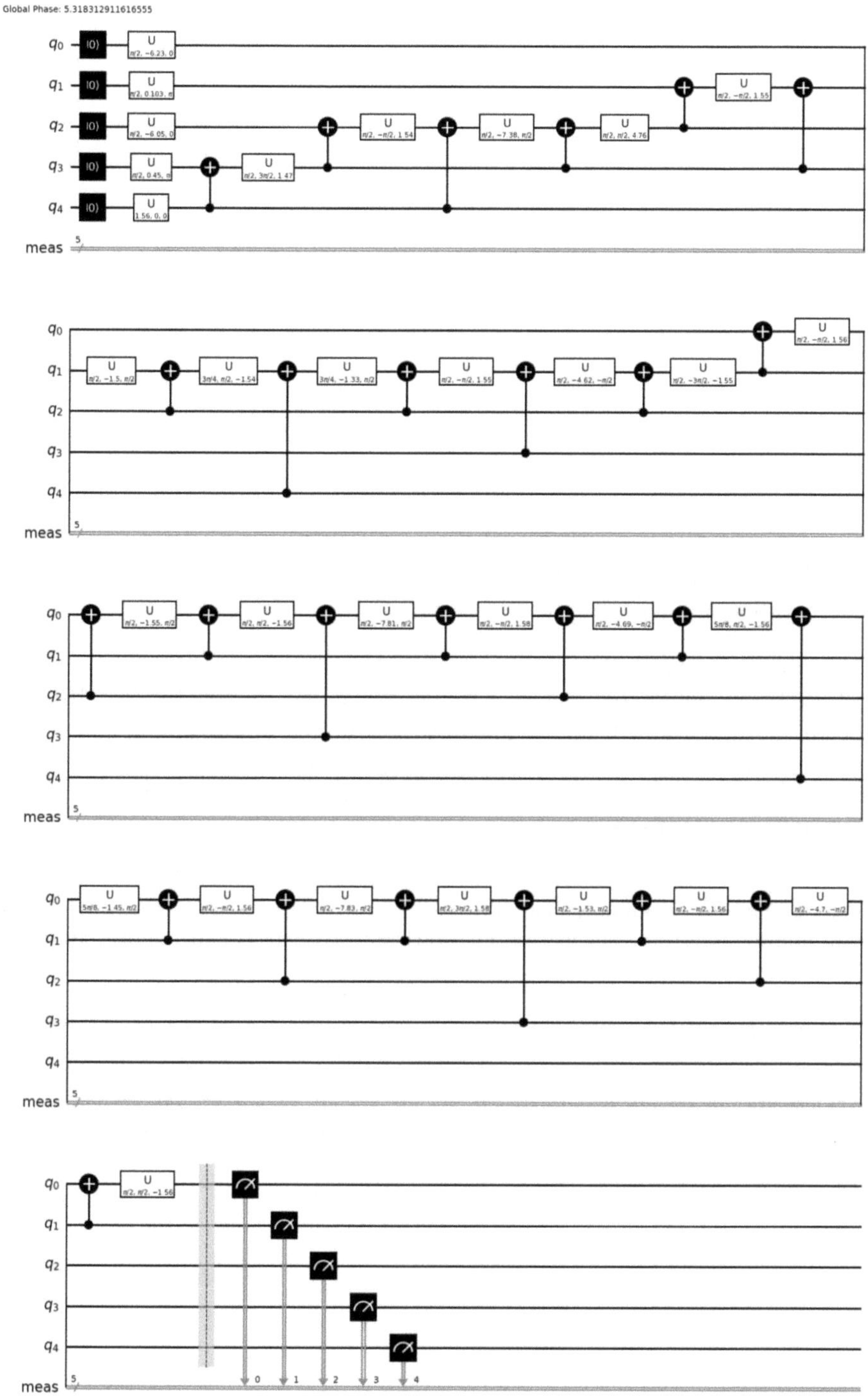

Fig. 2. A five-qubit circuit example for identifying two clusters.

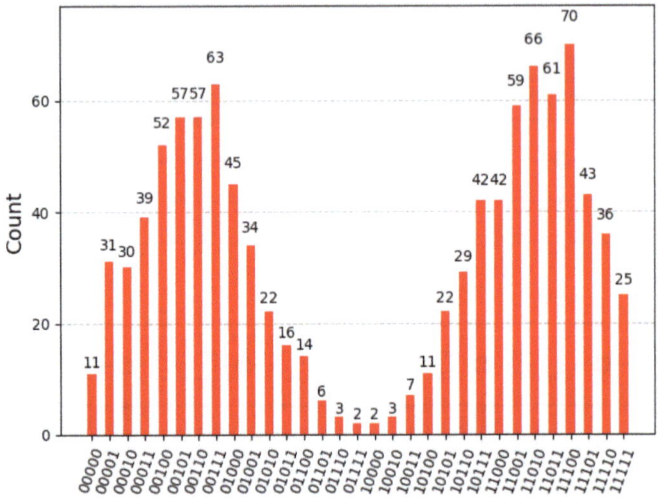

Fig. 3. The quantum state measurement revealed several dominant bitstrings, two of which $|11100\rangle$ and $|00111\rangle$ might correspond to the different encoded cluster centers. The simulation was done on Qiskit Aer with no noise model applied.

Based on our experiments, for larger qubit counts ($n \geq 10$), the `Initialize(`ψ`)` method failed because it requires a number of gates that increases exponentially with n. As a result, Qiskit refused to perform the synthesis and raised the error `Transpiler Error: HighLevelSynthesis is unable to synthesize 'state_preparation'` (Tables 1 and 2).

Table 1. ARI for one-dimensional two-cluster experiments as a function of the number of qubits under different noise configurations. Red entries indicate degradation in clustering performance.

Number of qubits	2	3	4	5	6	7	8	9
Single- and two-qubit depolarizing probabilities $(p_{1q}, p_{2q}) = (0.001, 0.01)$.	1.00	1.00	1.00	1.00	**0.18**	1.00	**0.31**	1.00
Single- and two-qubit depolarizing probabilities $(p_{1q}, p_{2q}) = (0.01, 0.05)$	1.00	1.00	1.00	1.00	1.00	1.00	**0.41**	**0.69**
No noise is applied	1.00	1.00	1.00	**0.63**	**0.32**	**0.12**	**0.05**	1.00

Table 2. Inter-cluster center's distances interpreted (postprocessing was made by selecting the top two bitstrings) for one-dimensional two-cluster experiments as a function of the number of qubits under different noise configurations. Red entries indicate cases with reduced cluster separation based on the ARI metric.

Number of qubits	2	3	4	5	6	7	8	9
Single- and two-qubit depolarizing probabilities $(p_{1q}, p_{2q}) = (0.001, 0.01)$.	11.6	12.4	11.6	12.3	1.70	11.5	0.80	10.5
Single- and two-qubit depolarizing probabilities $(p_{1q}, p_{2q}) = (0.01, 0.05)$	11.6	12.4	10.4	12.9	10.8	12.1	4.60	1.7
No noise is applied	11.6	12.4	10.4	1.1	1.70	0.70	0.40	10.1

The findings suggest that the drop in performance with the most basic post-processing approach was largely due to incorrectly identifying bitstrings of centers. In particular, it occurs when we mistakenly interpret that the top 2 bitstrings belong to different clusters, while in reality, they are part of the same one Fig. 4. Formally, this means that the following do not necessarily represent different individual distributions of the clusters:

$$\{\mathbf{b}_1, \mathbf{b}_2\} = \arg \overset{(2)}{\max_{\mathbf{b}}} H(\mathbf{b}) \nRightarrow \mathbf{b}_1, \mathbf{b}_2 \text{ represent different clusters centers.}$$

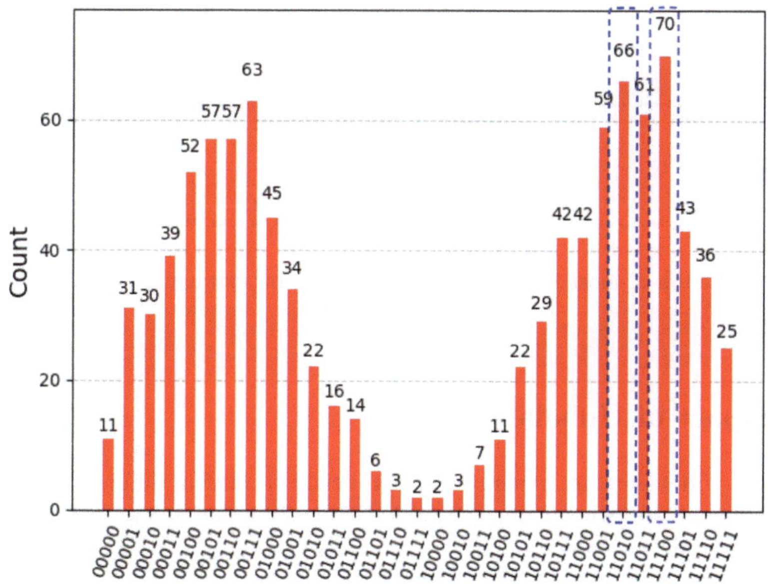

Fig. 4. Problem with blindly assuming that the top bitstring always belongs to different clusters.

3.1 Selecting top-k Bitstring Post-processing Strategy

More advanced post-processing might include the following steps: (1) measure the quantum circuit, obtaining a histogram of bitstrings; next, (2) identify the top-K most frequently occurring bitstrings (where K denotes the number of selected bitstrings, in this example, $K = 4$); after that, (3) the top-K bitstrings are grouped into a number of assumed clusters G_ℓ based on their proximity in the bitstrings; and finally, (4) select the bitstring with the highest count from each group (i.e., $\arg\max_{\mathbf{b}\in G_\ell} H(\mathbf{b})$ for group G_ℓ) and assume the resulting bitstrings might be mapped to cluster centers (Fig. 5).

$$\{\mathbf{b}_1, \mathbf{b}_2, ..., \mathbf{b}_K\} = \arg\max_{\mathbf{b}}^{(K)} H(\mathbf{b}),$$

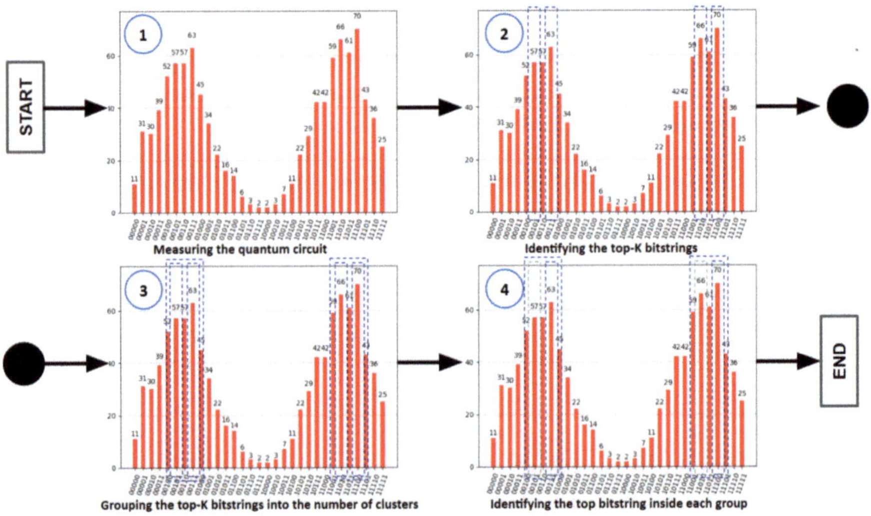

Fig. 5. The top-K most frequently occurring bitstrings strategy with subgrouping.

4 Impact of Unbalanced Data

Unbalanced datasets are generated by explicitly specifying different numbers of samples per cluster when calling the `make_blobs` function in a proportion of 75% to 25% (Fig. 6).

When the quantum state is formed as a sum of Gaussians, regions with higher data density have a stronger impact on the resulting wavefunction and, as a result, on the measurement probabilities. that is why balancing is a crucial part of the implementation.

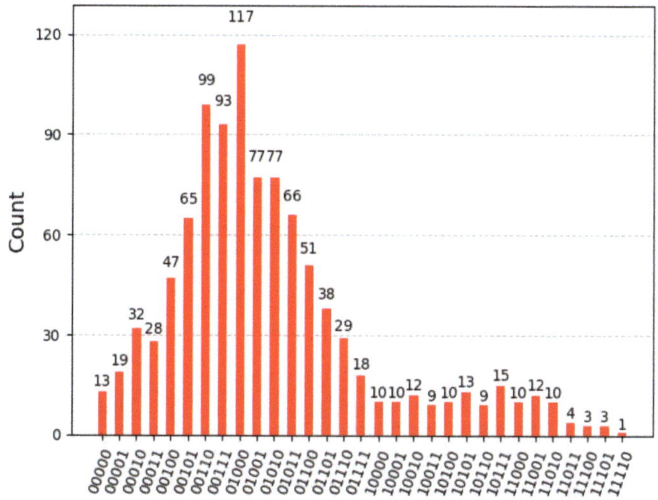

Fig. 6. Change in the imbalanced dataset toward a cluster containing a larger number of data points.

5 ODRA 5 Quantum Computer Experiments

The ODRA 5 quantum computer carries out universal quantum computations using qubits implemented with superconducting technology operating at cryogenic temperatures (20 mK, -273.13°C). The system is located at Wroclaw University of Science and Technology in Poland [13]. Schematically, the ODRA5 quantum processor is depicted in Fig. 7.

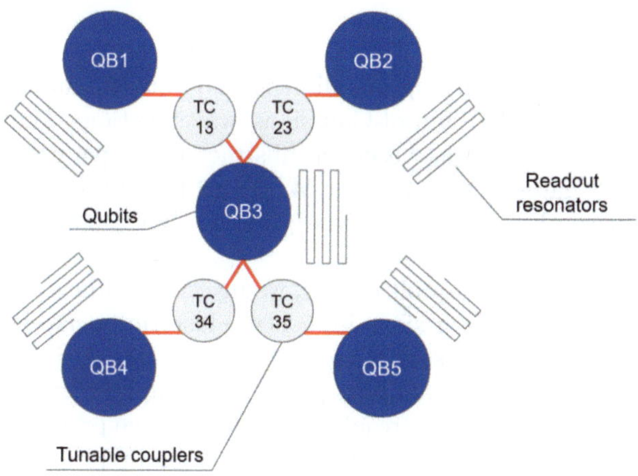

Fig. 7. ODRA5 quantum processor with 5 qubits.

For Odra5, a separate set of natively supported gates must be used. Otherwise, the following error will occur (Fig. 8):

```
ValueError: Invalid operation 'ry' found in IQMOptimize1QbDecomposition pass.
Expected operations: ['r', 'cz', 'move', 'if_else', 'measure', 'reset', 'delay',
'barrier']
```

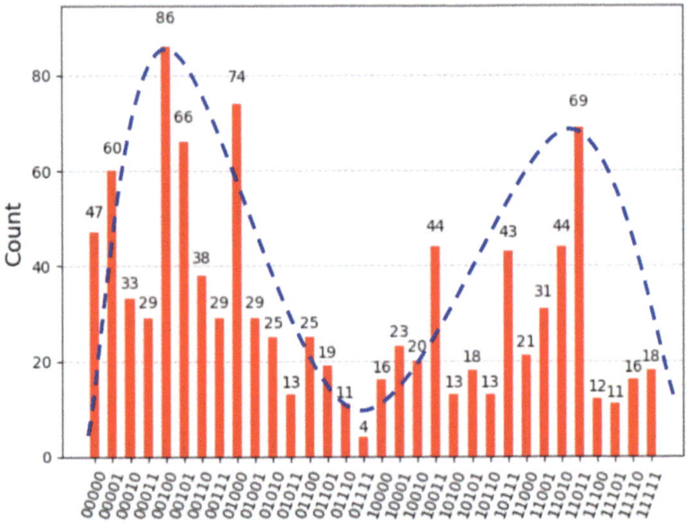

Fig. 8. 5-qubit gaussian clustering on ODRA5 for balanced data.

6 Conclusion

In this paper, we analyzed how a one-dimensional dataset can be grouped into two clusters by using the sum-of-Gaussians wavefunction. The implementation on the Aer simulator was based on abstract gate operations: Pauli rotations and CNOT gates $\{\text{rx}, \text{ry}, \text{rz}, \text{cx}\}$. While on the IQM ODRA5 quantum computer, we used the native hardware gate set $\{\text{r}, \text{cz}, \text{move}, \text{if\_else}, \text{measure}, \text{reset}, \text{delay}, \text{barrier}\}$.

By implementing a multi-qubit circuit, we investigated how representing the wavefunction as a sum of Gaussians influences clustering performance for both balanced and imbalanced datasets. A comparison using the Adjusted Rand Index (ARI) demonstrated how different post-processing strategies influence clustering performance as well. Overall, the results indicate that the practical usefulness of a Gaussian kernel in quantum computation may be limited (e.g., with respect to the number of clusters, data imbalance, etc.). Nevertheless, additional studies are required, particularly for encoding multi-dimensional datasets or using quantum clustering for larger and more complex cluster structures.

Two post-processing strategies were discussed for identifying clusterings: selecting the top bitstrings, and selecting the top-k bitstrings with subgrouping. A more efficient algorithm is required to encode a sum of Gaussians into a quantum state, without any classical pre-processing.

References

1. Initialize — qiskit api documentation (2026). https://quantum.cloud.ibm.com/docs/api/qiskit/qiskit.circuit.library.Initialize
2. Quantumcircuit — qiskit api documentation (2026). https://quantum.cloud.ibm.com/docs/api/qiskit/qiskit.circuit.QuantumCircuit
3. Bożejko, W., Trotskyi, S., Uchroński, M., Wodecki, M.: Optimizing two-machine scheduling in flexible manufacturing systems using autonomous AI and quantum computing. Neurocomputing **664**, 132067 (2026)
4. Carroll, M., Rosenblatt, S., Jurcevic, P., Lauer, I., Kandala, A.: Dynamics of superconducting qubit relaxation times. NPJ Q. Inf. **8**(1), 132 (2022)
5. Casaña-Eslava, R., et al.: A probabilistic framework for quantum clustering (2019). https://arxiv.org/abs/1902.05578
6. Horn, D., Gottlieb, A.: The method of quantum clustering. In: Proceedings of the 15th International Conference on Neural Information Processing Systems: Natural and Synthetic, NIPS'01, pp. 769–776. MIT Press, Cambridge, MA, USA (2001)
7. Kucharski, M.A.: A survey on quantum machine learning in speech acoustics. In: 2025 IEEE 25th International Symposium on Computational Intelligence and Informatics (CINTI), pp. 235–240. (2025)
8. Kucharski, M.A., Idzikowski, R.M., Kucharska, A., Pempera, K.: Real-hardware evaluation of quantum neural networks using a novel cornelian cherry dataset. SSRN Electron. J. (2026)
9. Preskill, J.: Quantum computing in the NISQ era and beyond. Quantum **2**, 79 (2018)
10. Rahman, M., Geiger, D.: Quantum clustering and gaussian mixtures (2016)
11. Scikit-learn documentation. sklearn.datasets.make blobs — scikit-learn 1.3.0 documentation. https://scikit-learn.org/stable/modules/generated/sklearn.datasets.make_blobs.html
12. scikit-learn documentation. User guide (2026). https://scikit-learn.org/stable/user_guide.html
13. Wrocław University of Science and Technology. Odra 5 – a quantum computer for polish science (2025). https://wit.pwr.edu.pl/en/news/odra-5-a-quantum-computer-for-polish-science-299.html
14. Xie, Y., Ben-Ami, N.: Efficient gaussian state preparation in quantum circuits (2025)

Developing Enterprise Architecture Reference Models for Assessing Digital Sovereignty of Data Centers

Wojciech Urbanczyk[1](✉) and Jan Werewka[2]

[1] Faculty of Computer Science, University of Vienna, Vienna, Austria
wojurbancz@gmail.com
[2] College of Economics and Computer Science, WSEI, Krakow, Poland
jwerewka@wsei.edu.pl

Abstract. The functioning of an organization can be assessed using quality attributes such as availability, performance, integrity, security, and digital sovereignty. The digital sovereignty is a relatively new quality attribute, and it requires dedicated analysis and scientific research, particularly in the context of data centers. At present, there is a lack of standards and systematic methods for maintaining and assessing digital sovereignty, whereas digital security has been extensively studied and standardized. Since security and digital sovereignty are highly correlated, it is legitimate to base solutions enhancing digital sovereignty on present security standards. This paper suggests an approach towards assessing the digital sovereignty of data centers through some combination of metrics, risk analysis, and holistic enterprise architecture models. This method allows for an assessment of the present state of sovereignty and a projection of possible loss of sovereignty using risk management techniques. Enterprise architecture models, developed using the ArchiMate graphical language, are used to describe the structure of data centers and to inventory organizational assets subject to digital sovereignty requirements. The proposed reference models provide a clear and systematic representation of data centers, supporting the identification of critical assets and architectural solutions necessary to ensure and assess digital sovereignty.

Keywords: Digital sovereignty · Data center · Enterprise architecture · Reference architecture · ArchiMate

1 Introduction

In the broad sense, organizations are increasingly confronted with ever-new challenges in a rapidly changing world. Traditionally, the quality of an organization's operations has been described through different quality attributes, such as security, efficiency, and availability. Over time, as new challenges emerge, new quality attributes have been introduced that address previously unexplored areas. One such relatively new concept is digital sovereignty.

W. Zamojski et al. (Eds.): DepCoS-RELCOMEX 2026, LNNS 2014, pp. 444–456, 2026.
https://doi.org/10.1007/978-3-032-27927-9_43

The issue of digital sovereignty has become a matter of national importance and is seen as a state responsibility. The EU has developed a set of standards, regulations, and frameworks. These legal acts [1] establish legal constraints and principles that enable the EU to implement digital sovereignty (i.e., regulate, secure, and manage digital technologies in accordance with EU values and law).

In the subject literature, the concept of digital sovereignty does not have a single, universally accepted definition. Most often, digital sovereignty is understood as the ability of a state, organization, or another entity to exercise effective control over its own digital environment [2]. This study examines digital sovereignty at the level of an IT organization and can be applied particularly to data centers.

Digital sovereignty of an organization refers to the ability to maintain full control over data, technologies, infrastructure, information flows, and the conditions governing data use, regardless of service providers, processing locations, or applicable jurisdictions.

According to [3], digital sovereignty—and the struggle to achieve it—affects everyone, including those who do not own a mobile phone or have never used an online service. Digital sovereignty encompasses the widely used concept of data sovereignty. Data sovereignty addresses questions such as where data are processed, who has access to them, and under what legal conditions they may be used. There is a need for systematic investigation into what data sovereignty is intended to encompass [4].

Data centers are organizations for which digital sovereignty is of critical importance, as it protects data, guarantees operational continuity, ensures legal compliance, and strengthens strategic independence and security. The overarching objective of the activities is to develop reference architecture and Enterprise Architecture (EA) models concerning data center (DC) sovereignty. Digital sovereignty varies between small and large data centers and essentially springs from the scope of responsibility, degree of risk, regulatory pressure, scale, and the real ability to retain control. The three pillars on which digital sovereignty sit are data sovereignty, operational sovereignty, and technological sovereignty.

The objectives of digital sovereignty vary depending on the nature of the data center. For large (governmental) data centers, sovereignty is understood strategically and in the long term, encompassing the protection of the public interest, state and economic security, critical infrastructure, and political and geopolitical stability. For smaller (corporate) data centers, sovereignty is primarily operational and economic, focusing on safeguarding business advantages, control over customer data and intellectual property, operational continuity, and compliance with sector-specific regulations. Research on digital sovereignty, particularly concerning data centers, is limited; a notable contribution is the review article [5].

2 Problems with Assessing Digital Sovereignty in an Organization

Digital sovereignty within an organization pertains to the systems owned and operated by that organization; however, not all systems require the same level of sovereignty. For some systems, loss of sovereignty may have strategic, legal, or operational consequences, while for others a lower level of sovereignty may be acceptable. As achieving full sovereignty for all systems is neither feasible nor economically justified, a structured approach to assessing and comparing sovereignty levels is required.

Below is a set of measurable indicators (KPIs - Key Performance Indicators) that allow for the assessment or comparison of systems in terms of sovereignty. The proposed indicators are normalized to a scale from 0 (lack of sovereignty) to 100 (full sovereignty) to enable comparison across heterogeneous systems and domains. Table 1 presents the types of KPIs: T—Technical, E—Economic, R—Regulatory, B—Security, O—Operational. These indicators are based on software quality models (ISO/IEC 25010), cloud interoperability standards (ISO/IEC 19941), security governance frameworks (ISO/IEC 27001, ISO/IEC 29147), and the European legal and policy context of digital sovereignty (GDPR, Schrems II, ENISA).

These indicators let you build an overall Software Sovereignty Index (SSI). Since not all indicators matter equally and some are related to each other, it makes sense to use a weighted index: $SSI_w = \Sigma(w_i \times KPI_i)$. You'd assign weights ($w_i$) based on what matters most to your organization and how critical each system is. The SSI_w index gives you a way to compare sovereignty across different systems and can be a useful tool when you're auditing sovereignty throughout your organization.

The selection of appropriate strategies enables the maintenance of sovereignty at a high level. For example, the study [7] presents an overview of strategies for maintaining digital sovereignty, as well as opportunities for further development based on emerging technologies. Good practice in companies is to maintain satisfactory quality attribute values by adhering to established standards. Many organizations seek to implement, for instance, the standard [8] concerning information security systems. Organizational audits allow for the assessment of security parameters' compliance with adopted norms. Standards related to digital sovereignty are still in an early stage of development and are not yet widely implemented. Standards and frameworks for data sovereignty include GAIA-X and the International Data Spaces Association (IDSA). The GAIA-X standard [9] encompasses Policy Rules, Trust Framework, and Federation Services, whereas IDS comprises the Reference Architecture Model, Certification Scheme, and Data Sovereignty Principles.

The KPI-based approach assesses the current level of digital sovereignty, while the risk-based analysis in the next section evaluates the likelihood and impact of its loss.

3 Risk Modeling of Digital Sovereignty Loss

Losing digital sovereignty can be costly, and these costs often emerge in unexpected ways. This uncertainty makes a structured approach to risk management essential. Much like cybersecurity, digital sovereignty can be managed using a risk-based approach, which is already well established in information security management systems [10]. However, the focus of the analysis differs. Cybersecurity primarily examines threats, vulnerabilities, and the potential impact of security incidents. Digital sovereignty, by contrast, is concerned with technological dependencies, legal and geopolitical constraints, and an organization's ability to retain effective control and decision-making autonomy. To develop a risk management model in this area, the first step is to clearly define what constitutes the risk of losing digital sovereignty.

Table 1. KPIs for the assessment of digital sovereignty of systems

KPI - Key performance indicator	Description	Type
SAI—Source access index	Percentage of the system available for audit	T
MS—Modifiability score	Degree to which the organization can modify the software independently	T
RLL—Reverse lock-in level	Inverted degree of vendor dependency (RLL = 100—LL, where LL denotes vendor lock-in level)	T/O
DPI—Data portability index	Degree of data portability based on open standards, APIs, export mechanisms	T
DPR—Domestic processing ratio	Percentage of data processed under a controlled jurisdiction	T/R
AS—Auditability score	Degree to which a full system audit can be performed (code, logs, architecture, configuration)	B/O
UCR—Update control ratio	Percentage of system updates that can be controlled prior to deployment	O
STI—Security transparency index	Degree of transparency of vendor-provided security vulnerability information (e.g., CVE disclosure, timeliness, completeness)	B/T
LIS—License independence score	Degree of license independence (modification, hosting)	E/T
RCPS—Reverse cost predictability score	Degree of stability and predictability of system maintenance and operation costs over time	E
LDP—Local deployment percentage	Percentage of components that can be run locally or in a sovereign cloud	O/T
VST—Vendor substitution time	Normalized time required to replace the vendor with an alternative provider within the permissible deadline	O
SSR—Skill self-sufficiency ratio	Degree to which the organization possesses the competence to maintain the system without the vendor	O/E

Digital Sovereignty Loss Risk (DSLR) is defined as the probability and potential impact of loss by an organization of effective autonomy, control, and decision-making capability relating to its digital assets, infrastructure, technologies, and data flows due to technological dependency or supplier constraint, legal obligation on them, enforcement of foreign jurisdiction on them, operational re-strictions on them or through geopolitical pressure imposed by external entities.

The risk of sovereignty loss is not determined by a single factor but arises from a combination of technological dependencies (D), active threats or coercion scenarios

(T), and jurisdictional and political exposure (J). Each of these elements increases the likelihood of losing decision-making and operational autonomy and can therefore be treated additively as components of the overall exposure to sovereignty loss. The DSLR risk model is defined as follows:

$$\mathrm{DSLR} = \mathrm{A} \cdot (\mathrm{DR} + \mathrm{TR} + \mathrm{JR}) \cdot \mathrm{V}$$

where:

- A—sovereignty-critical assets, referring to how important a given asset is from a digital sovereignty perspective
- DR—supplier dependency exposure, capturing reliance on specific vendors and technologies
- TR—threat exposure, including events that could trigger a loss of sovereignty
- JR—jurisdictional and geopolitical exposure, covering legal and political frameworks that may enable enforced loss of control
- V—sovereignty vulnerability, defined as the inverse of control capability

Table 2 presents exemplary metrics for the individual risk components DR, TR, JR, and V, together with their associated key performance indicators (KPIs). All components are assessed on a normalized scale from 0 (no risk) to 100 (maximal risk).

For each risk dimension (DR, TR, JR, and V), its value is calculated as the arithmetic means of the corresponding component metrics, i.e.,

$$\mathrm{DR} = \left(\sum \mathrm{DR_i}\right)/\mathrm{n},\ \mathrm{TR} = \left(\sum \mathrm{TR_i}\right)/\mathrm{n},\ \mathrm{JR} = \left(\sum \mathrm{JR_i}\right)/\mathrm{n},\ \text{and}\ \mathrm{V} = \left(\sum \mathrm{VR_i}\right)/\mathrm{n}.$$

This approach allows the model to be flexibly adapted to the characteristics of the analyzed asset and to the availability of measurement data.

DSLR Model and Risk Response Strategies. The DSLR model is not intended to assess the risk of technical failures or classical cybersecurity risks. Instead, it analyzes the extent to which external entities are capable of limiting or assuming control over an organization's digital resources, regardless of data location, implemented security mechanisms, or contractual arrangements. The model focuses on the risk of losing decision-making and operational autonomy.

The identified risk components require the definition of appropriate response strategies [11]. As in classical risk management, avoidance, mitigation, transfer, and acceptance strategies may be applied; however, their interpretation must reflect the specific nature of digital sovereignty.

Sovereignty Assets (A). Strategies define and identify assets critical to sovereignty. Limiting the areas of life where external bodies can wield their power over those assets is made possible by correct asset granularity, architectural segmentation—for instance, sovereign core vs. commodity IT—and application of a sovereignty-by-design approach [12] in which requirements for sovereignty are introduced already at the stage of system design. It also helps limit the degrees of freedom that external sources have over these assets.

Table 2. DSLR metrics related to KPIs.

Metrics	Description	KPI
DR_1—Risk of Vendor Lock-in	Degree of dependency on one or multiple suppliers	RLL
DR_2—Concentration Risk	Share of critical services provided by one or two suppliers	SAI, LDP
DR_3—Substitution Risk	Ability to substitute the supplier with an alternative provider	VST, SSR
TR_1—Threat Risk	Level of risk of digital sovereignty loss due to technical or operational threats	AS, UCR, STI
TR_2—Coercion/Influence Risk	Risk of political, economic, or technological coercion	DSS
TR_3—Disruption / Shutdown Threat	Risk of service shutdown imposed by an external entity	DPI, LIS
JR_1—Cross-Jurisdiction Exposure	Exposure of the asset to foreign legal systems and extraterritorial law	DPR
JR_2—Legal Sovereignty Misfit/Compliance Risk	Risk of non-compliance with local digital sovereignty requirements	SSR, LIS
JR_3—Arbitration & Dispute Vulnerability	Risk of enforced actions resulting from court rulings or arbitration decisions	DPI, VST
VR_1—Technical Dependency Fragility	Sensitivity to the lack of a viable technological alternative	MS, RLL
VR_2—Organizational Dependency Fragility	Lack of in-house capabilities to maintain and operate the system	SSR
VR_3—Operational Autonomy Deficit	Operational dependency on the supplier	LDP, OCPS

Supplier Dependency Risk (DR). The objective of these strategies is to reduce technological dependencies, vendor lock-in, and supplier concentration. This includes multi-vendor and multi-cloud architecture, the use of open standards and APIs, interoperability, modularity, and the development of in-house operational capabilities. At a strategic level, supplier diversification and limiting the dominance of individual vendors are also essential.

Threat and Coercion Risk (TR). Strategies in this area aim to limit the impact of political, economic, and technological pressure. They include designing architectures that are resilient to coercion, eliminating remote shutdown mechanisms and vendor-controlled kill switches, ensuring local control of cryptographic keys (sovereign key custody), implementing data and technology escrow mechanisms, restricting vendor-enforced automatic updates, and preparing sovereignty-oriented continuity plans (sovereign continuity plans). An important element is the concept of an operational sovereignty fallback, enabling a switch to alternative solutions without vendor consent.

Jurisdictional and Political Exposure (JR). These strategies aim to reduce exposure to foreign jurisdictions and geopolitical risks through local data processing, the use of local infrastructure operators, appropriate contractual clauses, and the selection of suppliers with low jurisdictional and political risk.

Vulnerability (V)/Control Capability. Strategies in this domain strengthen an organization's ability to independently operate and control digital systems through the development of internal technical competencies, local operational teams, comprehensive documentation, and the minimization of black-box solutions.

System-level strategies (meta-governance). These cross-cutting strategies include establishing governance structures for digital sovereignty, integrating sovereignty requirements into procurement and architectural processes, continuous monitoring of DSLR metrics, and treating sovereignty as an integral element of enterprise architecture governance.

Mitigating the risk of digital sovereignty loss is not based on increased technical levels of security but rather on systematic dependency relief, exposure limitation to jurisdictions, and in the strengthening of organization's effective decision-making and operational autonomy.

4 Data Center Assets Critical to Digital Sovereignty

Maintaining digital sovereignty in data centers requires the systematic identification of assets with respect to control over data, infrastructure, and operational processes. This forms the basis for evaluating the risk of loss of sovereignty and determining appropriate architectural as well as organizational measures that can be taken. It is in this respect that enterprise architecture models gain added importance since such modeling enables coherent yet comprehensive capturing of data center operation from a technical dimension and its organization as well as governance aspects.

Figure 1 presents digital sovereignty assets in the form of a landscape map, where services and capabilities provided by the data center are depicted along the horizontal axis, while functions and operational domains necessary to deliver them are represented along the vertical axis. This representation allows for a clear identification of assets of high, medium, and low significance from the perspective of sovereignty risk and facilitates the rapid identification of areas where dependencies are concentrated.

The practical significance of proper asset identification is illustrated by two cases:

Case 1. OVHcloud data center fire incident, Strasbourg, 2021 [13]. No formal rights of data ownership were violated, but victims discovered the absence of actual operational control over critical digital assets. Where physical data media, storage systems, and identity systems are lost, in most cases this results in the permanent loss of the sole authoritative version of the data. It shows that digital sovereignty is not a matter limited solely to legal aspects but rather includes the capacity to maintain real continuity of control over assets even in crisis situations.

Case 2. Schrems II—Loss of Control over Cross-Border Data Flows [14]. This case concerned Facebook's transatlantic data flows between EU-based data centers and U.S.-controlled backbone networks as well as their identity optimization services. The Court

Fig. 1. DC sovereignty digital assets presented as a landscape map

held that because U.S. surveillance laws permit access to data beyond contractual safeguards, network-level access control interferes with effective data sovereignty even when the location of the data is within EU-based data centers. Sovereignty requires control over jurisdiction, data flows, and network dependencies—not merely ownership or backup of data.

Table 3 shows two completely different ways your digital sovereignty can be at risk. With OVHcloud, the main issue was being locked into one provider without much control over your own operations. This led to a permanent loss of independence, even though foreign legal issues weren't really a factor. The Schrems II case tells a different story. Here, the problem was all about legal jurisdiction and government pressure. It shows that you can lose sovereignty even when your technology is running perfectly fine. This comparison confirms the necessity of a multidimensional sovereignty risk model that extends beyond traditional security and availability metrics.

The two cases show different, yet complementary, ways in which sovereignty can be lost. The OVHcloud SBG incident highlights that digital sovereignty can disappear entirely without any legal or geopolitical pressure, simply as a result of overreliance on a single supplier and a lack of real control. By contrast, the Schrems II ruling demonstrates that even infrastructures that are technically secure and operationally reliable can still fall short of sovereignty requirements when they are subject to foreign legal regimes that override contractual and regulatory protections. Taken together, these cases confirm the multi-dimensional nature of the DSLR model and underscore the need to address DR, JR, TR, and V in an integrated manner rather than treating them as separate concerns.

Table 3. OVHcloud SBG versus Schrems II comparison table, embedded in the DSLR model.

Dimension (DSLR)	OVHcloud SBG Fire	Schrems II case
Type of event	Physical infrastructure failure	Legal and jurisdictional enforcement
Primary loss driver	Loss of operational control over data and backups	Loss of legal and decision-making control over data flows
JR_1—Cross-Jurisdiction exposure	(Medium) EU-based infrastructure, EU law	(High) US surveillance law, extraterritorial reach
JR_3 - Arbitration/legal enforcement risk	(Low) No court-enforced data access	(High) CJEU ruling invalidating transfer mechanisms
TR_2—Coercion/influence risk	(Low) No political or legal coercion	(High) State-level surveillance and legal coercion
DR—Supplier dependency risk	(High) Single cloud provider, single location	(Medium) Dependency on US-based digital platforms and backbone services
V—Control capability deficit	(High) No autonomous backup, limited recovery capability	(Medium) Limited control over transatlantic data routing and access
Nature of sovereignty loss	Operational and technical	Legal, political, and strategic
Reversibility of impact	Largely irreversible (data permanently lost)	Partially reversible (new legal frameworks, re-architecture possible)
Key lesson	Sovereignty requires autonomous operational control, not only security	Sovereignty requires jurisdictional and legal control beyond contracts

Digital sovereignty is closely connected to other quality attributes of IT systems, especially security. Security measures—such as protecting data confidentiality and integrity, managing authentication and authorization, and handling cryptographic keys—are essential for sovereignty, but they are not sufficient on their own. The ENISA Threat Landscape for Data Centres report [15] highlights major risks facing data centers, including vendor lock-in, service concentration, supply chain vulnerabilities, and the loss of operational control. These risks align directly with the core dimensions of the Digital Sovereignty Loss Risk (DSLR) model.

The ISO/IEC 27001 standard does a solid job of helping you manage security for assets throughout a data center, but it's mainly focused on traditional information security concerns. When you look at the same setup through the lens of digital sovereignty, you start to notice what's missing—there aren't really standards that address things like which country's laws apply to your data, whether you're locked into specific vendors, or whether you can actually make your own technological decisions. This means we need to add sovereignty-focused methods to our existing security practices, not replace them.

For assets described this way, it is necessary to define the types of additional behaviors and actions that must be done to maintain sovereignty. Additional measures for security issues in the ISO 27001 standard are discussed in the paper Application of a Government Data Center (GDC) Reference Model for Security Management Analysis [16]. The paper [17] puts into perspective global governance and collaborative cybersecurity frameworks toward digital sovereignty. Cybersecurity and digital sovereignty relations are increasingly seen as complementary [18]. Work [19] adopts an Enterprise Architecture perspective, placing digital sovereignty and Enterprise Interoperability requirements seamlessly within International Data Spaces (IDS).

To model assets and map their significance for digital sovereignty, the ArchiMate language [20] is proposed, as it allows for the description of various DC layers:

- Motivation layer—goals, stakeholders, requirements, and principles, e.g., "Pre-vent Sovereignty Loss", "Vendor Independence Principle".
- Strategy layer—action strategies within the framework of available resources and capabilities, e.g., "Sovereignty Governance Team".
- Business layer—roles, processes, and functions for the maintenance of sovereignty, e.g., "Sovereignty Owner," "Sovereign Identity & Access Service."
- Application layer –user software components and their processes, e.g., "Sovereignty Compliance Engine"
- Technology layer—hardware, networks, operating systems, e.g., "Sovereign Hypervisor"
- Physical layer—facility infrastructure, e.g., power supply cooling
- Implementation & Migration layer—deployment projects. E.g. "Vendor De-pendency Reduction Program".

In Fig. 2, these layers are further subdivided into sublayers. The application layer comprises the Enterprise, Office, Middleware, and Database Application sublayers, while the technology layer is divided into DCIM and Virtualization.

Upon completing the mapping using the table, it is possible to infer which levels within the Data Center are most exposed to the risk of sovereignty loss. At least in those areas where digital sovereignty overlaps with security as defined in ISMS under ISO 27001. It is not possible to replace security aspects defined across the 14 domains of the ISO 27001 standard on a one-to-one basis with aspects of sovereignty. While correlations do exist between security aspects and digital sovereignty, these are limited to certain areas of the Data Center.

ArchiMate can be employed to map issues of digital sovereignty, particularly regarding technical, processual, and organizational aspects. However, for purely human or legal aspects, artifacts from the motivation layer or external annotations are often required to fully capture the nuances of digital sovereignty.

In conclusion, enterprise architecture modeling—and ArchiMate in particular—provides an effective tool for identifying and analyzing assets that are critical to the digital sovereignty of data centers.

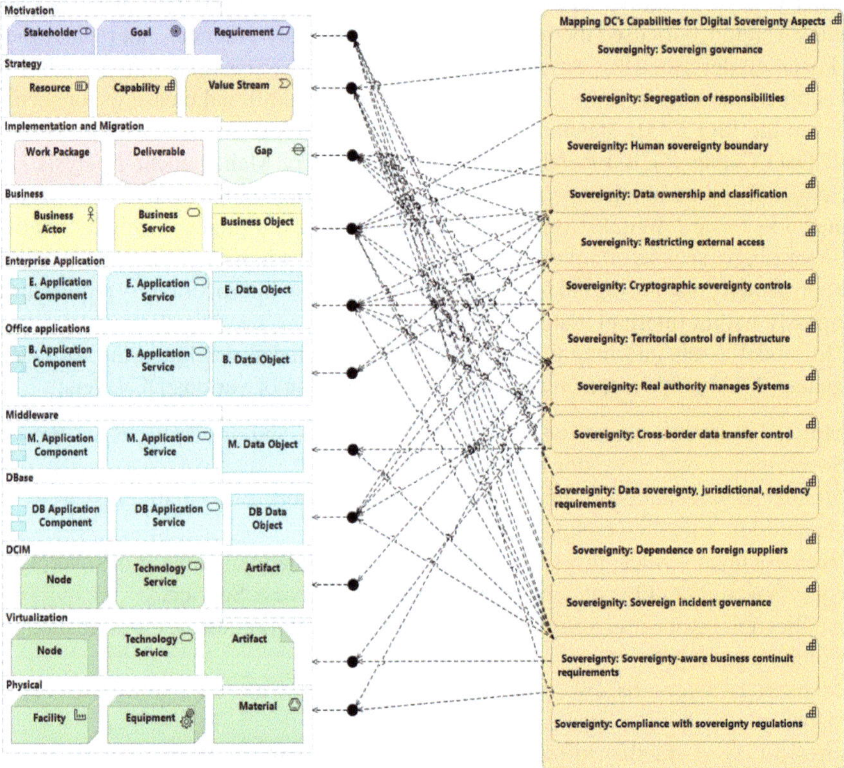

Fig. 2. Mapping digital sovereignty aspects to DC assets

5 Conclusion

In the context of rapidly evolving technological, regulatory, and geopolitical conditions, digital sovereignty has become a critical quality attribute for IT organizations, particularly data centers. Unlike classical quality attributes such as security or availability, digital sovereignty is not yet covered by a coherent and widely accepted set of standards and norms, which complicates its assessment and management.

This work lays out a practical, step-by-step way to analyze digital sovereignty. It starts with identifying your most important assets, then measures how much sovereignty you have using specific key performance indicators (KPIs) and finally maps out the risks of losing that sovereignty. In particular, it demonstrates that you can't manage sovereignty effectively unless you first figure out what resources your organization has, your data, technology, processes, and team skills—that let you make your own decisions and operate independently.

The risk-based approach enables the creation of a digital sovereignty risk register, where each asset's exposure to technological dependencies, operational threats, and jurisdictional and geopolitical factors is assessed. Based on this, priority intervention areas can be determined, and appropriate mitigation strategies—technical, organizational, and legal—can be defined.

When it comes to data centers, enterprise architecture models have turned out to be helpful for identifying which assets are crucial for maintaining digital sovereignty. These architectural maps make it easier to spot where you're at higher risk of losing control, and they connect those risky areas to actual projects and initiatives your organization can take to keep sovereignty where it needs to be. This approach could be a starting point for more hands-on research and for creating practical standards and frameworks that help IT organizations manage their digital sovereignty.

Future work will focus on the quantitative validation of the proposed sovereignty KPIs and the DSLR risk model, including their integration with enterprise architecture governance processes.

References

1. Knafo, S.: REPORT on European technological sovereignty and digital infrastructure. A10–0107/2025 European Parliament. Accessed: Jan. 18, 2026. https://www.europarl.europa.eu/doceo/document/A-10-2025-0107_EN.html
2. Pohle (Julia) and Thiel (Thorsten), Digital sovereignty. Accessed: Jan. 18, 2026. https://policyreview.info/concepts/digital-sovereignty
3. Floridi, L.: The fight for digital sovereignty: what it is, and why it matters, especially for the EU. Soc. Sci. Res. Netw., Rochester, NY: 3827089 (2021). https://doi.org/10.2139/ssrn.3827089
4. Hummel, P., Braun, M., Tretter, M., Dabrock, P.: Data sovereignty: A review. Big Data Soc. **8**(1), 2053951720982012 (2021). https://doi.org/10.1177/2053951720982012
5. Richardson, A. et al.: How sovereign is sovereign compute? A Review of 775 Non-U.S. Data Centers, (2025). arXiv: arXiv:2508.00932. https://doi.org/10.48550/arXiv.2508.00932
6. Stürmer, M.: Technological perspective on digital sovereignty. (2024). https://doi.org/10.48550/arXiv.2406.03266
7. Challenges and Opportunities: S. Misra, K. Barik, P. Kvalvik, Digital Sovereignty in the Era of Industry 5.0. Procedia Comput. Sci. **254**, 108–117 (2025). https://doi.org/10.1016/j.procs.2025.02.069
8. ISO - International Organization for Standardization, ISO/IEC 27001:2013, ISO. Accessed: Dec. 27, 2020. https://www.iso.org/cms/render/live/en/sites/isoorg/contents/data/standard/05/45/54534.html
9. Gaia-X context - Gaia-X Architecture Document - local Release. Accessed: Jan. 05, 2026. https://docs.gaia-x.eu/technical-committee/architecture-document/25.11/gaia-x_context/
10. Band, I, et al., Modeling enterprise risk management and security with the ArchiMate language, 2015
11. ISO/IEC 27005 Information Security Risk Management - Training Courses, PECB. Accessed: Jan. 19, 2026. https://pecb.com/en/education-and-certification-for-individuals/iso-iec-27005
12. Turobov, A., Farrand, B., Farrand Carrapico, H.: Digital sovereignty by design, academic articles and policy papers. (2023). Accessed: Jan. 19, 2026. https://reshare.ukdataservice.ac.uk/856533/
13. A. L. alawrence@uptimeinstitute.com Executive Director of Research, Uptime Institute, Learning from the OVHcloud data center fire, Uptime Institute Blog. Accessed: Jan. 19, 2026. https://journal.uptimeinstitute.com/learning-from-the-ovhcloud-data-center-fire/

14. Privacy Shield invalidated by CJEU but Standard Contractual Clauses remain—Byrne Wallace Shields LLP. Accessed: Jan. 26, 2026. https://byrnewallaceshields.com/news-and-recent-work/publications/privacy-shield-invalidated-by-cjeu-but-standard-contractual-clauses-remain.html
15. ENISA Threat Landscape 2021 | ENISA. Accessed: Jan. 18, 2026. https://www.enisa.europa.eu/publications/enisa-threat-landscape-2021
16. Urbanczyk, W. , Werewka, J.: Application of a government data center (GDC) reference model for security management analysis. In: IEEE International Conference on e-Business Engineering, ICEBE 2021, Guangzhou, China, IEEE, pp. 112–119 (2021). https://doi.org/10.1109/ICEBE52470.2021.00021
17. Ismail, S.: The Intersection of digital sovereignty and cybersecurity. In: IEEE International Conf. on Electro Information Technology (eIT), Valparaiso, IN, USA. pp. 098–102 (2025)
18. Sorrentino, E., Spagnuolo, A.F., Portaro, A., Taverniti, M., Cardillo, E.: Towards semantic coherence: understanding cybersecurity and digital sovereignty in the automotive landscape. In: 2025 IEEE 31st Int. Symposium on On-Line Testing and Robust System Design (IOLTS), Ischia, Italy, pp. 1–5 (2025)
19. Firdausy, D.R., De Alencar Silva, P., Van Sinderen, M., Iacob, M. -E.: Towards a reference enterprise architecture to enforce digital sovereignty in international data spaces. In: 2022 IEEE 24th Conf. on Business Informatics (CBI), Netherlands, pp. 117–125 (2022)
20. Azevedo, C.L.B., Iacob, M.-E., Almeida, J.P.A., van Sinderen, M., Pires, L.F., Guizzardi, G.: Modeling resources and capabilities in enterprise architecture: A well-founded ontology-based proposal for ArchiMate. Inf. Syst. **54**, 235–262 (2015). https://doi.org/10.1016/j.is.2015.04.008

On the Effectiveness of NLP Adversarial Attacks Beyond English

Paweł Walkowiak(✉)

Wroclaw University of Science and Technology, Wroclaw, Poland
pawel.walkowiak@pwr.edu.pl

Abstract. Adversarial example generation in natural language processing relies on a variety of techniques designed to induce incorrect model predictions while preserving the original meaning of the input. These techniques include character- and word-level perturbations, such as those employed by TextBugger, which introduce subtle and often imperceptible modifications that can significantly alter model behavior. Another prominent class of attacks focuses on replacing words with semantically similar synonyms, as exemplified by TextFooler, thereby maintaining semantic coherence while changing the predicted label. However, the majority of existing adversarial methods are developed and evaluated almost exclusively on non-inflectional languages, most commonly English, leaving their effectiveness in morphologically rich languages underexplored. In this work, we systematically investigate the performance of adversarial attacks in inflectional languages by evaluating four adversarial example generation methods: TextFooler, TextBugger, WordNetTextFooler, and BERT-Attack on BERT-based models. Furthermore, we analyse multiple approaches for assessing adversarial example quality, including BLEU, ROUGE, Natural Language Inference, and semantic similarity metrics, highlighting their strengths and limitations in this linguistic setting.

Keywords: Adversarial Examples · NLP · Language Models

1 Introduction

Transformer-based language models [26] currently represent the state of the art in text classification, yet their robustness can be significantly undermined by adversarial examples [24]. These strategically modified inputs are capable of misleading classifiers to the point of rendering them ineffective. Despite extensive literature on the generation and defense of such examples, their behavior within the context of inflectional languages remains largely unexamined. Furthermore, the methodology for selecting high-quality adversarial candidates, specifically how researchers filter them to remain indistinguishable from original text, is an under-investigated area. Current evaluation frameworks rely on diverse metrics, including n-gram statistics [14,20], semantic similarity [22], and natural language

W. Zamojski et al. (Eds.): DepCoS-RELCOMEX 2026, LNNS 2014, pp. 457–467, 2026.
https://doi.org/10.1007/978-3-032-27927-9_44

inference [1]. The inconsistent results produced by these varied approaches highlight a critical need to better understand their individual strengths and inherent limitations.

When choosing language models for particular tasks, decisions are typically driven mainly by performance measures such as accuracy. However, robustness is often neglected in this selection process. Given the broad spectrum of available models, both language-specific models [5,18] and multilingual models [4] may be taken into account. This leads to the question of whether certain properties of a model's training language, for example inflectional complexity, affect its robustness against text-based adversarial attacks. In this work, we seek to address this question through the construction of adversarial examples.

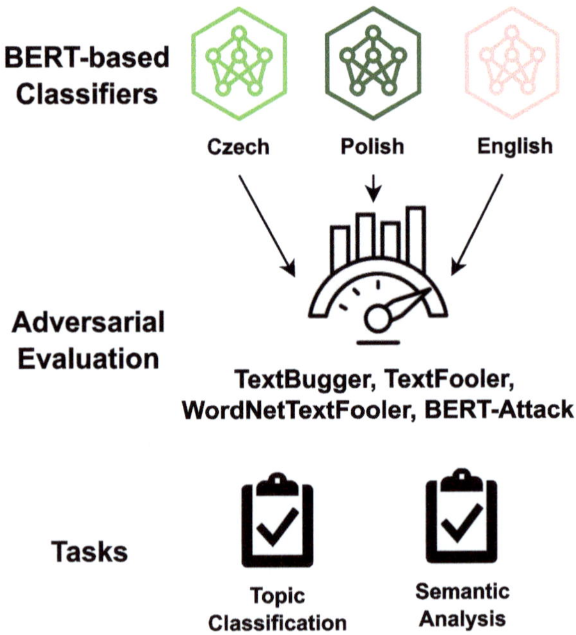

Fig. 1. Experimental pipeline for evaluation of adversarial examples generation methods.

To answer our research question, we evaluated the performance of various text attack methods [7,9,12,13] across three selected languages: two inflectional languages (Polish and Czech) and English, which serves as a widely used, non-inflectional reference language. In this paper, we present a comprehensive set of experiments aimed at understanding how the presence of inflectional morphology in a language influences the adversarial robustness of language models. By comparing models' responses to adversarial attacks in both inflectional (e.g. Polish and Czech) and non-inflectional (e.g. English) languages. The key contributions of this paper are:

- **A comprehensive evaluation and comparison of four adversarial example generation methods** (TextFooler [9], TextBugger [12], WordNet-TextFooler [7] and BERT-Attack [13]) across two cross-lingual NLP tasks, with a focus on topic classification and sentiment analysis in both highly inflectional (Polish and Czech) and non-inflectional (English) languages.
- Evaluation of widely used similarity assessment methods: **BLEU, ROUGE, Semantic Similarity and Natural Language Inference** for their effectiveness in measuring the quality of adversarial examples

2 Related Work

Text Classification. We consider a classification problem of mapping $X \rightarrow Y$, where X are input samples and Y are labels. Classification is the foundation for key NLP tasks, including sentiment analysis [25] and topic classification [3]. The Transformer model [26] marked a major advance in text classification, followed by BERT [5], which achieves state-of-the-art performance. BERT's pretraining on large corpora yields transferable language representations, enabling effective fine-tuning for classification tasks even with limited labeled data. Decoder-only models like GPT-3.5 [19] have recently shown strong performance across NLP tasks, raising questions about the continued dominance of BERT-based models for text classification. However, studies [15,27] indicate that BERT-based models can still outperform decoder-only architectures in this domain. These results highlight the importance of assessing the robustness of BERT-based approaches.

Adversarial Examples in Text Classification. Adversarial examples (AEs) are input samples that have been subtly perturbed in a way that is often imperceptible to humans but causes a model to produce incorrect predictions, deviating from the original correct output. They are currently considered a critical threat to the performance of the models and their trustworthiness. Since the influential work presented in [24], the field has seen the publication of tens of thousands of related studies. A substantial number of these contributions are catalogued in the continuously expanding list provided in [2]. Most adversarial example generation methods exploit vulnerabilities in model architectures by introducing subtle perturbations into the input text to induce misclassification. Among the growing number of adversarial examples generation methods, we focus on four of them specifically designed to target BERT-based models: TextFooler [9], TextBugger [12], WordNetTextFooler [7], and BERT-Attack [13].

Inflection. In the context of natural language processing, inflection refers to the variation in the form of a word to express grammatical features such as tense, number, gender, case, or person. These changes typically occur through the addition of affixes (such as suffixes). Inflection affects NLP tasks such as part-of-speech tagging, information retrieval, and machine translation. A key feature of inflectional languages in text generation is that a single morpheme (the smallest unit of meaning) can take on multiple functions through different suffixes.

Additionally, the relative freedom in word order allows paraphrasing while maintaining semantic content. These characteristics increase syntactic complexity and require more sophisticated approaches for attacks generation.

3 Adversarial Examples Generation Methods

TextFooler. (TF) introduced in [9], are methods to generate adversarial examples against text BERT-based classifiers. TextFooler identifies the most important words in a text-based model decision change on masked words. For each word, iteration is found in decreasing order of importance, defined as the word's impact on the model's decision marked as correct synonym using uncontextual embedding from the Glove model [21]. Adversarial samples are generated by replacing words with their substitutes in the original sentence.

TextBugger. (TB) is an adversarial example generation method introduced in [12]. The candidate word for attack is selected with the usage of Jacobian $J_F(i, y) = \frac{\partial F_y(X)}{\partial x_i}$, where $X = (x_1, x_2, ...x_n)$ is input text, which consists of words $\{x_1, x_2, ...x_n\}$, $F(X)$ is classification function and y model prediction. Based on the ranking, perturbations are introduced into important words in order to prepare an attack. The TextBugger proposes five perturbation methods: *insertion*, which involves adding a space within the perturbed word; *deletion*, which removes a random character; *swap*, which exchanges the positions of two adjacent characters; *Sub-C*, which substitutes characters with visually similar ones (e.g., replacing "l" with "1"); and *Sub-W*, which replaces a word with its nearest neighbour based on a GloVe model [21].

WordNetTextFooler. (WNTF) an adversarial example generation method, proposed in [7]. The method is based on the approaches introduced in TextFooler [9] and TextBugger [12]. However, it uses a different technique to generate word importance rankings and incorporates the aspect of inflectional variation of suggested synonyms, using Polish WordNet [6]. In contrast to the importance score estimation techniques used in previous methods, the authors of WordNet-TextFooler propose using a model explainability method Shapley values [16] to assess the importance of words. These values are computed per model on a designated subset of the data. The method replaces the candidate words with the appropriate synonyms drawn from WordNet.

BERT-Attack. (BA) [13] is an adversarial example generation method that takes advantage of a BERT-type model to generate possible word substitutions in a text. The approach is innovative in that it employs a pretrained language model to attack another language model, unlike earlier substitution methods that rely on simple perturbations or synonym replacements based solely on word vectors. The use of a contextualized language model allows for substitutions that take into account the surrounding text, yielding highly accurate replacements. The vulnerable words are selected by checking how the model prediction changes when masking each of the candidate words. The vulnerable words are replaced with candidate BERT generated tokens.

4 Experiments

The experiments are divided into two parts: (1) Evaluating the effectiveness of adversarial attacks on models trained for the same tasks across selected languages, to compare robustness as influenced by inflectional morphology; (2) Measuring the quality of adversarial examples generated for different languages.

4.1 Datasets

In the first part of our experiments, we assessed the effectiveness of adversarial attacks in reducing the robustness of BERT-based models across two NLP tasks: topic classification and sentiment analysis. For Polish, we used **Wiki_PL**[1], a 34-class Wikipedia-based dataset, and **MultiEmo** [10], a sentiment corpus. For Czech, we used **Czech Text Document Corpus v 2.0** [11], containing annotated news articles, and **CSFD**[2], a movie review dataset with sentiment labels. English benchmarks included **AG_News**[3] for topic classification and **IMDB** [17] for sentiment analysis. To perform the attacks, we trained classifiers using pre-trained language models: BERT[4] [5] for English, HerBERT[5] [18] for Polish, and RobeCzech[6] [23] for Czech. Dataset splits included training and test sets, with the latter used for evaluation and adversarial example generation. Dataset statistics and classification results are shown in Table 1.

Table 1. The dataset characteristics include the number of annotated classes, subset sizes, average word count, and classification accuracy (ACC) of the trained models.

Dataset	Language	Class number	Dataset size		Avg. words	Accuracy
			train	test		
AG_News	English	4	120,000	4,788	38	94.72
IMDB	English	2	8,000	1,260	211	95.17
Wiki_PL	Polish	34	6,885	1,859	186	94.88
MultiEmo	Polish	2	4,319	680	129	98.25
CTDC	Czech	37	11,955	1,574	247	83.33
CSFD	Czech	3	82,242	5,757	50	86.59

[1] http://hdl.handle.net/11321/222.
[2] https://corpora.kiv.zcu.cz/sentiment/csfd.zip.
[3] https://huggingface.co/sentence-transformers/distiluse-base-multilingual-cased-v1.
[4] https://huggingface.co/google-bert/bert-base-uncased.
[5] https://huggingface.co/allegro/herbert-base-cased.
[6] https://huggingface.co/ufal/robeczech-base.

4.2 Adversarial Examples Performance

In the experimental setup presented in Fig. 1, the selected adversarial examples generation methods were implemented: **TextBugger**, **TextFooler**, **WordNet-TextFooler** and **BERT-Attack**. We introduced a modification compared to the original methods, the unification of the identification of attack victim words. We used a masking algorithm that was applied to successive words in the texts, and changes in the confidence of the model, similar to the one proposed in the TextFooler method. For filtering adversarial example candidates to ensure their high quality, we used cosine similarity, with a threshold of 90%, on text embeddings generated by the multilingual SBERT model[7].

Table 2. Classification accuracy drop after attacks, for each attack method.

Dataset	Lang.	Δ Accuracy [%]			
		TF [%]	TB [%]	WNTF [%]	BA[%]
AG_News	ENG	1.90	**8.44**	2.78	2.36
IMDB	ENG	10.33	**19.69**	9.38	4.46
Wiki_PL	PL	5.10	**8.54**	5.10	2.03
MultiEmo	PL	7.66	**9.72**	4.13	2.81
CSFD	CZ	7.99	**22.62**	1.84	3.04
CTDC	CZ	5.38	**10.84**	0.86	4.23

The effectiveness of adversarial attacks was assessed on test subsets by comparing baseline classification accuracy with accuracy under attack. The results (baseline accuracy - attack accuracy) are shown in Table 2. Baseline models achieved over 94% accuracy in Polish and English datasets and over 83% on Czech. **TextBugger caused the largest performance drop**. Accuracy degradation aligns with attack success rates, with English sentiment analysis proving less robust. In topic classification, the English model showed higher robustness than the Polish and Czech models, while in sentiment analysis it suffered greater accuracy loss, indicating higher resilience of inflected-language models in this task.

4.3 How to Measure Adversarial Examples Quality?

A key requirement for adversarial examples is high similarity to the original text preserving meaning for humans while misleading the model. To assess this, we compare three common similarity metrics: **BLEU** [20], **ROUGE** [14], **Natural Language Inference (NLI)** [8], and **Semantic Similarity** [22]. NLI scores reflect entailment probabilities, while semantic similarity is measured via cosine

[7] https://huggingface.co/sentence-transformers/distiluse-base-multilingual-cased-v1.

Table 3. Similarity of adversarial examples to their originals, measured using four metrics. The results are presented by dataset and attack type. Values are calculated as the mean similarity of examples, with the standard deviation highlighted

DS	Attack type	Similarity Metric			
		BLEU [%]	ROUGE[%]	NLI[%]	Semantic[%]
AG_News	TB	76.54 ± 7.45	87.81 ± 3.61	**93.28** ± 13.35	96.51 ± 0.89
	TF	84.34 ± 8.45	92.81 ± 3.81	68.76 ± 34.58	**97.93** ± 1.32
	WNTF	**85.08** ± 7.83	**93.07** ± 3.43	79.79 ± 27.16	97.62 ± 1.21
	BA	82.75 ± 8.47	92.33 ± 3.8	75.56 ± 32.58	97.47 ± 1.23
IMDB	TB	94.14 ± 3.7	96.92 ± 1.98	**98.38** ± 3.02	98.2 ± 1.37
	TF	94.94 ± 3.52	97.47 ± 1.79	87.67 ± 23.3	**98.93** ± 1.11
	WNTF	**95.69** ± 2.98	**97.86** ± 1.49	94.55 ± 13.66	98.72 ± 1.2
	BA	94.75 ± 4.32	97.53 ± 2.24	75.25 ± 38.36	98.61 ± 1.27
Wiki_PL	TB	92.01 ± 8.0	95.82 ± 4.7	**96.44** ± 6.6	97.69 ± 1.3
	TF	92.93 ± 7.51	96.36 ± 3.56	72.22 ± 25.38	97.52 ± 1.43
	WNTF	92.93 ± 7.51	96.36 ± 3.56	72.19 ± 25.42	97.52 ± 1.43
	BA	**93.63** ± 7.59	**96.72** ± 4.12	93.73 ± 10.0	**98.35** ± 1.2
Multi_Emo	TB	88.45 ± 5.48	94.23 ± 2.9	**98.03** ± 1.55	98.01 ± 1.28
	TF	90.03 ± 5.02	95.41 ± 2.46	88.20 ± 15.36	98.25 ± 1.35
	WNTF	**93.39** ± 4.58	**96.50** ± 2.35	84.72 ± 19.81	98.64 ± 1.15
	BA	91.51 ± 5.13	95.9 ± 2.59	94.18 ± 11.58	**98.79** ± 1.21
CSFD	TB	71.07 ± 22.65	86.97 ± 11.50	**92.37** ± 15.13	97.53 ± 1.28
	TF	78.77 ± 16.04	90.12 ± 7.21	68.73 ± 29.26	97.66 ± 1.40
	WNTF	**90.94** ± 8.96	**96.27** ± 3.64	77.35 ± 29.16	**98.20** ± 1.48
	BA	81.8 ± 12.59	91.39 ± 6.23	84.23 ± 17.23	97.25 ± 1.30
CTDC	TB	94.31 ± 9.58	97.69 ± 3.45	**96.94** ± 5.97	98.25 ± 1.15
	TF	94.89 ± 5.1	97.69 ± 2.61	81.77 ± 22.2	98.14 ± 1.36
	WNTF	**98.90** ± 0.96	**99.56** ± 0.41	95.04 ± 9.74	**99.1** ± 1.22
	BA	96.37 ± 3.87	98.32 ± 1.86	92.86 ± 8.84	97.92 ± 1.32

similarity on SBERT embeddings. The similarity between the original and adversarial texts was computed and averaged per attack. The results, with standard deviations, are shown in Table 3. These metrics offer complementary perspectives: BLEU and ROUGE for statistical overlap, NLI for inferential agreement, and SBERT for semantic preservation.

More effective attacks, such as TextBugger, yield lower statistical similarity scores. For NLI-based confidence, cross-lingual differences are evident: TextBugger and WordNetTextFooler perform best for English, while TextBugger and BERT-Attack score highest for Polish and Czech. The high standard deviation in NLI scores indicates the metric instability between samples. In contrast, semantic similarity remains consistently high (above 96%) with low variance. In particular, WordNetTextFooler generates adversarial examples with the highest or second highest ROUGE and semantic similarity scores, highlighting its strength in producing semantically close adversarial texts.

To further analyze similarity metrics, we examined the distribution of NLI-based similarity scores for adversarial examples from the AG_News dataset with a high NLI standard deviation. The distributions by attack method are shown in Fig. 2. For AG_News, TextFooler, WordNetTextFooler, and BERT-Attack exhibit long-tailed distributions, aligning with their high variance. **The elevated standard deviation in NLI scores is driven by a subset of examples with very low similarity values.**

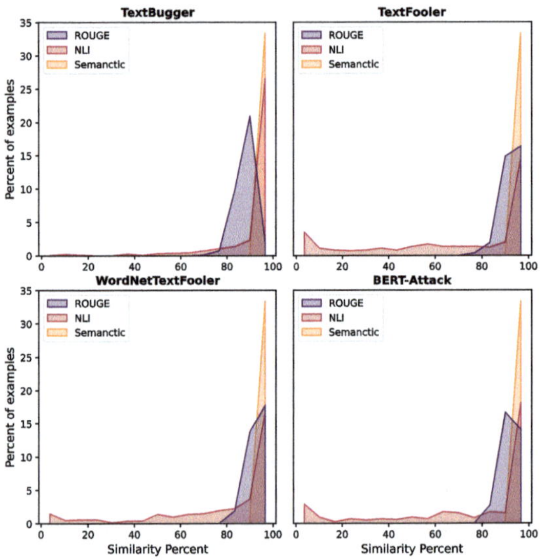

Fig. 2. Distribution of adversarial examples based on similarity to the original data for four evaluation metrics for the AG_News dataset.

5 Discussion

5.1 Conclusion

In this work, we evaluated how adversarial examples influence the robustness of the model in various language models. Incorporating inflexion in language model training corpus can increase its robustness against adversarial examples which focuses on bugging words (e.g. changing some of the letters). Among the set of evaluated adversarial example generation methods, TextBugger demonstrates the highest effectiveness, as measured by both the attack success rate and the corresponding decrease in model accuracy. This observation suggests that perturbation strategies based on character-level modifications, such as inserting, deleting, or replacing individual letters, may be more disruptive to language models than approaches that rely on substituting entire words, as employed in methods like TextFooler and WordNetTextFooler.

On the other hand, our analysis of adversarial example similarity levels indicated that TextBugger consistently produced statistically lower-quality attacks compared to the other evaluated methods. Among the similarity measurement techniques evaluated, Semantic Similarity and ROUGE demonstrated high stability. In contrast, the Natural Language Inference (NLI) method showed a high standard deviation in similarity scores across the dataset. Further investigation of the NLI results revealed that the method assigned scores close to zero for a subset of adversarial examples.

5.2 Limitations

One of the key limitations in researching less widely studied languages, such as Czech, is the limited availability of high-quality annotated datasets. This constraint is reflected in the performance of language models trained on Czech datasets such as CTDC and CSFD, where accuracy reaches respectively 83% and 86%. In contrast, models trained on Polish and English datasets consistently achieve accuracies exceeding 94%, highlighting a significant disparity.

Acknowledgments. The work was financed by CLARIN-PL: Common Language Resources and Technology Infrastructure (POIR.04.02.00-00C002/19, 2024/WK/01, FENG.02.04-IP.040004/24).

References

1. Camburu, O.M., Rocktäschel, T., Lukasiewicz, T., Blunsom, P.: e-snli: natural language inference with natural language explanations. In: Advances in Neural Information Processing Systems, vol. 31, pp. 9539–9549 (2018). https://papers.nips.cc/paper_files/paper/2018/hash/4c7a167bb329bd92580a99ce422d6fa6-Abstract.html
2. Carlini, N.: A complete list of all (arxiv) adversarial example papers (2019–2025). https://nicholas.carlini.com/writing/2019/all-adversarial-example-papers.html
3. Chen, J., Gong, Z., Liu, W.: A nonparametric model for online topic discovery with word embeddings. Inf. Sci. **504**, 32–47 (2019)
4. Conneau, A., et al.: Unsupervised cross-lingual representation learning at scale. CoRR arxiv:1911.02116 (2019)
5. Devlin, J., Chang, M., Lee, K., Toutanova, K.: BERT: pre-training of deep bidirectional transformers for language understanding. CoRR arxiv:1810.04805 (2018)
6. Dziob, A., Piasecki, M.: Implementation of the Verb Model in plWordNet 4.0. In: Proceedings of the 9th Global Wordnet Conference. pp. 113–122. Global Wordnet Association, Nanyang Technological University (NTU), Singapore (2018). https://aclanthology.org/2018.gwc-1.14

7. Gniewkowski, M., Syga, P., Klonowski, M., Walkowiak, P., Walkowiak, T.: Do Not Trust Me: Explainability Against Text Classification. In: ECAI 2023 - 26th European Conference on Artificial Intelligence, Kraków, Poland, 30 September–4 October 2023 - Including 12th Conference on Prestigious Applications of Intelligent Systems (PAIS 2023). Frontiers in Artificial Intelligence and Applications, vol. 372, pp. 875–882. IOS Press (2023). https://doi.org/10.3233/FAIA230356
8. He, P., Liu, X., Gao, J., Chen, W.: Deberta: decoding-enhanced bert with disentangled attention. In: International Conference on Learning Representations (2021). https://openreview.net/forum?id=XPZIaotutsD
9. Jin, D., Jin, Z., Zhou, J.T., Szolovits, P.: Is BERT really robust? Natural language attack on text classification and entailment. CoRR arxiv:1907.11932 (2019)
10. Kocoń, J., Miłkowski, P., Kanclerz, K.: MultiEmo: multilingual, multilevel, multidomain sentiment analysis corpus of consumer reviews. In: Paszynski, M., Kranzlmüller, D., Krzhizhanovskaya, V.V., Dongarra, J.J., Sloot, P.M.A. (eds.) ICCS 2021. LNCS, vol. 12743, pp. 297–312. Springer, Cham (2021). https://doi.org/10.1007/978-3-030-77964-1_24
11. Král, P., Lenc, L.: Czech text document corpus v 2.0. CoRR arxiv:1710.02365 (2017)
12. Li, J., Ji, S., Du, T., Li, B., Wang, T.: Textbugger: generating adversarial text against real-world applications. CoRR arxiv:1812.05271 (2018)
13. Li, L., Ma, R., Guo, Q., Xue, X., Qiu, X.: BERT-ATTACK: adversarial attack against BERT using BERT. In: Proceedings of the 2020 Conference on Empirical Methods in Natural Language Processing, EMNLP 2020, Online, 16–20 November 2020, pp. 6193–6202. Association for Computational Linguistics (2020). https://doi.org/10.18653/V1/2020.EMNLP-MAIN.500
14. Lin, C.Y.: ROUGE: a package for automatic evaluation of summaries. In: Text Summarization Branches Out, pp. 74–81. Association for Computational Linguistics, Barcelona, Spain (2004). https://www.aclweb.org/anthology/W04-1013
15. Lukito, J., Chen, B., Masullo, G.M., Stroud, N.J.: Comparing a BERT classifier and a GPT classifier for detecting connective language across multiple social media. In: Al-Onaizan, Y., Bansal, M., Chen, Y. (eds.) Proceedings of the 2024 Conference on Empirical Methods in Natural Language Processing, EMNLP 2024, Miami, FL, USA, 12–16 November 2024, pp. 19140–19153. Association for Computational Linguistics (2024). https://aclanthology.org/2024.emnlp-main.1067
16. Lundberg, S.M., Lee, S.I.: A unified approach to interpreting model predictions. In: Advances in Neural Information Processing Systems, vol. 30 (2017)
17. Maas, A.L., Daly, R.E., Pham, P.T., Huang, D., Ng, A.Y., Potts, C.: Learning word vectors for sentiment analysis. In: Proceedings of the 49th Annual Meeting of the Association for Computational Linguistics: Human Language Technologies, pp. 142–150. Association for Computational Linguistics, Portland (2011). http://www.aclweb.org/anthology/P11-1015
18. Mroczkowski, R., Rybak, P., Wróblewska, A., Gawlik, I.: HerBERT: efficiently pretrained transformer-based language model for Polish. In: Proceedings of the 8th Workshop on Balto-Slavic Natural Language Processing, pp. 1–10. Association for Computational Linguistics, Kiyv (2021). https://aclanthology.org/2021.bsnlp-1.1
19. OpenAI: Gpt-3.5 turbo technical overview (2023). https://platform.openai.com/docs/models/gpt-3-5. Accessed 03 May 2025
20. Papineni, K., Roukos, S., Ward, T., Jing Zhu, W.: BLEU: a method for automatic evaluation of machine translation, pp. 311–318 (2002)

21. Pennington, J., Socher, R., Manning, C.: GloVe: global vectors for word representation. In: Proceedings of the 2014 Conference on Empirical Methods in Natural Language Processing (EMNLP), pp. 1532–1543. Association for Computational Linguistics, Doha (2014). https://doi.org/10.3115/v1/D14-1162. https://aclanthology.org/D14-1162
22. Reimers, N., Gurevych, I.: Sentence-BERT: sentence embeddings using siamese BERT-networks. In: Proceedings of the 2019 Conference on Empirical Methods in Natural Language Processing. Association for Computational Linguistics (2019). http://arxiv.org/abs/1908.10084
23. Straka, M., Náplava, J., Straková, J., Samuel, D.: RobeCzech: Czech RoBERTa, a monolingual contextualized language representation model. In: Ekštein, K., Pártl, F., Konopík, M. (eds.) TSD 2021. LNCS (LNAI), vol. 12848, pp. 197–209. Springer, Cham (2021). https://doi.org/10.1007/978-3-030-83527-9_17
24. Szegedy, C., et al.: Intriguing properties of neural networks. In: 2nd International Conference on Learning Representations, ICLR 2014, Banff, AB, Canada, 14–16 April 2014, Conference Track Proceedings (2014). http://arxiv.org/abs/1312.6199
25. Tai, K.S., Socher, R., Manning, C.D.: Improved semantic representations from tree-structured long short-term memory networks. CoRR arxiv:1503.00075 (2015)
26. Vaswani, A., et al.: Attention is all you need. In: Advances in Neural Information Processing Systems 30: Annual Conference on Neural Information Processing Systems 2017, Long Beach, CA, USA, 4–9 December 2017, pp. 5998–6008 (2017). https://proceedings.neurips.cc/paper/2017/hash/3f5ee243547dee91fbd053c1c4a845aa-Abstract.html
27. Wang, Y., Qu, W., Ye, X.: Selecting between BERT and GPT for text classification in political science research. CoRR arxiv:2411.05050 (2024). https://doi.org/10.48550/ARXIV.2411.05050

Visualizing Authorial Style: How Chunk Size Affects UMAP and PaCMAP Representations

Tomasz Walkowiak(✉)

Wroclaw University of Science and Technology, Wrocław, Poland
tomasz.walkowiak@pwr.edu.pl

Abstract. A central objective of stylometry is authorship attribution, which focuses on mapping specific linguistic patterns to an individual's writing style. In this paper, we investigate how text chunk size influences the effectiveness of stylometric analysis in an unsupervised setting. Texts are represented using grammatical bigrams to capture stylistic information and are projected into a two-dimensional space using UMAP and PaCMAP for visualization and similarity analysis. Experiments were conducted on two datasets: a small corpus of Polish novels by two authors and a larger collection of 100 novels written by 34 authors. The quality of the resulting representations was evaluated using human judgment and quantitative metrics, including k-nearest neighbors accuracy and class separation measures. The results show that very small text chunks lead to unstable representations, while excessively large chunks reduce the number of available samples. Across both datasets and evaluation methods, chunk sizes between 2000 and 5000 words provide the best balance between stability and granularity. These findings offer practical guidance for selecting text segment sizes in unsupervised stylometric analysis.

Keywords: Authorship attribution · Text chunk size · Text similarity visualization

1 Introduction

From a machine learning perspective, authorship attribution [3] can be formulated as a supervised or unsupervised task. Supervised approaches rely on labeled training data to build a classifier, which is subsequently evaluated on a separate test set. In contrast, unsupervised approaches focus on discovering patterns in unlabeled data, typically through clustering or similarity analysis. The latter constitutes the main focus of this paper.

In general, the unsupervised workflow proceeds as follows. First, a collection of text documents is provided to the system. These documents are then transformed into multidimensional vector representations using an appropriate vectorization technique. The vectors are then projected into a lower-dimensional (typically two-dimensional) space, enabling visualization. Finally, similarities between

W. Zamojski et al. (Eds.): DepCoS-RELCOMEX 2026, LNNS 2014, pp. 468–477, 2026.
https://doi.org/10.1007/978-3-032-27927-9_45

texts are inferred by examining the spatial grouping of points in this projected space. This framework is highly general and can be applied to various notions of similarity, including semantic and stylometric similarity. A crucial component of this approach is the selection of suitable feature extraction and vectorization techniques that effectively capture the specific textual properties under investigation.

In our previous works [14,16], we demonstrated that in supervised classification settings grammatical bigrams, which encode stylistic information present in a text, can outperform other methods (including BERT-based representations and Large Language Models).

It is widely acknowledged that larger datasets enable more robust and reliable analysis. In the case of long texts, such as books, this can be achieved by dividing them into smaller segments (chunks) and analyzing these segments individually. However, this raises an important question regarding the optimal size of such chunks.

Feature representations based on frequency statistics, such as TF-IDF [11] and grammatical bigrams frequencies [16], require sufficiently large text samples to ensure stability. If the text segments are too short, the resulting frequency estimates become unreliable and highly variable. However, increasing the number of segments (i.e. using smaller fragments) provides more data points for analysis. This creates a trade-off between the stability of the representation and the size of the dataset.

Addressing this trade-off constitutes the main research question of this paper: what is the optimal text chunk size for reliable authorship attribution in literary texts?

2 Related Works

Stylometry is recognized as an interdisciplinary field that investigates the relationship between measurable textual features and meta-characteristics of higher-level, such as authorship, genre, or historical period [2,9,12]. Its applications span authorship attribution, author verification, and literary period detection. In particular, function-word frequencies, shallow syntactic patterns, and distributional features remain central to many approaches, as emphasized in studies by researchers such as Maciej Eder and Mike Kestemont [3,4,13].

Recent research trends have begun to explore the integration of traditional stylometric features with contemporary representation and visualization methods. Approaches based on TFâĂŞIDF-weighted [11] linguistic features—such as grammatical classes [8] or part-of-speech distributions—enable the capture of stylistic preferences while reducing topical bias. When combined with nonlinear dimensionality reduction techniques such as UMAP (Uniform Manifold Approximation and Projection) [6] or PacMAP [17], these representations allow for effective visualization and exploration of similarity structures in high-dimensional stylometric spaces. The problem of assessing the quality of the 2D projections obtained has been widely discussed in the literature, both in the context of document similarity analysis [5,15] and in evaluations of dimensionality reduction techniques [6,17].

3 Methods and Data Sets

3.1 Data Sets

The study was conducted on two datasets. The first, a smaller dataset, consists of five novels: two written by Eliza Orzeszkowa and three by Stefan Żeromski. The second, larger dataset comprises 100 Polish novels authored by 34 writers active in the late 19th and early 20th centuries [1].

To allow analysis at different levels of textual granularity, each book was divided into chunks of varying sizes, ranging from 100 to 50,000 words.

3.2 Text Representation

One of the most fundamental approaches to text vector representation is the bag-of-words model weighted with TF-IDF [11]. This method relies on counting word occurrences within a document and weighting them using term frequency and inverse document frequency (IDF), thereby emphasizing terms that are characteristic for a given document while being relatively rare across the corpus.

However, in authorship analysis, the selected features should reflect stylistic properties specific to the author rather than the semantic content of the text. This is particularly important because works by the same author may span diverse topics and domains. Consequently, the use of word-based features introduces the risk that a classifier will rely on semantically meaningful words—such as character names or domain-specific vocabulary—rather than stylistic patterns.

To mitigate this issue, we follow the approach proposed in [7,16], representing texts using bigrams of grammatical classes (i.e., sequences of two consecutive part-of-speech categories). Such representations largely eliminate semantic information while preserving stylistic and syntactic characteristics of the text.

Analysed texts were automatically processed using a morphosyntactic tagger [10], resulting in sequences of tokens corresponding to pairs of part-of-speech tags defined according to the National Corpus of Polish tagset [8]. In addition, a special tag was introduced to mark sentence boundaries. This procedure produced a vocabulary of 1,195 distinct grammatical bigrams. Grammatical bigram frequencies were weighted using TF-IDF to produce vector representations of the texts.

3.3 Dimensionality Reduction and Visualization

The vector representations of texts described in Sect. 3.2 were further analyzed using nonlinear dimensionality reduction techniques. In particular, we applied UMAP [6] and PaCMAP [17] to project high-dimensional document vectors onto a two-dimensional space.

These methods aim to preserve the underlying structure of the data, enabling exploration of similarities and differences between texts. The resulting 2D projections were used for visualization purposes, allowing us to inspect clustering patterns and potential grouping of texts with respect to authorship and stylistic properties.

The entire processing pipeline was implemented by the author and made publicly available as a web-based application called *WebSty*[1]. It enables users to upload their own corpora and interactively explore the resulting visualizations, facilitating qualitative analysis of stylistic similarities and differences between texts.

3.4 Evaluation Metrics

To evaluate the resulting two-dimensional projections, we employ three complementary metrics.

First, a human evaluator assigns a score to each visualization on a scale from 1 to 5, where 5 denotes the highest quality in terms of class separability and interpretability.

Second, we compute the k-nearest neighbors (kNN) accuracy. For each document (text chunk), we identify its k nearest neighbors in the projected space and verify whether they belong to the same author as the original document. The final accuracy is obtained by averaging this measure over all documents.

Third, we assess the degree of separation between classes (DS). For each document, we calculate the average distance to its k nearest neighbors of the same class and divide it by the average distance to its k nearest neighbors of different classes. A value greater than 1 indicates that the document is, on average, closer to texts written by the same author than to those written by others. To summarize this metric, we report both its mean value (DS_M) and the fifth percentile (DS_{5P}), which reflects the separation quality in the 5% most challenging regions.

4 Results

4.1 Small Dataset

For preliminary experiments, a data set consisting of only two authors (Orzeszkowa and Żeromski) was selected due to its relatively small size and the ease of visually analyzing the resulting projections. The corresponding UMAP visualizations are presented in Fig. 1. It can be observed that for small chunk sizes (up to 1000 words), the text segments from both authors largely overlap. As the chunk size increases, the texts become more clearly separable by author. However, for very large chunk sizes, the number of resulting data points decreases, which limits the usefulness of the visualization for analysis. For fragment sizes of 2250 and 4000 words, an interesting pattern emerges: texts written

[1] https://services.clarin-pl.eu/app/services/websty.

Table 1. Results for the small dataset (three books by Żeromski and two by Orzeszkowa) using UMAP projection. The table reports kNN accuracy, class separation (mean value and 5th percentile) computed for k = 5 and k = 10 nearest neighbors, as well as human-assessed visualization quality (on a scale from 1 to 5).

Chunk size	Number of chunks	kNN=5			kNN = 10			Human
		ACC	DS_M	DS_{5P}	ACC	DS_M	DS_{5P}	quality
100	2,852	0.28	2.06	0.43	0.13	1.90	0.46	1
300	950	0.77	5.28	0.74	0.66	4.44	0.72	1
500	571	0.88	8.51	1.69	0.79	6.97	1.66	2
1,000	286	0.99	14.47	4.41	0.98	10.32	3.74	2
1,500	191	0.99	16.56	5.18	0.98	11.74	4.11	4
2,000	143	**1.00**	**44.54**	**23.18**	**1.00**	**28.98**	**16.10**	4
2,250	128	**1.00**	19.40	9.03	**1.00**	13.72	6.42	**5**
2,500	115	**1.00**	30.19	18.36	**1.00**	20.60	13.88	4
2,750	104	**1.00**	25.82	16.82	**1.00**	17.93	12.29	4
3,000	96	**1.00**	29.23	16.08	**1.00**	19.46	12.48	4
4,000	73	**1.00**	30.29	17.55	**1.00**	19.23	13.10	**5**
5,000	59	**1.00**	28.55	19.58	**1.00**	18.26	11.35	4
10,000	30	**1.00**	7.61	4.74	**1.00**	5.29	3.02	2
30,000	11	0.18	1.79	1.37	0.00	1.84	1.38	1
50,000	7	0.00	1.57	1.17	0.00	1.57	1.17	1

by Orzeszkowa form distinct clusters corresponding to individual books, while texts by Żeromski remain grouped within a single cluster. The values of the evaluation metrics for UMAP are reported in Table 1. It can be noted that the best results according to analytical metrics based on k-nearest neighbors are achieved for a chunk size of 2000 words, whereas the human evaluator preferred the visualizations obtained for chunk sizes of 2250 and 4000 words. The corresponding results for PaCMAP are presented in Fig. 2 and Table 2. In this case, the human evaluator identified the best visualizations for chunk sizes in the range of 2000âĂŞ3000 words. However, it can be observed that the texts of Orzeszkowa tend to form elongated, line-like structures rather than compact clusters. In this respect, UMAP may be considered more effective, as it produces more coherent, group-like representations of texts originating from the same book. Regarding the analytical metrics, the best results are obtained for a chunk size of 2000 words when considering the 5th percentile of the degree of separation and for 5000 words when considering the mean degree of separation.

Taking into account both analytical and human evaluation metrics across the two projection methods, a chunk size of approximately 2000 words appears to be the most suitable choice. Nevertheless, chunk sizes up to 5000 words still yield relatively strong performance.

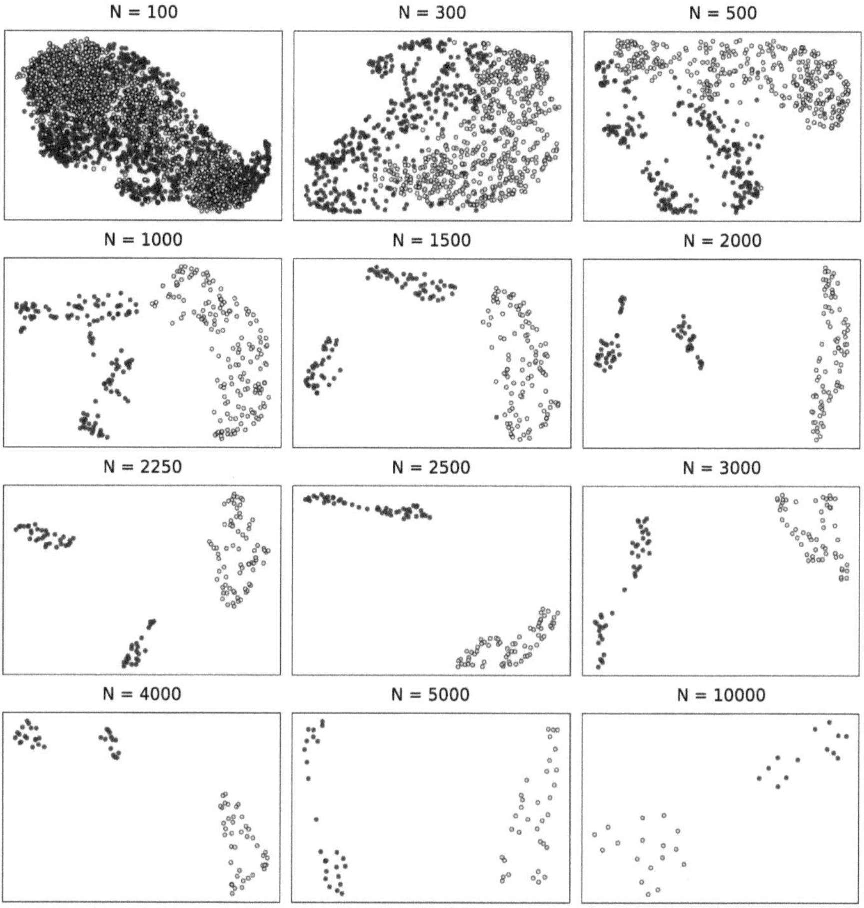

Fig. 1. UMAP visualization of the dataset for text chunks of varying sizes (from 100 to 50,000 words). Dark blue points represent texts by Orzeszkowa, while yellow points correspond to Żeromski. The figure illustrates how the two-dimensional representation evolves as the text chunk size increases.

4.2 100 Novels

In the second stage of the study, we performed an analytical evaluation on a dataset of 100 Polish novels. In this case, we did not assess the visualizations by humans, as representing 34 classes in a single plot is difficult due to the limited number of visually distinguishable colors. The values of analytical metrics are presented in Table 3. In addition to the metrics related to UMAP and PaCMAP visualizations, we also report classification accuracy for a supervised approach, as described in [14]. It can be observed that the best results for the unsupervised task are obtained for chunk sizes of 4000 and 5000 words, while the supervised classification achieves its highest accuracy for a chunk size of 3000 words. Nev-

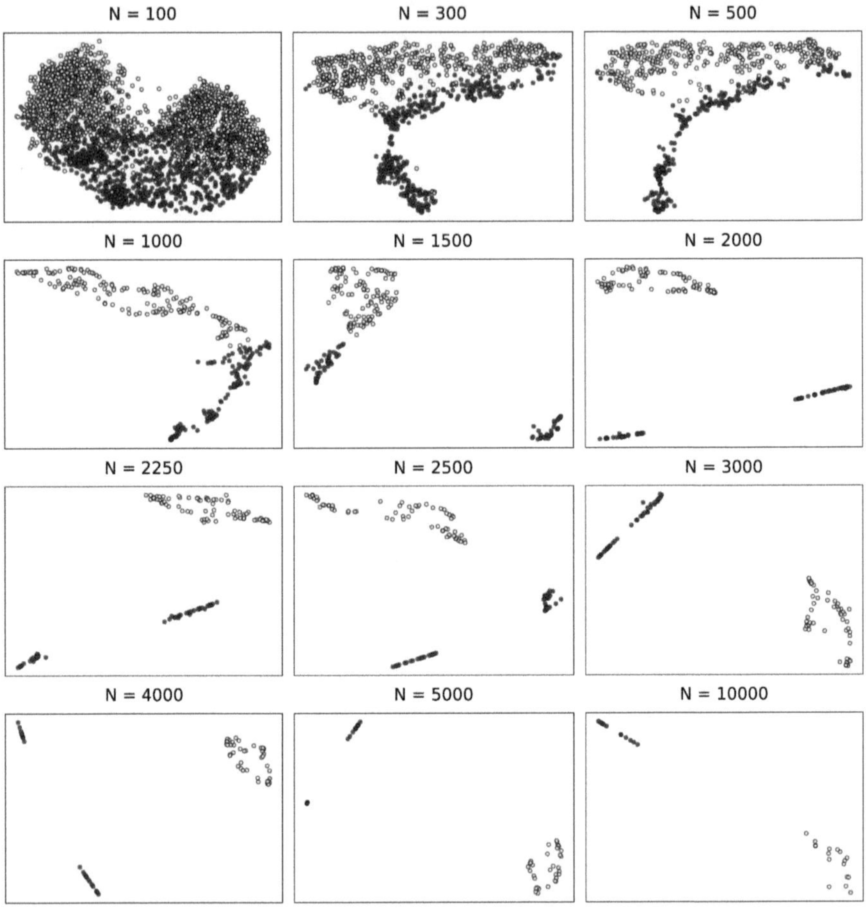

Fig. 2. PacMAP visualization of the dataset for text chunks of varying sizes (from 100 to 50,000 words). Dark blue points represent texts by Orzeszkowa, while yellow points correspond to Żeromski.

ertheless, results within the range of 2000 to 5000 words remain consistently strong across all evaluation methods.

These findings lead to the main conclusion of this paper: for visualization-based stylometric analysis of text similarity, the recommended text chunk size lies between 2000 and 5000 words.

5 Conclusion

This paper examined the effect of text chunk size on authorship attribution using stylometric features and unsupervised visualization methods (UMAP and PaCMAP). The results show that very small chunks lead to unstable representa-

Table 2. Results for the small dataset (three books by Żeromski and two by Orzeszkowa) using PacMAP projection. The table reports kNN accuracy, class separation (mean value and 5th percentile) computed for k = 5 and k = 10 nearest neighbors, as well as human-assessed visualization quality (on a scale from 1 to 5).

Chunk size	Number of chunks	kNN=5			kNN = 10			Human quality
		ACC	DS_M	DS_{5P}	ACC	DS_M	DS_{5P}	
100	2,852	0.44	2.84	0.46	0.29	2.75	0.48	1
300	950	0.76	7.20	0.70	0.67	7.71	0.73	2
500	571	0.87	17.68	1.85	0.81	13.31	1.69	2
1,000	286	0.98	37.21	4.93	0.97	23.37	4.14	2
1,500	191	**1.00**	37.13	7.72	0.98	22.47	6.12	3
2,000	143	**1.00**	82.14	**28.38**	**1.00**	44.54	**21.86**	**5**
2,250	128	**1.00**	64.71	18.28	**1.00**	37.18	13.49	**5**
2,500	115	**1.00**	58.56	19.27	**1.00**	29.80	12.33	**5**
2,750	104	**1.00**	71.87	25.54	**1.00**	41.76	17.08	4
3,000	96	**1.00**	72.11	23.65	**1.00**	38.73	18.36	4
4,000	73	**1.00**	86.33	22.36	**1.00**	39.74	17.43	4
5,000	59	**1.00**	**1,347.89**	24.46	**1.00**	**187.75**	18.81	3
10,000	30	**1.00**	41.38	14.40	**1.00**	17.59	8.95	3

Table 3. Results for the 100-novel dataset using UMAP and PaCMAP projections. The table shows kNN accuracy, class separation (mean and 5th percentile) computed for k = 5 nearest neighbors, and supervised classification accuracy (from [14]).

N	Number	UMAP			PACMAP			MLP
		ACC	DS_M	DS_{5P}	ACC	DS_M	DS_{5P}	ACC
100	48,724	0.03	0.82	0.22	0.05	0.92	0.22	0.33
300	16,026	0.15	1.47	0.23	0.20	1.73	0.23	0.59
500	9,237	0.28	3.29	0.24	0.35	5.04	0.25	0.67
1000	4,880	0.72	8.68	0.41	0.76	10.27	0.46	0.75
1,500	3,254	0.85	16.00	1.10	0.84	13.17	1.20	–
2,000	2,446	0.89	21.99	2.00	0.88	87.06	1.42	–
2,500	1,957	0.88	29.29	1.83	0.87	75.28	1.38	–
3,000	1,633	0.91	32.97	2.28	0.91	46.31	2.47	**0.82**
4,000	1,224	**0.93**	29.31	**2.84**	0.93	351.15	2.07	–
5,000	981	0.92	**40.09**	2.71	**0.94**	98.10	**2.88**	0.79
10,000	498	0.92	36.99	2.79	0.92	**2,104.44**	1.79	0.79
30,000	172	0.84	8.59	1.52	0.91	34.57	1.08	–
50,000	107	0.89	8.56	1.34	0.81	30.42	1.18	0.57

tions and poor separability, while increasing the chunk size improves both visual and analytical performance.

Experiments on both small and large datasets indicate that the best results are obtained for chunk sizes between 2000 and 5000 words. Smaller chunks lack statistical reliability, while larger chunks reduce the number of available samples, limiting analytical usefulness. The supervised classification results are consistent with this finding, with optimal performance observed around 3000 words.

Acknowledgements. The work was financed by CLARIN-PL: Common Language Resources and Technology Infrastructure (POIR.04.02.00-00C002/19, 2024/WK/01, FENG.02.04-IP.040004/24).

References

1. Eder, M., Rybicki, J.: Late 19th- and early 20th-century polish novels (2015). http://hdl.handle.net/11321/57, CLARIN-PL digital repository
2. Eder, M., Rybicki, J., Kestemont, M.: Stylometry with r: a package for computational text analysis. R J. **8**, 107–121 (2015). https://doi.org/10.32614/RJ-2016-007
3. Juola, P.: Authorship attribution. Found. Trends Inf. Retr. **1**(3), 233–334 (2006). https://doi.org/10.1561/1500000005
4. Kestemont, M.: Function words in authorship attribution. from black magic to theory? In: Feldman, A., Kazantseva, A., Szpakowicz, S. (eds.) Proceedings of the 3rd Workshop on Computational Linguistics for Literature (CLFL), pp. 59–66. Association for Computational Linguistics, Gothenburg, Sweden (Apr 2014). https://doi.org/10.3115/v1/W14-0908, https://aclanthology.org/W14-0908/
5. Marcińczuk, M., Gniewkowski, M., Walkowiak, T., Będkowski, M.: Text document clustering: Wordnet vs. TF-IDF vs. word embeddings. In: Proceedings of the 11th Global Wordnet Conference, pp. 207–214. Global Wordnet Association, University of South Africa (UNISA) (Jan 2021). https://www.aclweb.org/anthology/2021.gwc-1.24
6. McInnes, L., Healy, J., Melville, J.: UMAP: Uniform manifold approximation and projection for dimension reduction. arXiv preprint arXiv: 1802.03426 (2020)
7. Piasecki, M., Walkowiak, T., Eder, M.: Open stylometric system websty: integrated language processing, analysis and visualisation. Comput. Methods Sci. Technol. **24**(1), 43–58 (2018)
8. Przepiórkowski, A., Bańko, M., Górski, R.L., Lewandowska-Tomaszczyk, B. (eds.): Narodowy Korpus Jezyka Polskiego [Eng.: National Corpus of Polish]. Wydawnictwo Naukowe PWN (2012). http://nkjp.pl/settings/papers/NKJP_ksiazka.pdf
9. Päpcke, S., Weitin, T., Herget, K., Glawion, A., Brandes, U.: Stylometric similarity in literary corpora: Non-authorship clustering and Deutscher Novellenschatz. Digital Scholarship in the Humanities (Aug 2022). https://doi.org/10.1093/llc/fqac039
10. Radziszewski, A.: A tiered CRF tagger for Polish. In: Intelligent Tools for Building a Scientific Information Platform. SCI, vol. 467, pp. 215–230. Springer (2013). https://doi.org/10.1007/978-3-642-35647-6_16

11. Salton, G., B.C.: Term-weighting approaches in automatic text retrieval. Inf. Process. Manage. **24**(5), 513–523 (1988)
12. Sara El Manar El Bouanani, I.K.: Authorship analysis studies: a survey. Inter. J. Comput. Appli. **86**(12), 22–29 (2014). https://doi.org/10.5120/15038-3384
13. Stamatatos, E.: A survey of modern authorship attribution methods. J. Am. Soc. Inf. Sci. Technol. **60**(3), 538–556 (2009)
14. Walkowiak, T.: Authorship attribution of literary texts using named entity masking and maxlogit-based sequence classification for varying text lengths. In: Artificial Intelligence and Soft Computing: 22nd International Conference, ICAISC 2023, Zakopane, Poland, 18–22 June 2023, Proceedings, Part I, pp. 296–303. Springer-Verlag, Berlin (2023). https://doi.org/10.1007/978-3-031-42505-9_26
15. Walkowiak, T., Gniewkowski, M.: Evaluation of vector embedding models in clustering of text documents. In: Proceedings of the International Conference on Recent Advances in Natural Language Processing (RANLP 2019), pp. 1304–1311. INCOMA Ltd., Varna, Bulgaria (Sep 2019). https://doi.org/10.26615/978-954-452-056-4_149, https://www.aclweb.org/anthology/R19-1149
16. Walkowiak, T., Piasecki, M.: Stylometry analysis of literary texts in polish. In: Rutkowski, L., Scherer, R., Korytkowski, M., Pedrycz, W., Tadeusiewicz, R., Zurada, J.M. (eds.) Artificial Intelligence and Soft Computing, pp. 777–787. Springer International Publishing, Cham (2018). https://doi.org/10.1007/978-3-319-91262-2_68
17. Wang, Y., Huang, H., Rudin, C., Shaposhnik, Y.: Understanding how dimension reduction tools work: an empirical approach to deciphering t-sne, umap, trimap, and pacmap for data visualization. J. Mach. Learn. Res. **22**(201), 1–73 (2021). http://jmlr.org/papers/v22/20-1061.html

The Renaissance of Computer Piracy as a Result of the Evolution of Digital Content Distribution Models

Artur Wojciechowski and Kamil Dembowski(✉)

University of Information Technology and Management, Sucharskiego, Rzeszów, Poland
{awojciechowski,kdembowski}@wsiz.edu.pl

Abstract. The article analyzes the contemporary renaissance of computer piracy in the context of the evolution of digital content distribution models. The aim of the study is to show that piracy is not a temporary phenomenon, but a permanent element of the digital information system, subject to cyclical adaptation to technological, regulatory, and market changes. The paper presents an original model of the evolution of computer piracy, covering five stages of development—from physical piracy, through mass P2P networks, to contemporary functional piracy integrated with cloud solutions and service models. Particular attention is paid to the analysis of the technological infrastructure of torrent networks, the process of decentralization and anonymization, and the convergence of P2P and streaming models. It is pointed out that contemporary piracy is qualitative in nature—its essence is the professionalization of structures, the concealment of the P2P layer, and the shift of user motivations from ideological to functional. The article concludes that an effective anti-piracy policy requires a systemic approach combining legal instruments, the development of investigative skills, and the modernization of legal content distribution models.

Keywords: Computer piracy · Peer-to-peer networks · Torrent services · Copyright law · Cybercrime · Digital content distribution models · Streaming · Cybersecurity · Digital forensics · Information society

1 Introduction—Computer Piracy as a Structural Phenomenon, not a Temporary One

Computer piracy is one of the earliest identified phenomena accompanying the development of information technology. As early as the 1970s, it was pointed out that the process of digitising intangible goods would lead to a fundamental redefinition of the relationship between the technical possibility of copying and the construction of exclusive rights enjoyed by creators (Gates 1976). The promotion of this narrative intensified in the 1990s (Barta, Markiewicz 1993; Nowicka 1995). While in the analogue model the reproduction of works was associated with production costs, logistics and control of physical circulation, in the digital environment copying became almost cost-free, instantaneous and global (Goban-Klas 2005). This fundamental technological change created

W. Zamojski et al. (Eds.): DepCoS-RELCOMEX 2026, LNNS 2014, pp. 478–493, 2026.
https://doi.org/10.1007/978-3-032-27927-9_46

a lasting tension between the technological architecture of the internet and the normative structure of copyright law.

Copyright, whose contemporary form was shaped by the realities of the physical circulation of copies of works, has been transferred to an environment in which the concept of a "copy" has lost its physical, tangible obviousness (Barta, Markiewicz 2010). As a result, mechanisms for the protection of exclusive rights began to function in a space where control over the copying and distribution of works is technologically limited. Regulations concerning computer programmes and technical safeguards attempted to strengthen this protection (Nowicka 1995; Matlak 2007), but at the same time, tools enabling the circumvention of these safeguards were developed.

The literature on computer criminal law emphasises that cybercrime, including copyright infringement, is highly adaptable and capable of responding quickly to regulatory changes (Adamski 2000; Siwicki 2013). Unlike traditional forms of crime, crimes in cyberspace are decentralised, dispersed and transnational in nature. This means that strengthening repression in one country does not lead to the disappearance of the phenomenon, but to its migration to other jurisdictions or to technological transformation (Worona 2020). Computer piracy fits into this model—its history shows that every intensification of anti-piracy measures has led to a change in the form of illegal distribution, rather than its elimination.

Another important structural factor is the transformation of the model of social communication. The development of the information society has blurred the boundary between content producers and consumers (Goban-Klas 2005; Muchacki 2014). Internet users have become active participants in the circulation of information—not only recipients, but also senders and distributors. In such conditions, the model of centralised control over the distribution of works has ceased to correspond to technological realities (Szpor, Wiewiórowski 2012). Computer piracy can therefore be interpreted as the result of a structural shift in the way digital communication functions.

From the perspective of national security and cybersecurity, piracy is part of a broader environment of information threats (Rojszczak, Banasiński 2020; Górka 2014). Illegal content distribution uses the same technological mechanisms that are used in other areas of cybercrime—anonymity, encryption, distributed infrastructures, cross-border nature. Thus, the problem of piracy cannot be considered solely in terms of copyright; it requires a systemic approach that takes into account network architecture, international cooperation, and the analytical capabilities of law enforcement agencies (Adamski et al. 2015; Stefański 2017).

The economic dimension is also significant. Digitisation has led to the emergence of new business models based on subscriptions, time-based access and streaming services (Muchacki 2014). At the same time, the digital content market has become fragmented—access to the full catalogue of works requires the use of multiple parallel platforms. The literature on information resource management points out that excessive segmentation and lack of interoperability reduce the effectiveness of information systems (Pańkowska 2001). In such conditions, some users may perceive piracy as an alternative channel for integrating access to dispersed content.

The international nature of cyberspace further complicates law enforcement. Differences in national regulations, the lack of uniform standards and the need for cross-border cooperation make the prosecution of infringements time-consuming and costly (Worona 2020). Despite the establishment of dedicated institutions at European level (ENISA, EUIPO), cooperation in combating illegal content requires special coordination between public and private entities (Adamski et al. 2015), confirming that piracy operates in a complex infrastructural and institutional environment.

Considering the above conditions, computer piracy cannot be treated as a temporary effect of the "immaturity" of the Internet. Rather, it is a consequence of the structural properties of digital technologies, the global nature of the network, and the transformation of communication and economic models, as well as the newly emerging model of the information society (Barney 2008, Haber et al. 2006). Its persistence stems from the fact that it responds to specific systemic tensions: between technology and legal norms, between the global nature of the network and the territoriality of law, and between user expectations and the limitations of business models.

Recognising piracy as a structural phenomenon has important theoretical and practical consequences. Firstly, it requires a departure from the narrative that presents it as a pathology that can be eliminated. Secondly, it prompts an analysis of its transformations from an evolutionary perspective. Since the phenomenon is not disappearing but undergoing a transformation, it is necessary to understand its current stage of development. It is this need that leads to the concept of a "renaissance of computer piracy," understood as a qualitative change in the forms and functions of illegal distribution of digital content, which is the subject of the following sections of this study.

2 The Renaissance of Computer Piracy—a Qualitative Approach

The term “renaissance of computer piracy” may suggest a simple increase in the scale of copyright infringements, but in scientific terms it is more accurate to understand it as a qualitative change. It is not just about the number of downloads or the number of users using illegal sources, but about the transformation of the very structure and function of piracy in the digital information ecosystem. This phenomenon is part of broader processes of evolution in cybercrime, which, as indicated in the literature, are paradigmatic and adaptive in nature (Kosiński 2015; Pływaczewski ed. 2015).

In the early stages of the Internet’s development, piracy was often spontaneous and scattered. Users acted individually, driven by both economic and ideological motivations, referring to the concepts of "free culture" or free access to information (Williams 2003, Castells, Himanen 2009, Barta, Markiewicz 2010). Nowadays, however, we are seeing a gradual professionalisation and institutionalisation of illegal distribution structures. The literature on cybercrime emphasises that many forms of infringement take an organised form, with a clear division of roles, specialisation of functions and the use of advanced technological tools (Siwicki 2013; Kosiński 2015).

The renaissance of piracy therefore consists in a shift from mass and often chaotic activities to more orderly activities based on stable distribution structures. Contemporary platforms for the illegal exchange of content often operate in the form of closed communities, with invitation or registration systems, reputation mechanisms and internal regulations that also take into account the activities of legal creators (Digital Millennium

Copyright Act 1998, Digital Services Act 2022). Such solutions limit random participation and increase resistance to law enforcement, which is in line with the general trend towards the professionalisation of cybercrime (Pływaczewski ed. 2015).

An important factor contributing to this process is the evolution of legal models for the distribution of digital content. The development of streaming platforms, subscriptions and DRM (Digital Rights Management) systems was intended to be the market's response to the problem of piracy by offering convenient and relatively cheap access to content. However, over time, the market has become fragmented, with content divided among many competing platforms, often with territorial and licensing restrictions (Muchacki 2014). From the user's perspective, this means having to pay multiple fees and operate in an environment of limited interoperability. For this reason, pirate platforms offering complete digital content are regaining importance and are undergoing a period of transformation, commercialisation and professionalisation.

Regulations on digital rights management (DRM) were intended to strengthen control over the use of content, but in practice they often restrict the freedom of legal users, making it difficult to archive, copy for personal use or use content on different devices (Matlak 2007). As a result, some consumers began to perceive piracy not only as a "free alternative," but as a more functional solution, providing full access, no technical restrictions, and greater predictability of use.

This phenomenon can be interpreted in terms of information economics as a compensatory mechanism for market imperfections (Pańkowska 2001; Muchacki 2014). If the legal distribution system does not provide an optimal level of accessibility and usability, some users seek alternative channels of access. In this sense, piracy is not only a form of fee avoidance, but also a reaction to perceived system deficiencies.

At the same time, social and cultural dimensions must be taken into account. The information society is characterised by high expectations regarding immediate access to information and cultural goods (Goban-Klas 2005). Licence restrictions, regional blocks and delays in premieres in different countries are perceived as unjustified barriers in the global communication space. In this context, piracy can be seen as a manifestation of the tension between the global nature of the internet and the territoriality of copyright law (Worona 2020).

The resurgence of piracy is also linked to its partial commercialization. A documented example is the CDA.pl portal, now recognized as a fully legal VOD service operating on principles similar to those of YouTube or Netflix. Doubts about its legality stemmed mainly from the site's past, when it functioned as a hosting platform where users uploaded pirated content en masse. The portal also generated revenue from advertisements, often from popular companies and corporations, despite the questionable legality of the distributed film content. In 2018, the service was added to the Motion Picture Association of America's list of pirate sites, and since then, the company has changed its business policy to focus on the intensive promotion of verified content, and the CDA Premium service is now a fully legal VOD platform. In accordance with the U.S. Digital Millennium Copyright Act and European regulations—the Digital Services Act—the service provides publishers and distributors with the ability to report infringements and a tool for self-removal of copyright-infringing content. As demonstrated, in some cases, distribution structures generate revenue from advertising, intermediary services, or fees for access to technical infrastructure. Such activities resemble business models operating in the legal market, which further blurs the lines between criminal and quasi-market activities. This phenomenon is part of broader processes of professionalization and commercialization of cybercrime.

It is also worth noting that modern piracy increasingly operates on a service model. Users do not need to have advanced technical knowledge or directly use P2P protocols. All they need is access to an application or intermediary platform (CaaS—Crime-as-a-Service model) that performs the downloading and sharing process in the background. Such solutions meet the needs for convenience and anonymity, while at the same time making it difficult to identify the actual organisers of the infrastructure.

In this sense, the "renaissance" primarily means a change in the function of piracy. It ceases to be solely an act of law violation with a clear opposition to the market and begins to serve as an alternative distribution system—parallel to official models. Whereas in the past the rhetoric of opposition to the "monopoly" of the media industry dominated, today the main motivation is convenience, completeness of access and reduction of transaction costs.

Understood in this way, the renaissance of piracy is qualitative in nature: it consists in the integration of illegal distribution with modern technological solutions, the professionalisation of organisational structures and a shift in user motivation from ideological to functional. This phenomenon provides a direct context for analysing the evolution of technical infrastructure, in particular torrent services and P2P networks, which, despite their declared decline in importance, still play a key role in the distribution infrastructure of modern piracy.

3 The Evolution of Torrent Services and P2P Networks—the Technological and Investigative Dimension

The evolution of torrent services and peer-to-peer (P2P) networks is one of the most striking examples of the adaptive nature of computer piracy. From a technological and investigative perspective, the history of these solutions shows that every attempt to strengthen control over illegal distribution has led to a transformation of the system's architecture, rather than its demise.

3.1 The Original P2P Model—Decentralisation and Transparency

The first generation of P2P networks (e.g. Napster, Kazaa, eDonkey) was based on direct file sharing between users, using centralised or semi-centralised indexing servers. This model enabled mass and relatively easy content exchange, but at the same time generated a significant number of digital traces—IP addresses, server logs, file metadata—which could be used as evidence in criminal proceedings (Kalinowski 2011; Kasprzak 2015).

From the perspective of computer forensics, it was important that the end user remained directly visible on the network. File sharing involved disclosing the IP address and participating in a specific data transmission session. Law enforcement agencies could monitor public trackers and identify active participants in the exchange (Chojnowski

2020). During this period, a repressive strategy targeting both platform operators and end users prevailed.

However, with the intensification of anti-piracy activities and the development of legal instruments to combat cybercrime (Adamski 2000; Siwicki 2013), this model began to evolve.

3.2 Technological Transformation—Private Trackers and Dispersion

The next stage was the transition from public, open trackers to closed, private communities. Private trackers operate on the basis of an invitation system, the obligation to maintain a certain sharing ratio, and internal activity control mechanisms. Such solutions reduce the risk of infiltration by external entities and make it difficult to identify users.

From a technical point of view, the spread of magnet links (magnet URIs), which eliminate the need for a central torrent file, was also significant. The data identifying the resource is transmitted in the form of a cryptographic hash, which increases the system's resistance to the closure of indexing servers. This mechanism is part of the general trend towards decentralisation of internet infrastructure (Liderman 2003).

At the same time, there is an increasing use of data transmission encryption technologies and anonymising networks (e.g. VPN, TOR networks). From an investigative perspective, this means that it is more difficult to identify users unambiguously and that more advanced methods of network traffic analysis must be used (Chojnowski 2020; Luttgens, et al. 2016).

3.3 Hiding the P2P Layer—the "Background Infrastructure" Model

The current phase of torrent network evolution involves the gradual "hiding" of the P2P layer as infrastructure support. End users are increasingly less likely to use a torrent client directly; instead, they use applications or platforms that perform the download process in an external environment (so-called seedboxes, cloud services). This means separating the user from the technical act of file sharing.

From the perspective of criminal and procedural law, such solutions complicate the determination of the entity responsible for the dissemination of works. This raises the question of the scope of liability of the infrastructure operator, the intermediary service provider and the end user (Barta, Markiewicz 2010; Nieweęgłowski, Chrzanowski eds. 2016, Nowikowska 2022). This problem is part of the broader issue of the liability of internet intermediaries.

At the same time, the dispersion of infrastructure and the use of servers in different jurisdictions hinders effective law enforcement. Cyberspace is cross-border in nature, and differences in national regulations limit the ability of law enforcement agencies to act in a uniform manner (Worona 2020). International cooperation in combating illegal content requires coordination between public and private entities (Adamski et al. 2015), which significantly lengthens evidentiary procedures.

3.4 Evidentiary Challenges and Investigative Methodology

From the point of view of computer forensics, the proper securing and analysis of digital traces is of key importance. In a P2P environment, these traces include connection logs, file metadata, data from the operating memory, and information stored on intermediate servers (Kasprzak 2015; Chojnowski 2020, Jaworski, Grześkowiak 2026). However, the growing popularity of encryption and anonymisation services means that traditional methods of IP address identification are becoming less and less effective.

The literature on responding to security incidents indicates that the effectiveness of the response depends on the speed of data protection and cooperation between entities (Luttgens, Pepe, Mandia 2016, Opitek, Jaworski 2025). In the context of computer piracy, this means the need to develop analytical skills, including network traffic analysis, behaviour pattern identification and the use of data correlation tools (Liedel, Piasecka, Aleksandrowicz 2012).

At the same time, it should be noted that repressive measures targeting end users do not always lead to a reduction in the scale of the phenomenon. Closing individual services often results in users migrating to new, more technologically advanced platforms (Siwicki 2013). This phenomenon confirms the thesis about the adaptive nature of P2P infrastructure.

3.5 P2P as A Permanent Infrastructure Layer

Despite the narrative of the decline of torrents in the age of streaming, P2P technology has not disappeared—it has been transformed into an invisible infrastructure layer. Its effectiveness in the distributed distribution of large files makes it attractive for both legal and illegal applications. From a technical point of view, P2P offers scalability, fault tolerance and no central point of control—features that make it difficult to eliminate completely.

From an ICT security perspective, this means that P2P must be treated not as an incidental technology used exclusively for violations, but as an element of network architecture (Liderman 2003; Białas ed. 2002). The regulatory response cannot therefore be limited to penalising a specific tool, but should take into account the systemic nature of the infrastructure.

The development of torrent services and P2P networks shows that this technology is constantly adapting in response to regulatory and market pressures. There is also an element of torrenting that uses popular social media. The evolution from overt, mass platforms to distributed and partially hidden ecosystems confirms that computer piracy operates in a dynamic technological environment.

This infrastructural transformation provides the direct context for an attempt to systematise the phenomenon in the form of a model of the evolution of computer piracy, which will be presented in the next section of this study.

4 Model of the Evolution of Computer Piracy—Theoretical Approach and Analytical Diagram

4.1 Conceptualisation of the Phenomenon

An analysis of the changes in computer piracy to date leads to the conclusion that we are not dealing with a linear development of the phenomenon, but with a process of cyclical adaptation based on technological and social changes. The literature on cybercrime emphasises that forms of cybercrime are constantly being reconfigured in response to technological, regulatory and economic changes (Kosiński 2015; Siwicki 2013). This means that computer piracy should be analysed from a systemic perspective, as part of a dynamic system of relationships between technology, law, the market and society.

The theoretical model of the evolution of piracy is based on the assumption that this phenomenon functions within a three-factor structure:

1. Technology (T)—network architecture, degree of decentralisation, level of anonymisation, encryption tools (Liderman 2003; Chojnowski 2020).
2. Regulation (R)—intensity of penalties, scope of liability of intermediaries, effectiveness of international cooperation (Adamski 2000; Worona 2020).
3. Market (M)—availability, price and functionality of legal content distribution models (Matlak 2007; Muchacki 2014).

These three factors are complemented by a social variable:

4. User motivation (U)—economic, ideological, functional or compensatory (Barta, Markiewicz 2010; Muchacki 2014).

The relationship between these elements is feedback-based. A change in one area generates adaptation in the others. In this sense, computer piracy is a reactive and innovative phenomenon—it responds to system limitations and generates new technological solutions.

4.2 Model Structure—Four Analytical Dimensions

The model of piracy evolution can be presented in four interrelated dimensions:

1. Technological dimension

   It covers infrastructure architecture (centralisation vs. decentralisation), the level of anonymisation, the use of cryptography, and integration with cloud solutions. In the early stages, overt decentralisation (open P2P) dominated, whereas today we are seeing the P2P layer being hidden behind service interfaces (Chojnowski 2020; Luttgens, Pepe, Mandia 2016).
2. Regulatory dimension

   This concerns the scope and intensity of the state's response. The literature on computer criminal law indicates that each amendment to the regulations and extension of the liability of intermediaries has led to the migration of infrastructure towards a more dispersed and difficult to identify model (Adamski 2000; Siwicki 2013). The cross-border nature of the network further hinders uniform law enforcement (Worona 2020).
3. Economic dimension

   This concerns the structure of the digital content market. With the development of subscription models and DRM systems, control over distribution has increased, but at the same time market fragmentation has increased (Matlak 2007; Muchacki 2014). The literature on information management indicates that excessive segmentation limits the effectiveness of the system (Pańkowska 2001).
4. Motivational dimension

User motivations evolve with market and technological changes. In the mass P2P phase, the ideological component dominated (Barta, Markiewicz 2010). Today, functional motivation prevails—the pursuit of full access, no DRM restrictions and the integration of distributed content (Muchacki 2014).

4.3 Five-Stage Evolutionary Model

Based on the above dimensions, five stages of computer piracy development can be distinguished.

1. Stage I—Physical piracy

   Dominance of physical media, local trade, low scale. Regulation is based on classic instruments of criminal and civil law.
2. Stage II—Mass piracy (open P2P)

   Decentralised nature, global reach, visibility of users on the network. Partially ideological motivation (Barta, Markiewicz 2010).

   Stage III—Reaction to increased repression—migration to cyberlockers and intermediary services (Littlejohn, Tittel 2004). Intensification of regulation (R$\uparrow$) leads to technological transformation (T $\rightarrow$).
3. Stage IV—Institutionalised piracy

   Private trackers, reputation systems, professionalisation of structures (Kosiński 2015). Increase in internal control and specialisation of functions.
4. Stage V—Functional piracy (service model)

   Integration with cloud services, concealment of P2P as infrastructure. User motivation shifts from ideological to functional. Market deficits (M$\downarrow$ functionality) generate increased demand for alternative solutions (U$\uparrow$).

4.4 Adaptation Diagram of the Model

The model can be summarised as an adaptation cycle:

Technological change (T) → regulatory response (R) → infrastructure migration (T') → market model change (M) → redefinition of user motivation (U) → new regulatory pressure (R') (Fig. 1).

This diagram shows that piracy is neither a static nor a linear phenomenon. Each stage generates conditions for the next transformation. Adaptability is its fundamental structural feature.

4.5 Systemic Nature and Theoretical Implications

The model confirms that computer piracy functions as a "parallel system" to the official digital content market. It does not exist in a vacuum, but responds to changes in access, pricing and regulation (Muchacki 2014; Pańkowska 2001).

From a cybersecurity perspective, this means moving away from a purely repressive approach. The literature indicates that excessive penalisation without analysing user motivations leads to further professionalisation of the infrastructure (Siwicki 2013; Adamski 2000). An effective strategy requires influencing all four dimensions of the model—technological, regulatory, economic and social (Rojszczak, Banasiński 2020).

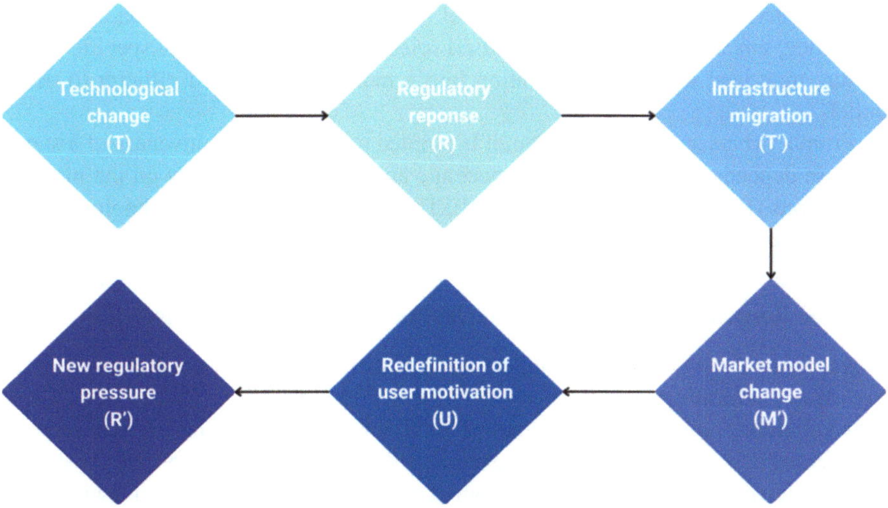

Fig. 1. Adaptation diagram of the evolutionary model.

4.6 Model Conclusions

1. Computer piracy is **cyclical and** adaptive in nature, not linear.
2. The most important factor in the current stage is **the convergence of P2P and service models.**
3. The functional deficits of legal models are a significant factor in strengthening user motivation.
4. Regulatory responses generate further technological innovations.
5. Piracy is part of the digital information system, not an external phenomenon.

This model provides a theoretical basis for further analysis of the implications for anti-piracy policies and regulatory strategies, which will be discussed in the next section of this paper.

5 Implications for Anti-Piracy Policies—a Systemic and Regulatory Approach

5.1 Limitations of the Repressive Approach

The demonstrated model of the evolution of computer piracy indicates that this phenomenon is cyclical and adaptive in nature. Historically, every intensification of repressive measures has led to a transformation of the technological infrastructure rather than to its disappearance. The literature on computer criminal law emphasises that criminalisation is an essential element of the legal protection system, but its effectiveness in cyberspace is limited by, among other things, the decentralisation and cross-border nature of the network (Adamski 2000; Siwicki 2013).

The closure of individual torrent platforms has resulted in users migrating to new, more technologically advanced solutions. This phenomenon is sometimes referred to as the displacement effect, characteristic of adaptive crime (Pływaczewski, ed. 2015). In practice, this means that actions focused solely on the elimination of specific infrastructure may lead to its reconfiguration into a more dispersed and difficult to detect model.

From a cybersecurity perspective, it is pointed out that the effectiveness of a state's response depends on its capacity for systemic analysis, and not solely on the intensity of sanctions (Rojszczak, Banasiński 2020). This means that both the technological and economic and social dimensions must be taken into account.

5.2 The Responsibility of Intermediaries and the Problem of Jurisdiction

An important area of regulation remains the liability of internet intermediaries—hosting providers, platform operators, cloud infrastructure administrators (Piech 2019). Copyright literature emphasises that extending the liability of intermediaries may increase the effectiveness of the protection of exclusive rights, but at the same time creates the risk of excessive interference with freedom of communication (Barta, Markiewicz 2010; Niewęgłowski, Chrzanowski eds. 2016).

The cross-border nature of cyberspace further complicates law enforcement. Differences in legal systems and the lack of uniform mechanisms for international cooperation hinder the effective prosecution of entities operating in multiple jurisdictions (Worona 2020). The literature on combating illegal content points to the crucial importance of cooperation between states and between the public and private sectors (Adamski et al. 2015). Without such coordination, repressive measures are fragmented and reactive.

5.3 The Importance of the Functionality of Legal Distribution Models

The model of piracy evolution indicates that an important factor in the current renaissance is the functional deficit of legal distribution models. The fragmentation of the streaming market, DRM restrictions, rising subscription costs and the constant reduction of product offerings affect the perception of the usefulness of legal services (Matlak 2007; Muchacki 2014). The literature on information resource management emphasises that an effective system should minimise access barriers and transaction costs (Pańkowska 2001).

From a public policy perspective, this means that countering piracy cannot be limited to repression, but should also include support for competitive, accessible and interoperable distribution models. If legal solutions offer a higher level of convenience and functionality, users' motivation to use illegal sources is weakened.

Modernized models of legal distribution in the age of digitization and changing consumer habits focus on a hybrid approach (online-offline), direct customer outreach, and the use of advanced technologies.

Countering digital content piracy could be based entirely on the legal distribution models proposed and discussed below.

A hybrid model that balances the convenience of online shopping with the physical experience. The customer can order goods via an app and pick them up or test them in a brick-and-mortar store that serves as a showroom. This model applies to the distribution of digital content on physical digital media, reinforcing the product's legality and giving the customer the opportunity for additional verification of its legal attributes.

Direct-to-Consumer Model. Digital goods producers are increasingly bypassing intermediaries such as traditional retailers, selling products directly through their own websites. This ensures a better margin, full control over the brand, and direct access to customer data.

Subscription Model. Allows for the legal sale of products via a subscription model, ensuring steady revenue and customer loyalty. This model is currently gaining popularity and works well not only in digital media but also for other consumer goods.

Selective and Exclusive Distribution Model. Limiting the number of points of sale to maintain a premium brand image. In its modern form, selectivity also applies to authorized e-commerce retailers, which helps control prices, distribution channels, and service quality.

Modernized models must also take into account sustainability principles and new international regulations governing relationships between manufacturers, distributors, and customers.

5.4 The Investigative Dimension and the Development of Analytical Skills

Modern P2P infrastructure and its integration with cloud services are increasing the complexity of evidence gathering. The literature on computer forensics indicates that effective securing and analysis of digital traces requires specialised tools and appropriate procedures (Chojnowski 2020; Luttgens, Pepe, Mandia 2016). The growing importance of encryption and anonymisation means that traditional methods of IP address identification are insufficient.

In this context, the development of scientific research in this area, the expansion of analytical competences and inter-institutional cooperation are of particular importance. Information analysis, data correlation and the use of digital intelligence tools (Konieczny 2022) are becoming key elements of an effective state response (Liedel, Piasecka, Aleksandrowicz 2012). The response to the renaissance of piracy therefore requires investment in technical and organisational capacity, not just legislative changes.

5.5 The Need for a Systemic Approach

According to the presented model of piracy evolution, an effective anti-piracy policy should address all four dimensions: technological, regulatory, economic and social. This means moving away from thinking in terms of eliminating a specific technology (e.g. P2P) in favour of risk management across the entire digital information system.

The literature on national security emphasises that threats in cyberspace are asymmetric and require flexible response strategies (Liedel ed. 2011; Górka ed. 2014). Computer piracy fits into this logic—its adaptability means that the state's response must be proportionate, balanced and take into account market dynamics.

5.6 Conclusions for Public Policy

Based on the model analysis, several key conclusions can be drawn:

1. **Repression is necessary but insufficient**—its excessive intensification may lead to further professionalisation of the infrastructure (Adamski 2000; Siwicki 2013.
2. **International cooperation is a prerequisite for effective regulation** in a cross-border environment (Worona 2020; Adamski et al. 2015).
3. **Improving the functionality of legal distribution models** is an important element in reducing user motivation (Muchacki 2014).
4. **The development of investigative and analytical skills** is crucial in the context of increasing anonymisation of infrastructure (Chojnowski 2020).
5. **A systemic approach** should replace selective and reactive measures (Rojszczak, Banasiński 2020, Jasworski 2024).

The renaissance of computer piracy does not therefore signify the failure of regulations but reveals the need for their modernisation. An effective anti-piracy policy should combine legal instruments, technological development, market model reform and analysis of social motivations. Only such an integrated approach can limit the adaptive potential of illegal distribution of digital content without upsetting the balance between the protection of exclusive rights and access to information.

6 Conclusion

The analysis confirms that computer piracy is not a temporary effect of the early stages of Internet development, but rather a phenomenon structurally embedded in the architecture of the information society. Its persistence stems from the fundamental tension between the technological ease of copying digital content and the structure of exclusive rights shaped by the realities of material trade (Barta, Markiewicz 2010; Nowicka 1995). The digitisation and globalisation of communication have not only increased the scale of potential infringements, but above all have changed the context in which copyright protection operates (Goban-Klas 2005; Muchacki 2014).

The article shows that the contemporary "renaissance" of piracy is qualitative in nature. It does not consist solely in an increase in the number of infringements, but in the transformation of technological infrastructure, the professionalisation of organisational structures and a change in user motivation. While in the initial phase the ideological component and the mass character of open P2P networks dominated, currently we are observing the integration of illegal distribution with service models, the concealment of the P2P layer in cloud infrastructure, and a shift in motivation towards functional compensation for market deficits (Kosiński 2015; Siwicki 2013).

A comparison of P2P and streaming models has shown that both systems respond to the same user needs but differ in their control architecture and scope of responsibility. The streaming model provides centralisation and formal legal compliance but often involves functional limitations and market fragmentation (Matlak 2007; Muchacki 2014). The P2P model offers decentralisation, permanent access to copies and greater interoperability, but operates outside the normative order. The contemporary convergence of both models—consisting in hiding the P2P infrastructure behind service interfaces—confirms the adaptive nature of piracy.

The model of computer piracy evolution proposed in the article—comprising five stages of development—indicates that this phenomenon operates in an adaptive cycle in which technological (T), regulatory (R), market (M) and social (U) changes remain in a feedback relationship. Each intensification of regulation led to a reconfiguration of infrastructure, and each change in the market model generated new motivations for users (Adamski 2000; Rojszczak, Banasiński 2020, Opitek, Jaworski 2025). In this sense, computer piracy should be treated as a parallel element to the legal content market, rather than as an external or completely marginal phenomenon.

An analysis of the implications for public policy leads to the conclusion that a purely repressive approach is insufficient. Penalisation and prosecution of infringements remain necessary, but without the simultaneous modernisation of distribution models, increased interoperability of services and the development of the analytical capabilities of law enforcement agencies, their effectiveness will be limited (Adamski et al. 2015; Chojnowski 2020). The cross-border nature of cyberspace also requires intensive international cooperation and coordination of regulatory activities (Worona 2020).

The renaissance of computer piracy should therefore be interpreted not as a regression in copyright protection, but as a sign of mature adaptation to digital realities. This phenomenon reveals structural tensions between the global nature of technology and the territoriality of law, between the subscription economy and the expectation of full access, and between the centralisation of control and the decentralisation of infrastructure.

In the perspective of further research, it seems reasonable to deepen the empirical analysis in three areas:

- the impact of streaming market fragmentation on user motivations,
- the effectiveness of cross-border cooperation models in combating illegal content,
- the role of cloud infrastructure in the transformation of the P2P model.

In summary, computer piracy is a permanent component of the digital information system, whose evolution reflects changes in technology, regulation and market economics. Understanding these dynamics—based on a systemic model—is a prerequisite for formulating rational and proportionate regulatory strategies that will be able to limit the negative effects of infringements without upsetting the balance between the protection of exclusive rights and access to information in the information society.

References

Adamski Andrzej: Prawo karne komputerowe. Warszawa: C.H. Beck, (2000)

Adamski Andrzej, Arkadiusz Lach, Kosiński Jerzy, Rocławska Monika, Bakalarz Sławomir: Współpraca w zwalczaniu nielegalnych treści w Internecie. Raport z badań. Toruń: Centrum Badań nad Cyberprzestępczością, (2015)

Barney Darin: Społeczeństwo sieci. Sic!: Warszawa, (2004)

Barta Janusz, Markiewicz Ryszard: Główne problemy prawa komputerowego. Warszawa: Wydawnictwa Techniczno-Naukowe, (1993)

Barta Janusz, Markiewicz Ryszard: Prawo autorskie. Warszawa: Wolters Kluwer, (2010)

Bezpieczeństwo XXI wieku. Asymetryczny świat. Pod red. Liedel Krzysztof, Piasecka Paulina, Aleksandrowicz Tomasz R. Warszawa: Difin, 2011

Chałubińska-Jentkiewicz Katarzyna, Karpiuk Mirosław: Prawo nowych technologii. Wybrane zagadnienia. Warszawa: Wolters Kluwer, 2015

Chojnowski Adam: Informatyka sądowa w praktyce. Gliwice: Helion, (2020)

Cyberbezpieczeństwo jako podstawa bezpiecznego państwa i społeczeństwa w XXI wieku. Pod red. Górka Marek. Warszawa: Difin. (2014)

Cyberbezpieczeństwo. Zarys wykładu. pod red. Banasiński Cezary. Warszawa: Wolters Kluwer, (2018)

Gierszewski Janusz: Organizacja systemu bezpieczeństwa społecznego. Warszawa: Difin, (2013)

Goban-Klas Tomasz: Cywilizacja medialna. Geneza, ewolucja, eksplozja. Warszawa: WSiP, (2005)

Grzybczyk Katarzyna, Auleytner Aleksandra, Kulesza Jan: Prawo w wirtualnych światach. Warszawa: Difin, (2013)

Grażyna Szpor, Gryszczyńska Agnieszka: Internet. Strategie bezpieczeństwa. Warszawa. C.H. Beck. (2017)

Haber Lesław Henryk, Niezgoda Marian: Społeczeństwo informacyjne. Aspekty funkcjonalne i dysfunkcjonalne. Kraków: Wydawnictwo UJ, (2006)

Hoffman Tomasz: Wybrane aspekty cyberbezpieczeństwa w Polsce. Poznań: FNCE, (2018)

Internet a prawo autorskie. Pod red. Niewęgłowski Adrian i Chrzanowski Mateusz. Lublin: Wydawnictwo UMCS, (2016)

Internet w społeczeństwie informacyjnym. Nowoczesne systemy informatyczne i ich bezpieczeństwo. Pod red. Pregiel Ryszard, Buchwald Paweł. Dąbrowa Górnicza: WSB, (2014)

Jaworski Damian Robert, Grześkowiak Zuzanna: Prawo wobec cyberzagrożeń i nowych technologii. Warszawa: Difin, (2026)

Kalinowski Artur M.: Metody inwigilacji i elementy informatyki śledczej. Kwidzyń: CHS, (2011)

Kasprzak Wojciech Andrzej : Ślady cyfrowe. Studium prawno-kryminalne. Warszawa: Difin, (2015)

Kosiński Jerzy: Paradygmaty cyberprzestępczości. Warszawa: Difin, (2015)

Kowalczuk-Szymańska Magdalena, Sztejnert-Roszak Olga: Naruszenia praw autorskich w internecie. Aspekty prawne i procedury dochodzenia roszczeń. Warszawa: Difin, (2011)

Kowalewski Jakub, Kowalewski Marian: Zagrożenia informacji w cyberprzestrzeni, cyberterroryzm. Warszawa: Oficyna Wydawnicza Politechniki Warszawskiej. (2017)

Liderman Krzysztof: Bezpieczeństwo teleinformatyczne. Warszawa: WSISiZ. (2003)

Liedel Krzysztof, Piasecka Paulina, Aleksandrowicz Tomasz R.: Analiza informacji. Teoria i praktyka. Warszawa: Difin. (2012)

Littlejohn Shnider Debra, Tittel Ed: Cyberprzestępczość. Jak walczyć z łamaniem prawa w Sieci. Gliwice: Helion, (2004)

Luttgens Jason, Pepe Matthew, Mandia Kevin: Incydenty bezpieczeństwa. Metody reagowania w informatyce śledczej. Gliwice: Helion, (2016)

Matlak Andrzej: Charakter prawny regulacji dotyczących zabezpieczeń technicznych utworów. Warszawa: Wolters Kluwer, (2007)

Matusiak Ireneusz: Gra komputerowa jako przedmiot prawa autorskiego. Warszawa: Wolters Kluwer. (2013)

McClure Stuart, Scambray Joel, Kurtz George: Vademecum hackingu. Skuteczna obrona sieci przed atakami. Gliwice: Helion. (2012)

Molski Marian, Opala Sebastian: Elementarz bezpieczeństwa sysytemów informatycznych. Warszawa: Micom. (2002)

Muchacki Mateusz: Cywilizacja informatyczna i Internet. Konteksty współczesnego konsumenta TI. Kraków: Impuls. (2014)

Nowicka Aurelia: Prawnoautorska i patentowa ochrona programów komputerowych. Warszawa: Dom wydawniczy ABC. (1995)

Opitek Paweł, Jaworski Damian Robert: Cyberprzestępczość w przwie karnym i kryminalistyce. Warszawa: Difin. (2025)
Pańkowska Małgorzata: Zarządzanie zasobami informatycznymi. Warszawa: Difin. (2001)
Podstawy bezpieczeństwa systemów teleinformatycznych. Podręcznik do szkoleń autoryzowanych przez Departament Bezpieczeństwa Teleinformatycznego Agencji Bezpieczeństwa Wewnętrznego. Pod red. Białas Andrzej. Gliwice: Wydawnictwo Pracowni Komputerowej Jacka Skalmierskiego, (2002)
Przestępczość XXI wieku. Zapobieganie i zwalczanie t.1 Problemy technologiczno-informatyczne, t.2 Problemy prawno-kryminologiczne. Pod red. Pływaczewski Emil W., Filipkowski Wojciech, Rau Zbigniew. Warszawa: Wolters Kluwer, (2015)
Rojszczak Marcin, Banasiński Cezary: Cyberbezpieczeństwo. Warszawa: Wolters Kluwer. (2020)
Regulation (EU) 2022/2065 of the European Parliament and of the Council of 19 October 2022 on the Digital Single Market and amending Directive 2000/31/EC (Digital Services Act)
Siwicki Maciej: Cyberprzestępczość. Warszawa: C.H.Beck. (2013)
Siwicki Maciej: Prawo karne wobec oszustw i innych związanych z nimi przestępstw w handlu internetowym oraz bankowości elektronicznej. Toruń: Wydawnictwo Naukowe Uniwersytetu Mikołaja Kopernika. (2018)
Stefański Ryszard A.: Metodyka pracy prokuratora w sprawach karnych. Warszawa: Wolters Kluwer. (2017)
Sobczak Jacek, Chałubińska-Jentkiewicz Katarzyna, Nowikowska Monika: Piractwo w sieci: Kobyłka, Silva Rerum. (2022)
Szpor Grażyna, Wiewiórowski Wojciech: Internet. Prawno-informatyczne problemy sieci, portali i e-usług. Warszawa. C.H. Beck. (2012)
Transsektorowe obszary bezpieczeństwa narodowego. Pod red. Liedel Krzysztof,. Warszawa: Difin, (2011)
The Digital Millennium Copyright Act of 1998 U.S. Copyright Office Summary
Waldemar, K.: Walka o informację w cyberprzestrzeni w XXI wieku. Rambler Press, Warszawa (2017)
Wojciech, W., Grzegorz, W.: Informatyka prawnicza. Trchnologia informacyjna dla prawników i Administracji publicznej. Kraków, Zakamycze (2006)
Williams Sam: W obronie wolnoćci. Krucjata hackera na rzecz wolnego oprogramowania. Helion: Gliwice, (2004)
Wojciechowska-Filipek Sylwia, Ciekanowski Zbigniew: Bezpieczeństwo funkcjonowania w cyberprzestrzeni jednostki—organizacji—państwa. Warszawa: CeDeWu.pl, (2016)
Worona Joanna: Cyberprzestrzeń a prawo międzynarodowe. Warszawa: Wolters Kluwer, (2020)

Cervical Cell Classification Using ResNet50V2 with Multiscale Feature Fusion for Enhanced Cervical Cancer Screening

Betelhem Zewdu Wubineh(✉), Andrzej Rusiecki, and Krzysztof Halawa

Faculty of Information and Communication Technology, Wroclaw University of Science and Technology, Wroclaw, Poland
{betelhem.wubineh,andrzej.rusiecki,krzysztof.halawa}@pwr.edu.pl

Abstract. Cervical cancer remains a major global health challenge, particularly in low-resource settings where early detection is limited. Traditional Pap smear analysis is effective but labour-intensive and prone to human error. This study presents a novel deep learning approach for automated cervical cell classification that enhances the ResNet50V2 architecture through a multiscale feature fusion mechanism. Unlike conventional CNN-based methods that rely on single-scale features, our model extracts and fuses intermediate and high-level representations using 1×1 convolutions, enabling the network to capture subtle morphological differences among cervical cell types. This architecture is lightweight and computationally efficient, requiring no data synthesis or ensemble models. Evaluated on the Cervix93 and Pomeranian datasets using five-fold cross-validation, the proposed model achieved 100% and 96% accuracy, respectively, outperforming existing methods while maintaining lower complexity. The key contribution of this work lies in introducing a simplified yet powerful multiscale fusion strategy within ResNet50V2 that significantly improves the discrimination of cervical cell types, offering a robust and scalable solution for early cervical cancer screening.

Keywords: Cervical cancer · Deep learning · ResNet50V2 · Multiscale feature fusion

1 Introduction

Cervical cancer is a significant global health concern, ranking as the fourth most common cancer among women worldwide [1, 2]. Each year, over half a million women are diagnosed with cervical cancer, and more than 300,000 die from the disease [3, 4]. Developing countries are particularly susceptible, with approximately 80% of cervical cancer deaths occurring in these regions due to limited medical infrastructure and a shortage of qualified pathologists [5–7]. The primary cause of cervical cancer is human papillomavirus (HPV), which damages cervical tissues and can lead to malignant cell growth. Cervical cells are generally classified as benign (precancerous, non-spreading) or malignant (cancerous, spreading) [8].

W. Zamojski et al. (Eds.): DepCoS-RELCOMEX 2026, LNNS 2014, pp. 494–503, 2026.
https://doi.org/10.1007/978-3-032-27927-9_47

The Pap smear test (Papanicolaou test) is the most widely used noninvasive screening method for cervical cancer [9]. This test involves collecting cell-based tissues from the cervix using a spatula or brush, staining them in a sealed container, and examining them under a microscope to identify abnormalities. However, the manual screening process is time-consuming, labor-intensive, and prone to human error and subjective bias [10]. It requires expert cytologists to examine thousands, or even millions, of cells per slide, which can vary significantly in size, color, and morphology. This can lead to high false negative rates and lower sensitivity for precancerous lesions.

To address the limitations of manual screening and improve diagnostic accuracy and efficiency, deep learning (DL) techniques, particularly Convolutional Neural Networks (CNNs), are extensively applied. DL algorithms, a subset of machine learning, use multilayered neural network models with interconnected processing units (neurons) to solve various problems [11]. CNNs are highly effective for image classification and recognition because they can directly process original images without sophisticated pre-processing [12], by integrating concepts like weight sharing, pooling, and local receptive fields, which significantly reduce training parameters [13, 14]. Additionally, transfer learning, where knowledge gained from training on large datasets such as ImageNet is transferred and fine-tuned for tasks in domains with limited data, such as medical imaging, has proven highly beneficial [15]. This approach effectively mitigates dataset limitations, improves the accuracy of the cervical cancer detection model, and plays a crucial role in supporting early detection and improving diagnostic precision in clinical practice.

Recent studies have applied transfer learning with deep convolutional neural networks (CNNs) for cervical cell classification to aid in the early detection of cervical cancer. Mulmule and Kanphade worked with AlexNet, which has been fine-tuned on the CERVIX93 dataset, achieving high accuracy, reaching 99.03% and a kappa score of 0.98 [16]. These approaches demonstrate the potential of deep learning as an assistive tool for automated cervical cancer screening from Pap smear images. Phoulady and Mouton conducted a study of a new cervical cytology dataset for nucleus detection and image classification using the cervix93 dataset. It achieves 86.1% precision and 89.5% recall in the CNN method [17]. Khan et al. propose CervixFormer, a multiscale Swin Transformer-based adversarial ensemble learning framework for the detection of cervical cancer on whole-slide images. It combines synthetic image generation using SAGAN, transformer-based ensemble learning, and a fusion model to improve the classification of pre-cancerous and cancerous lesions, achieving 95% accuracy, 95% recall, and 97% precision on six-class classification of a private dataset, and 100% accuracy on the Cervix93 dataset [18]. Recent work by Fang et al. proposed a deep integrated feature fusion (DIFF) framework that combines local features from CNNs with global features from transformers for cervical cytology image classification. The method achieved an accuracy of 85.06% and a precision of 85.47% [19].

Although these studies demonstrate the potential of deep learning in cervical cancer screening, several limitations remain. Most existing methods rely on single-scale feature extraction and standard classification layers, which may not capture the subtle morphological variations in cervical cytology images. Furthermore, these approaches often lack effective mechanisms to leverage multiscale information, which is crucial given the variability in cell size, shape, and texture. To address these limitations, our research proposes a multi-scale feature fusion strategy integrated with 1 × 1 convolutional layers to align and combine intermediate and high-level features from ResNet50V2. This approach enables richer feature representation by capturing complementary information across different abstraction levels, improving discrimination between visually similar cell types. By freezing the backbone and focusing on enhancing the classification head, our method also reduces computational overhead and mitigates overfitting on limited datasets. This novel architecture enhancement aims to advance the state-of-the-art in cervical cell classification, providing a more accurate and efficient diagnostic tool. The key contributions of this study are the design of effective multiscale feature fusion architecture for cervical cell classification to support more accurate and scalable automated screening systems.

The rest of the paper is organized as follows. Section 2 describes the materials and methods used in the study. Section 3 presents experimental results and discussion. Section 4 concludes the paper with final remarks.

2 Materials and Methods

In this section, we describe the dataset and its preprocessing, the proposed method, and the training details.

2.1 Dataset Collection and Preprocessing

In this study, the publicly available Cervix93 and a private dataset from the Pomeranian Medical University in Szczecin, Poland, were used for the classification of cervical cells. The Cervix93 dataset is a multicell image and consists of 93 image stacks along with their corresponding Extended Depth of Field (EDF) images. Each image has a resolution of 1280 × 960 pixels. Microscopic cervical cytology images are classified according to the Bethesda System into three categories: high-grade squamous intraepithelial lesions (HSIL), low-grade Squamous Intraepithelial Lesions (LSIL), and Negative. The second dataset is a private collection obtained from the Pomeranian Medical University. It is a single-cell image and contains a total of 419 images, categorized into HSIL, LSIL, and normal squamous intraepithelial lesion (NSIL) [20]. To prepare the images for training and testing, a series of preprocessing steps was applied. All images were resized to 224 × 224 pixels to meet the ResNet50V2 input requirements, and pixel values were normalized to the [0,1] range to stabilize and accelerate model convergence. The details of the dataset are depicted in Table 1.

To enhance the robustness of the model and reduce the risk of overfitting, data augmentation was applied using the ImageDataGenerator class of the Keras library. The following augmentation techniques were employed: shear transformation applied with a range of 0.2, zoom transformation with a range of 0.2, and horizontal flipping of images. The augmented images were generated on-the-fly during training, increasing the diversity of the dataset without requiring additional storage.

Table 1. Details of the dataset used for the study

No	Cervix93 dataset		Pomeranian dataset	
	Category	No. of images	Category	No. of images
1	HSIL	522	HSIL	124
2	LSIL	820	LSIL	61
3	Negative	260	NSIL	234
Total	–	1, 602	–	419

2.2 Proposed Method

To improve the feature representation capacity of the ResNet50V2 backbone for cervical cell classification, we propose a multiscale feature fusion approach. ResNet50V2, a widely used deep convolutional neural network architecture, incorporates identity mappings through residual connections, which help in alleviating the vanishing gradient problem and enable the training of deeper networks with improved performance and stability [21]. Its strong feature extraction capabilities make it an effective choice for various medical imaging tasks, including image analysis of cervical cytology.

However, conventional CNNs often focus predominantly on high-level semantic features of the final layers, potentially overlooking fine-grained texture patterns crucial for distinguishing visually similar cervical cell types. To address this, we introduce a multiscale feature fusion strategy that integrates both mid-level and high-level features. Multiscale feature fusion has been shown to improve classification accuracy by capturing complementary information at different abstraction levels, which is particularly valuable in complex visual recognition tasks [22, 23].

Specifically, we extract intermediate features from the layer known in the ResNet50V2 architecture as conv3_block4_out, which corresponds to the output of the third residual block (also referred to as Conv3_x in ResNet). To facilitate an effective combination, both feature maps are passed through separate 1×1 convolutional layers to reduce and align their channel dimensions. Global average pooling is then applied to each feature map to obtain compact, fixed-length feature vectors. These vectors are concatenated to form a fused feature representation that captures both mid-level texture patterns and high-level semantic context.

The fused vector is passed through fully connected layers with dropout regularization to prevent overfitting, followed by a softmax layer for final classification into HSIL, LSIL, and Negative categories. To further improve computational efficiency and prevent overfitting on the limited dataset, the ResNet50V2 backbone is frozen during training, while only the added classification layers are fine-tuned. This design allows the model to leverage multi-scale complementary information and enhances its ability to distinguish between subtle morphological differences in cervical cell images accurately. The proposed method is depicted in Fig. 1.

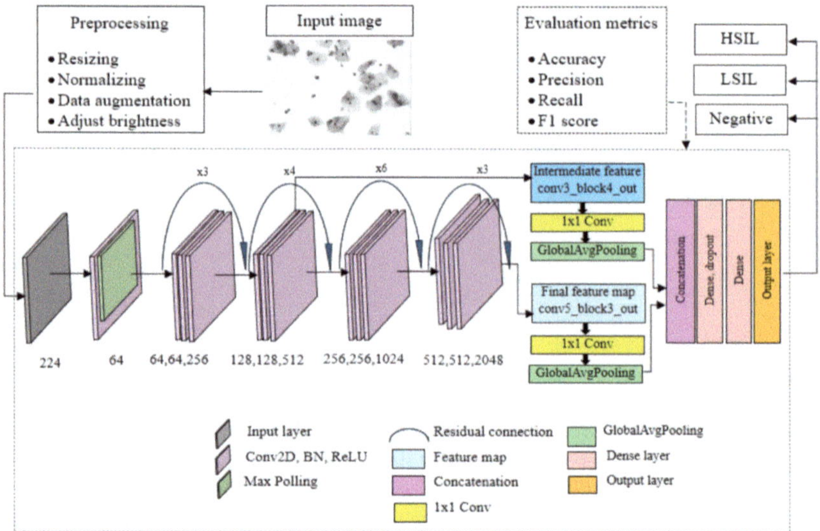

Fig. 1. The proposed ResNet50V2 multiscale feature fusion

2.3 Training Details

The proposed model was implemented using the TensorFlow and Keras deep learning frameworks. To ensure reliable, improved generalizability and unbiased evaluation, 5-fold cross-validation was used. The dataset was divided into 5 folds, and in each iteration, one fold was used for testing while the remaining folds were used for training. Training images were loaded with shuffling enabled, and testing images were loaded without shuffling to maintain consistency during evaluation. For every split, the ResNet50V2 backbone was initialized with ImageNet-pretrained weights, and all layers of the backbone were frozen to prevent overfitting and reduce computational complexity. Only the newly introduced multi-scale feature fusion layers and the classification head were trained.

The model was trained using the Adam optimizer with a categorical cross-entropy loss function suitable for multiclass classification. A batch size of 8 was used during training, and the maximum number of training epochs was set to 100. To avoid overfitting, early stopping was used with a patience value of 2 epochs, monitoring the validation accuracy. Additionally, model checking was used to save the best-performing model weights based on validation accuracy. Training was carried out in a GPU-enabled environment to accelerate computation.

2.4 Evaluation Metrics

For performance evaluation, the trained model was evaluated in the test set using multiple evaluation metrics: accuracy, precision, recall, and F1 score. All metrics were calculated using weighted averaging to account for class imbalance across the three categories. HSIL, LSIL, and Negative. Class predictions were obtained by applying the softmax activation function in the final layer, with the class corresponding to the highest predicted probability selected as the model's output. This comprehensive evaluation ensured that both the correctness (accuracy) and the robustness (precision, recall, F1 score) of the proposed method were thoroughly validated.

3 Results and Discussion

This section presents the experimental results obtained using the proposed multiscale feature fusion model based on ResNet50V2 for automated cervical cell classification on the Cervix93 dataset and the Pomeranian dataset. The model performance was evaluated using five-fold cross-validation, as well as class-wise analysis on the test set. Key evaluation metrics, including accuracy, precision, recall, and F1 score, were calculated to assess classification effectiveness and generalization. The results demonstrate that the proposed method consistently achieves outstanding performance across all folds and classes, with exceptionally high accuracy and balanced metric values. The performance of the proposed method using the test set is depicted in Table 2.

Table 2. Performance of the proposed model in the test set

Dataset	Category	Accuracy	Precision	Recall	F1 score	Support
Cervix93	HSIL	1.00	1.00	1.00	1.00	99
	LSIL	1.00	1.00	1.00	1.00	168
	NSIL	1.00	1.00	1.00	1.00	53
	Average	100%	100%	100%	100%	320
Pomeranian	HSIL	0.94	0.94	0.97	0.95	30
	LSIL	1.00	1.00	0.71	0.83	7
	NSIL	0.98	0.98	1.00	0.99	46
	Average	96%	96%	96%	96%	83

Table 2 provides a detailed class-wise breakdown of the model's performance on the test set, evaluating accuracy, precision, recall, F1 score, and the number of samples (support) per class. The model achieved perfect performance (100%) in all three categories: HSIL, LSIL, and negative class in the Cervix93 dataset. The uniform high performance across all classes confirms that the model is equally effective in distinguishing between different types of cervical cells, regardless of the class imbalance in the dataset. In addition, in the Pomeranian dataset, the model achieves 96% in all performance metrics. This suggests that the integration of multiscale feature fusion not only enhances global representation but also preserves the subtle morphological variations critical for accurate cell type differentiation. The confusion matrix for classifying cervical cells using the proposed method, obtained by aggregating the results from 5-fold cross-validation, is illustrated in Fig. 2.

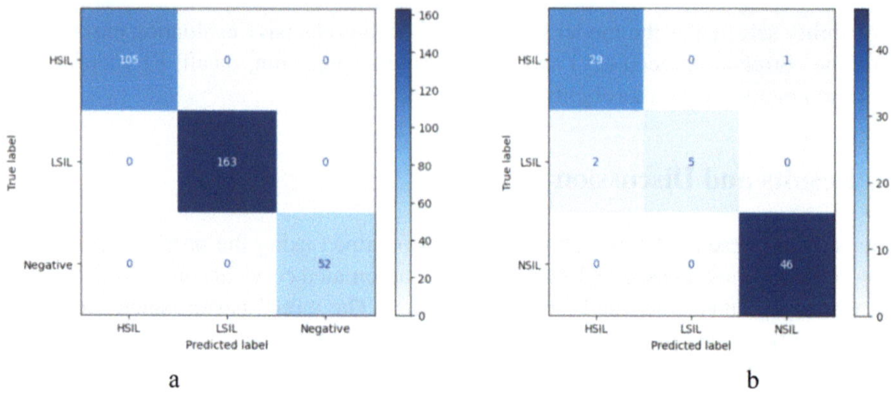

Fig. 2. Confusion matrix: **a** Cervix93 dataset and **b** Pomeranian dataset

The comparison of our results on the Cervix93 and the Pomeranian dataset is illustrated in Fig. 3. As we can see in Fig. 3, the model consistently maintains values of 100% for all metrics in the Cervix93 dataset and 96% on the Pomeranian dataset, reflecting outstanding generalization performance and reliability. The high precision confirms the model's ability to minimize false positives, while the high recall highlights its effectiveness in capturing true positive cases. Consistently elevated F1 scores indicate a balanced trade-off between precision and recall. Overall, these results validate the effectiveness and stability of the proposed ResNet50V2-based method for automated cervical cell classification. The comparison of our study with previous work is depicted in Table 3.

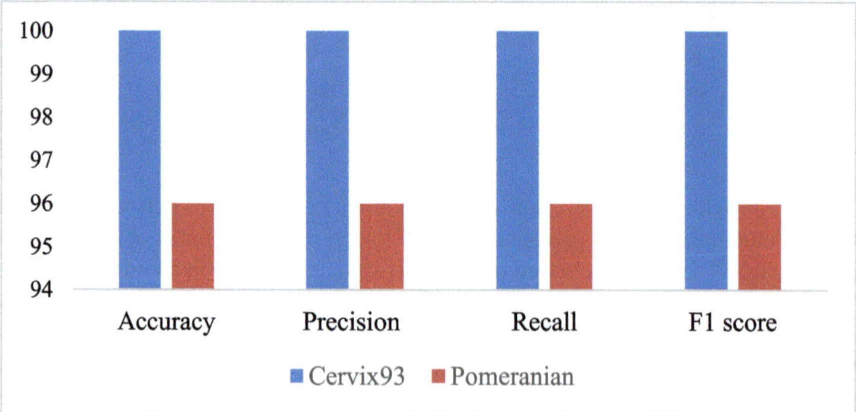

Fig. 3. Performance graph in five-fold cross-validation using the Cervix93 data set

Table 3. Comparison of our study with previous work

Ref	Dataset	Method	Class	Accuracy	Precision	Recall	F1-score
[16]	Cervix 93	AlexNet	3	99.03%	–	–	–
[17]	Cervix93	CNN	3	–	86.1%	89.5%	–
[18]	Private dataset	CervixFormer	6	95%	97%	95%	95%
	Cervix93	CervixFormer	3	100%	100%	100%	100%
[2]	Pomeranian	ResNet50V2	3	89%	89%	89%	89%
		Xception	3	94%	95%	94%	94%
Ours	Cervix93	ResNet50V2	3	100%	100%	100%	100%
Ours	Pomeranian	ResNet50V2	3	96%	96%	96%	96%

Table 3 presents the effectiveness of our proposed method is presented. We compared the results with several previous studies using different datasets and classification methods. Our approach, based on ResNet50V2, achieved superior performance with 100% accuracy, precision, recall, and F1 score in the Cervix93 dataset, outperforming earlier models such as AlexNet and CNN. While the CervixFormer approach also attained perfect classification on Cervix93, it relies on computationally intensive techniques, including synthetic image generation (SAGAN), transformer-based ensemble learning, and model fusion. In contrast, our method achieves equally strong performance using a single, lightweight model with a multiscale feature fusion approach, significantly reducing computational cost and complexity. On the Pomeranian dataset, the study [2] achieved accuracies of 89% and 94% using ResNet50V2 and Xception, respectively. Our approach outperformed these results, achieving 96% accuracy along with consistent precision, recall, and F1-score.

4 Conclusions

In conclusion, the proposed ResNet50V2-based method with multiscale feature fusion demonstrates outstanding performance in classifying cervical cell images, achieving perfect accuracy on the Cervix93 dataset and high accuracy on the Pomeranian dataset, consistently outperforming existing approaches. By effectively combining mid-level and high-level features, the model captures subtle morphological variations critical for accurate classification. Unlike more complex models that rely on synthetic data generation and ensemble techniques, our method maintains high effectiveness with a simpler, lightweight framework. These findings confirm the potential of the proposed approach to deliver reliable, scalable, and efficient automated cervical cancer screening, contributing to early detection, reduced diagnostic workload, and improved patient outcomes.

References

1. Rahaman, M.M., et al.: DeepCervix: a deep learning-based framework for the classification of cervical cells using hybrid deep feature fusion techniques. Comput. Biol. Med. **136**(July), 104649 (2021). https://doi.org/10.1016/j.compbiomed.2021.104649

2. Wubineh, B.Z., Rusiecki, A., Halawa, K.: Classification of cervical cells from the Pap smear image using the RES_DCGAN data augmentation and ResNet50V2 with self-attention architecture. Neural Comput. Appl. **0123456789** (2024). https://doi.org/10.1007/s00521-024-10404-x
3. Fang, M., Lei, X., Liao, B., Wu, F.X.: A deep neural network for cervical cell classification based on cytology images. IEEE Access **10**(December), 130968–130980 (2022). https://doi.org/10.1109/ACCESS.2022.3230280
4. Wubineh, B.Z., Rusiecki, A., Halawa, K.: Enhancing cervical cell classification with DenseNet201 and spatial attention mechanism. In: International Conference on Dependability of Computer Systems, pp. 229–236 (2025)
5. Wubineh, B.Z., Rusiecki, A., Halawa, K.: Segmentation and classification techniques for pap smear images in detecting cervical cancer: a systematic review. IEEE Access **12**(August), 118195–118213 (2024). https://doi.org/10.1109/ACCESS.2024.3447887
6. S. Patra and D. Panda, "Cervical cancer screening in developing countries: Report of a WHO consultation.," Program on Cancer Control
7. Habtemariam, L.W., Zewde, E.T., Simegn, G.L.: Cervix type and cervical cancer classification system using deep learning techniques. Med. Devices Evid. Res. **15**, 163–176 (2022). https://doi.org/10.2147/MDER.S366303
8. Zewdu Wubineh, B., Jeleń, Ł., Rusiecki, A.: DCGAN-based cytology image augmentation for cervical cancer cell classification. IEEE Trans. Med. Imaging **20**(50), 1003–1011 (2020). https://doi.org/10.1016/j.procs.2025.02.206
9. Yu, S., et al.: Automatic classification of cervical cells using deep learning method. IEEE Access **9**, 32559–32568 (2021). https://doi.org/10.1109/ACCESS.2021.3060447
10. Kaur, H., Sharma, R., Kaur, J.: Comparison of deep transfer learning models for classification of cervical cancer from pap smear images. Sci. Rep. **15**(1), 1–18 (2025). https://doi.org/10.1038/s41598-024-74531-0
11. Razzaq, K., Shah, M.: Machine learning and deep learning paradigms: from techniques to practical applications and research frontiers. Computers **14**(3) (2025). https://doi.org/10.3390/computers14030093
12. Wubineh, B.Z., Rusiecki, A., Halawa, K.: Data augmentation techniques to detect cervical cancer using deep learning: a systematic review. In: International Conference on Dependability of Computer Systems, Cham: Springer Nature Switzerland, pp. 325–336 (2024)
13. Huang, X., Pan, R., Wang, J.: A graphite ore grade recognition method based on improved inception-ResNet-v2 model. IEEE Access **13**(January), 48007–48018 (2025). https://doi.org/10.1109/ACCESS.2025.3550423
14. Lei, J., et al.: An intelligent network framework for driver distraction monitoring based on RES-SE-CNN. Sci. Rep. **15**(1), 1–13 (2025). https://doi.org/10.1038/s41598-025-91293-5
15. Hosseinzadeh Taher, M.R., Haghighi, F., Gotway, M.B., Liang, J.: Large-scale benchmarking and boosting transfer learning for medical image analysis. Med. Image Anal. **102**(August 2024), 103487 (2025). https://doi.org/10.1016/j.media.2025.103487
16. Mulmule, P.V., Kanphade, R.D.: Classification of cervical cytology overlapping cell images with transfer learning architectures. Biomed. Pharmacol. J. **15**(1), 277–284 (2022). https://doi.org/10.13005/bpj/2364
17. Phoulady, H.A., Mouton, P.R.: A new cervical cytology dataset for nucleus detection and image classification (Cervix93) and methods for cervical nucleus detection (2018). http://arxiv.org/abs/1811.09651
18. Khan, A., Han, S., Ilyas, N., Lee, Y.M., Lee, B.: CervixFormer: a multi-scale swin transformer-based cervical pap-smear WSI classification framework. Comput. Methods Programs Biomed. **240** (2023). https://doi.org/10.1016/j.cmpb.2023.107718

19. Fang, M., Fu, M., Liao, B., Lei, X., Wu, F.X.: Deep integrated fusion of local and global features for cervical cell classification. Comput. Biol. Med. **171**(February) (2024). https://doi.org/10.1016/j.compbiomed.2024.108153
20. Jeleń, Ł, Stankiewicz-Antosz, I., Chosia, M., Jeleń, M.: Optimizing cervical cancer diagnosis with feature selection and deep learning. Appl. Sci. **15**(3), 1–21 (2025). https://doi.org/10.3390/app15031458
21. Duklan, N., Kumar, S., Maheshwari, H., Singh, R., Sharma, S.D., Swami, S.: CNN architectures for image classification: a comparative study using ResNet50V2, ResNet152V2, InceptionV3, Xception, and MobileNetV2. SSRG Int. J. Electron. Commun. Eng. **11**(9), 11–21 (2024). https://doi.org/10.14445/23488549/IJECE-V11I9P102
22. Huo, X., et al.: HiFuse: Hierarchical multi-scale feature fusion network for medical image classification. Biomed. Signal Process. Control **87**(PA), 105534 (2024). https://doi.org/10.1016/j.bspc.2023.105534
23. Li, M., He, Z., Wang, H.: Heart sound classification based on multi-scale feature fusion and channel attention module. Bioengineering **12**(3) (2025). https://doi.org/10.3390/bioengineering12030290

Cyberattacks—Forecast of Their Role in Future Battlefields

Wiktor Ziętara(✉)

DSW Univeristy of Lower Silesia, Strzegomska Street 55, Wrocław, Poland
wiktor.zietara@dsw.edu.pl

Abstract. This article analyzes the application of radio communication systems and networks for special military purposes in military operations, using the Israel–Hamas conflict as a case study. It demonstrates how modern military telecommunications—including Software-Defined Radio (SDR) systems, Mobile Ad Hoc Networks (MANET), Unmanned Aerial Vehicles (UAVs), and integrated C4ISR systems—affect combat tactics, command processes, and the development of information superiority. The article is based on a qualitative methodology combining open-source intelligence, document analysis, and a comparative case-study approach. It includes a focused literature review on cyber warfare, military telecommunications, and network-centric warfare, and expands the discussion of operational implications by linking empirical observations to communication resilience, vulnerability to electronic attack, and cyber defense strategies. The article also discusses future development perspectives for military communication systems in the context of artificial intelligence, 6G technologies, satellite communications, and quantum communication. Israel is presented as an example of a state effectively adapting modern telecommunication technologies to the realities of asymmetric warfare.

Keywords: Military telecommunications · Radio communication systems · Asymmetric warfare · MANET networks · C4ISR · Cybersecurity

1 Introduction

1.1 Operational Context

The operational context of the Israel–Hamas conflict is defined by a combination of dense urban terrain, complex infrastructure, and multidomain engagement. These factors significantly influence the performance and requirements of communication systems deployed by military forces.

In such an environment, maintaining uninterrupted communication is essential not only for coordination but also for situational awareness and force protection. The need to operate under constant electromagnetic interference and cyber pressure requires highly adaptive and resilient communication architectures.

W. Zamojski et al. (Eds.): DepCoS-RELCOMEX 2026, LNNS 2014, pp. 504–516, 2026.
https://doi.org/10.1007/978-3-032-27927-9_48

Additionally, the integration of communication systems with intelligence, surveillance, and reconnaissance capabilities allows for the creation of a real-time operational picture, which is crucial for effective decision-making in fast-changing combat conditions.

This subsection frames the operational setting of the Israel–Hamas conflict as a dense urban and multidomain battlespace in which communications, cyber resilience, and electromagnetic control directly influence battlefield effectiveness.

1.2 Research Objective

From an operational perspective, the objective also includes identifying how communication systems contribute to shortening the decision-making cycle (OODA loop) and improving coordination between dispersed units. Particular attention is paid to the interaction between communication technologies and intelligence systems, which together enable real-time situational awareness.

Another important aspect of the research objective is to assess the resilience of communication systems in contested environments, including exposure to jamming, interception, and cyber intrusion. This allows the study to move beyond purely technical analysis and address operational effectiveness under real combat conditions.

The article examines how military radio communication systems support command continuity, information superiority, and operational adaptation under asymmetric-war conditions, and it links these issues to future capability development.

The Israel–Hamas conflict constitutes one of the most illustrative contemporary examples of how advanced military communication systems are employed under conditions of asymmetric warfare. Conducted predominantly in a densely urbanized environment and accompanied by extensive activity in the electromagnetic and cyber domains, this conflict provides a unique empirical context for analyzing the operational role of radio communication systems designed for special military applications.

Building upon the theoretical foundations presented in Part I, this article focuses on the practical implementation and operational consequences of modern military telecommunications. The Israel Defense Forces operate within an environment characterized by high threat dynamics, fragmented terrain, underground infrastructure, and continuous attempts by the adversary to disrupt information flows through electronic warfare and cyber operations.

The objective of this article is to analyze how the Israel Defense Forces employ radio communication systems—particularly Software-Defined Radio, Mobile Ad Hoc Networks, unmanned aerial platforms, and integrated C4ISR architectures—at both tactical and strategic levels. Special attention is devoted to the role of communications in supporting decision-making processes, enabling predictive analytics, and maintaining information superiority in a contested electromagnetic environment.

In addition to examining current operational practices, the article addresses challenges related to cybersecurity and electronic warfare, drawing on empirical data from recent years. It also explores future development trajectories for military telecommunications, including the integration of artificial intelligence, next-generation wireless technologies, satellite communications, and quantum communication.

2 Methodology

2.1 Research Design

The research design is structured to ensure methodological triangulation, combining multiple data sources and analytical perspectives. This approach increases the reliability of findings by validating observations across independent reports and publications.

Qualitative synthesis is used to identify recurring patterns in communication system deployment, cyber threats, and operational responses. This enables the construction of a coherent analytical framework linking technology with battlefield performance.

The study follows a qualitative case-study design supported by triangulation across open-source materials, analytical reports, and academic literature to reconstruct the operational role of communications in conflict.

2.2 Data Sources

The empirical basis includes think-tank studies, policy documents, institutional cyber reports, and peer-reviewed publications, selected to capture both technological and operational dimensions of the problem.

This study applies to qualitative research design based on multi-source analysis. The methodological core combines three complementary approaches: open-source intelligence, analysis of strategic and analytical documents, and a case-study method focused on the Israel–Hamas conflict.

The research procedure consisted of identifying recurring technological, operational, and strategic patterns in the available sources, comparing them across document types, and synthesizing them into an analytical framework linking telecommunications with battlefield effectiveness. The study does not rely on classified intelligence.

The Israel–Hamas conflict is treated as an illustrative and exploratory case study rather than as a universal model for all future wars. It was selected because it combines three analytically valuable features: intensive fighting in a dense urban environment, the use of advanced military communications by a technologically sophisticated state actor, and the presence of an asymmetric opponent employing cyber, information, and electronic means.

3 Literature Review

3.1 Telecommunications Research

Research on military telecommunications emphasizes the transition from centralized to distributed communication architectures. This shift reflects the need for flexibility, scalability, and resilience in dynamic operational environments.

Studies highlight that modern communication systems must support not only data transmission but also integration with command systems, sensors, and decision-support tools, making them a core element of multidomain operations.

Telecommunications literature emphasizes distributed command, resilient networking, and the growing integration of communications with C4ISR and machine-assisted decision-making.

3.2 Cyber Warfare Research

A parallel strand of research shows that cyber operations, disinformation, and digital disruption have become inseparable from modern combat operations and now directly affect military communications.

The current state of research indicates that military telecommunications are no longer examined solely as a technical support function, but as an integral component of command effectiveness and multidomain operations. RAND analyses emphasize the growing importance of distributed command architectures, C4ISR integration, and machine-assisted decision-making in future conflicts [1, 4, 12].

A second major stream of literature concerns cyber warfare, information operations, and the use of digital technologies by state and non-state actors. Byman and McCaleb [2], Chang et al.

A third body of literature addresses the technological evolution of military communications themselves. Studies on SDR, machine learning for command support, satellite systems, and emerging quantum communications [7, 11, 12] indicate that future communication networks will need to be adaptive, software-driven, and tightly integrated with sensing, analytics, and cyber protection.

4 Operational Role of Radio Communication Systems

4.1 Tactical Layer

At the tactical layer, communication systems must operate under severe constraints, including limited visibility, infrastructure damage, and electromagnetic interference. Their effectiveness directly influences the ability of units to coordinate maneuvers and respond to threats.

The integration of adaptive technologies such as SDR and MANET ensures continuity of operations even when parts of the network are degraded or destroyed, which is critical in urban combat scenarios.

At the tactical level, communication systems determine whether dispersed units can coordinate maneuver, fires, and situational awareness under pressure and in degraded urban terrain.

4.2 Systems Integration

Operational effectiveness depends not on isolated devices but on the integration of radios, networks, unmanned platforms, and command systems into one resilient information architecture.

At the tactical level, radio communication systems are not limited to the transmission of voice and data; they function as the connective layer of combat operations. In urban combat, where line-of-sight is constrained, terrain is fragmented, and units are highly dispersed, communication resilience becomes a prerequisite for effective command.

Concrete operational examples help clarify the practical value of these technologies. SDR-enabled radios support frequency hopping, encryption, and adaptive waveform management, making them suitable for operations in contested electromagnetic environments.

These examples show that the practical importance of communication technologies lies in their ability to preserve command continuity under pressure. Their value is therefore measured not only by bandwidth or range, but by resilience against jamming, redundancy of transmission paths, and compatibility with intelligence and fire-control systems.

5 Strategic Significance and Network-Centric Warfare

5.1 Command Architecture

Command architecture in modern armed forces is increasingly dependent on integrated communication networks that connect all levels of command. These networks enable rapid dissemination of information and synchronization of actions across multiple domains.

The effectiveness of such architecture depends on interoperability, data fusion capabilities, and the ability to maintain secure and uninterrupted communication flows under adversarial conditions.

At the strategic level, military communications operate as an element of command architecture, shaping how information is collected, fused, distributed, and converted into decisions.

5.2 Network-Centric Warfare

The network-centric model treats each platform and unit as a node in a wider information ecosystem, thereby increasing tempo, interoperability, and the capacity for distributed action.

At the strategic and operational levels, the role of telecommunications extends beyond transmission and enters the domain of command architecture. Network-centric warfare assumes that each platform, sensor, and unit becomes a node in a larger information ecosystem.

Within the Israeli case, this logic is visible in the integration of communication networks with intelligence collection, predictive analytics, and targeting processes. The case study thus serves a dual purpose in this article.

6 Cybersecurity and Electronic Warfare

To quantitatively illustrate the scale of cyber threats during the Israeli Palestinian conflict, selected empirical data concerning the number of incidents and cyberattacks in 2022–2024 are presented in the following tables and figures. These data are analytically important because they connect the technological discussion with concrete operational challenges: communication resilience, vulnerability to electronic attack, and the growing requirement for integrated cyber defense in military communication architectures.

The dynamic development of advanced telecommunication systems, which form the basis of contemporary C4ISR architectures, leads to a significant expansion of the

Table 1. Cyber incident reports and cyber incidents in Israel (2022–2024)

Indicator	Period	Value
Reports to INCD/119 (estimated	2022	~9,120
Reports to INCD/119	2023	13,040
Reports during the "Iron Swords" war period	X–XII 2023	~8,870
Recorded cyberattacks	07.10–31.12.2023	3,380
Incidents with significant damage potential	07.10–31.12.2023	~800
Reports to INCD/119 (verified)	2024	17,078
Documented ransomware attacks (cybercrime)	2024	>300

Source author's own elaboration based on reports from the Israel National Cyber Directorate (INCD) and CSIS [3, 11, 13, 14]

threat spectrum in cyberspace and the electromagnetic domain [3, 6, 11]. Empirical data confirming this trend is presented in Table 1.

Commentary on Table 2: In 2024, the number of verified reports to the 119-center increased to 17,078, confirming the sustained elevated level of cyber pressure following the escalation of the conflict in 2023. It is also worth noting the parallel increase in cybercriminal activity (including ransomware), which indicates a blurring of the boundaries between state-sponsored and criminal threats and highlights the necessity of multi-layered protection (SOC, network segmentation, zero-trust) also within military communication systems (Figs. 1 and 2).

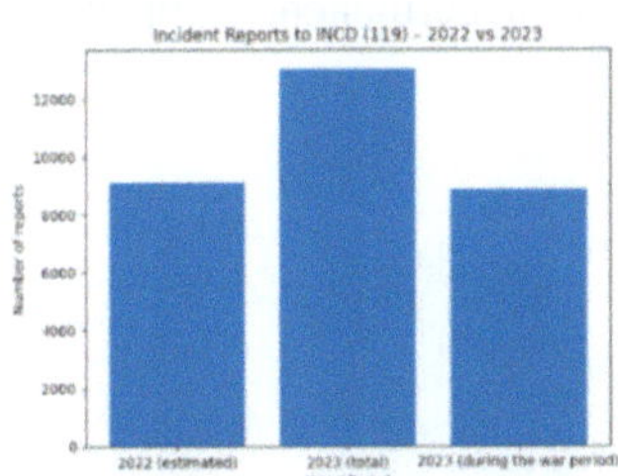

Fig. 1. Cyber incident reports in Israel in 2022–2024—number of reports submitted to INCD/119, with reference to the period October–December 2023

The 2022–2024 trend shows an increase in the number of incident reports, with the period from October to December 2023 accounting for a significant share of the annual activity. From an operational perspective, this means that communication infrastructure (including C4ISR nodes) must be designed with the assumption of continuous engagement in cyberspace—not only during phases of kinetic escalation.

In asymmetric conflict environments, cybersecurity and electronic warfare are becoming elements of the battlefield equal in importance to kinetic operations. The

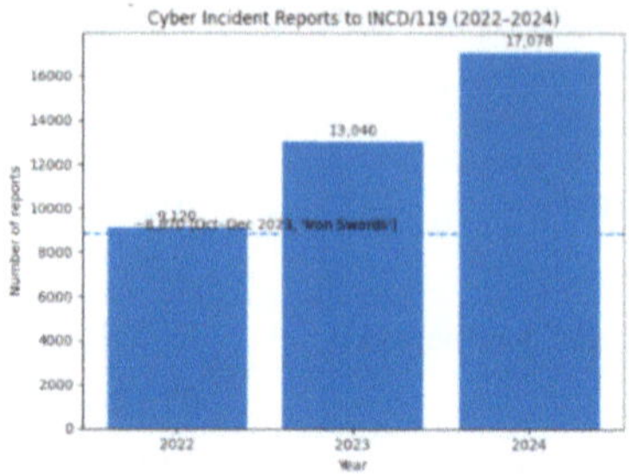

Fig. 2. Cyber incident reports to INCD/119 (2022–2024)

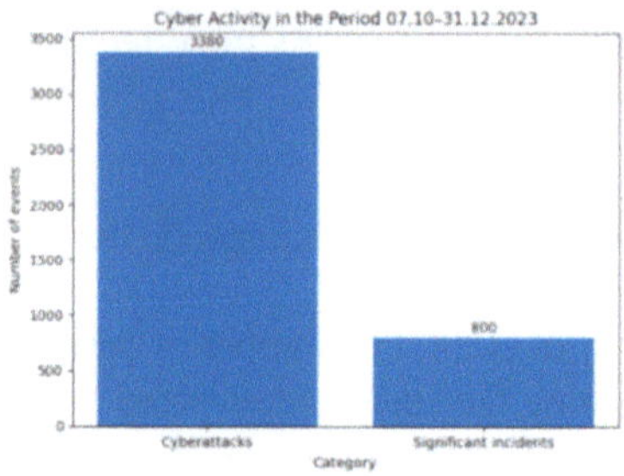

Fig. 3. Scale of cyber activity in the period 07.10–31.12.2023—comparison of the number of cyberattacks and incidents with significant damage potential

information space—encompassing ICT networks, radio systems, and digital infrastructure—has now become an area of permanent competition between the parties to the conflict (Fig. 4).

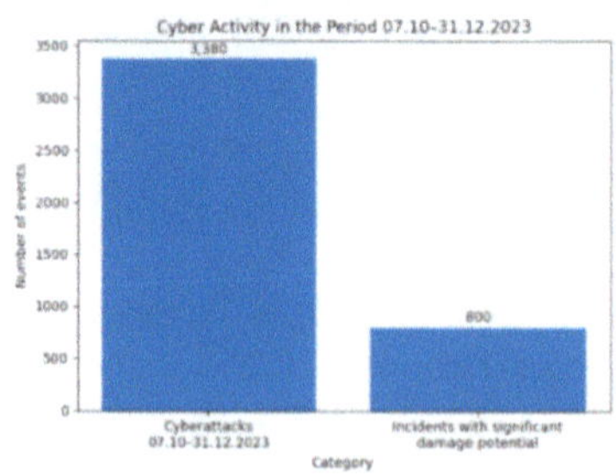

Fig. 4. Cyber activity in the period 07.10–31.12.2023

The ratio of the number of cyberattacks to incidents with significant potential damage indicates strong event selectivity—part of the activity is mass-scale and disruptive in nature, while a smaller proportion consists of high-impact actions. For radio communication systems, resilience mechanisms (LPI/LPD, FHSS) and rapid detection of network anomalies are of key importance.

In the Israel–Hamas confrontation, the adversary—despite limited technological resources—intensively employs electronic warfare and cyber means to disrupt the IDF's information superiority. These activities include attempts to jam radio transmissions in the VHF/UHF bands, locate emission sources, intercept unencrypted signals, and exploit readily available civilian technologies to conduct improvised jamming operations.

A key element in protecting radio communication systems is the use of advanced cryptographic methods and dynamic electromagnetic spectrum management techniques. Israeli communication systems employ, among others, frequency hopping, adaptive modulation, and low probability of intercept/low probability of detection (LPI/LPD) techniques.

In parallel with protecting its own systems, the IDF conducts active electronic warfare operations, including jamming, deception, and degradation of the adversary's communication systems. Interception and analysis of Hamas radio traffic enable the identification of command structures, routes of fighter movement, and the locations of key communication nodes.

Cybersecurity constitutes a second, equally important pillar of protection for modern military communication systems. Hamas activities in cyberspace include disinformation campaigns, psychological operations conducted via social media, and attempts to attack Israeli ICT infrastructure.

A particularly important role in this area is played by specialized units such as Unit 8200, responsible for conducting cyber operations and signals intelligence. This unit performs both defensive tasks—related to the protection of critical infrastructure and military ICT systems—and offensive tasks, involving the infiltration and neutralization of the adversary's information systems [6].

A significant challenge remains the growing complexity of telecommunication systems and their dependence on software and digital components. This increases the risk of vulnerabilities to supply-chain attacks, malware, or data manipulation.

Cybersecurity and electronic warfare no longer function as separate domains of activity but constitute an integral part of an integrated operational environment. The ability to protect one's own communication systems, simultaneously disrupt and disorganize the adversary's systems, and conduct information operations in cyberspace is one of the key factors determining operational and strategic superiority in contemporary armed conflicts.

7 Prospects for the Development of Military Telecommunications

Prospects for the development of military telecommunications are closely linked to the ongoing digitalization of the battlefield, the increasing autonomy of combat systems, and the integration of activities within multidomain operations [4, 10, 12]. In the coming decades, military communication systems will evolve toward greater bandwidth, increased resistance to interference, and the capability for autonomous management of information traffic in a highly contested electromagnetic environment.

One of the key development directions is the integration of telecommunication systems with artificial intelligence and machine learning. AI algorithms will increasingly be used for automatic signal classification, detection of anomalies in network traffic, prediction of interference, and dynamic reconfiguration of communication networks.

Emerging 6G technologies mark the next stage in the evolution of military wireless communications. Although they currently remain in the research and testing phase, they already indicate potential operational capabilities such as ultra-low latency, very high throughput, and the ability to integrate vast numbers of sensors, unmanned platforms, and wearable systems within a single combat network.

Another important trend is the development of next-generation satellite communications, including low Earth orbit (LEO) satellite constellations. The scale of this phenomenon is illustrated by the data presented in Table 2 and Fig. 3.

Table 2. Number of operational and active satellites—selected reference points (2023–2025)

Year/reference date	Number of satellites
05.2023	~7,560
12.2023	~9,000
05.2025	~11,700
2025 (ESA—active payloads)	~11,000

Source author's own elaboration based on data from UCS and the ESA Space Environment Report 2025 [4, 15]

The increase in the number of satellites—particularly LEO constellations—enhances the resilience of strategic communications through diversification of transmission paths, while simultaneously increasing the risk of orbital congestion and the need for space traffic management. For military applications, this translates into the growing importance of multiband terminals, automatic handover mechanisms, and connection planning under interference conditions (Figs. 5 and 6).

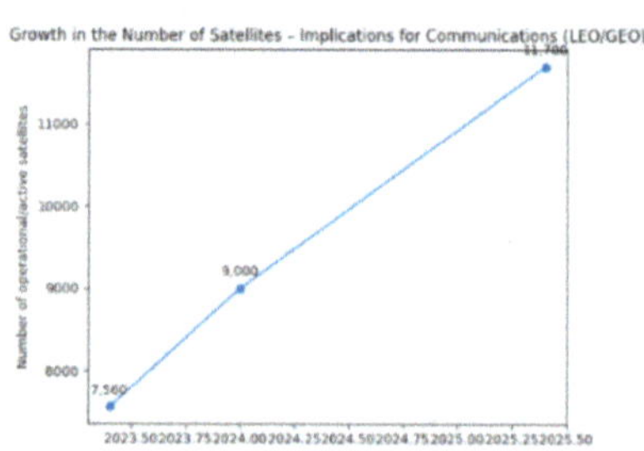

Fig. 5. Growth in the number of operational and active satellites—implications for communications

The increasing number of active satellites/payloads strengthens SATCOM potential (greater redundancy, lower latency in LEO), but also raises requirements related

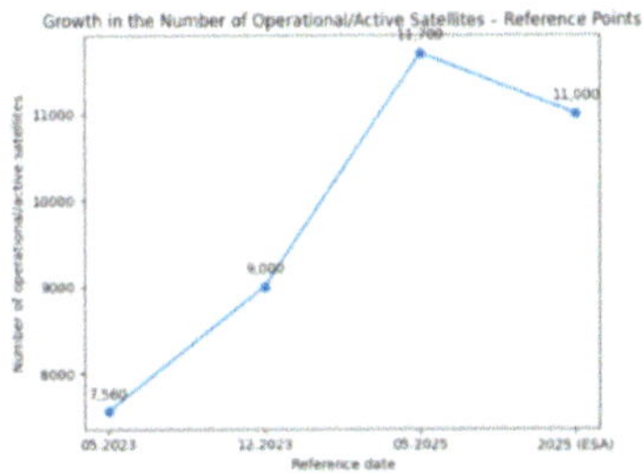

Fig. 6. Growth in the number of operational and active satellites—selected reference points (2023–2025)

to coordination and security (collision risk, cyberattacks on the ground segment, signal jamming). Consequently, the role of physical-layer resilience and cryptography is growing.

Another breakthrough development direction is quantum communication, perceived as a potential response to the increasing threats associated with interception and the breaking of cryptographic protections. The use of quantum-mechanical phenomena enables the creation of communication channels that are practically immune to eavesdropping.

An analysis of IDF experience and technological trends indicates that radio communication systems for special applications are becoming the core of the network-centric warfare architecture—integrating SDR/MANET networks, UAV platforms, and C4ISR nodes into a single, coherent data-exchange ecosystem. At the same time, the growing scale of activity in cyberspace (17,078 verified reports to the INCD 119 center in 2024) necessitates the design of communications by design with built-in mechanisms for cyber protection, monitoring, and rapid service restoration in the event of degradation.

At the strategic level, a parallel increase in the importance of satellite communications is evident—particularly within the LEO model, where the growth in the number of active satellite payloads (approximately 11,000 according to ESA) increases redundancy and the availability of beyond-line-of-sight (BLOS) channels. These benefits, however, are counterbalanced by the rising risk of orbital congestion and the need for space traffic management, which in practice translates into the requirement for multiband terminals, automatic handover, and resistance to jamming and interception.

From a market development perspective, forecasts for 2025–2030 confirm a steady increase in investments in military communications, driven by multidomain operations, interoperability requirements, and the integration of security with the transport layer. A key conclusion for capability planning is the necessity of parallel development of: (1) tactical SDR/MANET networks, (2) satellite communications, and (3) cyber resilience and spectrum management—to shorten the decision-making loop and maintain command continuity in heavily contested environments.

The projected growth of the military communications market reflects the transition of armed forces toward multidomain architectures (MDO) as well as the increased demand for interference-resilient communications (SATCOM-on-the-move,

Table 3. Military communications market – forecast for 2025–2030.

Year	Market value (USD billion)
2025	35.62
2026	37.83
2027	40.18
2028	42.68
2029	45.33
2030	48.17

Source author's own elaboration based on data from Mordor Intelligence and CSIS [5, 12]

mesh/MANET, SDR). In practice, this translates into accelerated deployment of open-architecture solutions and the integration of security mechanisms with the transport layer (encryption, zero-trust, monitoring) (Tables 3 and 4).

Table 4. Structure of the military communications market—selected shares (2024) and dynamics (2025–2030)

Indicator (Mordor intelligence)	Year/period	Value (%)
Share of ground-based systems in the market	2024	36.34
Share of C2/C3 applications in revenues	2024	39.49
Share of land platforms (land forces)	2024	41.67
Forecast market CAGR	2025–2030	6.22

Source author's own elaboration based on Mordor Intelligence (Military Communications Market) [12]

The dominance of the ground segment and C2/C3 applications confirms that communications modernization is driven primarily by the need for a coherent situational picture and a rapid decision-making cycle (sensor-to-shooter). At the same time, the projected growth dynamics indicate the increasing importance of security solutions and interoperability in a multidomain environment (Figs. 7 and 8).

Armed forces that are the first to successfully integrate artificial intelligence, next-generation technologies, and advanced information protection mechanisms will gain a lasting operational and strategic advantage in future armed conflicts.

The slope of the curve (CAGR of approximately 6.22%) suggests a stable, multi-year investment cycle. From a military user's perspective, this implies parallel funding of: (1) modernization of radios and tactical networks (SDR/MANET), (2) satellite communications and multiband terminals, and (3) cybersecurity tools and spectrum management. Operational success will depend on the integration of these layers within a coherent C4ISR architecture.

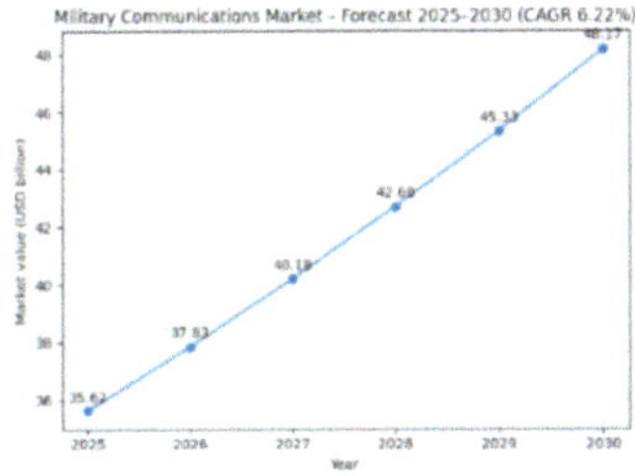

Fig. 7. Forecast of the growth of the military communications market in 2025–2030 (CAGR 6.22%)

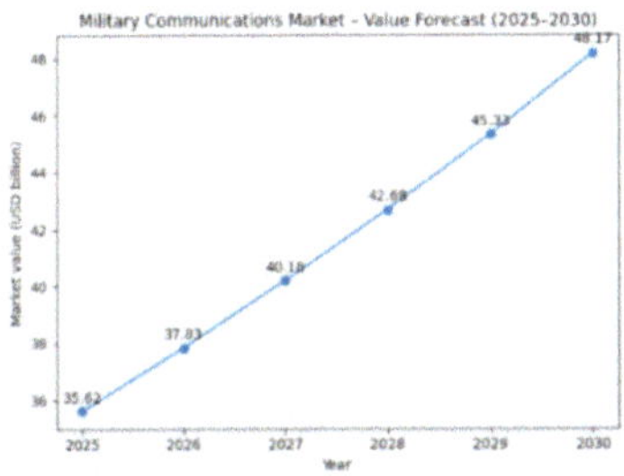

Fig. 8. Military communications market—value forecasts for 2025–2030

8 Conclusion

The conducted analysis clearly indicates that contemporary armed conflicts cannot be considered in isolation from information and telecommunications infrastructure, which has become one of the fundamental elements of the modern battlefield. Radio communication systems for special military applications no longer serve solely as tools for transmitting orders and data; instead, they constitute an integrated component of the operational architecture, determining the effectiveness of command, the tempo of decision-making, and the ability to achieve information superiority.

The case study of the Israel–Hamas conflict makes it possible to capture the practical dimension of these processes. Operating in an asymmetric, urbanized, and heavily electromagnetically contested conflict environment, the Israel Defense Forces demonstrated the ability to effectively adapt advanced telecommunication systems to the realities of the modern battlefield.

The analysis shows that the key factor behind Israel's operational advantage is not only a high level of technological sophistication, but above all a doctrinal and organizational approach to managing information as a strategic resource. In this model, military telecommunications have been fully integrated with command, reconnaissance, predictive analytics, and operations in cyberspace and the electromagnetic domain.

Of particular importance in the context of future conflicts is the growing role of cybersecurity and electronic warfare. The study demonstrates that the ability to protect one's own communication systems, simultaneously disrupt and disorganize those of the adversary, and conduct information operations constitutes one of the key determinants of advantage at both tactical and strategic levels.

The development perspectives for military telecommunications outlined in the article indicate further evolution toward autonomous, intelligent, and highly network-centric systems. The integration of artificial intelligence, next-generation communication technologies (6G), satellite systems, and quantum communication will gradually redefine the conduct of military operations.

The conclusions drawn from the analysis of Israel's experience are universal in nature and may serve as an important source of knowledge for other armed forces, including those of NATO and the European Union. Investments in modern military telecommunications should be regarded not as a supplement to combat capabilities, but as their foundation—one that conditions operational effectiveness in future conflicts.

References

1. Black, J., et al.: Command and control in the future. RAND. N/A (2024)
2. Byman, D., McCaleb, A.: Understanding Hamas's uses of information technology. CSIS. N/A (2023)
3. Chang, Y., Keblis, M., Li, R., Iakovou, E., White, C.: Misinformation and disinformation in modern warfare. Oper. Res. (2022). https://doi.org/10.1287/opre.2022.2263
4. Cohen, R., et al.: The future of warfare in 2030. RAND. N/A (2020)
5. Cohen, R., et al.: From cast lead to protective edge. RAND. N/A (2017)
6. Cordey, S.: Israeli unit 8200 study. ETH Zurich. N/A (2019)
7. Dickson, J., Harding, E.: Quantum potential. CSIS. N/A (2024)
8. Harding, E.: Israeli intelligence analysis. CSIS. N/A (2023)
9. INCD Annual Report. N/A (2025)
10. INCD Cybercrime Report. N/A (2025)
11. Jeliazov, P.M., Marinov, M.S.: Military radar systems. In: IEEE Conference. N/A (2004)
12. Madison, A., et al.: Machine learning for C2. ICMCIS. N/A (2024)
13. Stanescu, G.: Informational war. N/A (2023)
14. Wirtschafter, V.: Information ecosystem. N/A (2023)

Author Index

W. Zamojski et al. (Eds.): DepCoS-RELCOMEX 2026, LNNS 2014, pp. 517–518, 2026.
https://doi.org/10.1007/978-3-032-27927-9

MIX
Papier aus verantwortungsvollen Quellen
Paper from responsible sources
FSC® C105338
FSC
www.fsc.org

If you have any concerns about our products,
you can contact us on
ProductSafety@springernature.com

In case Publisher is established outside the EU,
the EU authorized representative is:
Springer Nature Customer Service Center GmbH
Europaplatz 3, 69115 Heidelberg, Germany

Printed by Libri Plureos GmbH
in Hamburg, Germany